Lecture Notes in Computational Science and Engineering

32

Springer
Berlin
Heidelberg
New York
Hong Kong
London
Milan
Paris
Tokyo

Preface

The workshop on "computational physics of transport and interfacial dynamics" was held in Dresden, Germany from February 25 to March 8, 2002. The Max Planck Institute for the Physics of Complex Systems has sponsored the workshop and the preliminary lecture-based seminar.

The workshop has closely pursued the recent progress of research in computational physics and materials science, particularly in modelling both traffic flow phenomena and complex multi-scale solidification. These branches of science have become topics of considerable diversity linking disciplines as different as physics, mathematical and computational modelling, nonlinear dynamics, materials sciences, statistical mechanics and foundry technique. The international workshop brought together experts from different fields in order to enhance the exchange of knowledge, to assess common interests and to provide closer cooperation between different communities of researchers. The workshop intended to create a comprehensive and coherent image of the current research status and to formulate various possible perspectives for joint future activities. Special emphases laid on exchanging experiences concerning numerical tools and on the bridging of the scales as necessary in a variety of scientific and engineering applications. An interesting possibility along this line was the coupling of different computational approaches leading to hybrid simulations. In this sense, we explicitly addressed researchers working with different numerical schemes as diverse as cellular automata, coupled maps, finite difference and finite element algorithms for partial differential equations (e.g. phase-field computations). The following scientific topics were treated in the contributions:

- Fundamentals and modelling of microstructure formation: Free boundary problems, phase transitions, solidifiation, interfacial dynamics, Ginzburg-Landau equations, Cahn-Hilliard equations, phase-field modelling.
- Numerical simulations on different time and length scales: Fluid dynamic computations, adaptivity, finite element-discretizations, discrete modelling techniques such as molecular dynamics and Monte Carlo.
- Multi-scale problems in crystal growth, porous media, solidification and traffic flow.

The workshop was attended by 87 participants from 22 different countries. This volume contains 36 refereed papers having been drawn on from the oral presentations and the posters at the workshop. The proceedings book is organized in two main chapters, the first chapter on interface dynamics and solidification and the second chapter on transport processes. The first chapter contains contributions on phenomenology and modelling as well as on numerical simulations.

The editors gratefully acknowlegde the granting of the workshop by the Max-Planck Gesellschaft and thank the Max-Planck Institute in Dresden for providing space and equipment and for being a very kind host. We wish to heartily thank all members of the local organizing committee for their professional and invaluable help.

Heike Emmerich Britta Nestler Michael Schreckenberg

Table of Contents

I *INTERFACE DYNAMICS:* Phenomenology and Modelling

Part I

INTERFACE DYNAMICS: Phenomenology and Modelling

Melting Kinetics of Prolate Spheroidal Crystals

M.E. Glicksman[1], A. Lupulescu[1], and M.B. Koss[2]

[1] Rensselaer Polytechnic Institute, Materials Science & Engineering Department, Troy, NY, 12180-3590 USA.
[2] Department of Physics, College of the Holy Cross, Worcester, MA 01610-2395 USA.

Abstract. The melting kinetics of a pivalic acid (PVA) dendritic mushy zone was observed for the first time under convection-free conditions. Video data show that PVA dendrites melt into fragments that shrink at accelerating rates to extinction. Individual fragments follow a characteristic time-dependence derived here for the diminishing length scales within a melting mushy zone. The melting kinetics against which the experimental observations are compared is based on the conduction-limited quasi-static process of melting under shape-preserving conditions. Agreement between analytic theory and experiment was found for the melting of a prolate spheroidal crystal fragment with an aspect ratio of $\mathcal{C}/\mathcal{A} = 12$.

1 Introduction

1.1 Dendritic Growth

Many theoretical [1], [2], [3], [4] and experimental [5], [6], [7], [8] dendritic growth studies were reported over the past 40 years. Dendrite growth, as a pattern-forming process, is much simpler to describe as free crystallization from a pure molten phase. The thermodynamic driving force for crystallization in this case is the melt supercooling, $\Delta T \equiv T_m - T_\infty$, where T_m is the melting point of the material, and T_∞ is the temperature of the melt far from the heat emitting solid-liquid interface. The crystallization process for free growth requires that the latent heat released be transported away from the solid-liquid interface by thermal conduction to the cooler (supercooled) melt.

1.2 Isothermal Dendritic Growth Experiment (IDGE)

Theories of dendritic growth in pure materials are best tested under conditions of conduction-controlled heat transfer, a situation possible only where gravity is reduced to nearly zero, where buoyancy-induced natural convection is eliminated [10], [11], [12], [13]. The Isothermal Dendritic Growth Experiment, or IDGE, is a basic science experiment, developed under NASA support, that provides both terrestrial and microgravity data on the kinetics and morphology of dendritic solidification. Before the advent of IDGE in 1994, it was not possible to test quantitatively either Ivantsov's solution [14] for the steady-state transport field surrounding a dendrite, or evaluate the so-called interfacial "stability constant," $\sigma^* \cong 0.02$, predicted from several physics theories of near-equilibrium solid-liquid interfaces [2], [3], [4].

The IDGE flew in near-Earth orbit three times, and provided the first compelling evidence that Ivantsov's heat-conduction solution closely describes steady-state dendritic growth for the test materials succinonitrile (SCN) and pivalic acid (PVA). These cubic organic crystals offer three major experimental advantages compared to metals: 1) they are transparent, 2) easily purified to high levels, and 3) have conveniently low melting points. IDGE instruments, on board the space shuttle *Columbia*, provided CCD images (telemetered to Earth during each flight), NASA-processed 35-mm film negatives (available postflight), and, for the first time, on USMP-4, real-time streaming of video data telemetered to our laboratory via the K-band (high-frequency) antenna on the shuttle.

1.3 Mushy Zones

A mushy zone is defined as a finely-mixed two-phase region, consisting of a dendritic solid phase embedded within its melt phase. Mushy zones are known to occur in ingots, castings, and welds, and are thought to play a rôle in the evolution of the Earth's semi-solid core, and planetesimals. Mushy zone, in brief, represent "active" regions in a solidifying body, in which the microstructure forms and evolves [15], [16]. In order to simplify the behavior of a melting mushy zone, we studied pure PVA (4-9's purity) under nearly gravity-free conditions. The melting of a mushy zone in a pure substance under convection-free conditions should be kinetically controlled by conduction-limited heat transfer, provided that convection is absent. The kinetics of melting is, of course, normally studied in the presence of relatively strong buoyancy-induced melt convection. Convection alters in complicated ways the thermal fields surrounding dendrites, plus, as an added complexity, sedimentation often takes place because the solid and its melt usually differ (by a few percent) in their mass densities. Conduction-limited freezing and melting processes are of importance in both terrestrial and planetary processes. For example, mushy-zone evolution in castings and ingots, and in fusion welding, can occur where the length scales available for thermal buoyancy are limited, and little convection occurs. Similar mushy-zone behavior may occur in elementary Earth's core processes [17], [18], [19], [20],[21], [22], and in meteoritic and planetary genesis [23], [24], [25], all of which occur either where gravity, *pe se*, is nearly absent, or reduced below terrestrial levels.

2 Video Melting Data

When melting begins, the dendritic side-arms first shorten, and then larger secondary branches detach from the primary stem. Figure 1 shows several video frames representative of a few stages of the total melting sequence. In microgravity, it is important to note that the individual crystalline elements of the mushy zone remain motionless. At the late stages of melting, the remaining dendritic fragments present a large variety of shapes, from nearly spherical blobs to elongated needle-like crystals, all of which eventually melt away.

The mushy-zone melting events, recorded every 1/30th second as video frames, provide to our knowledge the first quantitative data compilation of convection-free melting over a significant range of length scales: from about 10^{-2}m down to about 50×10^{-4}m. The large dynamic range of these length scales allows a detailed kinetic analysis of the convection-free melting process.

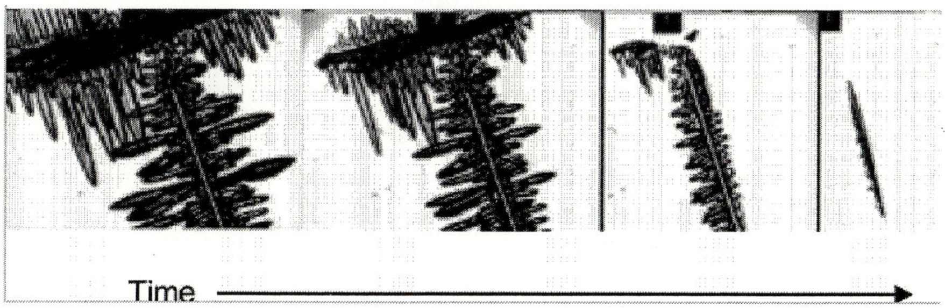

Fig. 1. Video sequence (frame height ~ 1.5 cm) showing the progressive melting of PVA dendrites in microgravity. As the dendrites fragment, individual crystallites remain motionless within the melt. Video data were analyzed frame-by-frame for the melting of the needle crystal shown in the last panel and reported in Figure 6.

3 Data Analysis

3.1 Image Processing

The shapes of the melting dendritic fragments, as analyzed from the IDGE video data, were determined using commercial image analysis software. The fragments, particularly toward the later stages of melting, are best described as prolate spheroids. Figure 2 shows the shape of a melting dendrite stem outlined using this software tool. The equivalent ellipse of the two-dimensional image of this fragment is calculated on the basis of its major axis, $\mathcal{C}$, and minor axis, $\mathcal{A}$. The physical particles are, of course, three-dimensional objects, approximated over time as prolate spheroids of decreasing size. Image analysis permitted measurement of the major and minor axes of the equivalent ellipsoidal fragment as a function of time. Sampling the $\mathcal{C}/\mathcal{A}$ ratio over time towards the end of the melting cycle shows that the values of the $\mathcal{C}/\mathcal{A}$ ratio for a particular fragment do not vary much. For the particular crystal measured, the initial length was $\mathcal{C}_0 = 0.76$cm. Its $\mathcal{C}/\mathcal{A}$ ratio varied from a starting value of about 11, rose to a maximum of about 14, and then decreased to about 10, with almost 95% of the crystal's volume melted. Thus, the $\mathcal{C}/\mathcal{A}$ ratio for this melting fragment may be approximated by its volume-weighted average, which in this instance was $\langle \mathcal{C}/\mathcal{A} \rangle = 12.0$. As shown in Figure 2, this crystal is needle-like in appearance.

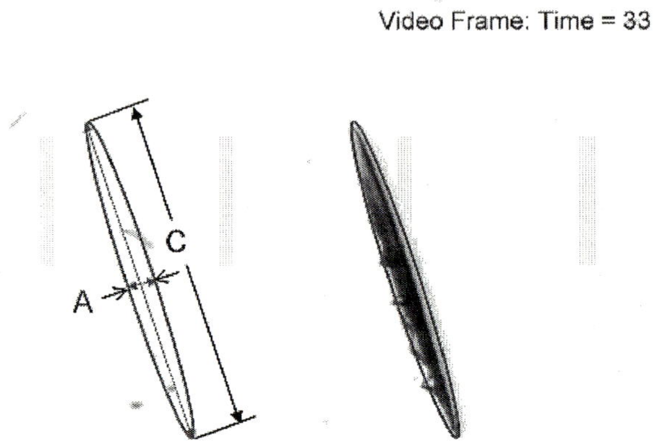

Fig. 2. Sample digitized video frame for image analysis of a melting PVA dendritic fragment. The image analyzer calculates the equivalent ellipse for the fragment, allowing determination of its major axis length, $\mathcal{C}(t)$, and its $\mathcal{C}/\mathcal{A}$ ratio.

4 Quasi-static Theory

4.1 Background

Quasi-static and moving boundary solutions are well known [26] [27] for the growth of spherical crystals for which ($\mathcal{C}/\mathcal{A} = 1$). For the more general cases of freezing ellipsoids and hyperboloids, moving boundary solutions were developed by Ham [28]. However, Ham's similarity solutions for ellipsoids apply *only* to the case of crystal growth. Ham clearly states in his paper that the similarity transformations used in his moving boundary solutions are inapplicable to cases of *melting* ellipsoidal particles. Lacking an analytical moving boundary solution, we developed a quasi-static model for the conduction melting of prolate spheroids. Subsequently, we will demonstrate that the melting kinetics, without convection, of needle-like dendritic fragments may be accurately described as a quasi-static conduction-limited process.

4.2 Potential Theoretic Formulation

The quasi-static melting or freezing of a crystal is based on determining the heat current entering or leaving its *static* solid-melt interface, and then formulating kinematic equations for the crystal's shrinkage or growth. The rate of change of the particle's volume is assumed to occur slowly enough so that the surrounding transport field has sufficient time to adjust to the changing particle. The thermal transport field used only approximates the time-dependent solution to the conduction equation as a quasi-static solution based on Laplace's equation. The key condition determining the applicability of such approximations is that the Stefan number—the intensity measure for melting or freezing—is small, i.e., $St \ll 1$.

We first choose cylindrical coordinates, (r, z, φ), to describe a prolate spheroidal crystal with major axis, $\mathcal{C}$, and minor axis $\mathcal{A}$. The crystal's major axis is aligned with the z-axis. The plane $z = 0$ is the equatorial plane that divides the crystal symmetrically, so that its poles $(r = 0)$ occur at the semi-major axis locations, $z = \pm\mathcal{C}/2$. The axis-symmetric temperature field, $T(r, z)$, is non-dimensionalized as a thermal potential, ϑ, defined as

$$\vartheta(r, z) \equiv \frac{T(r, z) - T_m}{T_\infty - T_m}, \tag{1}$$

where T_∞ is the temperature of the melt infinitely far away from the particle. The propensity and rate of melting (or freezing) is determined by the sign and magnitude of the denominator in Eq.(1). A dimensionless parameter based on that temperature difference is the Stefan number, defined here as

$$St \equiv \frac{T_\infty - T_m}{\Delta H_f / C_\ell^p}, \tag{2}$$

where ΔH_f is the molar enthalpy of fusion, and C_ℓ^p is the molar specific heat of the melt at constant pressure. It is convenient to express the potential defined in Eq.(1) in an orthogonal, curvilinear coordinate system in which the solid-melt interface of the prolate spheroid is an *isothermal* coordinate surface labeled $\xi = \Xi$, for which the potential is $\vartheta(\Xi) = 0$. Such a coordinate system, (ξ, η, φ), is discussed by Morse and Feshbach [29], for which the metric coordinates ξ and η are scaled to the focal length of the prolate spheroid, $a/2$, where the interfocal length $a = \mathcal{C}/\Xi = \mathcal{A}/\sqrt{\Xi^2 - 1}$. Confocal prolate spheroidal coordinates with $a = 2$ are shown as the $\varphi = 0$ projection in Figure 3. In brief, heat flow to or from a prolate spheroid occurs through the melt along the hyperboloidal surfaces η =const., whereas the isotherms of the melt's temperature field are confocal prolate ellipsoids, $\xi \geq \Xi$=const. Specifically, the Laplacian potential surrounding a prolate spheroid with the fixed interfacial potential $\vartheta(\Xi) = 0$, embedded in a melt where the far-field potential $\vartheta(\infty) = 1$, is given in these curvilinear-coordinates as

$$\vartheta(\xi, \eta) = 1 - \frac{\log \frac{\xi+1}{\xi-1}}{\log \frac{\Xi+1}{\Xi-1}}. \tag{3}$$

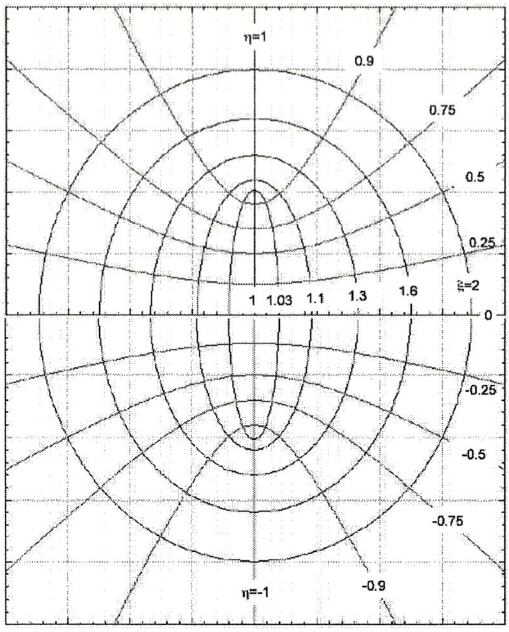

Fig. 3. Confocal prolate spheroidal coordinates (ξ, η, φ) on the $\varphi = 0$ plane. The solid-melt interface has the profile $\xi = \Xi$, where its aspect ratio $\mathcal{C}/\mathcal{A} = \Xi/\sqrt{\Xi^2 - 1}$. The isotherms surrounding this crystal are surfaces $\xi \geq \Xi = const.$, and the heat flow follows surfaces of the form $\eta = constant$. The foci for these ellipsoids are located at $\xi = 1, \eta = \pm 1$, (φ arbitrary) so the interfocal length is $a = 2$.

The gradient field expressed in prolate spheroidal coordinates associated with the thermal potential given in Eq.(3) is easily found by taking the partial derivative of the potential, $\vartheta(\xi, \eta)$, with respect to ξ, and dividing the result by the system's ξ-coordinate scale factor, $h_\xi = (a/2)\sqrt{(\xi^2 - \eta^2)/(\xi^2 - 1)}$. Carrying out these mathematical operations yields the desired gradient field

$$\frac{1}{h_\xi}\frac{\partial\vartheta(\xi,\eta)}{\partial\xi} = \nabla\vartheta(\xi,\eta) = \frac{2}{a}\sqrt{\frac{\xi^2-1}{\xi^2-\eta^2}}\frac{2}{(\xi+1)(\xi-1)}\left(\log\frac{\Xi+1}{\Xi-1}\right)^{-1}, \quad (4)$$

which may be simplified to

$$\nabla\vartheta(\xi,\eta) = \frac{4}{a}\left(\sqrt{(\xi^2-1)(\xi^2-\eta^2)}\log\frac{\Xi+1}{\Xi-1}\right)^{-1}. \quad (5)$$

The relationship of the gradient field, Eq.(5), to the actual thermal gradient in the melt may be found by differentiating Eq.(1) with respect to ξ, and again dividing through by the curvilinear coordinate scale factor, h_ξ,

$$\nabla\vartheta(r,z) = \frac{1}{h_\xi}\frac{\partial T(r,z)}{\partial\xi}\left(\frac{1}{T_\infty - T_m}\right). \quad (6)$$

Rescaling the potential gradient through the characteristic temperature of the material, $\Delta H_f/C_\ell^p$, and substituting the definition of the coordinate scale factor, h_ξ, yields

$$\frac{\Delta H_f}{C_\ell^p}\nabla\vartheta(\xi,\eta)=\frac{\Delta H_f}{C_\ell^p}\cdot\frac{2}{a}\sqrt{\frac{\xi^2-1}{\xi^2-\eta^2}}\cdot\frac{\partial T(r,z)}{\partial\xi}\left(\frac{1}{T_\infty-T_m}\right). \tag{7}$$

If the temperature gradients are evaluated at the solid-melt interface, where $\xi=\Xi$, then Eq.(7) may be written in the compact form

$$\frac{\Delta H_f}{C_\ell^p}\nabla\vartheta(\Xi,\eta)=\frac{1}{St}\nabla T_\Xi. \tag{8}$$

Here the definition of the Stefan number, Eq.(2), has been substituted into Eq.(7), and the symbol ∇T_Ξ denotes the dimensional thermal gradient in the melt evaluated at the solid-melt interface, $\xi=\Xi$.

4.3 Enthalpy Flux

The enthalpy flux, $\mathbf{J}$, for thermal conduction through the melt normal to the solid-melt interface may be found using Fourier's law, $\mathbf{J}=-k_\ell\nabla T_\Xi$, where k_ℓ is the thermal conductivity of the melt. The gradient at the solid-melt interface, Eq.(8), can be substituted into Fourier's law to yield the enthalpy flux,

$$\mathbf{J}=-k_\ell\frac{\Delta H_f}{C_\ell^p}St\cdot\nabla\vartheta(\Xi,\eta). \tag{9}$$

Inserting the expression for the potential gradient, Eq.(5), evaluated on the solid-melt interface, $\xi=\Xi$, into Eq.(9), gives the thermal flux normal to the solid-melt interface, as

$$\mathbf{J}=-k_\ell\frac{\Delta H_f}{C_\ell^p}St\cdot\frac{4}{a}\cdot\left(\sqrt{(\Xi^2-1)(\Xi^2-\eta^2)}\log\frac{\Xi+1}{\Xi-1}\right)^{-1}. \tag{10}$$

A nondimensionless flux, $\mathcal{J}$, may be defined as the last term in Eq.(10), namely,

$$\mathcal{J}\equiv-\left[\sqrt{(\Xi^2-1)(\Xi^2-\eta^2)}\log\frac{\Xi+1}{\Xi-1}\right]^{-1}. \tag{11}$$

The flux entering the ellipsoidal crystal at its poles, $\eta=\pm1$, is found from Eq.(11) to be

$$\mathcal{J}_{pole}=-\left[(\Xi^2-1)\log\frac{\Xi+1}{\Xi-1}\right]^{-1}, \tag{12}$$

whereas the flux entering the ellipsoid at the equatorial plane, $\eta = 0$, is

$$\mathcal{J}_{equator} = -\left[\sqrt{(\Xi^2 - 1)(\Xi^2)}\log\frac{\Xi + 1}{\Xi - 1}\right]^{-1}. \tag{13}$$

The ratio of the polar-to-equatorial fluxes is

$$\frac{\mathcal{J}_{pole}}{\mathcal{J}_{equator}} = \frac{\Xi}{\sqrt{\Xi^2 - 1}} = \frac{\mathcal{C}}{\mathcal{A}}. \tag{14}$$

Equation (14) proves that the the flux ratio between the poles and the equator is equal to the $\mathcal{C}/\mathcal{A}$ ratio. This result is consistent with the requirement that the imposed quasi-static melting rates maintain the particle's shape by keeping the $\mathcal{C}/\mathcal{A}$ ratio fixed in time. Note too that the η-dependent terms in Eqs.(10) and (11) are responsible for what is often termed the "point-effect" of diffusion. As a prolate spheroid increases its $\mathcal{C}/\mathcal{A}$ ratio, it becomes more needle-like, and the fluxes around the poles increase compared to the fluxes in the equatorial region. See Figure 4.

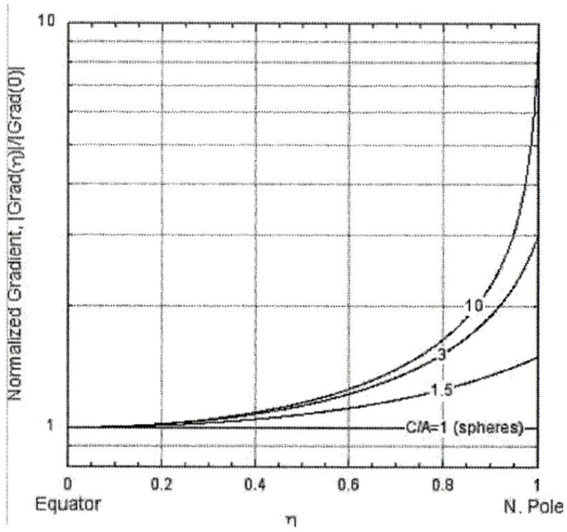

Fig. 4. "Point-effect" of diffusion as described by Eq.(11). The flux (log-scale) at any latitude, η, is normalized to the equatorial flux. As the spheroid becomes more needle-like, the fluxes in the polar regions ($\eta \sim 1$) increase relative to those in the equatorial region ($\eta \sim 0$).

4.4 Enthalpy Current

The total heat current conducted to or from a *static* prolate spheroid may be expressed as a surface integral of the enthalpy flux,

$$\dot{Q}_{tot} = \oint \mathbf{J} \cdot \mathbf{n} dS, \tag{15}$$

where the differential element of solid-melt interfacial area on the prolate spheroid, $\xi = \Xi$, may be expressed in the curvilinear coordinates as

$$dS = r d\sigma d\varphi = r h_\eta d\eta d\varphi, \tag{16}$$

where $r = a/2\sqrt{(\Xi^2 - 1)(1 - \eta^2)}$ is the radial distance from the major axis of the spheroid to an arbitrary point, η, on its solid-melt interface; $d\sigma = h_\eta d\eta$ is the differential of the arc length; and h_η is the η-coordinate scale factor needed to relate the arc length, σ, along the interface in curvilinear coordinates defined by Morse & Feshbach as $h_\eta \equiv a/2\sqrt{(\Xi^2 - \eta^2)/(1 - \eta^2)}$.

Substituting these transformations into Eq.(16) yields the differential element of interfacial area on the prolate spheroid, $\xi = \Xi$,

$$dS = \frac{a^2}{4}\sqrt{(\Xi^2 - 1)(\Xi^2 - \eta^2)} d\eta d\varphi. \tag{17}$$

Inserting Eq.(10) for the normal flux vector, and Eq.(17) for the differential interfacial area, into Eq.(15) yields an integral expression for the total enthalpy current,

$$\dot{Q}_{tot} = -\frac{a k_\ell \Delta H_f St}{C_\ell^p}\left(\log\frac{\Xi + 1}{\Xi - 1}\right)^{-1}\int_{-1}^{1} d\eta \int_0^{2\pi} d\varphi, \tag{18}$$

where the integration limits $-1 \le \eta \le 1$ and $0 \le \varphi \le 2\pi$ correspond to the total interfacial area from pole-to-pole on the ellipsoid $\xi = \Xi$. Carrying out the closed surface integration indicated in Eq.(18) introduces the expected total spherical image, 4π, and yields the total heat current entering the prolate spheroid, $\xi = \Xi$, with an interfocal length a,

$$\dot{Q}_{tot} = -\frac{4\pi}{\log\frac{\Xi+1}{\Xi-1}}\frac{a k_\ell \Delta H_f}{C_\ell^p} St. \tag{19}$$

4.5 Kinematic Formulation

Heat entering (or leaving) the spheroidal crystal results in its shrinkage (or growth) caused by the $S \rightleftharpoons L$ phase transformation. The time-dependent volume of a melting or freezing prolate spheroid may be expressed, applying our notation, in terms of its time-dependent major- and minor-axes through the formula [30],

$$V(t) = \frac{\pi}{6}\mathcal{A}^2(t)\mathcal{C}(t) = \frac{\pi}{6}\left(\frac{\mathcal{A}}{\mathcal{C}}\right)^2 \mathcal{C}^3(t), \tag{20}$$

where the time-dependence of the spheroid's fixed aspect ratio, $\mathcal{C}/\mathcal{A}$, is explicitly suppressed here in the notation used for the right-hand equality. Note that in the limiting case of a sphere of diameter $\mathcal{D}$, the aspect ratio $\mathcal{C}/\mathcal{A} = 1$, so that $\mathcal{C} = \mathcal{A} = \mathcal{D}$, and the familiar result obtains from Eq.(20). If the crystalline spheroid melts (or freezes) and is subject to a time-rate of change of its volume, then the time-rate of change of its major-axis length is related kinematically to the volume derivative as

$$\dot{V}(t) = \frac{\pi}{2}\left(\frac{\mathcal{A}}{\mathcal{C}}\right)^2 \mathcal{C}^2(t)\,\dot{\mathcal{C}}(t), \tag{21}$$

where the overdot notation indicates differentiation with respect to time. The rate of enthalpy change associated with the volume change may be found by multiplying Eq.(21) by the volumetric enthalpy, $\Delta H_f/\Omega$,

$$\dot{H}_{tot} = \frac{\pi}{2}\frac{\Delta H_f}{\Omega}\left(\frac{\mathcal{A}}{\mathcal{C}}\right)^2 \mathcal{C}^2(t)\,\dot{\mathcal{C}}(t), \tag{22}$$

where Ω is the molar volume. This enthalpy expression ignores the imperceptible advection occurring in the melt near the interface caused by the small ($\approx 3\%$) density differences of the phases. Coriell and McFadden have shown that this effect is generally negligible in the case of steady-state dendritic crystallization [31]. The kinematic energy balance required under quasi-static conditions is therefore $\dot{H}_{tot} = \dot{Q}_{tot}$. Equating Eq.(22) with Eq.(19) gives

$$\frac{\pi}{2}\frac{\Delta H_f}{\Omega}\left(\frac{\mathcal{A}}{\mathcal{C}}\right)^2 \mathcal{C}^2(t)\,\dot{\mathcal{C}}(t) = -\frac{4\pi}{\log\frac{\Xi+1}{\Xi-1}}\frac{ak_\ell \Delta H_f}{C_\ell^p}St. \tag{23}$$

Equation (23) is the O.D.E. for the quasi-static evolution of a prolate spheroidal crystal of constant aspect ratio contacting a superheated, or supercooled, melt of Stefan number, St. Equation (23) may be solved for the measurable quantity $\mathcal{C}(t)$ by substituting the transformation for the interfocal length $a = \mathcal{C}/\Xi$, and then separating the variables, after which one obtains

$$\int_{\mathcal{C}_0}^{\mathcal{C}(t)} \mathcal{C}d\mathcal{C} = -\frac{8}{\Xi\log\frac{\Xi+1}{\Xi-1}}\left(\frac{\mathcal{C}}{\mathcal{A}}\right)^2 \alpha_\ell St\int_0^t dt'. \tag{24}$$

Carrying out the integrations indicated in Eq.(24) yields, after a few steps of algebra, the kinetic expression for quasi-static melting or freezing of prolate spheroids, namely

$$\mathcal{C}(t) = \sqrt{\mathcal{C}_0^2 - \frac{16\alpha_\ell St}{\Xi\log\frac{\Xi+1}{\Xi-1}}\left(\frac{\mathcal{C}}{\mathcal{A}}\right)^2 t}. \tag{25}$$

Equation (25) may be non-dimensionalized as

$$\frac{\mathcal{C}(t)}{\mathcal{C}_0} = \sqrt{1 - K_{prol}\cdot St\cdot Fo}, \tag{26}$$

where the kinetic coefficient, K_{prol}, is defined as

$$K_{prol} \equiv \frac{16}{\Xi \log \frac{\Xi+1}{\Xi-1}} \left(\frac{\mathcal{C}}{\mathcal{A}}\right)^2, \tag{27}$$

and the Fourier number (dimensionless transformation time) is defined as

$$Fo \equiv \frac{\alpha_\ell}{\mathcal{C}_0^2} t. \tag{28}$$

It is convenient for the current experimental purposes to relate K_{prol} to the crystal's $\mathcal{C}/\mathcal{A}$ ratio. Equation (27) is cross-plotted in Figure 5 against the aspect ratio $\mathcal{C}/\mathcal{A} = \Xi/\sqrt{\Xi^2 - 1}$. If one sets aside questions of interfacial stability, Eq.(25) provides an analytic expression of the quasi-static kinetic law for the growth ($St \geq 0$) or melting ($St \leq 0$) of a prolate spheroidal crystal.

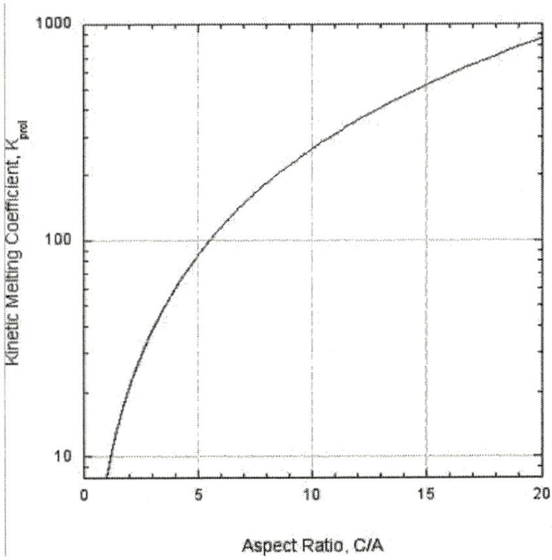

Fig. 5. Quasi-static kinetic coefficient, K_{prol}, versus the aspect ratio, $\mathcal{C}/\mathcal{A}$. Note that this kinetic coefficient approaches a value of 8 as the $\mathcal{C}/\mathcal{A}$ ratio approaches unity (spheres), but rises rapidly for the case of ellipsoidal needle-like crystals.

5 Results

5.1 Experimental Melting Rates

The PVA crystal that we analyzed in microgravity formed originally as a part of a dendritic mushy zone created in IDGE growth cycle 4. These dendrites grew from pure molten pivalic acid that was supercooled 0.40K. The mushy zone consisted of approximately 4% dendritic solid when melting was initiated. The total time needed to melt this mushy zone completely was approximately 12 minutes. The last large fragment to melt was the one selected for quantitative analysis. (See again the final panel shown in Figure 1.) This dendrite fragment (of initial length $C_0 = 0.76$cm) was the residual crystallite from a melting primary dendrite stem, and exhibited a nearly constant aspect ratio $\mathcal{C}/\mathcal{A} \cong 12$ throughout its final 40s melting sequence. To calculate the melting kinetics of this dendritic fragment, one must know, in addition to the $\mathcal{C}/\mathcal{A}$ ratio, the thermal diffusivity and the Stefan number of the melt. The thermistors arranged within the IDGE thermostat were optimized for measuring the supercooling *prior* to dendritic growth quite precisely (±0.002K). The subsequent temperature history within the growth chamber, especially during the melting portion of an experimental growth cycle, is, unfortunately, less certain. Specifically, temperature uncertainties arose because of a lack of precise synchronization between the video clock and the mission clock. In view of this situation, one may instead adopt a self-consistent method of estimating the Stefan number in the melt during the last 40s of the video melting sequence. The basis of the value of the self-consistent Stefan number is just the observed extinction time of the fragment. The extinction time (in this instance, 40s) is the interval needed for the melting crystal to just disappear. If the left-hand side of Eq.(25) is set equal to zero, a relationship may be established among the unknown Stefan number, the starting size of the crystal, and the crystal's observed extinction time, $t^{\dagger} = 40s$, namely

$$St = \frac{C_0^2}{\alpha_\ell \cdot K_{prol}} \cdot \frac{1}{t^{\dagger}}. \tag{29}$$

The extinction time can be measured precisely to within 1 video frame (±0.033s), and the thermal diffusivity of molten PVA at its melting point is known to be $\alpha_\ell = 7.0 \times 10^{-4}\mathrm{cm}^2/\mathrm{s}$ [13]. Inserting these data and the value for $K_{prol} = 360.6$ into Eq.(29), (See again Figure 5 for $\mathcal{C}/\mathcal{A} = 12.0$.) yields the self-consistent Stefan number, $St = 0.057$, based on the observed extinction time of 40s . The crystal melting data reported in Figure 6 were fit in their entirety using the self-consistent Stefan number, which corresponds to the melt being superheated 0.63K above its equilibrium melting point, $T_m = 35.97K$.

The overall prediction of the melting kinetics during the lifetime of this dendrite fragment, using the quasi-static theory developed in Section 4, appears to be in good agreement with experiment. Scrutiny of the video data indicates that only small disparities with the predictions occur over the melting interval. Particularly as the extinction time is approached, and the speed of melting accelerates, capillary and interfacial attachment effects may enter the kinetics and alter the melting rate from that calculated by quasi-static conduction theory. The departures observed here from quasi-static theory are, however, minor, and will not be pursued further in this paper.

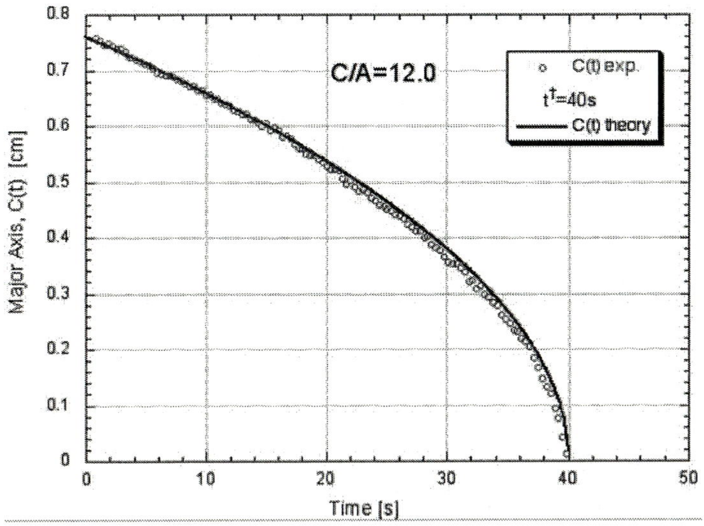

Fig. 6. Melting kinetics of a dendritic pivalic acid (PVA) crystal fragment ($\mathcal{C}/\mathcal{A} = 12$) in microgravity. Experimental melting data (open symbols) obtained through digital analysis of 230 digitized video frames measure $\mathcal{C}(t)$ versus time. Solid curve is the theoretical prediction from Eq.(25) fitted with a Stefan number equal to 0.057. The fit with quasi-static theory is based on the observed extinction time $t^{\dagger} = 40$s.

5.2 Thermal Data

Thermal data telemetered from the IDGE thermostat indicate independently that the *maximum* Stefan number recorded by the high-resolution thermistors internal to the growth chamber at the end of this experimental cycle was about 0.16. Melting of the last dendritic fragments clearly occurred at Stefan numbers much less than 0.16, but the precise range of values over which the observed crystal melted remains imperfectly known. As an alternative to obtaining direct measurements of the melt temperature within the chamber, one may recognize that the rate of temperature increase in the growth chamber depended on two countervailing factors: 1) the temperature difference between the exterior thermostat temperature (fixed at 37.85C during melting) and the interior melt temperature, and 2) the quantity of crystals remaining within the chamber that actively absorb latent heat. When melting of the mushy zone begins near the chamber walls, the temperature difference is ~1.8K, and the chamber contains ~4% dendritic crystals. The rate of temperature rise is, however, initially slow, because most of the heat added is absorbed as the latent heat of melting. As the dendrites melt and their total mass decreases, the rate of latent heat absorption also decreases, and thus the rate of temperature rise increases. The rate of temperature rise within the IDGE growth chamber should reach its maximum value at about the time at which the last crystals disappear, i.e., near the time when our detailed melting data were taken. It is at that point where all latent heat absorption ceases within the chamber, yet the temperature difference between the interior and the exterior is still relatively large (~1.2K in this case). We emphasize that using a heating-rate versus time correlation is an approximate procedure. Nonetheless, when the time derivative of the Stefan number versus time is calculated for the experiment, a distinct peak occurs about half-way through the total melting period at t = 720s. The time at which the peak rate occurs corresponds to a Stefan number of 0.055. During the 40s period for the melting sequence of the prolate spheroid we selected, the value of the maximum time derivative of the Stefan number is $dSt/dt \cong 2 \times 10^{-4}\text{s}^{-1}$. Thus, the change of Stefan numbers over the 40s observation period is estimated to be less than ~ 0.008. The average Stefan number during melting based on the IDGE thermal data is 0.055 ± 0.004—a value in agreement with the self-consistent Stefan number ($St = 0.057 \pm 0.001$) determined from Eq.(29) by fitting the kinetic data to the crystal's observed extinction time, $t^\dagger$ = 40s.

Possible melting behaviors for prolate spheroidal crystals with $\mathcal{C}/\mathcal{A} = 12$, subjected to different Stefan numbers are shown for comparison in Figure 7.

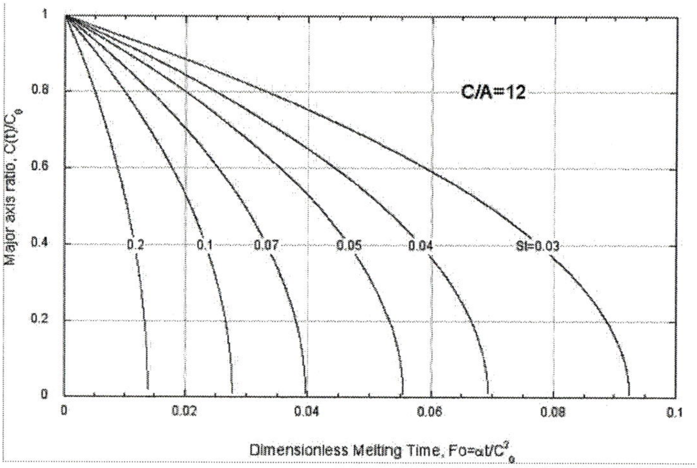

Fig. 7. Major axis length scaled to initial major axis length, $\mathcal{C}(t)/\mathcal{C}_0$, versus melting time, predicted by Eq.(25) for $\mathcal{C}/\mathcal{A} = 12$. These curves show the sensitivity to the the Stefan number of the melting kinetics for needle-like prolate spheroidal crystals.

These curves are predicted using Eq.(25) with the indicated Stefan numbers, which were selected to cover the full range of values encountered during the 12 minute melting portion of the experimental cycle. Note that the melting kinetic curves for $\mathcal{C}/\mathcal{A} = 12$ change substantially with only modest variations in the Stefan number. It is interesting to point out that as the aspect ratio of a crystal decreases towards unity, its kinetic melting coefficient, K_{prol}, rapidly falls toward the limiting value for spheres of 8. Extinction times, as noted in Figure 7, are rather sensitive to a crystal's aspect ratio.

6 Conclusions

1. Dendritic mushy zones were formed in supercooled high-purity PVA melts in microgravity. At the end of each experimental cycle the mushy zone was progressively melted by raising the temperature of the thermostat. Video data at 30 fps of the melting process for one cycle (0.4K supercooling) are reported here. Upon raising the temperature, the mushy zone melts into numerous fragments, many of which may be characterized as prolate spheroids.
2. Individual dendritic fragments remain motionless in the low-gravity environment, progressively melting toward extinction at nearly constant $\mathcal{C}/\mathcal{A}$ ratios.
3. Moving boundary solutions, such as those proposed by Ham for ellipsoidal crystals, are not applicable to the melting process. We developed instead a more broadly applicable quasi-static analytic theory, using curvilinear coordinates, in lieu of a numerical solution to the moving boundary problem. The theory describes the kinetics of conduction melting of prolate spheroids of arbitrary $\mathcal{C}/\mathcal{A}$ ratios. The quasi-static analytic results are, however, limited to small Stefan numbers ($St \ll 1$).
4. The observed melting kinetics in microgravity of a prolate spheroidal crystal with an aspect ratio $\mathcal{C}/\mathcal{A} = 12$ was found to be described accurately by the quasi-static analysis. Predictions of the crystal size as a function of time require knowledge of the Stefan number and thermal diffusivity of the melt, along with the kinetic parameter derived here from the theory that accounts for the crystal's aspect ratio.
5. As the $\mathcal{C}/\mathcal{A}$ increases, and a spheroid becomes more needle-like in shape, the polar flux increases relative to the equatorial flux, in accord with the so-called "point effect" of diffusion.
6. The Stefan number required to compare theory and experiment was estimated using the observed extinction time of the crystal. Measured heating rates yield an independent Stefan number found to be in good agreement with the self-consistent value derived from the extinction time. The predicted convection-free melting kinetics accurately describe the observations in microgravity that are reported here.

7 Acknowledgements

The authors gratefully acknowledge the support provided for this work by the National Aeronautics and Space Administration, Division of Biological and Physical Sciences Research, Code U, Washington, D.C., under Grant NAG 8-1696.

References

1. Glicksman, M.E., Marsh, S.P., " The Dendrite." *Handbook of Crystal Growth.* Ed. D.T.J. Hurle, Elsevier Science Publishers, Amsterdam (1993) 1,075-1,122.
2. Langer, J.S., Muller-Krumbhaar, H.," Theory of dendritic growth. I. Elements of a stability analysis." *Acta Metallurgica,* Vol. 26 (1978) 1,681-1,687.
3. Langer,J.S., " Nonequilibrium physics and the origins of complexity in nature." *Princeton Series in Physics. Critical Problems in Physics: Chapter 2,* Princeton University Press 1996.
4. Mullins,W.W., Sekerka S.F., "Morphological stability of a particle growing by diffusion of heat flow." *J. Appl. Phys.*, Vol. 34 (1961) 323-329.
5. Glicksman, M.E., Schaefer, R.J., Ayers, J.D., "Dendritic Growth-A Test of Theory." *Metall. Trans. A*, Vol. 7A (1976) 1,747-1,759.
6. Glicksman, M.E., "Fundamentals of Dendritic Crystal Growth." *Crystal Growth of Electronic Materials. Chapter 5.* Ed. E. Kaldis, Elsevier Science Publishers (1985) 57-69.
7. Huang, S.C., Glicksman, M.E., "Fundamentals of dendritic solidification-I. Steady-state tip growth." *Acta Metallurgica,* Vol. 29 (1981) 701-715.
8. Trivedi, R., Mason, J.T., "The effects of interface attachment kinetics on solidification interface morphologies." *Metall. Trans. A*, Vol. 22A (1991) 235-249.
9. Lipton, J., Glicksman, M.E., Kurz, W., "Equiaxed Dendrite Growth in Alloys at Small Supercooling." *Metall. Trans. A*, Vol. 18A (1987) 341-345.
10. Glicksman, M.E., Koss, M.B., Bushnell, L.T., Lacombe, J.C., Winsa, E.A., "The Isothermal Dendritic Growth Experiment." *Materials Science Forum,* Vol. 215-216 (1996) 179-190.
11. LaCombe, J.C., Koss, M.B., Fradkov, V.E., Glicksman, M.E., " Three-dimensional dendrite-tip morphology." *Physical Review, E,* Vol. 52, No. 3 (1995) 2,778-2,786.
12. LaCombe, J.C., Koss, M.B., Corrigan, D.C., Lupulescu, A., Tennenhouse, L.A., "Implications of the interface shape on steady-state dendritic crystal growth." *Journal of Crystal Growth,* Vol. 206 (1999) 331-344.
13. Lupulescu, A., LaCombe, J.C., Frei, J.E., Glicksman, M.E., Koss, M.B., "USMP-4 Microgravity PVA Dendritic Growth Data -Preliminary Analysis-." *Microgravity Symposium: Proceedings TMS Meeting,* Nashville, 2000, CD-ROM.
14. Ivantsov, G.P., "The Temperature Field Around Spherical , Cylindrical, and Needle-Shaped Crystals which Grow in Supercooled Melts." *Dokl. Akad. Nauk. USSR,* Vol. 58, No. 4 (1947) 567-569.
15. Marsh, S.P., Glicksman, M.E., "Overview of geometric effects on coarsening of mushy zones." *Metallurgical and Materials Transactions A*, Vol. 27A (1996) 557-567.
16. Paradies, C.J., Smith, R.N., Glicksman, M.E., "The influence of convection during solidification on fragmentation of the mushy zone of a model alloy." *Metallurgical and Materials Transactions A*, Vol. 28A (1997) 875-883.
17. Fearn, D.R., Loper, D.E., Roberts, P.H., "Structure of the Earth's inner core." *Nature,* Vol. 292 (1981) 232-233.
18. Loper, D.E., "Structure of the inner core boundary." *Geophys. Astrophys. Fluid Dyn.* Vol. 25 (1983) 139-155.

19. Wenk, H.R., Takeshita, T., Jeanloz, R., Johnson, G.C., "Development of texture and elastic anisotropy during deformation of hcp metals." *Geophysical Research Letters*, Vol. 15 (1988) 76-79.
20. Cormier, V.F., "Inner core structure inferred from body waveforms." *EOS*, Vol. 75 (1994) 67.
21. Souriau, A., Romanowicz, B., "Anisotropy in inner core attenuation: a new type of data to constrain the nature of the solid core." *Geophysical Research Letters*, Vol. 23 (1996) 1-4.
22. Bergman, M.I., "Measurements of elastic anisotropy due to solidification texturing and the implications for the Earth's inner core." *Nature*, Vol. 389 (1997) 60-63.
23. Rushmer, T., Minarik, W.G., Taylor, G.J., "Physical Processes of Core Formation." *Origin of the Earth and Moon*, Eds. R.M. Canup, K. Righter, The University of Arizona Press, Tucson, (2000) 227-245.
24. Haack, H., Scott, E.R.D., "Asteroid Core Crystallization by Inward Dendritic Growth." *Journal of Geophysical Research*, Vol. 97 (1992) 14,727-14,734.
25. Esbensen, K.H., Buchwald, V.F., "Planet(oid) core crystallization and fractionation- evidence from the Agpalilik mass of the Cape York iron meteorite shower." *Phys. Earth. Planet. Inter.*, Vol. 29 (1982) 218-232.
26. Sekerka, R.F., Jeanfils, C.L., Heckel, R.W., "The Moving Boundary Problem." *Lectures on the Theory of Phase Transformations*, Ed. H.I. Aaronson, The Metallurgical Society, AIME, New York (1982) 117-169.
27. Glicksman, M.E., *Diffusion in Solids: Field Theory, Solid-state Principles, and Applications.* John Wiley and Sons, New York, 2000.
28. Ham, F.S., "Shape-Preserving solutions of the time-dependent diffusion equation." *Quart. J. Appl. Math.*, Vol. 17 (1959) 137-145.
29. Morse, P.M., Feshbach, H., *Methods of Theoretical Physics*, Vol. 2, McGraw-Hill, New York (1953) 1,284-1,292.
30. Eshbach, O.W., *Handbook of Engineering Fundamentals*, 2nd edition, John Wiley & Sons, New York (1961) 2-60.
31. McFadden, G.B., Coriell, S.R., "The Effect of Fluid Flow due to the Crystal-Melt Density Change on the Growth of a Parabolic Isothermal Dendrite." *Journal of Crystal Growth*, Vol. 74 (1986) 507-512.

Deterministic Behaviour in Sidebranching Development

Ricard González-Cinca[1] and Yves Couder[2]

[1] Departament de Física Aplicada, Universitat Politècnica de Catalunya, Campus Nord UPC, Ed. B5, J. Girona Salgado s/n, E-08034 Barcelona, Spain
[2] Laboratoire de Physique Statistique, Ecole Normale Supérieure, 24 rue Lhomond, 75231 Paris CEDEX 05, France

Abstract. The development of sidebranching in solidifying dendrites in a regime of finite diffusion length is studied both experimentally and by means of a phase–field model. The growth rate of each sidebranch shows a power–law behaviour from the early stages of its life. From their birth, branches which finally succeed in the competition process of sidebranching development have a greater growth exponent than branches which are stopped.

1 Introduction

Dendritic patterns in diffusion–controlled systems have received much attention in the past decades [1–3]. In crystal growth, the study of sidebranches that appear at the flanks of a growing dendrite has been developed more recently. Most of the work devoted to sidebranching has been focussed in the scenario of selective amplification of fluctuations in the region close to the tip in which linear theories are valid [4–7]. However, the shape of the dendrite further down from the tip and, in particular, the development of sidebranches there have been scarcely studied [8,9].

In this paper we present an experimental and numerical study on the growth process of sidebranches at initial stages, which turns out to be of crucial importance in the selection of the final shape of a two–dimensional dendrite. We consider solidification systems where growth is controlled by a finite–range diffusive field. Numerical simulations were carried out by means of a phase–field model.

In the next Section we present some experimental results on the behaviour of competing sidebranches. In Sec. 3, the phase–field model employed in this work is presented. The numerical procedure as well as the simulation results and comparison with experiments are presented in Sec. 4.

2 Experimental Results

We grew ammonium bromide crystals in conditions similar to those described in Refs. [10,11], but from a solution with different concentration. The system is driven by mass diffusion, and we limit to the case when the diffusion length

$l_D = D/V$, D and V being the diffusion coefficient and the dendrite velocity respectively, is small when compared to the region occupied by the dendrite.

In Fig. 1 a portion of the considered dendrite is shownin which fully developed sidebranching can be observed. We focus in the smaller region shown in Fig. 2 and we choose three sidebranches (labeled a, b and c in the figure) growing from the main dendrite.

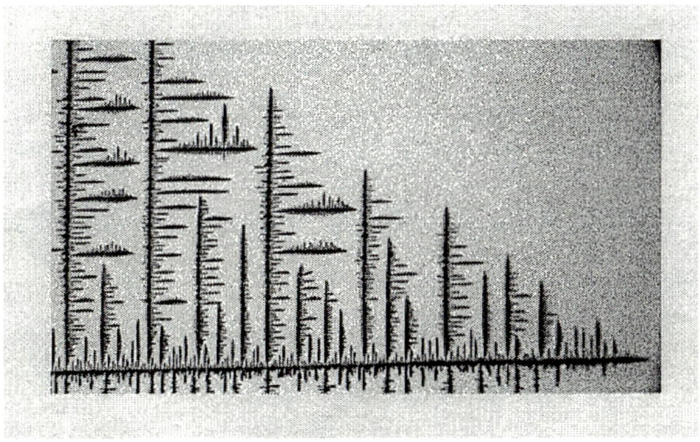

Fig. 1. Ammonium bromide dendrite grown in a regime of finite diffusion length.

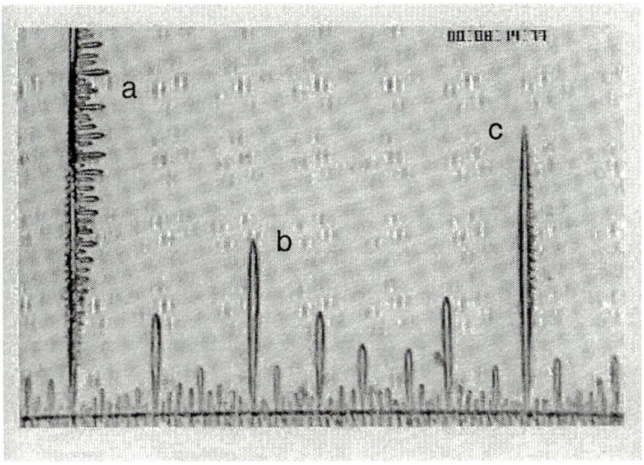

Fig. 2. Region of the dendrite containing selected sidebranches a, b and c.

We have recorded the shortest distance of the tip of each sidebranch to the axis of the main dendrite, y, as a function of time from the birth of the sidebranch. The plot of the three curves in Fig. 3a shows the saturation in the growth of branch b caused by the competition process taken place with a and c. Fig. 3b corresponds to a log–log plot of the same curves. By eliminating the saturation part and fitting each of them to $y \sim t^n$, it is found $n = 1.24, 0.82$ and 1.01 for branches a, b and c, respectively. Thus, the growth rate of sidebranches follows a power–law behaviour and the exponent is smaller in those branches which are screened off in the competition process.

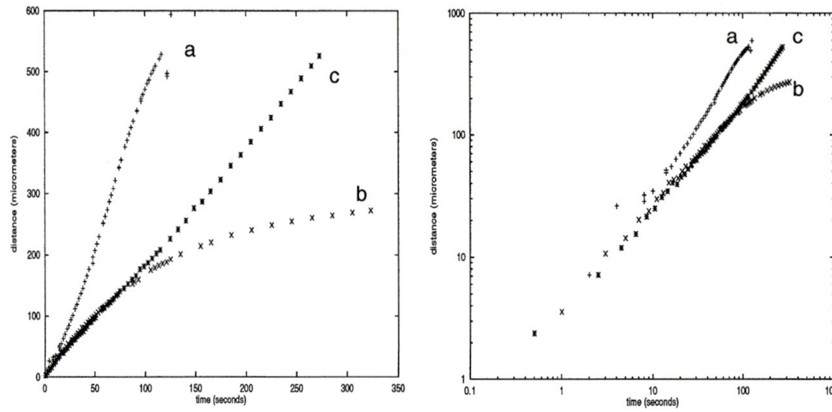

Fig. 3. a) Distance between the tip of each sidebranch and the axis of the main dendrite *vs.* time; b) log–log plot.

3 Model

We have performed simulations of dendritic growth by means of a phase–field model for solidification [12,13]. The corresponding equations for the time evolution of the phase field ϕ and the dimensionless diffusive field $u(\mathbf{r}, t)$ can be written in the following form [12]:

$$\epsilon^2 \tau(\theta) \frac{\partial \phi}{\partial t} = \phi(1-\phi) \left(\phi - \frac{1}{2} + 30\epsilon\alpha\Delta u \phi(1-\phi) \right)$$
$$-\epsilon^2 \frac{\partial}{\partial x} \left[\eta(\theta)\eta'(\theta) \frac{\partial \phi}{\partial y} \right] + \epsilon^2 \frac{\partial}{\partial y} \left[\eta(\theta)\eta'(\theta) \frac{\partial \phi}{\partial x} \right] + \epsilon^2 \nabla \left[\eta^2(\theta) \nabla \phi \right] \quad (1)$$

$$\frac{\partial u}{\partial t} + \frac{1}{\Delta} \left(30\phi^2 - 60\phi^3 + 30\phi^4 \right) \frac{\partial \phi}{\partial t} = \nabla^2 u + \psi(x, y, t), \quad (2)$$

where Δ is the dimensionless undercooling. Lengths are scaled in some arbitrary reference length ω, while times are scaled by ω^2/D. In these equations

θ is the angle between the x-axis and the gradient of the phase field and $\eta(\theta) = \sigma(\theta)/\sigma(0)$, $\sigma(\theta)$ being the surface energy. The kinetic coefficient has been taken as isotropic, $\beta(\theta) = \beta_o$, which leads to $\tau(\theta) = m\eta(\theta)$ with constant $m = c_p D\beta_o/Ld_o$, c_p being the specific heat per unit volume, L the latent heat per unit volume and d_o the capillary length. α is equal to $\frac{\sqrt{2}\omega}{12d_o}$. The parameter ϵ controls the interface thickness and by $\epsilon \to 0$ the classical sharp–interface model is recovered. The term ψ introduces an external noise that induces sidebranching [7].

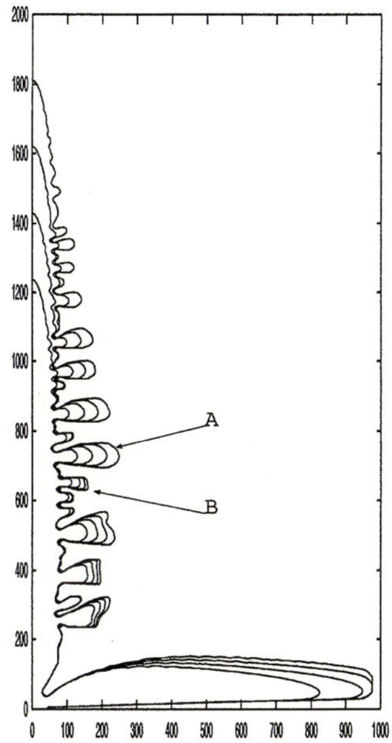

Fig. 4. Dendrite at different times obtained with a phase–field model.

4 Numerical Procedure and Results

The phase-field model equations have been solved on rectangular grids with mesh spacing Δx using first-order finite differences. A four-fold surface tension anisotropy $\sigma = \sigma(0)(1 + \gamma\cos(4\theta))$ has been considered. The noise term is evaluated at each uncorrelated cell of lateral size Δx simply as $I \cdot r$, where I denotes the amplitude of the noise, and r is a uniform random number in the interval $[-0.5, 0.5]$. We have used a set of parameters that gives rise to a needle without sidebranching when $I = 0$. The employed parameters are $\alpha = 400$, $m = 20$, $\gamma = 0.06$, $\epsilon = 0.003$, $I = 19$, $\Delta t = 10^{-4}$, $\Delta x = 0.01$ and

$\Delta = 0.42$. In Fig 4 a dendrite is shown which is obtained with this set of parameters at different times. Labels A and B indicate the sidebranches that we will consider. It can be observed that growth of branch B saturates at later times.

A more detailed picture of the region containing the considered branches is shown in Fig. 5. We have computed the position y of the tips of both branches as a function of time, similarly to what we did with the selected sidebranches among those grown in the experiment.

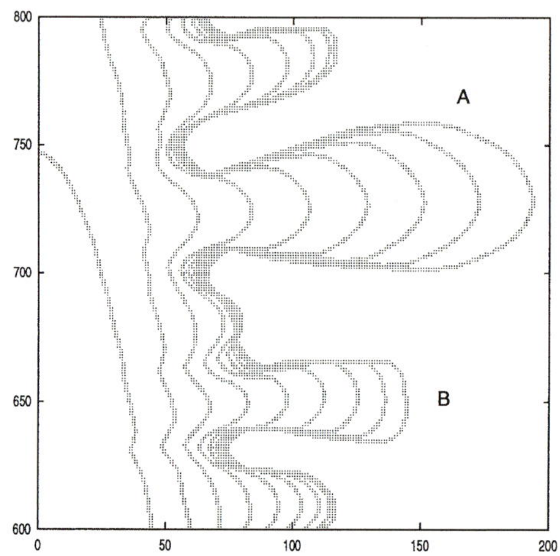

Fig. 5. Region containing branches A and B at different times.

In Fig. 6 the points (t,y) for branches A and B as well as fitting curves $y \sim t^n$ for both data sets are represented. The value of the growth exponent is 1.06 and 0.93 for A and B respectively, which indicates that the winner branch grows from the initial stages of its life faster than the looser branch. This conclusion is obtained systematically when comparing different branches.

The results presented in this paper suggest that the noise mechanism plays a less important role in the determination of the final sidebranching structure than expected. We have found that the competition process between branches is fully determined from the beginning and the branch which starts growing faster succeeds.

We acknowledge A. Hernández–Machado for fruitful discussions.

References

1. Langer, J.S., in *Chance and Matter*, Souletie, J., Vannimenus, J., Storra, R., Eds. North Holland, (1987) 629

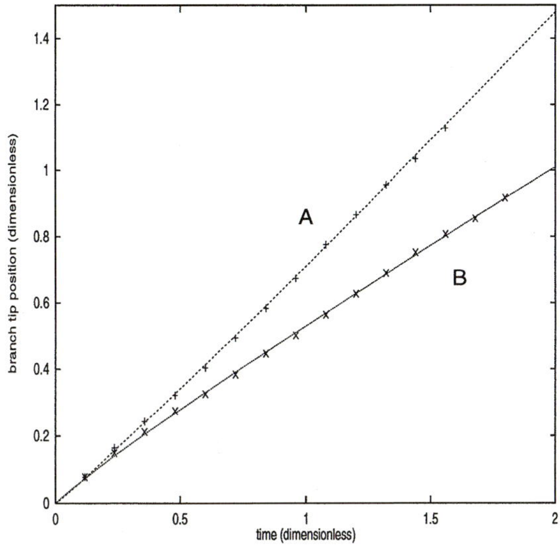

Fig. 6. Position of the tip of branches A and B as a function of time (points) and the corresponding fitting curves $y \sim t^n$ (lines).

2. *Dynamics of Curved Fronts*, Pelcé, P., Ed. Academic Press (1988)
3. *Growth and Form. Nonlinear Aspects*, Ben Amar, M., Pelcé, P., Tabeling, P., Eds. Plenum Press (1988)
4. Brener, E., Temkin, D.: Noise–induced sidebranching in the three–dimensional nonaxisymmetric dendritic growth. Phys. Rev. E **51** (1995) 351–359
5. Karma, A., Rappel, W.-J.: Phase–field model of dendritic sidebranching with thermal noise. Phys. Rev. E **60** (1999) 3614–3625
6. Pavlik, S.G., Sekerka, R.F.: Fluctuations in the phase–field model of solidification. Physica A **277** (2000) 415–431
7. González–Cinca, R., Ramírez–Piscina, L., Casademunt, J., Hernández–Machado, A.: Sidebranching induced by external noise in solutal dendritic growth. Phys. Rev. E **63** (2001) 051601-1–9
8. Bisang, U., Bilgram, J.H.: Shape of the tip and the formation of sidebranches of xenon dendrites. Phys. Rev. E **54** (1996) 5309-5326
9. Li, Q., Beckermann, C.: Evolution of the sidebranch structure in free dendritic growth. Acta Mater. **47** (1999) 2345-2356
10. Maurer, J., Bouissou, P., Perrin, B., Tabeling, P.: Europhys. Lett. **6** (1989) 67
11. Couder, Y., Argoul, F., Arnéodo, A., Maurer, J., Rabaud, M.: Statistical properties of fractal dendrites and anisotropic diffusion–limited aggregates. Phys. Rev. A **42** (1990) 3499–3503
12. Wheeler, A.A., Murray, B.T., Schaefer, R.J.: Computation of dendrites using a phase–field model. Physica D **66** (1993) 243–262
13. McFadden, G.B., Wheeler, A.A., Braun, R.J., Coriell, S.R., Sekerka, R.F.: Phase–field models for anisotropic interfaces. Phys. Rev. E **48** (1993) 2016–2024

Growth Dynamics during Solidification of Undercooled Melts

Dieter Herlach[1]

Institut fuer Raumsimulation, DLR, D–51170 Institut fuer Raumsimulation, DLR, D-511 Koeln, Germany

Abstract. Electromagnetic levitation technique is applied for containerless undercooling of liquid metals, metallic alloys and semiconductors as well. The velocity of the solidification front is directly measured as a function of undercooling. Measurements on pure metals reveal growth velocities ranging up to 100 m/s at the largest undercooling. The dependence of the growth velocity on undercooling is analysed within current theories of dendrite growth. The experimental findings are well described without any free parameters assuming the validity of the collision limited growth model and deviations from local equilibrium at the solid-liquid interface. Solid solutions behave similar as pure metals with the difference that also deviations from the chemical equilibrium at the interface have to be taken into account. As tested on binary alloys of strongly partitioning elements the deviations from chemical equilibrium are well understood by solute trapping described by a velocity dependent partitioning coefficient. A phenomenon analogous to solute trapping is disorder trapping during solidification of superlattice structures in intermetallics. Examples of measurements on intermetallic alloys are presented. Results of experiments on quasicrystalline and polytetrahedral alloys are presented. The sluggish growth behaviour in such systems with complex crystalline structure is understood by a growth dynamics, which is not collision but rather diffusion limited. The transition from faceted to non-faceted growth is discussed on the basis of experimental investigations of the growth dynamics of semiconducting materials.

1 Introduction

Most of all materials are prepared from the liquid state as their parent phase. Therefore, solidification is one of the most important phase transformations in materials science. If a melt is cooled one may distinguish between two different cases. Under the conditions of directional solidification a positive temperature gradient is applied externally, which stabilizes a planar solid/liquid interface. Crystal growth dynamics is externally controlled by the change of the temperature gradient. The dynamics of the crystal growth behaves very sluggish preserving local thermodynamic equilibrium at the solidification front. This means the velocity with which the interface is proceeding is very small compared to the atomic diffusion speed in the liquid. On the other hand if a melt is undercooled below its equilibrium melting temperature a negative temperature gradient is arising. Such a negative temperature gradient is destabilizing a planar interface. Any instantaneous perturbation tends to grow since a curved interface is more favourable for the heat redistribution at the solid/liquid interface. In case of alloys a concentration gradient is built up because of the different solvability of the solute in liquid and solid solvent. This concentration gradient enhances the instability of the solid/liquid interface caused by the negative temperature gradient in undercooled melts.

The gradients in front of the interface during solidification of undercooled melts give rise to dendritic crystal growth. The morphology of a dendritic solidification front ensures a high interface area to volume ratio that is more favourable for heat and mass redistribution compared to a planar interface. As a consequence the velocity of the dendritic interface is higher compared with the kinetics of a planar interface. Provided large undercoolings are achieved, deviations from local equilibrium are occurring, which lead to a kinetic undercooling of the interface and a reduction of the concentration profiles in alloys.

In the present work electromagnetic levitation technique is applied to process melts in diameter of several mm in containerless state. Owing to the complete avoidance of heterogeneous nucleation on container walls large undercoolings are accessible ranging up to about 25 % of the melting temperatures of the respective metals [1]. A freely suspended drop offers the additional benefit that it is directly observable for rapid solidification by using proper diagnostic means. In particular, different techniques are applied to measure quantitatively the dendritic growth velocity of levitation processed melts of metals and semiconductors as a function of undercooling. Experimental results will be presented for pure metals, solid solutions, dilute non-miscible alloys, intermetallics and quasicrystals as well. In addition, also the growth dynamics of levitation undercooled semiconductors is investigated. The experimental findings are discussed within dendrite growth theory and stability analysis of the solidification front.

2 Experimental

Electromagnetic levitation is a very efficient tool to process bulk melts in containerless state. A schematic view of an electromagnetic levitation chamber for containerless undercooling and solidification experiments is shown in Fig. 1. This technique gives the extra benefit that the freely suspended drops are not only accessible for direct diagnostics but also for external manipulation of the crystallization process. The levitation coil together with the sample in diameter of about 6 mm are placed within an ultra-high-vacuum chamber which can be evacuated to pressures of $p \approx 10^{-8}$ mbar before backfilling with gases such as He or He-H_2 mixture. The gases are purified by an oxysorb system and, additionally, by passing them through a liquid nitrogen cold trap. The sample is processed within the levitation coil which is connected to a high frequency generator. The maximum power output of the rf-generator is 24 kW and the frequency can be changed in the range between 300 kHz and 1.2 MHz. Temperature control in a limited range is possible by varying the sample position along the symmetry axis of the levitation coil and by using forced convection due to cooling gases. The temperature of the sample is measured by means of a two-colour pyrometer with an absolute accuracy of ± 3 K and a sampling rate of 100 Hz. However, such a frequency is not sufficient to monitor the rapid temperature rise during recalescence. Therefore, a high speed photosensing device has been developed to record relative temperature changes during rapid solidification with a frequency of 1 Mhz . Solidification of the undercooled melt can be externally initiated by touching the lower bottom of the sample with a crystallization trigger needle.

Owing to the avoidance of heterogeneous nucleation on container walls and ist far reduction on free surfaces by foreign phases by e.g. oxides thanks to processing in an ultra-high-vacuum chamber under high purity environment large undercoolings are achieved. Thus, the growth dynamics can be studied as a function of temperature over extended ranges of undercooling. To measure the growth velocity a rapid photosensing device was developed. It measures the time $\Delta t = t_2 - t_1$ that is needed by the solidification front to propagate through the observation window of the photosensing device. Because of the fact that solidification is externally stimulated by touching the lower bottom of the sample with the sample holder the geometry of the solidification pathway is well defined. The difference of the solidification path $\Delta d = d_2 - d_1$ is therefore easily determined and the growth velocity V is obtained from the ratio $V = \Delta d/\Delta t$ [2]. As checked by a double sensor the growth velocity is independent on the location of the solidification front. Solidification in

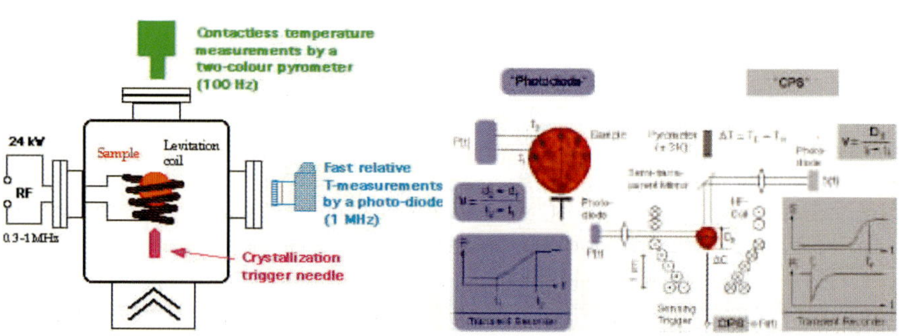

Fig. 1. Left: Electromagnetic levitation chamber for containerless undercooling and solidification of metals.
Right: Experimental techniques to measure the growth dynamics of the solidification front in levitation undercooled samples: The Photodiode method (left hand side) and the capacity sensor technique, CPS (right hand side). More details are given in the text.

undercooled melts is by dendritic growth (see chapter 3). The rapid photosensing technique is based upon the assumption that the solidification front is approximated by the envelope of the tips of all dendrites, which start to propagate at the external nucleation point. This assumption is fulfilled in the medium and large undercooling range at which many small dendrites are nucleated at the same location at sample surface. But the assumption may be violated at small undercoolings where only a few fat dendrites are formed which slowly migrate through the volume of the undercooled melt. In order to measure the growth velocity even in this undercooling range the capacity sensor technique (CPS) was developed. The nucleation sensing trigger is made of a nickel needle that is part of an electrical resistance Ũ capacity (RC) circuit. The frequency of the RC circuit depends on the capacity C that is changed if the trigger needle is touching the metallic liquid drop and initiates externally crystallization. The signal R recorded as a function of time on a transient recorder marks the time at which the needle touches the sample by a sharp cusp at time t_i. At the nucleation point a central dendrite is formed which propagates from the lower end of the sample to its upper end. As soon as the dendrite is arriving at the top of the sample the pyrometer measures a rapid increase of pyrometer signal S at time t_f. The growth velocity of this central dendrite is then easily determined by the ratio of the time difference and the sample diameter D_0, $V = (t_f - t_i)/D_0$ [3].

3 Description of Dendrite Growth in Undercooled Melts

Solidification of undercooled melts takes place by free dendritig growth. The release of the heat of crystallization leads to an increase of the temperature at the solid-liquid interface and the undercooled melt acts as a heat sink itself. As a consequence a negative temperature gradient arises that destabilizes a planar solidification front. In pure metals the negative temperature is the physical origin of the dendritic growth morphology. The strongly curved interface of a dendrite is most favourable for the heat redistribution at the interface during solidification and causes rapid crystal growth, which is much faster than any crystallization kinetics of a planar interface during e.g. directional solidification. Classical theory for free dendrite growth is basing upon the solution of the heat and mass transport equations for liquid and solid phase [4]. Accordingly, the total undercooling of a melt, ΔT, which can be measured by experiment is consisting of several contributions:

$$\Delta T = \Delta T_t + \Delta T_r + \Delta T_k + \Delta T_c \tag{1}$$

The thermal undercooling ΔT_t is calculated by the solution of the heat transport equation resulting in

$$\Delta T_t = \frac{\Delta H_f}{C_p^l} \cdot Iv(P_t) \qquad P_t = \frac{V \cdot R}{2a_l} \tag{2}$$

with ΔH_f the heat of fusion, C_p^l the specific heat of liquid phase, $Iv(P_t)$ the Ivantsov function in terms of the thermal Péclét-number P_t, V the velocity of the solidification front, R the curvature radius at the tip of a dendrite, and a_l the thermal diffusivity. Due to strong curvature at the tip of a dendrite the equilibrium melting temperature is reduced because of the Gibbs-Thomson effect that leads to a local interface undercooling ΔT_r of the solid-liquid interface. The curvature undercooling ΔT_r is given by

$$\Delta T_r = 2 \cdot \frac{\Gamma}{R} = 2 \cdot \frac{\sigma/\Delta S_f}{R} \tag{3}$$

with Γ the Gibbs-Thomson coefficient, σ the interfacial energy and ΔS_f the entropy of fusion.

At growth velocities comparable or larger than the atomic diffusive speed V_d, local deviations at the interface are occurring due to a kinetic undercooling ΔT_k. The kinetic undercooling is estimated within rate theory by calculating the transition rates of atoms going from liquid to solid state and vice versa [5]. It results

$$\Delta T_k \approx V/\mu \qquad \mu = f \cdot \frac{\Delta H_f}{k_B T_L^2} \cdot V_0 \tag{4}$$

μ is the kinetic growth coefficient, T_L the liquidus temperature and k_B Boltzmannś constant and f the atomic side factor. The velocity V_0 is determined by $V_0 = \delta\, \Phi\, exp(-\Delta G_a/k_B T_i)$. δ represents an interatomic spacing, Φ a typical atomic frequency, ΔG_a the activation energy for atomic diffusion and T_i the interface temperature. In the case of collision limited growth ($\Delta G_a = 0$) V_0 is approximated by the speed of sound in liquid state V_s. In pure metals and solid solutions the condition of collision limited growth is evidenced [6]. However, the situation may change if complex systems with large unit cells are considered. Here, short-range diffusion processes are required to built up the solid from the liquid. In such a case V_0 should approximate the atomic diffusive speed, V_d, rather than V_s. Since the speed of sound, V_s, is about three orders of magnitude larger than V_d, fast growth dynamics is expected in simple metallic systems while in alloys with complex structure a rather sluggish growth kinetics should be observed. The atomic side factor f will depend on the similarity of short-range order in undercooled liquid and solid. Provided there is a similar short-range order in both phases f is given by unity, $f = 1$, i. e. each atomic jump from the liquid to the solid is successful in the sense that the atom is directly integrated into the solid. But if the short-range order differs from liquid to solid f should be smaller than unity, $f < 1$, since each atom needs several jumps to find the equilibrium lattice side. In case of alloys a constitutional undercooling, ΔT_c, has to be taken into account. The solubility of the alloy solute is often smaller in solid than in liquid. This causes a pile up of solute in front of the interface and establishes a concentration gradient, which leads to the constitutional undercooling ΔT_c

owing to the local change of liquidus und solidus temperature in front of the solid-liquid interface. It also destabilizes the interface in addition to the negative temperature gradient. ΔT_c reads

$$\Delta T_c = m_l c_0 \left[\frac{1}{1-(1-k_E)\cdot Iv(P_c)} - 1 \right] \tag{5}$$

m_l denotes the slope of the liquidus line, c_0 the nominal concentration of the alloy, k_E the equilibrium partition coefficient, $Iv(Pc)$ the Ivantsov function in terms of the chemical Péclét number $P_c = (VR)/2D$ with D the chemical diffusion coefficient. Similar as in case of the kinetic undercooling, deviations occur from the local chemical equilibrium if the solidification velocity becomes comparable or even exceeds the atomic diffusive speed. Solute is trapped into the solid beyond the chemical equilibrium. This non-equilibrium effect is taken into account in dendrite growth theory by a velocity dependent partition coefficient and a velocity dependent liquidus slope [7],[8]. The velocity dependent partition coefficient $k(V)$ is determined within the solute-trapping model of Aziz [9] as:

$$k(V) = \frac{k_E + V/V_d}{1 + V/V_d} \tag{6}$$

Equation 1 describes the undercooling ΔT as a function of growth velocity V and curvature radius R in terms of the Péclét numbers P_t and P_c, respectively. In order to find a unique correlation between these quantities an additional relation has to be found. An independent equation correlating the growth velocity V with the tip radius R is inferred from morphological stability analysis. According to the hypothetical criterion of marginal stability [10] an equation is found for the dendrite tip radius, R, as

$$R^2 = \frac{\Gamma/\sigma^\star}{m_l G_c - G_t} \tag{7}$$

G_c and G_t denote the chemical and thermal gradient in front of the interface, and $\sigma^\star$ is the stability constant. The marginal stability analysis leads to an expression for the $\sigma^\star$:

$$\sigma^\star = \frac{2 d_0 a_l}{R^2 V} = \frac{1}{4\pi^2} \tag{8}$$

with $d_0 = \sigma_{sl} T_l C_P^L / \Delta H_f^2$ the capillary length (σ_{sl}: solid-liquid interfacial energy). The criterion of marginal stability is a hypothetical one. The solvability theory leads to a self-consistent solution of the problem by introducing the anisotropy of the interface as a parameter that dissolves the degeneration of the solutions of the transport equation [11],[12],[13]. According to solvability theory, the stability constant differs from the expression of equ. (8) by an anisotropy parameter ϵ, which is however close to unity for systems with metallic bonding.

4 Experimental Results on Dendrite Growth in Undercooled Melts

4.1 Pure Metals

Figure 2 shows the dendrite growth velocity V as a function of undercooling ΔT for pure Ni, in a double logarithmic scale. The closed squares represent early measurements by using the photo-sensing device while the open squares give results of measurements particularly in the low undercooling range. The dots correspond to measurements using the capacity sensor. The solid line gives the prediction of dendrite growth theory taking into account a kinetic undercooling of the interface [3]. In the medium undercooling range 100 K $< \Delta T <$ 200 K there is a good agreement between experiment and theory. At large undercoolings $\Delta T >$ 200 K deviations between theory and experiment are observed, which lead to a break in the temperature dependence of the growth velocity-undercooling relation at a critical undercooling $\Delta T^{\star} \approx$ 175 K [14]. At small undercoolings $\Delta T <$ 100 K the measured growth velocity is systematically larger than the theoretically predicted values. This finding becomes obvious by the measurements using the photo-sensing device even though the data are showing a considerable scatter. But using the capacity sensor the accuracy of the measurements is essentially improved. It is clearly discriminated that at velocities $V <$ 1 m/s the experimentally determined values of the growth velocities are enhanced to the predictions of theory. It is interesting to note that these deviations are observed at growth velocities that are comparable with the fluid flow velocities induced externally in electro-magnetically processed melts [15]. Fluid flow will certainly cause changes in the heat and mass transport in front of the solid-liquid interface and therefore will lead to a corresponding change in the growth dynamics. This will be subject of future research also including experiments on board of the International Space Station. These experiments make use of reduced natural convection and electromagnetically induced stirring in melts, which are processed containerlessly in the Electro-Magnetic Levitator of the Materials Science Laboratory in the ESA module (Columbus) [16]. The positioning forces needed for containerless processing in space are by about three orders of magnitude smaller than the forces required to levitate a sample on Earth against the gravitational forces. The reduced positioning fields, in turn, guarantee a very quiescent state of the melt whose solidification kinetics is not very much disturbed by externally stimulated heat and mass transport in the liquid.

The experimental data in the medium undercooling range are well described within dendrite growth theory provided the condition of local equilibrium is relaxed by taking into account a kinetic undercooling. Such kinetically caused deviations from local equilibrium at the solid-liquid interface are not surprising at growth speeds as high as $V \approx$ 70 m/s at the largest undercooling ΔT = 330 K. Such a growth velocity is by about one order of magnitude larger than the atomic diffusive speed V_d in metallic systems [17]. In calculating the kinetic undercooling ΔT_k according to equ. (4) a value for the speed of sound of liquid Ni, $V_s = 2000$ m/s, has been taken. It leads to a satisfactory agreement with the experiment, i. e. the atomic side factor is equal to unity, $f = 1$. This situation differs from that of pure Fe melts.

Figure 2, right hand side, illustrates the dependence of the dendrite growth velocity of undercooling for pure Fe. The open circles represent measured data applying the photosensor technique, and the closed circles correspond to measurements using the capacity sensor. The solid lines give the predictions of dendrite growth theory if the prefactor V_0 of the kinetic undercooling is decreased from the $V_0 = V_s = 4000$ m/s, the speed of sound for liquid Fe to a value of $V_0 = 250$ m/s. Only such a value being by more than one order of magnitude smaller than V_s is capable to describe the experimental data. But it is hard to understand that V_0 should be essentially smaller than the speed of sound V_s for pure Fe [18]. According to equ. (4) a larger kinetic

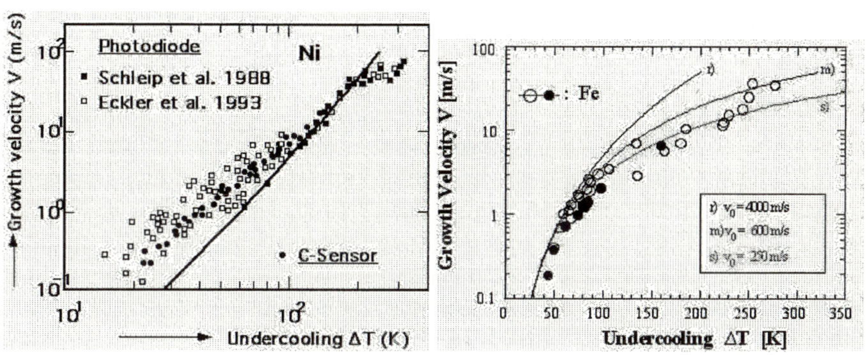

Fig. 2. Left hand side: Dendrite growth velocity V as a function of undercooling ΔT for Ni. Closed squares: measured by the photo-sensing device [2], open squares: measured by the photo-sensing device with special emphasis placed on the small undercooling range, solid circles: measured by the capacity sensor technique [3], solid line: prediction of growth theory.
Right hand side: Dendrite growth velocity V as a function of undercooling ΔT for pure Fe. Open circles: measured by the photosensing device, closed circles: measured by the capacity sensor. The solid lines give the predictions of dendrite growth theory for different values of the prefactor V_0.

undercooling may also be caused by a change of the atomic side factor f. If the difference between the prediction of dendrite growth theory assuming the validity of the collision limited growth model for pure Fe and the experimental data is entirely attributed to f, it corresponds to a change from $f = 1$ to $f \approx 0.1$. This means that each atom in the liquid boundary adjacing to the interface needs several jumps (in average 10) to find the correct lattice place in the solid. A possible source of such a situation is a different short-range order in the liquid and the solid. Recent neutron scattering experiments on pure Fe melts both in the stable and undercooled melt regime indicate a coordination number of $Z \approx 12$ [19], which is close to $Z = 12$ for fcc structure but different from $Z = 8$ for bcc structure of the solid. Undercooling and nucleation experiments on pure Fe reveal that primary solidification of bcc phase is energetically favoured compared to fcc because of an interfacial energy σ_{sl}(bcc), which is smaller than the interfacial energy of σ_{sl}(fcc) [20]. Hence, there must be some rearrangements of atoms to occur, which attempt to go from the liquid to the solid. This effect becomes even more pronounced in systems as Ge or Si which behave metallic like in liquid and show semiconducting behaviour in the solid state (see chapter 4.5).

4.2 Solid Solutions

In alloys constitutional effects have to be taken into account. The constitutional undercooling leads to a reduction of the growth dynamics because it arises due to redistribution of solute at the interface, which requires energy. On the other hand the concentration gradient in front of the interface leads to a narrowing of the dendrite. Since thin dendrites propagate faster than fat dendrites this effect of stability counteracts the effect of constitutional undercoolings. The stability effect however, only dominates in the dilute regime of alloys containing a solute with a large diffusion coefficient as e.g. C in Ni [21]. Usually, the addition of solute to solvent leads to a reduction of the growth dynamics. Figure 4 shows the growth velocity V as a function of undercooling ΔT for $C_{70}Ni_{30}$, (left) and $Co_{50}Pd_{50}$ (right), alloys with complete miscibility of the components in solid state. For both alloys the growth velocity increases with undercooling according to a power law up to a critical undercooling $\Delta T^{\star}$. At this critical undercooling there is a discontinuous change of the slope of velocity-undercooling relationship. The critical undercooling $\Delta T^{\star}$ amounts to 200 K for $Cu_{70}Ni_{30}$ and to 150 K for $Co_{50}Pd_{50}$ respectively. At undercoolings $\Delta T > \Delta T^{\star}$ the velocity linearly rises with undercooling. The solid lines are calculated within dendrite growth theory [7],[8]. In the case of $Cu_{70}Ni_{30}$ alloy an equilibrium partition coefficient $k_E = 0.81$ leads to an underestimation of the growth dynamics, while the assumption of complete partitionless solidification ($k = 1$) results in an overestimation of growth dynamics. Assuming the validity of the solute trapping model by Aziz [9] enables a satisfactory description of the experimental results up to the critical undercooling. The break in the temperature dependence of V at $\Delta T^{\star}$ is not well understood so far. A special result has been achieved for $Co_{50}Pd_{50}$. This alloy possesses a small heat of fusion, thus, a relatively small hypercooling limit $\Delta T_{hyp} = \Delta H_f / C_p^l = 268$ K, which only amounts to 65 % of the averaged value of the ideal solution [22]. However, no anomalous behaviour is observed if the velocity V exceeds the hypercooling limit.

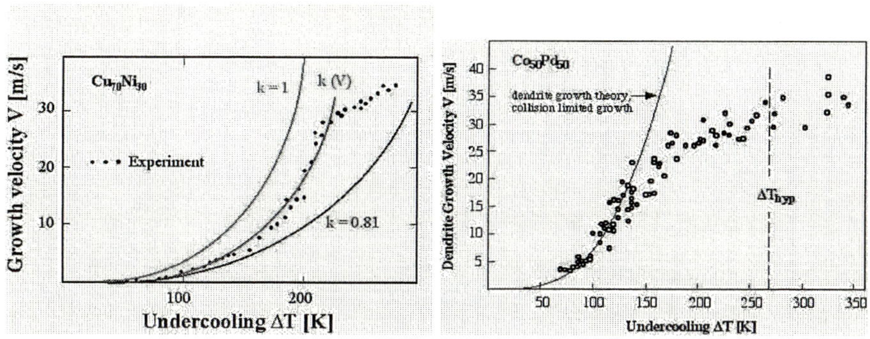

Fig. 3. Dendrite growth velocity V as a function of undercooling ΔT for solid solutions; the dots give results of measurements while the solid lines represent the predictions of dendrite growth theory.
Left side: $Cu_{70}Ni_{30}$ alloy, the various lines correspond to growth velocities under the assumption of local chemical equilibrium (equilibrium partition coefficient k_E = 0.81), a velocity dependent partition coefficient $k(V)$ (according to the solute trapping model by Aziz) and for a partition coefficient $k = 1$ (complete partitionless solidification) [14].
The **right hand side** gives equivalent results for $Co_{50}Pd_{50}$ alloy [23]. Similar to $Cu_{70}Ni_{30}$ the experimental results are well described by theory, but a break in the temperature dependence is observable at a critical undercooling $\Delta T^\star = 200$ K ($Cu_{70}Ni_{30}$) and $\Delta T^\star = 150$ K ($Co_{50}Pd_{50}$), respectively. In case of $Co_{50}Pd_{50}$ alloy undercoolings have been achieved exceeding the hypercooling limit ΔT_{Thyp}. No change in the Velocity-undercooling relation is found.

4.3 Dilute Multiphase Alloys

Ni-B and Ni-Zr alloys show eutectic points in their phase diagrams and belong to the glass forming alloys at their eutectic compositions. Another characteristic feature is the steep fall of liquidus temperature with B and Zr composition and the very small equilibrium partition coefficient k_E on the Ni-rich side. Therefore, strong partitioning is expected during solidification of dilute Ni-B and Ni-Zr alloys, which should lead to a correspondingly growth dynamics. Figure 4 illustrates the dependence of the growth velocity V on undercooling for dilute Ni-B (left) [24] and one dilute Ni-Zr (right) alloy [25]. The experimental results are represented by the closed symbols. In case of Ni-B two dilute alloys were investigated, containing 0.7 at% B (squares) and 1.0 at% B (circles). For comparison the results for pure Ni [14] are given in the right hand diagram. The solid lines correspond to predictions of dendrite growth theory. For both dilute alloy systems the growth dynamics is very much reduced compared to pure Ni at undercoolings below a critical undercooling $\Delta T_c^\star$. The sluggish growth in this undercooling range is interpreted to be chemical diffusion limited growth. According to the equilibrium partition coefficient k_E solute is redistributed in front of the interface in the regime

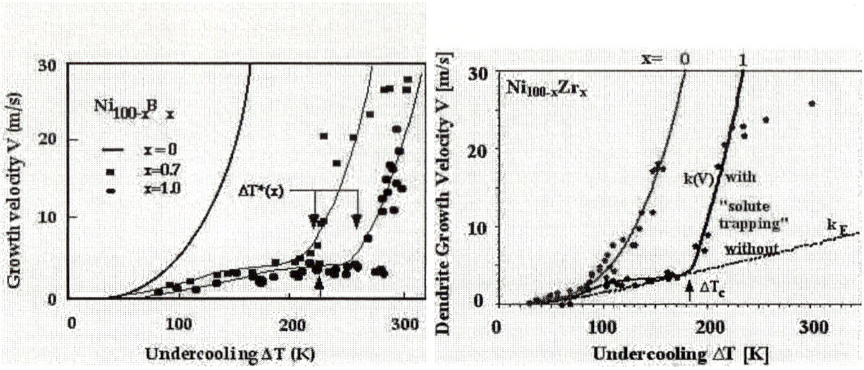

Fig. 4. Dendrite growth velocity V as a function of undercooling ΔT for dilute Ni-B [24] and Ni-Zr [25] alloys. Closed symbols: experimental results, solid lines: predictions of dendrite growth theory. For comparison, the growth velocity of pure Ni is also given (x=0). In both alloys, the growth velocity of the alloys is very much reduced compared to pure Ni at undercoolings less than a critical undercooling $\Delta T_c^\star$, which depends crucially on the alloy system and its concentration. If the undercooling reaches the critical undercooling the growth dynamics is well described by theory taking into account a velocity dependent partition coefficient $k(V)$ according to the solute trapping model by Aziz [9].

of very small undercoolings. As soon as the undercooling reaches the critical undercooling the growth velocity steeply rises with increasing undercooling. In this regime the growth dynamics is well described by theory provided a velocity dependent partition coefficient $k(V)$ according to the solute trapping model by Aziz [9] is taking into account. In particular in this range of solute trapping the description of the growth velocity-undercooling relationship crucially depends on the parameters of the chemical diffusion coefficient D and the atomic diffusive speed V_d. Therefore, both of these parameters were independently determined by laser surface re-solidification experiments in combination with Rutherford Backscattering experiments on $Ni_{99}Zr_1$ alloys [25]. As can be seen in the right hand diagram of Fig. 4 the experimental results obtained from Ni-Zr are in good agreement with the dendrite growth theory described without any free adjustable parameter. At largest undercoolings a break in the temperature dependence of the growth velocity is observed similar as for pure Ni, whose physical origin is subject of on going research [26]. The critical undercooling depends essentially on the concentration of the solute. It changes from a value of $\Delta T_c^\star = 223$ K for $Ni_{99.3}B_{0.7}$ to $\Delta T_c^\star = 256$ K for $Ni_{99}B_1$. This means, only a slight variation of the concentration by 0.3 at% B changes the critical undercooling $\Delta T_c^\star$ by 33 K! Also, the critical undercooling depends on the solute element. At equal concentration of 1 at% B and Zr, $\Delta T_c^\star$ is altered from $\Delta T_c^\star = 256$ K for $Ni_{99}B_1$ to $\Delta T_c^\star = 185$ K for $Ni_{99}Zr_1$ alloy.

4.4 Intermetallics with Superlattice Structures

A phenomenon quite analogous to solute trapping is disorder trapping. It occurs during rapid solidification in melts of intermetallic compounds, which form superlattice structures under equilibrium solidification conditions. Disorder trapping means that the growth velocity is too high to build up the superlattice of the intermetallic phase. It is theoretically described by a model of Boettinger and Aziz [27]. Figure 5 shows the results of measurements on the equiatomic intermetallic compounds FeSi and CoSi [28]. The squares represent the experimental results. Dendrite growth theory is used to analyse the experiments. First, it is assumed that the interface undercooling $\Delta T_k = V/\mu$ is determined from the model of collision-limited growth [6]. This leads to the dashed line in Fig. 5. Apparently, the assumption of the validity of collision-limited growth overestimates the velocity. Alternatively, diffusion-limited growth is supposed. Under these conditions, the atomic diffusive speed is the limiting factor, and μ is small, which results in a large interface undercooling ΔT_k. The experimental data can be fitted under the assumption of diffusion-limited growth. The best fit is obtained for $\mu = 0.021$ m/sK, which is much less than $\mu = 4.2$ m/sK for collision-limited growth. The resulting conclusion is that at undercoolings below a critical undercooling $\Delta T^\star \approx 310$ K growth appears to be diffusion limited. This may be understood by the fact that the build-up of a superlattice structure requires atomic diffusion over at least two interatomic spacings. However, if the undercooling

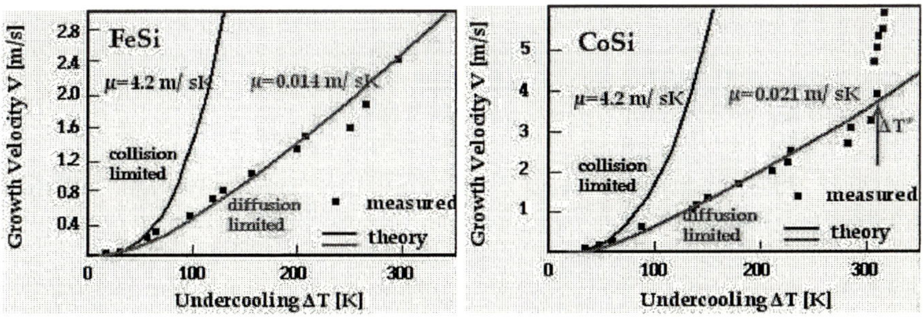

Fig. 5. The dendrite growth velocity V as a function of undercooling ΔT for the intermetallic compound FeSi (left) and CoSi (right), respectively. The squares give the measured velocities. The solid lines represent the theoretical predictions assuming either collision-limited growth or diffusion-limited growth [28].

exceeds a critical value $\Delta T^\star \approx 310$ K for CoSi alloy, the velocity V starts to rise rapidly. The slope of the $V(\Delta T)$ curve beyond the critical undercooling is comparable to that of the curve calculated under the assumption of collision-limited growth. This transition is understood by the onset of disorder trapping at $\Delta T^\star \approx 310$ K when the growth velocity becomes comparable to, or even larger than the atomic diffusive speed. The atoms can no longer sort themselves out onto the various sublattices of the intermetallic phase. This leads to a disordered structure. In the present levitation experiments, the cooling rate is too small to avoid reordering of the as-solidified samples. Therefore, the disordered structure, primarily solidified at the largest undercooling, cannot be revealed by, e.g., diffraction studies on the as-solidified material cooled down to room temperature. However, it is evidenced in Ni-Al based intermetallic compounds by investigations both of the solidification velocity versus undercooling relation [29] and transmission electron microscopy (TEM) studies as well [30]. The TEM micrographs clearly show the existence of antiphase domains [31].

Even in the regime of sluggish growth of intermetallic alloys dendritic growth can be collision limited [32]. Based upon a study of the changes of interface undercooling due to non-equilibrium effects as solute trapping and solute drag, the interface undercooling ΔT_i is expressed as sum of various terms:

$$\Delta T_i = \Delta T_{sd} + \Delta T_{st} + \Delta T_{dt} + \Delta T_k \tag{9}$$

with ΔT_k the kinetic undercooling (cf. equ. (4)), ΔT_{st} a contribution by solute trapping, ΔT_{sd} a term, which is originating from solute drag and ΔT_{dt} the part by disorder trapping. The various contributions of equ. (9) have been calculated on the basis of the concept of maximum driving force [32]. Accordingly, ΔT_{sd} only prevails at small growth velocities $V/V_d < 10^{-2}$. If the velocity is approaching the atomic diffusive speed V_d the terms of solute and disorder trapping, ΔT_{st} and ΔT_{dt} become dominant, but they vanish if V exceeds V_d. At $V > V_d$, the true kinetic undercooling ΔT_k becomes most prominent. Based upon this analysis the $V - \Delta T$ relations measured for Ni-Al intermetallic alloys are interpreted. Figure 6 shows V as a function of ΔT for $Ni_{73.7}Al_{26.3}$ and $Ni_{76.5}Al_{23.5}$ alloy, respectively. According to the metastable phase diagram of Ni-Al as calculated by Ansara and Dupin [33] a phase competition of various phases exists depending on concentration and undercooling: β the ordered bcc B2 compound (NiAl), γ' the ordered ccp L12 compound (Ni_3Al) and γ the disordered ccp solid solution (Ni). NiAl melts congruently at about 46 at% Al, Ni_3Al incongruently in a peritectic reaction $\gamma' \rightarrow \gamma + L$ at about 24 at% Al.
Due to the significant undercoolings and consequently high measured growth velocities, the phase selection behaviour is assumed to be mainly nucleation-controlled. Once a critical nucleus has formed, the first stage of solidification, involving recalescence, is expected to be completed without further nucleation events occurring. The closed circles represent the experimentally determined growth velocities while the lines give the predictions of theory for the growth dynamics of the various phases. For the composition Ni-26.3 Al (left hand side), the growth velocity versus undercooling relation measured in levitation experiments are understood if primary crystallization of bcc phase is supposed. It is well known from undercooling experiments on Fe-Ni [34] and

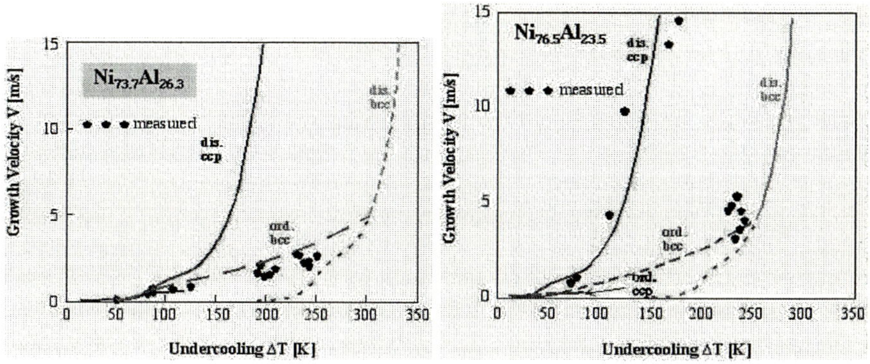

Fig. 6. Growth velocity V as a function of undercooling ΔT for $Ni_{73.7}Al_{26.3}$ (left) and $Ni_{76.5}Al_{23.5}$ (right) intermetallic alloy. The thicker dashed curve has been calculated for the ordered bcc-phase, the thinner dashed line for the disordered bcc-phase, the dotted line for the unstable regime of the disordered bcc-phase, the thinner solid line for the disordered ccp-phase and the thicker solid line for the ordered ccp-phase [31].

Ni-V [35] alloys that, due to the lower energy of the solid/liquid interface, bcc crystal structures are often favoured over ccp with respect to nucleation. The theoretical transition point from partially ordered to completely disordf the ccp phase are somewhat larger than for an alloy Ni-25 Al. The velocities measured for the ordered bcc structure are located just near the transition point to disorder trapping [31]. Apparently, the studies in Ni-Al reveal that even in case of ordered superlattice structures in intermetallic compounds the atomic attachement kinetics at the liquid-solid interface is described by collision limited growth provided several contributions to the interface undercooling ΔT_i are taken into consideration. But, we assume that a transition from collision limited growth (with the speed of sound as the upper velocity limit) to diffusion limited growth (with the atomic diffusion speed as the upper limit) should occur if the complexity of the unit cell of crystalline structure is enhanced.

4.5 Complex Structures of Polytetrahedral Phases

Alloy systems as Al-Fe showing large unit cells with polytetrahedral ordering are belonging to the class of so called approximant phases. The size of the unit cell with polytetrahedral order differs essentially depending on the alloy system. In case of μ-phase in intermetallic compound Al_5Fe_2 the unit cell consists of 15 atoms, while in case of λ-phase in intermetallic compound $Al_{13}Fe_4$ it is increased to a size of 102 atoms. An infinitely large unit cell of polytetrahedral order is present in quasicrystalline alloys. To built up large unit cells of polytetrahedral order atomic migration is needed not only over one interatomic spacing as for e.g. simple cubic structure but atoms have to jump over more than one interatomic distance to sort them out selves for the construction of a large unit cell with polytetrahedral order. Therefore, growth dynamics in undercooled melts of such alloys with a complex unit cell structure is expected to be diffusion controlled rather than collision limited. Figure 7 illustrates such diffusion controlled growth on a binary Al_5Fe_2 alloy with a unit cell of polytetrahedral order containing 15 atoms. In addition, also the growth dynamics of an $Al_{72}Pd_{21}Mn_7$ alloy was measured, which forms a stable icosahedral phase with quasiperiodicity in three dimensions [36]. The alloys were undercooled and the growth velocity was measured by the photo-sensing device. In addition, high speed video camera technique was also applied to record the propagation of the solidification front through the undercooled melt [37]. The growth velocities of the polytetrahedrally ordered

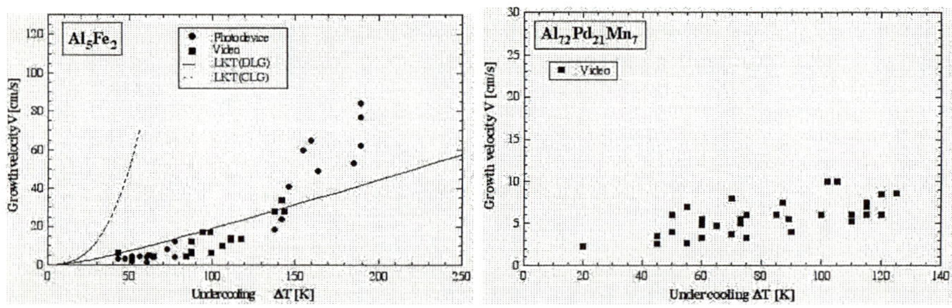

Fig. 7. Growth velocity V as a function of undercooling ΔT for the polytetrahedral μ-phase formed from an Al_5Fe_2 melt (left) and for an icosahedral I phase formed from an $Al_{72}Pd_{21}Mn_7$ melt. The symbols represent experimental results as obtained by measurements with the photo-sensing device (closed circles) and with the high speed video camera (closed squares), respectively. The lines give the predictions of dendrite growth theory assuming collision-limited growth (dashed line) and diffusion-limited growth (solid line) for the binary Al_5Fe_2 alloy. The dendrite growth theory allows for calculations of the velocity - undercooling relation only for pure metals and binary alloys. For the ternary alloy, a quantitative analysis is therefore lacking [37]

phases are smaller by one to two orders of magnitude if compared with growth velocities measured in pure metals and solid solutions (cf. Fig. 2 and Fig. 3). The growth dynamics is even more sluggish than that of the intermetallic compounds (cf. Fig. 5 and Fig. 6). In case of Al_5Fe_2 alloy there is a fair description of the results by assuming diffusion limited growth, at least in the undercooling range $\Delta T < 150$ K. At larger undercoolings a steep rise of the velocity is observed somehow similar as for CoSi alloy (cf. Fig. 5). The temperature dependence of the growth velocity at undercoolings $\Delta T > 150$ K becomes comparable to the dashed curve calculated for collision limited growth. This behaviour may indicate the onset of disorder trapping during rapid solidification of this alloy. For the ternary icosahedral alloy a monotonic increase of V is recorded without any change in ist temperature derivative. The maximum growth velocity is smaller than 0.1 m/s even at the largest undercooling of $\Delta T \approx 130$ K. The sluggish growth of the polytetrahedral phases is understood by the chemical ordering in the construction of the large unit cells with polytetrahedral order. The atoms in the liquid cannot occupy each crystal site but have to find their correct sites in solid state. This requires short-range diffusion of the atoms at the solid-liquid interface and consequently the growth kinetics is controlled by a diffusion limited growth process. The dynamics of growth is the more sluggish the larger the unit cell, and consequently slowest for quasicrystalline alloys with icosahedral order.

4.6 Faceted and Non-Faceted Growth of Semiconductors

From a more general perspective, there are two idealized mechanisms for crystal growth from the melt. (i) Systems with rough solid-liquid interface are thought to grow by the continuous growth mechanism, in which atom transfer from liquid can occur at any site on the interface [38]. (ii) Strongly faceting materials of high entropy of fusion exhibit edgewise growth, in which the interface is atomically smooth except for the presence of ledges or steps, and in which atom transfer can occur only at a few special sites on the interface. Considering solidification into an undercooled melt, Cahn et al. [39] have suggested a model featuring a transition from edgewise growth (EG) to continuous growth (CG) when the driving force for crystallization is raised. Figure 8 gives an overview of the various microstructures formed in pure Ge upon solidification of differently undercooled melts. At a small undercooling of $\Delta T = 40$ K (a), a faceted structure becomes apparent which is indicative for edgewise growth; at a medium undercooling of $\Delta T = 316$ K (b), a coarse grained microstructure is substituted for lamellar twins that are ubiquitous in Ge crystals grown under normal conditions (dendritic substructure is visible at moderate undercoolings if introducing minute amount of tin, e.g., in the resultant grain structure of $Ge_{99.61}Sn_{0.39}$ at $\Delta T = 277$ K, not shown). At the largest undercooling, $\Delta T = 426$ K (c), an equiaxed, grain refined structure is observed [40]. By investigations on a whole series of samples of pure Ge, a critical undercooling $\Delta T_c^\star$ was determined that separates two growth regimes: $\Delta T < \Delta T_c^\star \approx 300$ K, edgewise growth and a faceted structure; $\Delta T > \Delta T_c^\star$, dendritic growth resulting in coarse grains. According to Evans et al. [41], the first microstructural conversion from twins to coarse grains is explained as follows: the critical radius of two-dimensional nucleation becomes comparable with the step size or with the so-called correlation length at the solid-liquid interface when ΔT exceeds the critical value, suggesting that the step at the interface will loose its "identity". Regarding the second microstructural transition to refined grains it has been recognized that the origin of grain refinement in undercooled materials is due to dendrite break up. The physical mechanism of this microstructural transformation in undercooled melts has been investigated by theoretical analysis [42] and experiments as well [43]. The models (LKT/BCT) of dendrite growth [7],[8] have

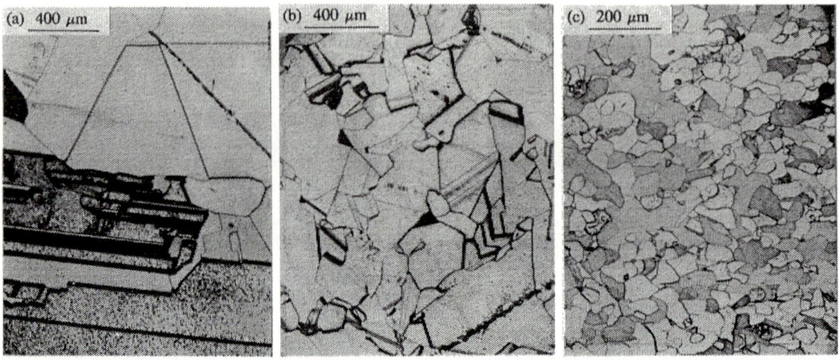

Fig. 8. Morphological transformation of pure Ge with increasing undercooling ΔT: (a) $\Delta T = 40$ K, faceted structure, (b) $\Delta T = 316$ K, coarse grains with some dendritic elements, (c) $\Delta T = 426$ K, grain refined structure [40].

been developed to describe continuous crystal growth. Two points merit consideration before using these models for faceting materials. The first concerns the transition from edgewise to continuous growth. The LKT/BCT models have been proven valid for dendritic solidification (see previous chapters). For the Ge-based systems they should be applicable in the regime $\Delta T > \Delta T_c^\star$ (dendritic growth), provided an offset of the total (bath) undercooling ΔT by a critical interface undercooling $\Delta T_i^\star$ for the EG $\leftarrow$ CG transition is taken into account. Evans et al. [41] gave a plot of the interface temperature at the onset radius for instability as a function of the bulk undercooling ΔT for Ge, according to their computation on stability of spherical crystal growth. From the computed curve, one derives $\Delta T_i^\star \approx 153$ K at $\Delta T_c^\star = 300$ K for pure Ge. The numerical value of $\Delta T_i^\star \approx 153$ K should be independent on alloy concentration in dilute Ge-Sn alloys. The total undercooling eq. (1) reads:

$$\Delta T - \Delta T_i^\star = \Delta T_t + \Delta T_r + \Delta T_k + \Delta T_c \qquad (10)$$

To test this equastion the growth velocity was measured as a function of undercooling for pure Ge and dilute Ge-Sn alloys [40]. The results of the measurements are shown in Fig. 9. Two salient features are visible for pure Ge: first, in the range $\Delta T < \Delta T_c^\star \approx 300$ K, the growth velocity is very small and rises slowly. In this region, the crystal grows in the edgewise mode. It results in the faceted structure and twins (cf. Fig. 8a). Second, at $\Delta T_c^\star =$ 300 K, a change takes place in the temperature dependence of the growth behaviour. In the range $\Delta T > \Delta T_c^\star$, the velocity rapidly increases with ΔT. Such a behaviour is typical for the continuous growth mode of dendritic solidification as described by Peteves et al. [44]. Adding a small amount of 0.39

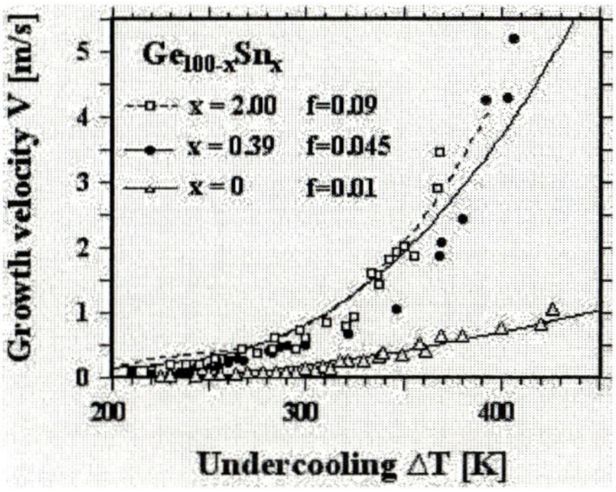

Fig. 9. Growth velocity, V, as a function of undercooling, ΔT, for pure Ge (triangles), a very dilute $Ge_{99.61}Sn_{0.39}$ alloy (closed circles), and a dilute $Ge_{98}Sn_2$ alloy (squares). The symbols and curves correspond to measured values and predictions, based on the modified dendrite growth model, respectively. f denotes the fraction of interfacial sites at which growth can occur. The critical undercoolings $\Delta T_c^\star$ for the microstructural transition from faceted to non-faceted growth are $\Delta T_c^\star \approx 300$ K (pure Ge), $\Delta T_c^\star \approx 255$ K ($Ge_{99.61}Sn_{0.39}$), $\Delta T_c^\star \approx 230$ K ($Ge_{98}Sn_2$).

at% Sn to pure Ge leads to a drastic enhancement of the crystal growth velocity by a factor of five at the largest undercooling. Further increasing the Sn concentration up to 2 at% Sn slightly reduces this effect.

The experimental results were analysed within the frame of dendrite growth theory. Concerning the Ge-based systems the interface undercooling merits special attention since it is known that semiconductors grow with large interfacial undercooling. According to equ. (4), the interfacial undercooling is given by $\Delta T_k = V/\mu$ with μ the kinetic growth coefficient. In case of collision limited growth Vo is given by the speed of sound, V_s, and for metals, the atomic side factor $f = 1$. Since liquid Ge is metallic, it is realistic to assume the model of collision limited growth model. However, solid Ge is a semiconductor of strongly directional bonds. Because of the differences of bonding in metallic like liquid Ge and semiconducting solid Ge, the factor f may be less than unity. The factor f is used as the only free parameter to describe the growth velocities measured at $\Delta T > \Delta T_c^\star$ within dendrite growth theory. The results of the calculations are represented by the solid lines in Fig. 9. The numerical values of the factor f as inferred from the fits are changing from $f \approx 0.01$ for pure Ge to $f \approx 0.045$ for $Ge_{99.61}Sn_{0.39}$ and $f \approx 0.09$ for $Ge_{98}Sn_2$. This means that the fraction of interface sites at which atomic attachment can occur increases by the addition of the metallic component with the consequence of acceleration of the growth dynamics. This effect prevails the counteracting effect of the contribution of the constitutional undercooling ΔT_c, which causes a decrease of the growth velocity. This finding is supported by the results of the alloy containing 2 at% Sn. Even though f is further increased in comparision with the very dilute alloy, the growth velocity is slightly reduced. This indicates that with increasing concentration of Sn, the constitutional undercooling becomes more important.

5 Summary and Conclusions

Electromagnetic levitation technique in combination with proper diagnostic means to obserbe the propagation of the solidification front into undercooled melts was applied to measure the growth dynamics of containerlessly processed samples as a function of undercooling. Deviations from local equilibrium at the solid-liquid interface were evidenced owing to the large driving forces for rapid crystallization at the undercoolings achieved by containerless undercooling. The deviations are discussed within a kinetically undercooled interface on pure metals and solid solutions, solute and disorder trapping in alloys of solid solutions, dilute Ni-B/Ni-Zr alloys, intermetallics and diffusion controlled growth in alloys with unit cells showing polytetrahedral order. Finally, the transition from edgewise to continuous growth in Ge and dilute Ge-Sn alloys was discussed.

Acknowledgements

This work benefited from serious collaboration with my coworkers H. Assadi, M. Barth, K. Eckler, D. Holland-Moritz, D. Li, J. Schroers, M. Schwarz, T. Volkmann, B. Wie and R. Willnecker. The author expresses his sincere thanks to all of them for the always pleasant and fruitful cooperation. The author is very much indebted to the Deutsche Forschungsgemeinschaft (DFG) for financial support of the present work within several projects of the priority program SPP 729 "Undercooled Metallic Melts".

References

1. Herlach, D. M.: Annual Review of Materials Science **21** (1991) 23.
2. Schleip, E., Willnecker, R., Herlach, D. M. and Görler, G. P.: Mat. Sci. Eng. A **98** (1988) 39.
3. Eckler, K. and Herlach, D. M.: Mat. Sci. Eng. A **178** (1994) 159 - 162.
4. Kurz, W. and Fisher, D. J.: *Fundamentals of Solidification*, (Trans Tech Publications Switzerland-Germany-UK-USA 1989).
5. Christian, J. W.: *The Theory of Transformations in Metals and Alloys*, Pergamon, Oxford, 1975.
6. Turnbull, D.: Acta Metall. **30** (1982) 2135.
7. Trivedi, R., Lipton, J. and Kurz, W.: Acta Metall. **35** (1987) 965.
8. Boettinger, W. J., Coriell, S. R. and Trivedi, R. in Mehrabian, R. and Parrish, P. A.: (eds.), *Rapid Solidification Processing: Principles and Technologies IV*, Claitor's, Baton Rouge, LA, 1988, p. 13.
9. Aziz, M. J.: J. Appl. Phys. **53** (1982) 1158.

10. Langer, J. S., and Müller-Krumbhaar, H.: Acta Metall. **79** (1978) 1689, 1697.
11. Pelce, P. and Pomeau, Y.: Stud. Appl. Math. **74** (1986) 245.
12. BenAmar, M. and Pomeau, Y.: Europhys. Lett. **2** (1986) 307.
13. Barbieri, A., Hong, D. C. and Langer, J. S.: Phys. Rev. A, **35** (1986) 1802.
14. Willnecker, R., Herlach, D. M. and Feuerbacher, B.: Phys. Rev. Lett. **62** (1989) 2707.
15. El-Kaddah, N. and Szekely, J. in: Mat. Res. Soc. Proc. Symp. "Materials Processing in the Reduced Gravity Environment of Space", ed.: G.E. Rindone, 9 (1982), p. 191.
16. Piller, J., Seidl, A., Stauber, M. and Dreier, W. in: *Solidification 1999*, eds. Hofmeister, W. H., Rogers, J. R., Singh, N. B., March, S. P., Vorhees, P. W.: Proc. TMS Fall Meeting (Rosemont 1998) p. 3.
17. Smith, P. M. and Aziz, M. J.: Acta Metall. Mater. **42** (1994) 3515.
18. Eckler, K., Gärtner, F., Assadi, H., Norman, A. F., Greer, A. L. and Herlach, D. M.: Mat. Sci. Eng. A **226-228** (1997) 410.
19. Schenk, T., Holland-Moritz, D., Simonet, V., Bellissent, R. and Herlach, D. M.: to be published.
20. Barth, M., Eckler, K., Herlach, D. M. and Alexander, H.: Mat. Sci. Eng. A **133** (1991) 790.
21. Eckler, K., Hamerton, R. G., Herlach, D. M. and Greer, A. L.: Mat. Sci. Eng. A **133** (1991) 730.
22. Wilde, G., Görler, G. P. and Willnecker, R.: Appl. Phys. Lett. **69** (1996) 2995.
23. Volkmann, T., Wilde, G., Willnecker, R. and Herlach, D. M.: J. Appl. Phys. **83** (1998) 3028.
24. Eckler, K., Cochrane, R. F., Herlach, D. M., Feuerbacher, B. and Jurisch, M.: Phys. Rev. B **45** (1992) 5019.
25. Arnold, C. B., Aziz, M. J., Schwarz, M. and Herlach, D. M.: Phys. Rev B. **59** (1999) 334.
26. Karma, A.: private communication, and this issue.
27. Boettinger, W. J. and Aziz, M. J.: Acta Metall. **37** (1989) 3379.
28. Barth, M., Wei, B. and Herlach, D. M.: Phys. Rev. B **51** (1995) 3422.
29. Goetzinger, R., Barth, M., Herlach, D. M., Hunziker, O. and Kurz, W.: Mat. Sci. Eng. A **226-228** (1997) 415.
30. Assadi, H., Barth, M., Greer, A. L. and Herlach, D. M.: Materials Science Forum **215-216**, 3 (1996) 7.
31. Assadi, H., Barth, M., Greer, A. L. and Herlach, D. M.: Acta Mater. **46** (1998) 491.
32. Assadi, H. and Greer, A. L.: J. Crystal Growth **172** (1997) 249.
33. Dupin, N.: Ph.D. thesis, Institut Nastional Polytechnique de Grenoble, 1995.
34. Schleip, E., Herlach, D. M. and Feuerbacher, B.: Europhys. Lett. **11** (1990) 751.
35. Notthoff, C., Feuerbacher, B., Frans, H., Herlach, D. M. and Holland-Moritz, D.: Phys. Rev. Lett. **86** (2001) 1038.
36. Tsai, A. -P., Inoue, A. and Masumoto, T.: Philos. Mag. Lett. **62** (1990) 95.
37. Schroers, J., Holland-Moritz, D., Herlach, D. M. and Urban, K.: Phys. Rev. B **61** (2000) 14 500.
38. Jackson, K. A.: Mat. Sci. Eng. **65** (1984) 7.
39. Cahn, J. W., Hillig, W. B. and Sears, G. W.: Acta Metall. **35** (1964) 1421.
40. Li, D. and Herlach, D. M.: Phys. Rev. Lett. **77** (1996) 1801.

41. Evans, P. V., Vitta, S., Hamerton, R. G., Greer, A. L. and Turnbull, D.: Acta Metall. Mater. **38** (1990) 233.
42. Karma, A.: Int. J. Non-Equilibrium Processing **11** (1998) 201.
43. Schwarz, M., Karma, A., Eckler, K. and Herlach, D. M.: Phys. Rev. Lett. **73** (1994) 1380.
44. Peteves, S. D. and Abbaschian, R.: Metall. Trans. **22A** (1991) 1259.

Thermodynamics of Diffuse Interfaces

G.J.Schmitz[1]

ACCESS e.V. , D-52056 Aachen, Germany

1 Introduction

Classical thermodynamics describe phase-equilibria between bulk phases rather than kinetics of phase transitions and interface dynamics. Especially, classical thermodynamics do not comprise the definition of a specific length scale, as all energy formulations in thermodynamics hold on any length scale if it is only large enough to allow for reasonable statistics.

As no specific length is defined in the free energy formulations, there is no hope to deduce information about the evolution of structures–which do have length scales–by minimization of classical thermodynamic free energy functionals. Accordingly, classical thermodynamics yields only information about the fractions of the bulk phases being in equilibrium, but not about their spatial distribution the microstructure.

Structures may be defined as interfaces/transitions between different, homogeneous bulk areas i.e. between regions not revealing any structure themselves. The inclusion of contributions of interfaces to the total energy of the thermodynamic system therefor is mandatory if the evolution of structures shall be accessible at all.

As interfacial energies are several orders of magnitude smaller than the energies of the respective bulk phases they often however are not considered. If interfacial energies are considered, they are included into the systems free energy description as mathematically sharp surfaces contributing a related surface energy. It is important to note, that the thickness of the interface represents a characteristic length scale in the system, which however is degenerate in the case of mathematically sharp interfaces.

Interfaces in physical systems often have a finite extension of several monolayers and therefor the extension of thermodynamics including sharp interfaces towards a more general thermodynamic description involving diffuse interfaces seems meaningful.

Scope of the present paper is to derive a free energy functional for diffuse interfaces starting from a thermodynamic approach. The principle idea of the proposed approach is most easily demonstrated for a pure substance undergoing a phase transition e.g. from solid "s" to liquid "l". The total Gibbs free energy G^{tot} for such system is commonly defined as the sum of the Gibbs free energies of the individual phases G^l and G^s:

$$G^{tot} = \phi_s^{tot} G^s + \phi_l^{tot} G^l = \iiint\limits_{vol} \phi_s(\vec{x},t) G^s dV + \iiint\limits_{vol} \phi_l(\vec{x},t) G^l dV \quad (1)$$

where indices l, s denote liquid and solid respectively and ϕ_l, ϕ_s denote the volume fractions of these phases as function of space $\vec{x}$ and the time t. For a fixed ratio ϕ_l/ϕ_s several possible arrangements of the two phases are depicted in Fig. 1a which, according to equation (1), all have the same total energy G^{tot}. However, these arrangements obviously vary in the total interfacial area

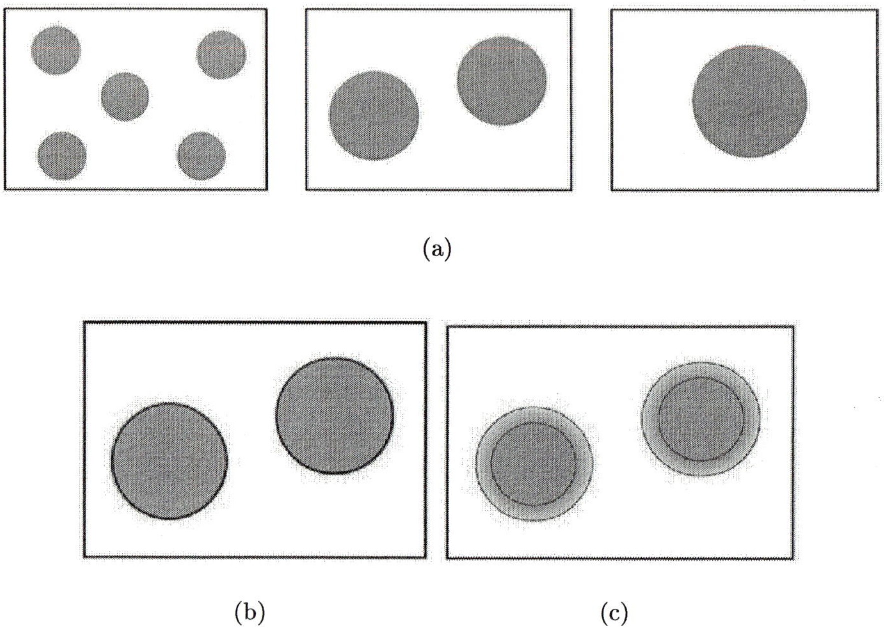

Fig. 1. Several arrangements of two phases coexisting in a finite volume. Note that the volume fraction is in all cases the same. The arrangements however differ significantly with respect to the amount of interface between the two phases which may be a sharp interface (b) or a diffuse interface (c).

between the two phases involved. This maybe accounted for by adding a surface energy term γA to the bulk free energy of the system, Fig. 1b:

$$G^{tot} = \iiint_{vol} \phi_s G^s dV + \iiint_{vol} \phi_l G^l dV + \iint_{sur} \gamma dA \tag{2}$$

A diffuse interface, however, is not characterized by a two dimensional surface element dA but by a finite volume in which the two phases coexist. As a consequence, the surface integral in equation (2) turns into a volume integral, its integrand γA then being some function of ϕ_s and ϕ_l, Fig. 1c:

$$G^{tot} = \iiint\limits_{vol} \phi_s'(\vec{x},t) G^s dV + \iiint\limits_{vol} \phi_l'(\vec{x},t) G^l dV \tag{3}$$
$$+ \iiint\limits_{vol} f\left(\phi_s'(\vec{x},t), \phi_l'(\vec{x},t)\right) dV$$

The ϕ_s, ϕ_l have been primed here to distinguish them from from their sharp interface counterparts. The primes " ' " will however be neglected in the following. Scope of the present work is to identify the contribution of the interface i.e. the function $f(\phi_s(\vec{x},t), \phi_l(\vec{x},t))$ and the amount of renormalization necessary for the fractions of the bulk phases, i.e. to give relation between ϕ and ϕ'. Within this paper the identification of this function and an overall free energy functional shall proceed by:

- general considerations
- theories of crystal growth (Kossel-model, Jackson-model, Temkin-model)
- theories of phase transitions (Ginzburg-Landau theories)
- statistical considerations

2 General Considerations

As the desired function f is a measure of interface resp. interface density, it has to vanish in either of the bulk phases, which means:

$$f(\phi_s, \phi_l) \equiv 0 \quad \text{for} \qquad \phi_s = 1 \text{ (and } \phi_l = 0) \tag{4}$$
$$\text{and for} \qquad \phi_l = 1 \text{ (and } \phi_s = 0) \tag{5}$$

for any point $\vec{x}$ being located in the bulk regions and at any time t.

A simple ansatz fullfiling this requirement for are products:

$$f \sim \phi_l^n * \phi_s^m \tag{6}$$

The exponents n and m have to be positive but not necessarily integers- as otherwise f would not be defined in the bulk regions where either ϕ_l or ϕ_s are 0. More general would be a linear combination:

$$f = \sum_{n,m} a(n,m) \phi_l^n \phi_s^m \tag{7}$$

($n, m \neq 0$ for all interface terms, n or m being 0 will correspond to the other term to represent the bulk phase)

The inclusion of other terms vanishing in the bulk regions (ϕ_s = 1 and ϕ_l = 0 and vice versa) without being products like $\phi_s \ln(\phi_s)$ or $\phi_l \ln(\phi_l)$ or linear combinations of such logarithmic terms like e.g. occurring for concentrations in ideally mixed thermodynamic systems is also possible.

As within the bulk phases the values of the ϕ-fields do not vary and thus gradients of these fields only do occur at interfaces, the function f may also contain terms proportional to $\nabla\phi_s$ and/or $\nabla\phi_l$. Most general would be a combination of any of above terms–products, logarithms, gradients.

3 Thermodynamic Approaches to Interfacial Energy

Classical thermodynamics are based on Gibbs formulation of the following variational problem [1]:

$$\delta\left[-\sum\phi_i\ln(\phi_i)-\lambda\sum E_i\phi_i-\mu\left(\sum\phi_i-1\right)\right]=0 \tag{8}$$

with λ,μ being Lagrange multipliers accounting for the constraints of energy and probability conservation. The solution of this variational problem yields the well known relation for the probability of specific state ϕ_i:

$$\phi_i\sim\exp(-\lambda E_i)/S \qquad S=\sum\exp(-\lambda E_i) \tag{9}$$

with S being the thermodynamic sum being necessary to normalize the probabilities.

It is interesting to note instead of the variation δ also the full functional derivative $\delta/\delta\phi=[\nabla(\partial/\partial(\nabla\phi))-\partial/\partial\phi]=0$ being used to derive the equations of motion for the phase field leads to this result as there are no explicit dependencies on gradients in the-equivalent-formulation:

$$[\nabla(\partial/\partial(\nabla\phi_i))-\partial/\partial\phi_i]\left[-\sum\phi_i\ln(\phi_i)-\lambda\sum E_i\phi_i-\mu\left(\sum\phi_i-1\right)\right] \tag{10}$$

Replacing the logarithmic terms $\sum\phi_i\ln(\phi_i)$ by gradient terms/ scalar products $(\vec{d}\nabla\phi)\ln(\vec{d}\nabla\phi)$ as physically motivated in the following chapters drastically changes that situation. No more exponential terms are expected in the solution for the ϕ and under some simplifying assumptions a gradient energy expression $|\vec{d}|^2\,|\nabla\phi|^2$ being the gradient energy known from phase–field models occurs instead:

$$[\nabla(\partial/\partial(\nabla\phi_i))-\partial/\partial\phi_i]\left[|\vec{d}|^2\,|\nabla\phi|^2-\lambda\sum E_i\phi_i-\mu\left(\sum\phi_i-1\right)\right]=0 \tag{11}$$

To motivate this replacement of the logarithmic terms by gradient terms it is helpful to consider models of crystal growth.

4 Crystal Growth Models

Thermodynamics of interfaces are especially important in the case of growing crystals, where the transition of atoms from the liquid to the solid state at the interface largely determines the structure and the properties of the grown crystal. More information about the function f describing the interfacial energy density may therefor be expected from models of crystal growth. Two models describing the crystal growth of a so called Kossel–crystal, see Fig. 2, will be analyzed in more detail:

- the Jackson model of faceted growth and [7]
- the Temkin model for the growth of a diffuse interface [6]

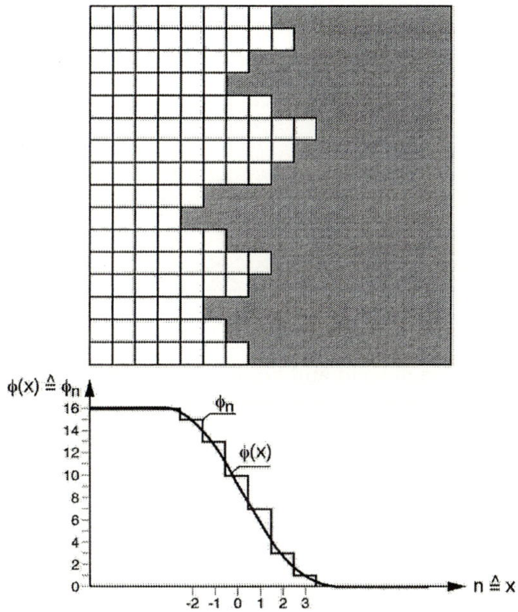

Fig. 2. Kossel's model of the growing crystal. Atoms attach to the interface in layers. The model assumes that the atoms may only adhere on the top of already solid atoms ('solid on solid'). A smooth transition with a finite interface thickness is identified when averaging the fraction of solid atoms parallel to the interface.

Both models allow for a region of interface between the pure solid and the pure liquid, which in case of the Jackson model is restricted to one monolayer and extends across several layers in the Temkin model.

The spatial distribution of the two phases in this interface region is a priori unknown, Fig. 3(top). At first order ideal mixing between the solid and liquid phases may be assumed generating a spatially homogeneous distribution of solid and liquid in the entire interface volume, Fig.3(middle). In classical thermodynamics an ideal mixture is described by logarithmic terms, which in case of deviations from ideal mixture are supplemented by additional terms (e.g. Redlich-Kister polynomials in the case of mixed solute).

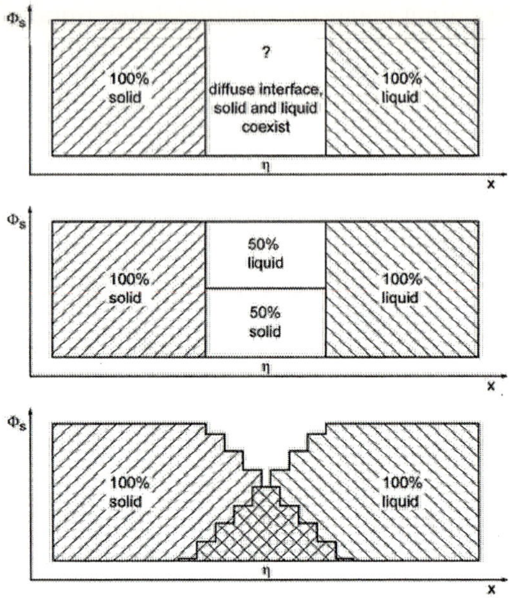

Fig. 3. Spatial distribution of solid and liquid phases coexisting in a region between the pure bulk phases. Nothing is a priori known about this distribution (top). Introducing ideal mixing in this region corresponds to the model of Jackson introduced to describe faceted growth by assuming ideal mixing in a monolayer between bulk solid and bulk liquid (middle). Gradual mixing across several layers was first introduced by Temkin in a multilayer extension of Jacksons approach.

For a phase transition in a unary system this situation basically corresponds to the conditions given by Jackson in his theory of faceted growth, were he considered the mixture between a pure solid and a pure liquid to be restricted to one monolayer. For this monolayer, the Gibbs free energy G^{int} is (adapted from Woodruff [5]):

$$G^{int} = kT\left[\phi\ln(\phi) + (1-\phi)ln(1-\phi) + \alpha\phi(1-\phi)\right] \tag{12}$$

where ϕ denotes the fraction of solid atoms in the interface layer and $(1-\phi)$ represents the fraction of atoms in the liquid state respectively. $\alpha = \Delta S_f/kT$ is the so called Jackson-factor enabling differentiation between diffuse ($\alpha \leq 2$) and faceted ($\alpha > 2$) growth. Note, that in the case of a sharp interface where ϕ_s and ϕ_l are either identical 0 or 1 (obeying $\phi_s + \phi_l = 1$) all of the terms in f vanish. They only contribute to the the total energy G^{tot} in region of the diffuse interface where $\phi, 1-\phi \equiv \{\phi \in \mathbf{R} : 0 < \phi < 1\}$ and thus are related to the interfacial energy.

It seems reasonable to assume that the transition from solid to liquid across a diffuse interface is not a "two step" process, (100% $\rightarrow$

50% and 50% → 0%) but involves at least several steps, Fig.3(bottom), and may eventually be described by a monotonous, continuous function. A respective ansatz has been given by Temkin in his multilayer model assuming a finite number n of transition layers between solid and liquid to describe the diffuse interface of a Kossel- crystal. According to Temkin the Gibbs free energy related to this "diffuse" interface reads:

$$\Delta G = \Delta\mu \left(\sum_{n=-\infty}^{0} (1 - \phi_n^s) - \sum_{n=1}^{\infty} \phi_n^s \right) + 4\omega \sum_{n=-\infty}^{\infty} \phi_n^s (1 - \phi_n^s) \tag{13}$$
$$+ kT \sum_{n=-\infty}^{\infty} (\phi_{n-1}^s - \phi_n^s) \ln (\phi_{n-1}^s - \phi_n^s)$$

where the term in brackets with $\Delta\mu$ (difference in chemical potential) is related to the bulk phases ϕ_s and ϕ_l (i.e. : $1 - \phi_s$) and represents the driving force for a transition. In equilibrium ($\Delta\mu = 0$) we get:

$$G^{int} = 4\omega \sum_{n=-\infty}^{\infty} \phi_n^s (1 - \phi_n^s) \quad + kT \sum_{n=-\infty}^{\infty} (\phi_{n-1}^s - \phi_n^s) \ln (\phi_{n-1}^s - \phi_n^s)$$

This formulation recovers Jackson's model as a limiting case for a single interface layer as can be easily verified by setting $0 < \phi_n < 1$ for only a single value of n.

Besides obviously representing a more general formulation, Temkin's model reveals following interesting features:

- summation (resp. integration) across all layers (i.e the entire volume)
- description of a coupling to the bulk phases for non-vanishing $\Delta\mu$
- introduction of a gradient and accordingly of a length scale

Although the gradient is not immediately obvious in above formulation it may identified by $\Delta\phi = \phi_{n-1}^s - \phi_n^s = d\phi/dz * \Delta z$ in one dimension. Generalization to three dimensions then yields $\vec{a}^s \nabla\phi^s$ with $\vec{a}^s$ representing a vector characterizing the metric imposed by the crystal structure (its components a_x, a_y, a_z representing the lattice spacings in x, y, z directions.)

Reformulating Temkin's expansion in a continuous way accordingly yields:

$$G = kT \iiint_{vol} (\nabla\phi^s \vec{a}^s) \ln (\nabla\phi^s \vec{a}^s) + \iiint_{vol} 4\omega\phi^s(1 - \phi^s) \tag{14}$$

giving the first hints how the desired function f might look like.

Remarkable is the loss of symmetry as no equivalent terms $\vec{a}^l \nabla\phi^l$ do appear, which may be a peculiarity of the second phase in Temkin's model being a liquid with no characteristic length scale $\vec{a}^l$. In case of two interacting solids s_1 and s_2 with different crystal structures $\vec{a}^{s_1}$ and $\vec{a}^{s_2}$ the situation corresponds to stressed interfaces and a more general -because more symmetric-formulation reads:

$$G = kT \iiint_{vol} \left[(\nabla\phi^{s_1}\vec{a}^{s_1}) \ln(\nabla\phi^{s_1}\vec{a}^{s_1}) + (\nabla\phi^{s_2}\vec{a}^{s_2}) \ln(\nabla\phi^{s_2}\vec{a}^{s_2}) \right] + \iiint_{vol} 4\omega\phi^{s_1}\phi^{s_2} \tag{15}$$

This formulation can be rewritten under specific assumptions (see appendix) as:

$$G = kT \iiint_{vol} \left[\left(|\vec{a}^{s_1}|^2 + |\vec{a}^{s_2}|^2 \right) |\nabla\phi^{s_1}|^2 + (|\vec{a}^{s_2}| - |\vec{a}^{s_1}|) |\nabla\phi^{s_1}| \right] + \iiint_{vol} 4\omega\phi^{s_1}\phi^{s_2} \tag{16}$$

ω is a constant related to bond energies and the difference $|\vec{a}^{s_2}| - |\vec{a}^{s_1}|$ in the latter part of the sum corresponds to a lattice mismatch between the two solid phases and can be neglected in case of solid/liquid phase transition as the spacing of fictitious liquid lattice would immediately adapt to the solid lattice leading to a vanishing difference $|\vec{a}^{l}| - |\vec{a}^{s}|$. A detailed treatment of this term will however be interesting in solid-solid phase transitions like e.g. in Solid-Phase-Epitaxy (SPE) or in precipitation phenomena in solids. Neglecting these stress-related terms by approximating $|\vec{a}^{l}| \sim |\vec{a}^{s}| \sim |\vec{a}|$ yields a free energy formulation similar to the well known double-obstacle formulation of the phase-field free energy density functional:

$$G = kT \iiint_{vol} 2|\vec{a}|^2 |\nabla\phi^{s_1}|^2 + \iiint_{vol} 4\omega\phi^{s_1}\phi^{s_2} \tag{17}$$

To summarize the implications of above interpretation of Temkin's formulation it should be noted that logarithmic terms known from classical thermodynamics have evolved into and thus have been related to gradient terms known from Time- dependent Ginzburg-Landau theory. This fact introduces a length scale into thermodynamics and has strong impact on the variational minimization of the respective functional.

To finish the interpretation of Temkin's model it is moreover interesting to have a closer look at the formulation of the driving force term related to the bulk energies:

$$G_{bulk} = \Delta\mu \left(\sum_{n=-\infty}^{0} (1 - \phi_n^s) - \sum_{n=1}^{\infty} \phi_n^s \right)$$

with $\Delta\mu$ representing the driving force for the transition, which has not been treated further by Temkin.

For a single component system undergoing a transition from liquid to solid $\Delta\mu$ corresponds to $\Delta G = G^l - G^s$. The extraction of ($\Delta G = G^l - G^s$) however obviously produces some artifacts relating the Gibbs energy of one phase with the fraction of the respective counter phase:

$$G_{bulk} = G^s \sum_1^{\infty} \phi^s + G^l \sum_{-\infty}^{0} \phi^l - G^l \sum_1^{\infty} \phi^s - G^s \sum_{-\infty}^{0} \phi^l$$

Neglecting these artifacts (i.e the terms $G^l \sum \phi^s$ and $G^s \sum \phi^l$) extending the sums to infinity yields

$$G_{bulk} = G^s \sum_{-\infty}^{\infty} \phi^s + G^l \sum_{-\infty}^{\infty} \phi^l$$

which is in compliance with equation (1) when written in continuous form in three dimensions:

$$G_{bulk} = \iiint_{vol} G^s \phi^s + \iiint_{vol} G^l \phi^l$$

5 Approaches from Theories of Phase Transitions

Time and space resolved modeling of phase-transitions has gained more and more importance in modeling of solidification phenomena and evolving microstructures in the recent years. The fundamental difference between sharp interface models and diffuse interface theories is the finite interface thickness yielding an additional degree of freedom for the system. The evolution of this continuous parameter in time may be used to describe as well the first formation of interfaces-i.e. nucleation-as also their stability during propagation in a continuum model. Actual diffuse interface theories of phase-transitions make only use of a stationary value of the interface thickness. They do not address fluctuations or evolution of this parameter in the time.

Shape and thickness of the transition layer are largely determined by the choice of the potential energy formulation in the phase-field free energy functional. Especially the interface thickness corresponds to the height of the energy barrier and various potentials lead to the same sharp interface limit. As by now the thickness of the interface, its internal dynamics and its mesoscopic shape in the direction normal to the interface have not in depth been considered, the choice of the potential actually entering the phase-field free energy formulation has been of minor interest. Actual phase-field models accordingly treat the interface thickness as a fixed, stationary parameter not exhibiting its own dynamics and not having a physical meaning at all. Interfaces in physical systems-e.g. in thin film deposition or in solidification of many metals - however often have a finite thickness of several monolayers, the number of which varies as a function of external conditions like e.g. temperature.

The general structure of the phase-field free energy formulation, from which the equations of motion are derived by functional differentiation, consist of three contributions:

- a gradient energy being related to gradients of the order parameter
- a potential energy (in general a double-well-potential with minima either at $+/-1$ or -in more recent work at 0 or 1. Sometimes a double obstacle potential is used.)
- a driving force term describing the deviation from equilibrium and in general being proportional to an undercooling ΔT

The common aspect of all formulations [2–4,8] is identified in the driving force terms, which all involve deviations from an equilibrium state described by an undercooling ΔT. This corresponds to a Taylor expansion around an equilibrium state:

$$G^{tot} = G^{equi} + (\partial G/\partial T)\Delta T$$

Thus by now all the free energy functionals are parameterized by this value. A free energy formulation *not referencing* to the equilibrium state, but only being based on the energies of the phases involved, their volume fractions, their interfacial area (being proportional to gradients of the fraction solid/liquid respectively) and interfacial energy σ would represent a more comprehensive description:

$$G^{tot} = G^{tot}(G^s, G^l, \sigma, \phi_s, \phi_l, \nabla\phi^s, \nabla\phi^l)$$

To approach a respective formulation, the potential and driving force parts of the free energy functional used by Steinbach et al [4]. may serve as an instructive example when being compared to Temkin's model. As in Temkin's model, terms relating the Gibbs energy of one phase to the fraction of the counter phase appear when expressing ΔG as $G_i - G_j$:

$$(\phi_i\phi_j)^2 + G_i(1/3\phi_i^3 + \phi_i^2\phi_j - \phi_i\phi_j^2 - 1/3\phi_j^3) - G_j(1/3\phi_i^3 + \phi_i^2\phi_j - \phi_i\phi_j^2 - 1/3\phi_j^3)$$

which, when considering the pure bulk contributions separately i.e neglecting terms $\sim \phi_i\phi_j$, yields:

$$G_{bulk} = 1/3G_j\phi_j^3 + 1/3G_i\phi_i^3 - 1/3G_i\phi_j^3 - 1/3G_j\phi_i^3$$

This expression looks similar to Temkin's expression for the bulk energy contributions besides the fact of different powers of ϕ occurrs in either expression. This fact - meaning that superposition of the bulk energies might not be linear in phase-fraction in mixed phase systems- and the fact that different forms of potentials (double-well, double obstacle) may equivalently be used to describe the phase transition leads to more detailed considerations about the origin of the potential in the following chapter.

6 Statistical Approach

To get a statistical approach to the problem, the fractions ϕ_s are now considered as the probabilities of finding an atom in the solid state (i.e. being in an environment typical for a solid state) when being drawn out of a control-volume centered around $\vec{x}$ at time t. Following very general statement can be made for any finite size of this control volume at any length scale:

$$\phi_s(\vec{x},t) + \phi_l(\vec{x},t) = 1 \tag{18}$$

and accordingly (since $1^n = 1$)

$$\phi_s^2 + 2\phi_s\phi_l + \phi_l^2 = 1 \tag{19}$$

$$\phi_s^3 + 3\phi_s^2\phi_l + 3\phi_s\phi_l^2 + \phi_l^3 = 1 \tag{20}$$

$$\phi_s^4 + 4\phi_s^3\phi_l + 6\phi_s^2\phi_l^2 + 4\phi_s\phi_l^3 + \phi_l^4 = 1$$

and so forth

Here the extension has been continued up to fourth order as to allow for comparison with the classical Ginzburg-Landau theory that uses the double-well fourth order polynom.

This procedure can be considered as defining interface states (i.e. product terms $\phi_s\phi_l$) between the two states ϕ_s and ϕ_l and simultaneously renormalizing the probabilities for occurrence of the bulk phases in the interface regions. In contrast to renormalization group procedures, however, where starting from a large number of degrees of freedom irrelevant degrees of freedom are averaged out of the system, here the starting point is a low number of macroscopic states and necessary additional degrees of freedom are introduced:

$$\phi_s^2 + 2\phi_s\phi_l + \phi_s^2 = 1 \text{ may be reinterpreted as } p_s + p_{if} + p_l = 1 \tag{21}$$

with p_s being the probability of an atom being in a solid environment, p_l being the probability of an atom being in a liquid environment and p_{if} being the probability of an atom to belong to the interface. The definition of products terms $\phi_s\phi_l$ as a probability of both bulk phases to be present in the interface region has already been stated in the chapter on general considerations. It is interesting to note that such products also appear in both Jackson and Temkin models as also in classical and Time–dependent Ginzburg– Landau theories.

The physical meaning of these interface terms may be best explored using the picture of a three dimensional cubic crystal with a sixfold coordination ($Z = 6$) revealing 5 (resp. 7) different types of configurations, Fig. 4, each occurring with different probability. These probabilities can be calculated by the binomial coefficients of $(\phi_s + \phi_l)^Z$:

$$1 = \phi_s^6 + 6\phi_s^5\phi_l + 15\phi_s^4\phi_l^2 + 20\phi_s^3\phi_l^3 + 15\phi_s^2\phi_l^4 + 6\phi_s\phi_l^5 + \phi_l^6 \tag{22}$$

Fig. 4. Three dimensional view of a Kossel crystal. The occurrence of each of the cubes has a different probability and is related to a specific energy level ..(see text).

It is easily seen, that e.g. the term $\phi_s\phi_l^5$ may be identified with a cube facing the liquid with 5 sides and the solid with 1 side (e.g. cube 1 in Fig. 4). This state is sixfold degenerate as no face of the attached cube is preferred. The degree of degeneracy- indicated by the binomial coefficients will be less in the case of anisotropic systems. The degree of degeneracy of the other cubes may be verified accordingly.

Given the probabilities for the possible states from above considerations, the determination of the systems energy becomes possible by weighting each state with its specific energy:

$$\phi_s + \phi_l = 1 \qquad \phi_s G^s + \phi_l G^l = G^{tot} \tag{23}$$

$$(\phi_s + \phi_l)^2 = 1 \qquad \phi_s^2 G^s + 2\phi_s\phi_l G^{sl} + \phi_l^2 G^l = G^{tot} \tag{24}$$

$$(\phi_s + \phi_l)^3 = 1 \qquad \phi_s^3 G^s + 3(\phi_s S^s \phi_l S^l)\phi_s\phi_l + \phi_l^3 G^l = G^{tot} \tag{25}$$

$$(\phi_s + \phi_l)^4 = 1 \qquad \phi_s^4 G^s + 4(\phi_s^2 S^s \phi_l^2 S^l)\phi_s\phi_l + 6\phi_s^2\phi_l^2 W^{sl} + \phi_l^4 G^l = G^{tot} \tag{26}$$

This procedure does not affect the pure solid bulk regions where $(\phi_s)^4 = (\phi_s)^3 = (\phi_s)^2 = \phi_s = 1$ (and thus $G^{tot} = G^s$) and $\phi = 0$ (or vice versa for the liquid bulk regions) and thus recovers the classical energy formulation given

8 Acknowledgments

This work is based on joint discussions with the ACCESS microstructure simulation group and their work mainly being funded by: BMBF, DFG and the Corus group.

Thanks are due to Toni Ivas for commenting on the manuscript and getting it into a readable form.

References

1. Jaynes, E.T: Probability Theory: The logic of Science, chapter **11**, (internet: http://omega.albany.edu:8008/JaynesBook.html)
2. Kobayashi, R.: Modeling and numerical simulations of dendritic crystal growth. Physica D **63** (1993) 410–423
3. Caginalp, G., Fife, P.C.: Phys. Rev. B **33** 11 (1986) 7792
4. Steinbach, I., Pezzola, F., Nestler, B., Seesselberg, M., Prieler, R., Schmitz, G.J., Rezende, J.L.L.: A Phase field concept for multiphase systems. Physica D **94** (1996) 135-147
5. Woodruff, D.: The solid liquid interface. Cambridge University Press, 1973
6. Temkin, D.E.: Crystallization Processes (ed. N.N. Sirota, F.K. Gorskii and V.M. Varikash). English translation published by consultants Bureau, New York (1966)
7. Jackson, K.A.: Liquid Metals and solidification ASM. Cleveland Ohio, (1958) 174
8. Karma A., Rappel, W.J.: Numerical Simulation of Three dimensional Dendritic Growth. Phys. Rev. Lett. vol.**77** No. **19** (1996) 4050-4053
9. Fix, G.: Phase field models for free boundary problems. in Free boundary problems Vol. **II**, (ed. A.Fasano, M.Primicerio) Piman Boston (1983)
10. Wang, S.L., Sekerka, R.F., Wheeler, A.A., Murray, B.T., Coriell, S.R., Braun, R.J., McFadden, G.B.: Thermodynamically-consistent phase-field models for solidification. Physica D **69** (1993) 189-200

Table 1. The different free energy formulations addressed in this paper. Note the arrows indicating the introduction of a length scale into entropy

Model	Terms related to field gradients		Interface terms (vanish in bulks)		Bulk terms; driving forces	Constraints
	$\|\nabla\phi\|^2$	$(\vec{a}\nabla\phi)\ln(\vec{a}\nabla\phi)$	$\phi_i \ln \phi_i$	$\phi_i\phi_j$	ϕ_i, ϕ_i^2, etc	
Gibbs/Shanon			$\sum \phi_i \ln \phi_i$		$\sum E_i\phi_i$	$\sum \phi_i - 1 = 0$
Jackson			$\sum \phi_i \ln \phi_i$	$\phi_i\phi_j$	not found	
Temkin		$(\vec{a}\nabla\phi)\ln(\vec{a}\nabla\phi)$ ←		$\phi_i\phi_j$	$\Delta G(\phi_i - \phi_j)$	
Ginzburg/Landau				$A\phi_i\phi_j + B(\phi_i\phi_j)^2$		
Time-dependent Ginzburg-Landau	↓ $\|\nabla\phi\|^2$			$A\phi_i\phi_j + B(\phi_i\phi_j)^2$		
Phase-field (double-well)	$\|\nabla\phi\|^2$			$(\phi_i\phi_j)^2 + \Delta G(1/3\phi_i^3 + \phi_i^2\phi_j - \phi_i\phi_j^2 - 1/3\phi_j^3)$		
Phase-field (double-obstacle)	$\|\nabla\phi\|^2$			$\phi_i\phi_j$		$\sum \phi_i - 1 = 0$
Proposed form		$(\vec{a}\nabla\phi)\ln(\vec{a}\nabla\phi)$		$\phi_i\phi_j$	$G_i\phi_i^2 + G_j\phi_j^2$	$(\sum \phi_i)^2 - 1 = 0$

Computer Investigation of the Influence of the Internal Structure Topology on the Percolation Process in Two- and Three-Dimensional Inhomogeneous Systems

Aliaksei Konash and Sergey Bagnich

Institute of Molecular and Atomic Physics, Minsk 220072, Belarus
Email: konash@imaph.bas-net.by

Abstract. In our work the percolation process in two- and three-dimensional inhomogeneous lattices is studied. The inhomogeneous lattice is simulated by a random distribution of obstacles differing in size and number. The influence of obstacles on the parameters (critical concentration, average number of sites in finite clusters, percolation probability, critical exponents, and fractal and spectral dimensions of a finite cluster) characterizing the percolation in the system is analyzed. It is demonstrated that all these parameters essentially depend on the linear size and relative area of the obstacles.

1 Introduction

Nowadays developing of the computer technique for imitating such complex processes as the process of percolation in disordered systems and related phenomena is an actual task. Rock fracture and fragmentation [1], conducting in a random resistance grating [2], strongly inhomogeneous media [3], propagation of forest fires [4] and doped semiconductors [5] were successfully analyzed in terms of the percolation theory. And in [6] Kopelman *et. al.* developed a cluster formalism for describing the electronic excitation energy transfer in inhomogeneous systems. In such a presentation to characterize a condition of investigating a system, the function of the percolation probability P_∞ and the average number I_{AV} of sites in the cluster were chosen. The dependence of them on the concentration C is determined by next scaling relationships [7]

$$I'_{AV} \propto |C/C_c - 1|^{-\gamma} \tag{1}$$

$$P_\infty \propto |C/C_c - 1|^{\beta} \tag{2}$$

where C_C is the critical concentration of the sites and β and γ are the critical exponents(depend only on space dimension).

An application of this theory for transport of electronic excitation energy in the mixed molecular crystals [8] and solutions of organic compounds in low-molecular vitrifying solvents [9] has demonstrated good correspondence

Table 1. The values of the critical exponents

Critical exponents	Percolation theory		Isotopically mixed molecular crystals	Chemically mixed molecular crystals	Solid solution of benzaldehyde in ethanol	Ethanol solution of benzaldehyde in porous glass
	2D	3D				
β	0.14	0.41	0.13	0.13	0.41	0.25
γ	2.1	1.6	2.1	2.09	1.7	1.95

between experimental results and the theoretical predictions of the percolation theory for the critical exponents (tabl.1).

However, recent experimental studies [10] of similar processes in porous matrices revealed an influence of the internal inhomogeneous structure of the system on the topology of energy transport.

In our work, we performed the computer simulation of the percolation process on a square and cubic lattices with obstacles differing in size and relative area (volume) in order to elucidate the possible effect of these obstacles on the critical concentration C_c, the average number $I_{AV}$of sites in a cluster, the percolation probability P_∞, the fractal dimension of an infinite cluster and the critical exponents.

2 Computational Technique

We will solve the site percolation problem, because it is the most important one from the viewpoint of energy migration in heterogeneous systems. Let us consider a lattice formed by a set of sites and bonds. It is assumed that C is the part of sites painted black in a random manner. An aggregate of black sites connected to one another either directly or through chains of connected black sites is referred to as a cluster. At C=1, black clusters are absent in the system. At C«1, black clusters consist of a small number of sites: single sites, pairs, triads, etc. However, as the percolation threshold is approached, particular clusters merge together and their average size increases. The average number of sites in finite clusters is defined by the expression

$$I_{AV} = \frac{\sum\limits_{m} i_m m^2}{\sum\limits_{m} i_m m}, \tag{3}$$

where i_m is the number of clusters containing m sites. The analytical dependence of $\boldsymbol{I}_{AV}$ on the fraction C is unknown. Numerical calculations showed that, at $C \to C_c - 0$, the quantity I_{AV} goes to infinity [see relationship (1)]. According to [11], the percolation probability is defined as the ratio between the number of sites forming an infinite cluster and the total number of sites in the lattice. In the numerical simulation, the number of sites contained in the maximum cluster (m_{max}) is calculated and the percolation probability is estimated from the formula

$$P_{\max} = \frac{m_{\max}}{L \times L} \quad , \tag{4}$$

where l is the linear size of the lattice. Extensive simulations and theoretical considerations show that, near $C \rightarrow C_c + 0$, the percolation probability decreases according to the power law (2).

All the results presented in this work were obtained from simulations of the percolation process on 200*200 and 100*100*100 lattices for two- and three- dimension cases respectively. Despite the currently available methods of computer reconstruction of Vycor porous glasses (see, for example, [12]), the inhomogeneous system was initially simulated by introducing square obstacles of a specified size into the lattice, as was done by Bujan-Nunez *et al.* [13]. For this purpose, a new lattice with a cell size equals to the linear size of obstacles was constructed on the primary lattice. In the simulation, some of the cells were randomly chosen as obstacles and all the primary lattice sites lying within the chosen obstacles were eliminated (removed) from consideration. Fig. 1 displays variants of the model matrices with obstacles differing in linear size and relative area.

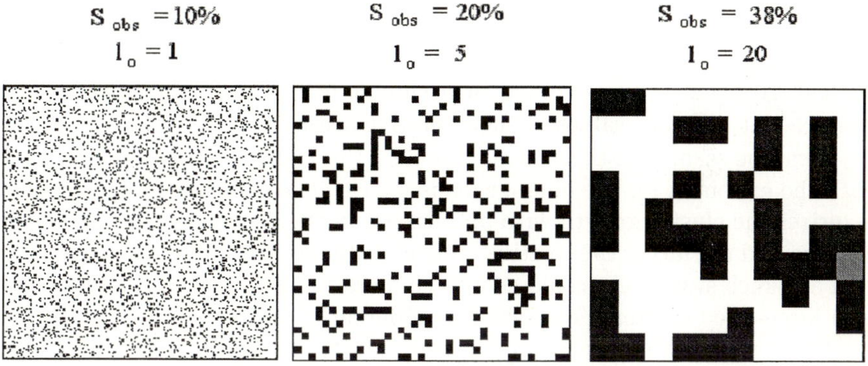

Fig. 1. Variants of matrices with obstacles differing in linear size and relative area.

3 Results and Discussion

First and foremost, we analyzed haw the inhomogeneity of the matrix affects the critical concentration. In each case, the value of C_c was determined by two methods. According to Hoshen *et al.* [14], the critical concentration can be determined from the position of the maximum in the dependence of the reduced average number I'_{av} of sites in clusters on the site concentration C:

$$I'_{AV} = \frac{\sum\limits_{m} i_m m^2}{\sum\limits_{m} i_m m} - \frac{m^2_{\max}}{\sum\limits_{m} i_m m}. \tag{5}$$

The dependence obtained by averaging over 200 realizations (2D space) exhibit a rather smeared maximum, even though the concentration in the course of the experiment was changed with the step $\Delta C = 0.001$. For this reason, the critical concentration for each realization was taken as the concentration corresponding to the onset of the percolation between opposite sides of the lattice. The critical concentration determined from these data agrees closely with the value obtained using other methods for a square lattice [4,5,7,11]. The introduction of obstacles into the lattice considerably affects the critical concentration: its value increases substantially (Fig.2).

The dependence of the average number of sites in a cluster on the reduced concentration C/C_{AV} of occupied sites is well described by the power law predicted by formula (1). The deviation from the power dependence near the critical concentration is caused by the finite sizes of the lattice. In fact, as follows from formula (1), this quantity should increase to infinity at the critical point, which, in principle, is impossible in systems of finite size. The critical exponent γ determined from this dependence coincides with the value obtained by the same method in [14] and is slightly less than the exponents derived from other techniques [7]. Figure 3 displays the dependencies of the critical exponent γ (determined in a similar manner) on the linear size of obstacles at different values of their relative areas.

It can be seen, that the critical exponent for obstacles with l_0=1 coincides with the exponent for the homogeneous lattice. The critical exponent γ characterizes the cluster growth with an increase in the concentration. The larger the size and the larger the relative area of obstacles, the higher the concentration at which sites begin coalesce into clusters and small-sized clusters merge into large-sized clusters. To state this differently, an increase in the average cluster size with an increase in the concentration is more pronounced than that in the system with a homogeneous matrix; in fact this corresponds to an increase in the critical exponent γ. The observed decrease in the exponent γ for matrices with obstacles of size l_0=20 can be explained by the finite sizes of the lattices. As can be seen from Fig.1, an increase in the size of the obstacles brings about the transformation of the inhomogeneous matrix into the matrix with a specially confined structure.

This is essentially pronounced for lattices with the maximum relative area of the obstacles used in the computer experiment. Numerical simulation revealed that, for 2D matrices at l_0=20 and S_{obs} = 38%, an increase in the lattice size to 400*400 is accompanied by increase in the critical exponent γ

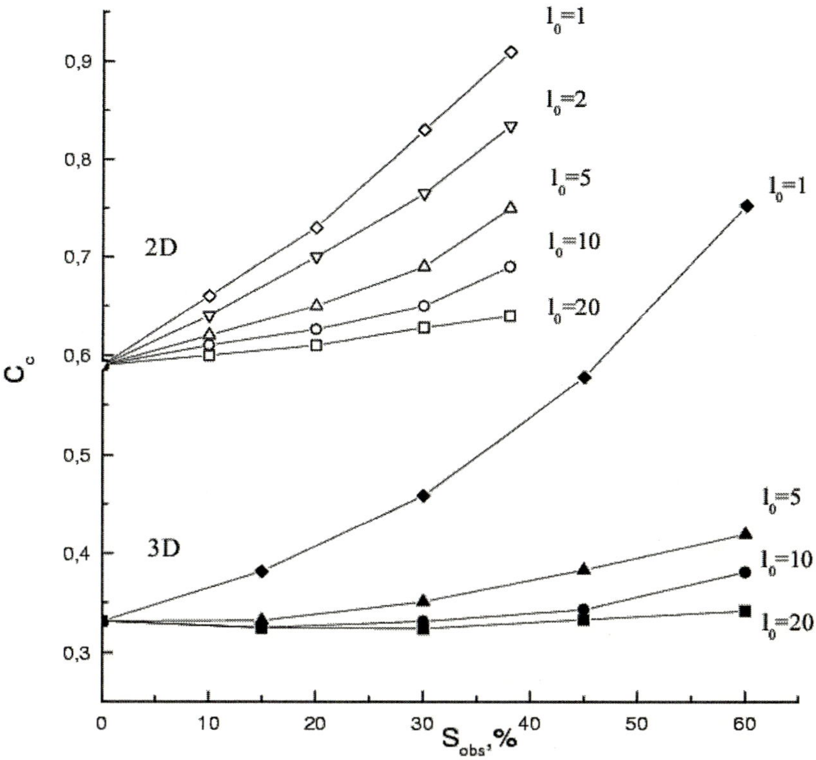

Fig. 2. Dependence of the critical concentration C_c on the homogeneity fraction S_{obs} in the matrix at different linear sizes l_0 of obstacles.

to 2.94(for 200*200 lattice, γ=2.50). Note that an increase in the size of the obstacles to 40 results in a further decrease in the exponent.

Figure 4 depicts the dependence of the percolation probability on the reduced concentration for the lattices with different fractions of the obstacles. It is clearly seen that the introduction of obstacles into the lattice is attended by a more rapid increase in the percolation probability with an increase in the concentration.

The fractal dimension d_f is a principal characteristic of the infinite cluster at the critical point. Mandelbrot [15] was the first to introduce the notion of a fractal. Subsequently, he specified the tentative concept and defined the fractal as a structure consisting of parts that, in some sense, are similar to a unity [16]. An infinite cluster at the critical point exhibits a statistical self-similarity [11]. The fractal geometry of the infinite cluster and its statistical self-similarity are interrelated. This interrelation leads to the following

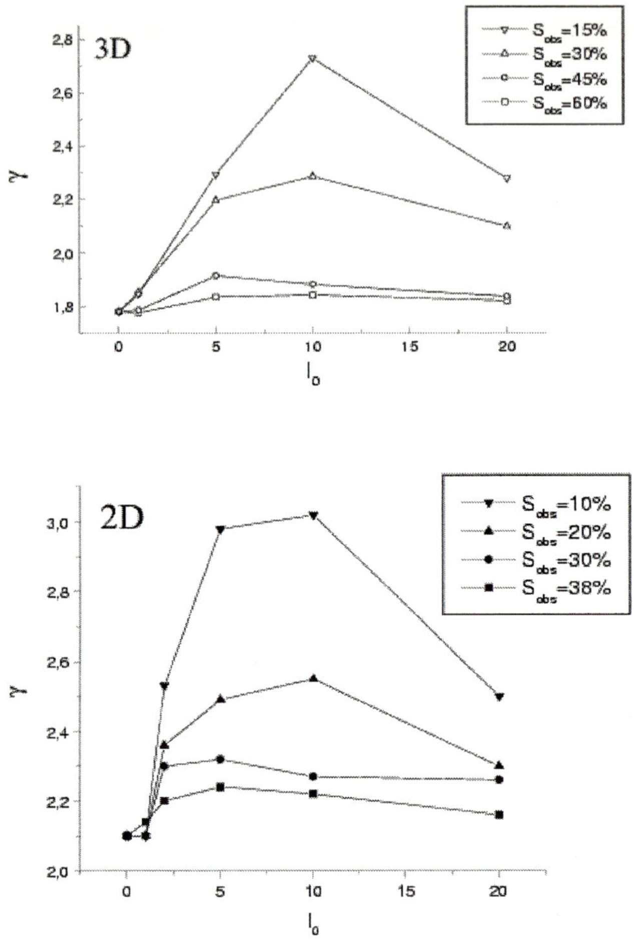

Fig. 3. Dependence of the critical exponent γ on the linear obstacle size l_0 in lattices for different values of their relative areas.

relationship between the mass and the linear size of the cluster:

$$M = l^{d_f} \tag{6}$$

Sokolov in [17] showed that the fractal dimension in virtually all physical problems is defined as the exponent in the relationship (4). At present, the fractal dimension is determined using different methods. One of them is the embedded square method proposed in [11]. Forrest and Witten [18] proposed to bring the central point into coincidence with the center of gyration of the studied object in order to improve the reproducibility of the results. In our work, we also determined the fractal dimension of the infinite cluster by using the embedded square method. To accomplish this, among all the possible realizations, we chose clusters whose centers of gyration were close to $L/2$. The results presented below were obtained by averaging over 300 different clusters. For lattices with large–sized obstacles, additional averaging was performed over 20 configuration of distribution of the obstacles in these lattices.

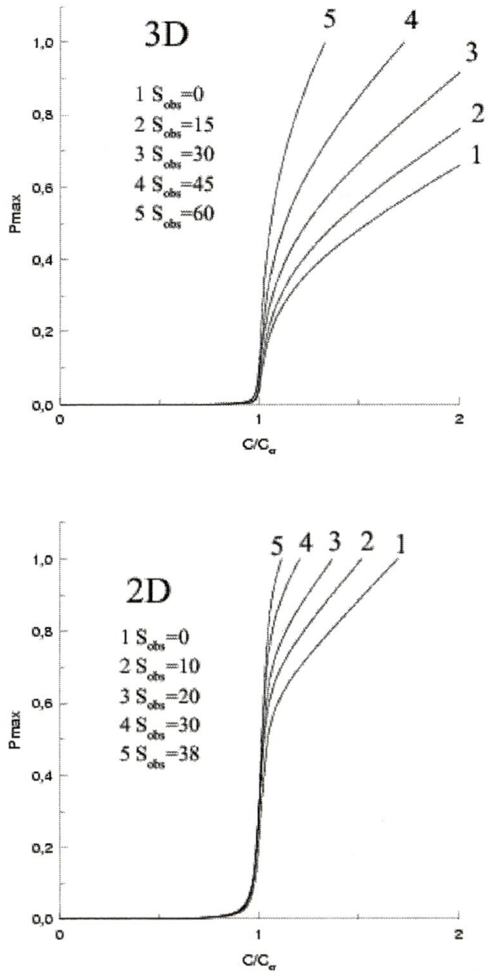

Fig. 4. Dependence of P_{max} on the reduced concentration C/C_c for lattices with obstacles of the linear size l_0=1 at different fractions of the obstacles in the matrix.

The dependence of the cluster mass (the number of sites) exhibits a linear behavior beginning with square sizes of the order of 20*20. The fractal dimension determined from the slope of this dependence is equal to 1.8. This value is slightly less than the exact dimensional $d_f = 91/48$, which was calculated in [4,7] in terms of the scaling theory. This difference can arise for two reasons. First, the fractal dimension determined by the embedded square method is underestimated compared to that obtained by other methods. Second, the underestimated value of d_f can be dictated by finite size of the lattice. In any case, our prime concern is with the influence of the inhomogeneous properties of the lattice on the fractal dimension rather than in its absolute value. This influence is illustrated by the data shown in Fig.5.

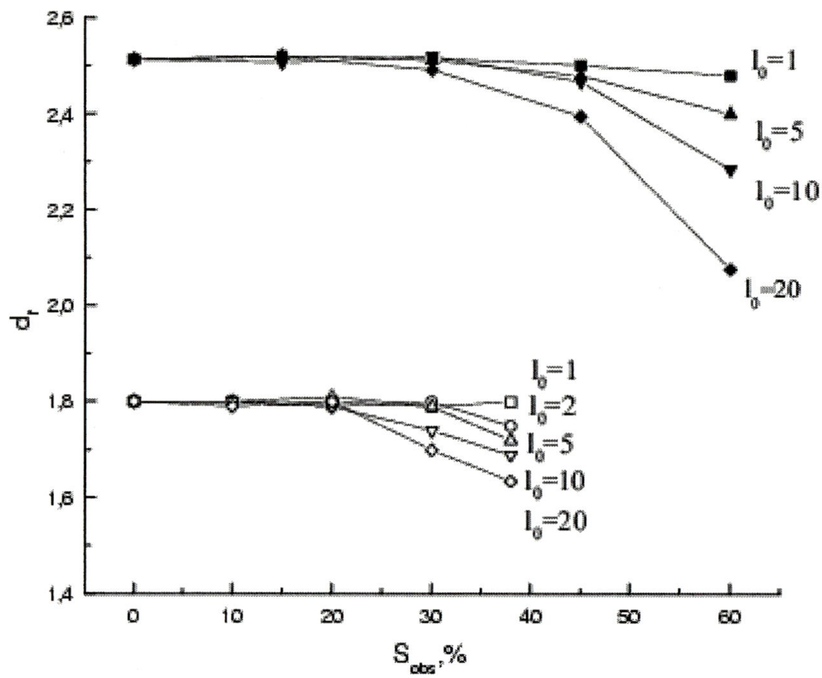

Fig. 5. Dependence of the fractal dimension of a percolation cluster on the fraction of the obstacles S_{obs} in the matrix at different linear sizes l_0 of the obstacles.

Another dimension associated with random walks on a cluster is a spectral dimension d_S. In [19], it has been shown that at the percolation threshold the number S_N of distinct sites visited during an N-step random walk on an infinite cluster varies asymptotically as:

$$S_N \propto N^{ds/2}, d_s = 4/3 \tag{7}$$

in all Euclidean dimensions (Alexander-Orbach hypothesis) [20].

In our investigation we obtained the values of spectral dimension closed to 4/3 for lattices differing in relative area and linear size of the obstacles. This result is in a good agreement with the Alexander-Orbach hypothesis.

4 Conclusion

The results obtained in this work confirm the assumption made earlier (on the basis of the available data on the energy transfer in disordered system, specifically, in matrices with different structures on the microscopic level) that the inhomogeneous properties of matrices substantially affect the percolation process. The introduction of the obstacles causes strong influence on all percolation parameters in both two- and three- dimensional spaces with next laws. Increase of the relative area of the obstacles leads to increase of the critical concentration value, increase of the critical exponent γ value, increase of the growth rate of percolation probability with increase of the concentration and decrease of the fractal dimension of percolation cluster. In turn, increase of the linear size of the obstacles (with the same the relative area of them) results in a fall of the critical concentration value, increase of the critical exponent γ value, a fall of the growth rate of percolation probability with increase of the concentration and decrease of the fractal dimension of percolation cluster. The spectral dimension of percolation cluster does not depend on the presence of the obstacles in lattice what accords with the conception of super universality for this parameter.

References

1. Fragmentation Form and Flow in Fractured Media, Eds. R. Engelman and Z. Jaeger (IPS, Bristol, 1986).
2. *Percolation Structures Processes*, Ann. Isr. Phys. Soc., Vol.. Eds. G. Deutsher, R. Zallen and J. Adler, (Hilger, Bristol, 1983), **5**.
3. B. I. Shklovskiy and A. L. Efros. Percolation theory and conductivity of highly inhomogeneous media, Sov. Phys. Usp. **18** (1975) 845-878.
4. D. Stauffer. Introduction to Percolation Theory, (Taylor & Francis, London, 1985) 420.
5. B. I. Shklovskiy, A. L. Efros. Electronic Properties of Doped Semiconductors, (Springer –Verlag, New York, 1984).
6. R. Kopelman. in Spectroscopy and Exitation Cynamics in Condenced Molecular Systems, Ed. by V.M. Agranovich and R.M. Hochstrasser (Amsterdam, North-Holland, 1983).
7. D. Stauffer. Scaling theory of percolation cluster, Phys. Rep., **54** (1979) 1-74.
8. Borczyskowski, von C., Kirski, T., Bunsenges, Ber.; Dispersive energy transport in disordered molecular crystals, Phys. Chem. **93** (1989) 1373-1377.
9. S. A. Bagnich. Triplet excitation migration for compound molecules in solid solutions, Chem. Phys. **185** (1994) 229-236.
10. S. A. Bagnich. Migration of benzaldehyde triplet excitation in porous matrices, SPIE Proc. **3176** (1997) 212-218.

11. J. Feder. Fractals (Plenum, New York, 1988) 322.
12. M. E. Kainourgiakis, E. S. Kikkinides, A. K. Stubos, N. K. Kanellopoulos. Simulation of self-diffusion of point-like and finite-size tracers in stochastically reconstructed Vycor porous glasses, J. Chem. Phys. **111** (1999) 2735-2743.
13. M. C. Bujan-Nunez, A. Miguel-Fernandez, M. A. Lopez-Quintela. Diffusion-influenced controlled reaction in an inhomogeneous medium: Small concentration of reagents, J. Chem. Phys. **112** (2000) 8495-8501.
14. J. Hoshen, R. Kopelman, E. M. Monberg. Percolation and cluster distribution. II. Layers, variable-range interaction, and exciton cluster model, J. Stat. Phys. **19** (1978) 219-242.
15. B. B. Mandelbrot. The Fractal Geometry of Nature, (W.H. Freeman, New-York, 1983) 486.
16. B. B. Mandelbrot. Fractals Encyclopedia of Physics and Technology, **5** (1987) 579.
17. I. M. Sokolov. Dimension and other geometrical critical exponents in the percolation theory, Sov. Phys. Usp. **29** (1986) 924-948.
18. S. R. Forest, T. A. Witten. Long-range correlation in smoke-particle aggregates, J. Phys. A **12** (1979) L109 - L117.
19. R. Rammal, G. Toulose. Random walks on fractal structures and percolation clusters, J. Phys. Lett. **44** (1983) 13-22.
20. S. Alexander, R. Orbach. Density of states on fractals: "tractons", J. Phys. Lett. (Paris) **43** (1982) 625-631.

Electron Transport of Nanoperm Alloys

K. Pekala

Department of Physics, Warsaw University of Technology Chodkiewicza 8,
PL-02-524 Warsaw, Poland.

Abstract. Electrical resistivity and thermoelectric studies of nanocrystallizing $Fe - Zr - Cu - B$ alloys allow to separate the surface and bulk crystallization processes. Nucleation mechanisms specific for both crystallization channels are correlated to different effects of boron and copper atoms. A temperature variation of crystalline fraction confirms a strong influence of grain sizes and grain boundaries on electrical resistivity of the nanocrystallizing alloy. This first thermoelectric study of Nanoperm alloys exhibits a weak dependence of TEP on temperature and chemical composition of amorphous alloys. TEP of the nanocrystalline alloy is determined mainly by the α - Fe phase.

1 Introduction

The recently studied Nanoperm alloys are distinguished by excellent magnetic properties, such as high saturation magnetization, small coercive field, high magnetic permeability. Numerous experimental approaches, mainly magnetic, were applied in order to unfold the microscopic structure which entails such specific physical properties. The Moessbauer spectroscopy allowed to separate the fine processes of surface crystallization [1]. Transport investigations of Nanoperm alloys are very limited [2,3]. The present work is aimed at studying the unresolved change in temperature variation of resistivity appearing at 500 K. Electrical resistivity is supplemented for a first time by thermoelectric power, which is also a sensitive probe of structural transformations.

2 Experiments

Alloys with compositions $Fe_{89}Zr_7B_4$, $Fe_{87}Zr_7B_6$, $Fe_{83}Zr_7B_{10}$, $Fe_{87}Zr_6 Cu_1B_6$ and $Fe_{85}Zr_7Cu_2B_6$ were produced in argon atmosphere by melt spinning technique. Details are described separately [4]. Electrical resistivity was measured using the dc four probe configuration with the accuracy of 0.1%. Temperature variation of thermoelectric power was recorded by a differential method. The error was smaller than 0.2 mV/ K.

3 Results and Discussion

3.1 Electrical Resistivity

All our alloys exhibit a broad minimum of electrical resistivity located around 350 K, which is close to the Curie temperature of the alloys (Fig. 1 and 2). A similar temperature variation of resistivity in the vicinity of the para- to ferromagnetic transition is reported for $Fe-Zr$ alloys [2,3] and also for other alloys, such as e.g. $Co-Si-B$ [2,5] . Above this minimum, the temperature variation of resistivity exhibits two ranges of approximately linear variation with a slope change of 500 K - 630 K depending on composition. An abrupt drop of resistivity is observed above the threshold temperature of 800 K, when crystallization processes start. The decrease of resistivity above 900 K reveals further crystallization stages.

The observed temperature variation of electrical resistivity is built by a superposition of the structural and magnetic components. According to the Ziman model the structural part above the Debye temperature varies proportionally to the temperature [6]. The magnetic part related to the electron scattering on disordered spins becomes constant above the Curie temperature [7].

Analysis of the resistivity data allows us to distinguish subtle phenomena in crystallization. It is well known that due to additional topological and chemical disorder, experienced by atoms located at surfaces, the threshold surface crystallization temperature is lower as compared to the bulk crystallization threshold. Thus a change in the slope of resistivity versus temperature observed at 500 K may be ascribed to the surface crystallization affecting a small part of the sample volume (Figs. 1 and 2). The structural rearrangements at the surface were also detected at this temperature by Moessbauer spectroscopy [8]. An abrupt resistivity change around 800 K reflects the crystallization processes occurring in a bulk. They result in a precipitation of nanocrystals a - Fe phase. This iron phase was confirmed by the X-ray and Moessbauer data. It is noteworthy that the relative change in resistivity during crystallization is greater than in Finemet and aluminium based alloys [9,10]. Moreover, crystallization occurs in a relatively narrow temperature interval.

Alloys with higher boron content exhibit higher threshold temperatures both for the surface and for bulk processes. Fig. 1 confirms the stabilizing role of boron atoms which introduce covalent bonds $Fe-B$. It was checked that the local surface crystallization temperature is suppressed when boron atoms are displaced from a surface layer. On the other hand, the copper atoms affect only the bulk crystallization but not the surface crystallization. Previous observations lend a firm support to the idea that copper atoms form clusters being nucleation centres [11]. This in turn explains that the raising copper content diminishes the threshold bulk crystallization temperature.

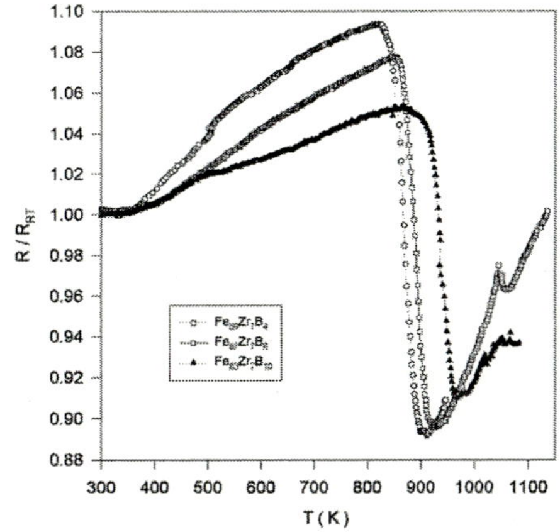

Fig. 1. Temperature variation of normalized electrical resistivity of $Fe - Zr - B$ alloys for various boron contents.

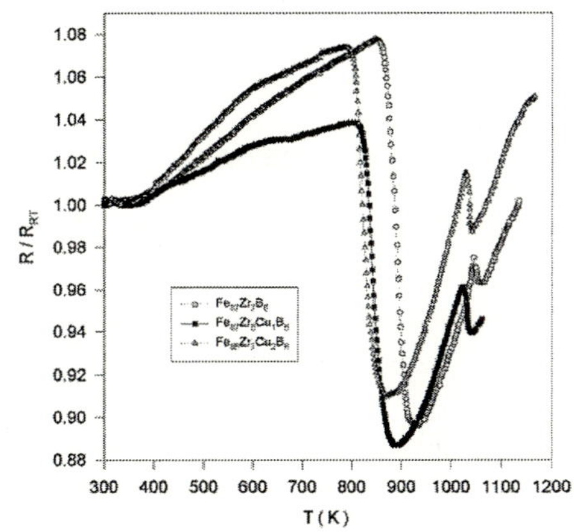

Fig. 2. Temperature variation of normalized electrical resistivity of $Fe{-}Zr{-}Cu{-}B$ alloys for various copper contents.

3.2 Crystalline Fraction

Kinetics of crystallization may be quantitatively evaluated, when evolution of a crystalline fraction X is determined as a function of temperature or time. Transport studies are highly sensitive to structural rearrangements and offer electrical resistivity measurements as the precise probe for crystallization processes . The so called Maxwell - Garnett model is applied for numerical calculations of the crystalline fraction X, [12]. This model describes electrical resistivity $\rho(T)$ considering an alloy as a mixture of the crystalline and amorphous phases characterized by different $\rho_c(T), \rho_a(T)$ electrical resistivities, respectively. Then

$$\rho(T) = \rho_a(T)\left[\frac{1+3X\beta}{1-X\beta}\right],$$

where $\beta = [\rho_a(T) - \rho_c(T)]/[\rho_a(T) + 2\rho_c(T)]$. It was checked that an assumption of the linear temperature dependence of electrical resistivity of the amorphous phase is well substantiated. Electrical resistivities of the amorphous and crystalline phases are used as experimental parameters in calculation. The temperature dependence of electrical resistivity for the crystalline α - Fe phase taken from [15] is used only in the relatively narrower temperature interval, in which a crystallization occurs. The calculated evolution of the crystalline fraction with temperature is plotted in Fig. 3. It shows that during the first crystallization stage below 830 K only 5 % of the sample volume transforms to the nanocrystalline phase. Keeping in mind that the surface crystallization starts at 500 K one may conclude that this initial 5 % nanocrystalline volume results from surface crystallization. It is very plausible that in a temperature interval just below 830 K also the intense nucleation processes occurs. The nucleation centers created inside the whole sample in this interval facilitate bulk crystallization which proceeds rapidly above 830 K (Fig. 1 and 2).

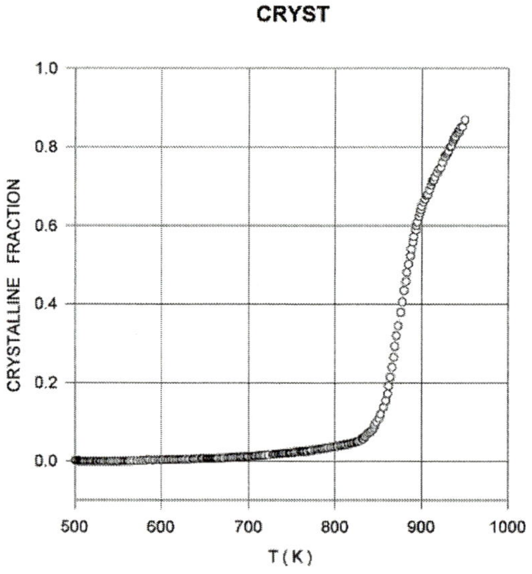

Fig. 3. Crystalline fraction X as a function of temperature calculated from electrical resistivity data for $Fe_{89}Zr_7B_4$ alloy.

During the second crystallization stage the nanocrystalline grains developing around the nuclei grow up quickly - leading to final 87% fraction of nanocrystalline phase in the alloy. This is in agreement with the crystalline fraction determined from magnetization data by Malkinski et al. [13] Electron scattering on grain boundaries is well known for enhancing strongly electrical resistivity when grain sizes are reduced [15,16]. It is found that calculations of the crystalline fraction optimally reproduce the measured electrical resistivity when applying the electrical resistivity of nanocrystalline phase is 2.3 times higher than the resistivity of the polycrystalline α - Fe phase with micrometer size grains. Such a high ratio of electrical resistivities in the nano- and microcrystalline phases is not exceptional and was reported e.g. for the nanocrystalline Cu, Ni and Fe [17,18].

3.3 Thermoelectric Power

Thermoelectric power (TEP) is negative for the amorphous alloys studied, as shown in Fig. 4 and 5. It varies slowly and proportionally to the temperature. Per analogiam to electrical resistivity, a change in TEP slope is found at 500-600 K due to the surface crystallization. TEP is only insignificantly influenced by chemical composition of amorphous alloys. The abrupt deviations appear at about 800 K, when the bulk crystallization starts. After crystallization, the TEP shifts towards positive values (Fig. 4 and 5). For a comparison, TEP of pure polycrystalline iron is plotted in Fig. 4 [14]. A similarity in TEP temperature variations of iron and crystalline alloys proves that TEP of these alloys is overwhelmingly determined by the atomic structure of the α - Fe phase. A similar behavior of TEP was reported for Al-Y-TM alloys [4]. The values of TEP in nanocrystalline alloys are relatively smaller compared to polycrystalline alloys of the same composition.The same ratio of TEP magnitudes in polycrystalline and in nanocrystalline samples was reported for the Cu, Ni, Fe samples [18].

4 Conclusion

Electron transport parameters are strongly correlated to the structural evolution of solids. The advanced analysis of temperature variations of resistivity of the alloys allows to distinguish intervals of the surface crystallization starting from 500 to 630 K, depending on composition, and of the bulk crystallization above 800 K. The threshold temperatures of both crystallization processes reveal the stabilizing role of the boron atoms creating covalent $Fe-B$ bonds. The negligible influence of copper atoms on the surface crystallization, along with the noticeable suppression of the bulk crystallization temperature, suggests specific nucleation mechanisms for both types of crystallization. The reduced boron content in a surface layer suppresses the local threshold crystallization temperature [7]. On the other hand, the copper clusters form bulk nucleation centers [11]. Temperature variation of crystalline fraction, as well as a comparison to magnetic studies of a crystallization kinetics [13], confirm a strong influence of the grain sizes and grain boundaries on electron

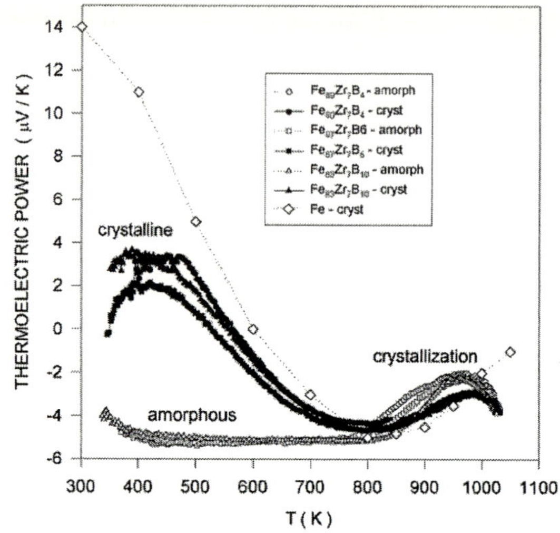

Fig. 4. Temperature variation of thermoelectric power of $Fe - Zr - B$ alloys for various boron contents.

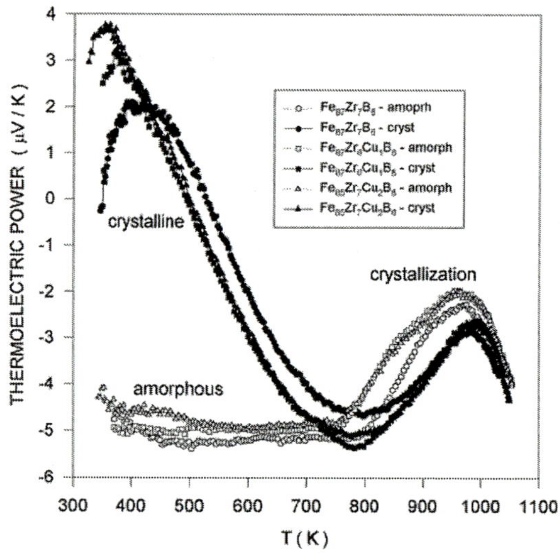

Fig. 5. Temperature variation of thermoelectric power of $Fe - Zr - Cu - B$ alloys for various copper contents.

scattering in alloys composed of the bcc - Fe nanocrystals embedded in the amorphous matrix. Temperature variations of thermoelectric power of the amorphous alloys depend weakly on temperature and on alloy composition. Thermoelectric power is less affected by the surface and bulk crystallizations, as compared to the electrical resistivity. TEP of the nanocrystalline alloys is overwhelmingly determined by the atomic structure of the α- Fe phase.

Acknowledgement

This work is supported in parts by grants No 7 TO8A02121 of KBN and 503 G022 of Faculty of Physics.

References

1. M. Kopcewicz, A. Grabias, J. Appl. Phys. 80 (1996) 3422.
2. J.M. Barandiaran, L. Fernandez Barquin, , J.C. Gomez Sal, P.Gorria, A. Hernanando, Sol. St. Commun, 88 (1993) 75.
3. P. Duhaj, I. Matko, P.Svec, J. Sitek, D. Janickovic, Mater.Sci. Eng. B39 (1996) 208.
4. K. Pekala, J. Non-Crystalline Solids 287 (2001) 183.
5. U. Mizutani, Progress in Mater. Sci. 2 (1983) 97.
6. P.J. Cote, L. Meisel, Phys. Rev. Lett. 40 (1978) 1586.
7. G. Bergmann, P. Marquardt, Phys. Rev. 17 (1978) 1355.
8. M. Kopcewicz, A. Grabias, J. Appl. Phys. 80 (1996) 3422.
9. T. Kulik, T. Horubala, H. Matyja, Mater. Sci. Eng. A157 (1992) 107.
10. P. Jalkiewicz, K. Pekala, J. Latuch, Nano-Structured Materials 11 (1999) 733.
11. Y. Zhang, K. Hono, A. Inoue, A. Makino, T. Sakurai, Acta Mater. 44 (1996) 1497.
12. R. Landauer, in Electrical Transport and Optical Properties of Inhomogeneous Media, AIP Conf. Proc. No. 40, ed. J.C. Garland, D.B. Tanner (AIP, New York, 1978), 1.
13. L. Malkiwski, A.Slawska -Waniewska, Mater. Sci. Eng. 226 (1997) 716.
14. W. Fulkerson, J.P. Moore, D.L. McElroy, J. Applied Physics 37 (1966) 9644.
15. A.F. Mayadas, M. Shatzkes, Appl.Phys.Lett 14 (1969) 345.
16. J. Ederth, L.B. Kish, E. Olsson, C.G. Granqvist, J. Applied Physics 88 (2000) 6578.
17. R.K. Islamgaliev, K. Pekala, M. Pekala, R.Z. Valiev, Phys. Status Sol (a)162 (1997) 559.
18. K. Pekala, M. Pekala, Nanostructured Materials 6 (1995) 819.

Self-Organized Formation of Fractal and Regular Pores in Semiconductors

Jens Christian Claussen[1,2], Jürgen Carstensen[1], Marc Christophersen[1], Sergiu Langa[1], and Helmut Föll[1]

[1] Chair for general materials science, CAU University of Kiel, Kaiserstr. 2, 24143 Kiel, Germany http://www.tf.uni-kiel.de/matwis/amat/
[2] Theoretische Physik und Astrophysik, Universität Kiel, Leibnizstr. 15, 24098 Kiel, Germany http://www.theo-physik.uni-kiel.de/~claussen/

Abstract. Electrochemical etching of semiconductors, beside technical applications, provides an interesting experimental setup for self-organized structure formation capable of regular, diameter-modulated, and branching pores. The underlying dynamical processes governing current transfer and structure formation are described by the Current-Burst-Model: all dissolution processes are assumed to occur inhomogeneously in time and space as a Current Burst (CB); properties and interactions between CB's are described by a number of material- and chemistry- dependent ingredients, like passivation and aging of surfaces in different crystallographic orientations, giving a qualitative understanding of resulting pore morphologies.

1 Electrochemical Etching: Basic Experimental Setup

The solid - liquid junction of Silicon and HF - containing electrolytes exhibits a number of peculiar features, e.g. a very low density of surface states, i.e. an extremely well "passivated" interface [1]. If the junction is biased, the IV - characteristics (Fig. 1) in diluted HF is quite complicated and exhibits two current peaks and strong current- or voltage oscillations at large current densities (for reviews see [2,3]). These oscillations have been decribed quantitatively by the Current-Burst-Model [4–7].

The PC controlled experimental setup is shown in Fig. 1. Using a four electrode arrangement, a potentiostat/galvanostat is contacting the sample and the electrolyte, allowing for a well defind potential resp. current for the electrochemical dissolution reactions. Backside contact, front- and/or backside illumination and electrolyte pumping can be varied as well as cell size (from under 0.3 cm up to wafers of 6 in), semiconductor material (Si, InP, GaAs, GaP) including various doping levels and crystallographic orientations. In addition, the electrolytes (e.g. HF, HCl, H_2SO_4) and their concentrations and temperature can be varied. With this setup a rich variety of porous semiconductors [8,9] can be generated, from nanopores to mesopores and macropores, and with various regular and branching structures.

Most of the phenomenonology can be well understood within the framework of the Current-Burst-Model which seems to reflect a number of quite general properties of semiconductor electrochemistry.

2 The Current-Burst Model

The Current Burst Model [4–6,10] states that the dissolution mainly takes place on small spots in short events, starting with a direct Si-dissolution, and possibly followed by an oxidizing reaction (see Fig. 2).

After these two short processes, the oxide hump undergoes dissolution, a time-consuming process which ensures at the location of the current burst a dead-time of fixed length during which no new burst can start. However, immediately after dissolution the Si surface has the highest reactivity, resulting in a maximal probability of annother current burst. Due to H-termination the surface becomes passivated, and the probability for bursts decays until it reaches the properties of a completely passivated Si surface, comparable to the situation before the nucleation of the first Current Burst.

Smoothing Effect of Oxide Layers and Global Current Oscillations Depending on the regime in the IV-curve current bursts and their interaction play different roles: For extremely high currents, in the oscillation regime, one has a permanent oxide coverage, and due to the high forcing of the system Current Bursts are started at all locations where the oxide layer is sufficiently thin for a breakthrough. In this regime, a detailed Monte Carlo study [5–7] including the lateral interaction between CBs (overlap of the oxide humps of neighboring CBs) has shown that the CB Model quantitatively can explain the experimental observations of globally oscillating etching current. Due to a phase synchronization of neighboring CBs oxide domains are formed. The size of this domains increases with increasing oxide generation/reduced oxide dissolution. At a percolation point only one oxide domain exists on the sample surface with a synchronized cycle of oxide growth and dissolution, resulting in a macroscoping oscillation of the external current. The size of the domains as well as the oscillation time can be controlled by the chemical parameters. Even without global oscillations for all regions of the IV-curve where oxide is formed one finds domains of synchronized oxide growth which define the length scale for the roughness of the electrochemically polished sample surface and thus lead to a smoothing of the surface.

Passivation Effects: The Aging Concept While at high current densities the semiconductor surface is completely covered with oxide, at low current densities most of the semiconductor surface will be in direct contact to the electrolyte. It is well known [1] that after chemical dissolution the free surface is passivated, i.e. the density of surface states reduces as a function of time which increases the stability of the surface against further electrochemical attack. Schematically the perfection of the surface passivation and the resulting reduction of the probability for a chemical attack as a function of time are plotted in Fig. 3. In silicon, the speed and the perfection of passivation of the (111) crystallographic surface is larger than for the (100) surface. This selective aging of surfaces leads to a self amplifying dissolution of (100) surfaces (which will become pore tips) and a preferential passivation of (111) surfaces (which will become pore walls). Under optimized chemical conditions with an extremely large passivation difference between (111) and (100) surfaces a self organized growth of octahedral cavities occurs. The octahedra consists of (111) pore walls. As soon as the complete surface of the octahedra reaches a critical value, it is easier to start a new cavity at a (100) tip of the old cavity, since the current density in the new, small cavity is larger and no surface passivation will occur until the surface again becomes to large. This growing mechanism leads to an oscillation of both the current through each pore and the diameter of each pore as a function of time. As in the case of oxide dissolution an internal time constant is related to the pore growth.

If a global current density smaller than the average current density in a current burst is applied, the aging concept becomes essential to explain the generation of macropores: At passivated surfaces, the nucleation probability of new current bursts is lower than at sites where a current burst has taken place before. Therefore the situation shown in Fig. 2(c) is much more unlikely than the situation in Fig. 2(d), so that the whole surface separates in two phases with current-carrying pores and passivated surface without contributions to the total current.

3 Consequences on Pore Geometries

As described above, depending on the electrochemical reactions the interaction of Current Bursts, either surface smoothing due to the dissolution of an isotropic oxide layer or strongly anisotropic dissolution of the semiconductor (due to the crystallography dependent aging of surfaces which is one of the most important reasons for pore formation) will dominate. Thus one obtains completely different surface morphologies when changing the electrochemical etching conditions; e.g. in a Si-HF-organic electrolyte system only by increasing the HF concentration (i.e. faster oxide dissolution and thus reduced influence of oxide) the electrochemical etching changes from electropolishing (strong oxide smoothing) over macropore formation (pores with diameters of several μm and smooth pore walls) to mesopore formation (strongly anisotropic, narrow pores with diameters of less than 400nm) [11].

Silicon has one of the most stable oxides of all semiconductors. So generally electrochemical etching of III/V compounds leads to rougher surfaces and smaller pore diameters. In addition the surface aging of III-V compounds is more complicated, since there exist two different (111) surfaces; e.g. in GaAs only the {111}A planes (Ga-rich planes) appear as stopping planes. So in most III-V compounds not octahedra (eight (111) surfaces as stopping planes) as in silicon are etched but tetrahedra with only four (111)A surfaces as stopping planes, and the {111}B planes (As-rich) serve as preferential growing directions like the (100) directions in Si (Fig. 4).

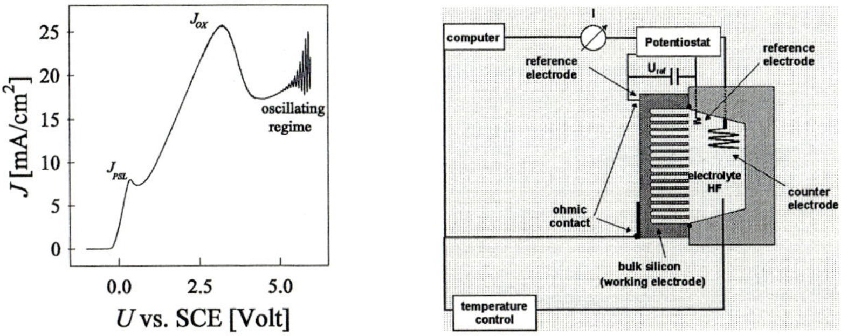

Fig. 1. Left: The IV- characteristics of the silicon-hydrofluoric acid contact shows different phenomena from generation of a porous silicon layer (PSL), oxidation and electropolishing (OX) and electrochemical oscillations at higher anodic bias. Right: The experimental set-up used for electrochemical anodization of semiconductors.

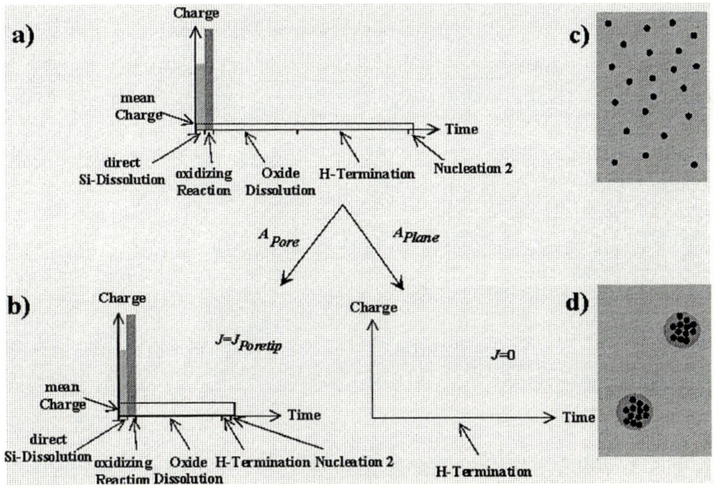

Fig. 2. Applying a global current density smaller than the average current density in a current burst will either require very long time constants to keep the mean charge density passed in a burst small and to cover the surface completely with current lines (a), a statistical arrangement of current lines with the optimized (smaller) time constant of the system, leaving parts of the surface without current and creating nanopores (b, c), or induces a phase separation by rearranging the current lines in areas corresponding to macropores (d).

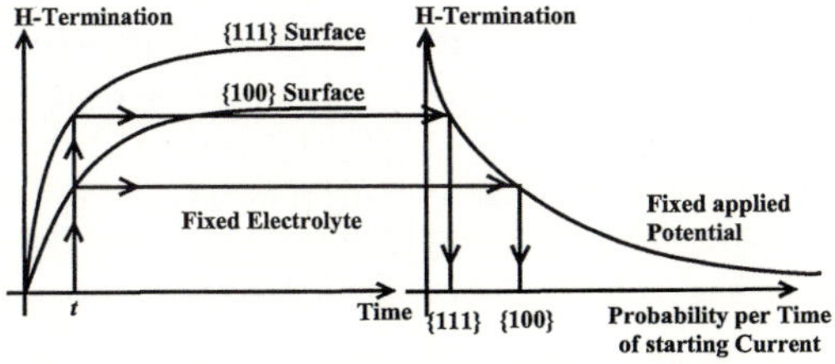

Fig. 3. Time constant and perfection of H-termination differ strongly for (100) and (111) Si surface orientation with higher nucleation probability on (100) surface.

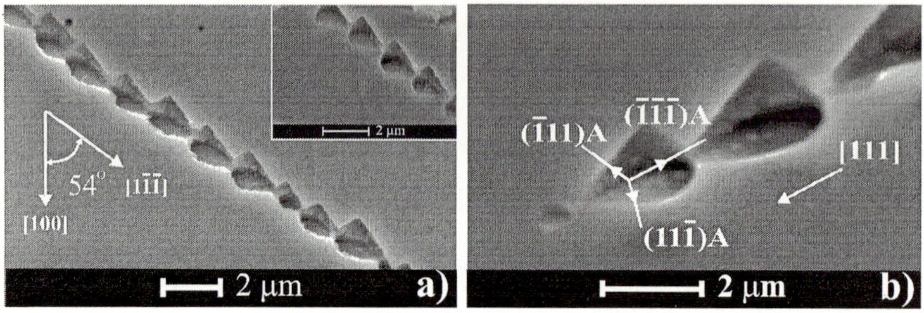

Fig. 4. Tetrahedron-like pores oriented along $< 111 >$ directions obtained in (100)-oriented GaAs at high current densities in HCl electrolytes ($n = 10^{17}$ cm^{-3}).

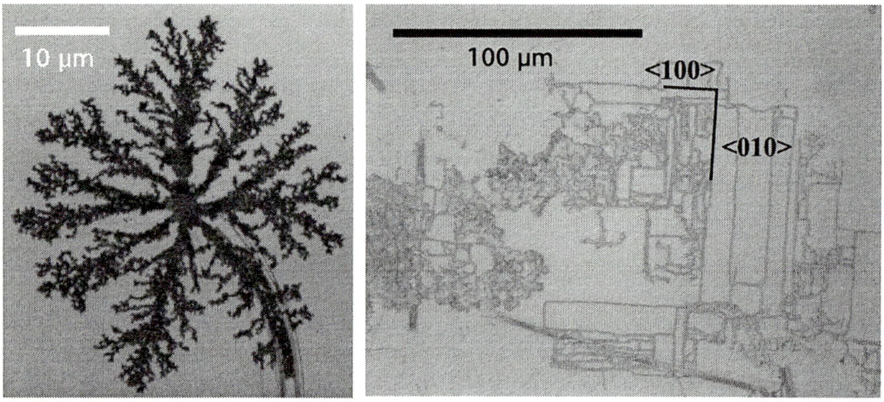

Fig. 5. Left: Without crystallographic preferentiation, surface fractals similar to dendritic structures occur. Right: Under certain etching conditions lateral growth along lines in crystallographic direction occurs which also seems to generate some self-similar structure.

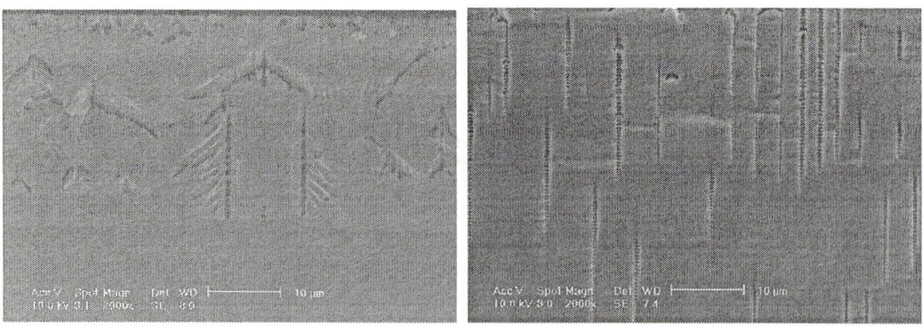

Fig. 6. Branching of pores (left, for DMF, 4wt-% HF, constant current) can be suppressed by an open-loop control method triggering to the system-inherent time scale (right, period 0.5 min).

Lateral Growth of Current Bursts: Generation of Surface Fractals For extremely low current densities and low HF concentration, the total number of active current bursts is so low that most of the surface is passivated or covered with oxide. CB nucleation becomes a severe problem. The neighborhood of a just terminated CB is therefore a preferential site for new CB nucleation, leading to the formation of surface fractals as shown in Fig. 5 (left). A further decrease of the current density allows for a partial passivation of the surface area of a terminated CB. According to Fig. 3 nucleation in the 100 direction of the surface near CBs will be preferential (Fig. 5).

Sidebranching Pores and Suppression by Open-Loop-Control For technical applications, it would be desirable to generate pores with a diameter of less than 500nm. However, at all etching conditions investigated up to now, in this diameter range always pores always show intensive side-branching. By modulation of the intensity of illumination or by modulation of the etching current with a modulation frequency meet the intrinsic time scale of the pore formation, the side pore formation is hindered by the reduction of the etching current at the right phase. (Fig. 6).

4 Conclusions

The Current Burst model, a local stochastic and highly nonlinear model, and the Aging concept describing the time- and orientation-dependence of the passivation behavior, provide on an intermediate level of abstraction a general approach to explain pore geometries, oscillations and synchronization with a minimum of material-dependent, but experimentally accessible, parameters. The lateral interaction of Current Bursts gives rise to synchronization phenomena, and a percolation transition to global ordering. Under special conditions lateral growth of fractal structures is obtained. An open-loop control can be successfully applied to suppress sidebranching of pores. To clarify quantitatively the dynamical processes in this system, further experimental and theoretical efforts have to be made.

References

1. Yablonovich, E., Allara, D. L., Chang, C. C., Gmitter,T., Bright, T. B.: Phys. Rev. Lett., 57, 249 (1986)
2. Föll, H., Appl. Phys. A **53**, 8 (1991)
3. Smith, R. L., Collins, S. D., J. Appl. Phys., **71**, R1 (1992)
4. Carstensen, J., Christophersen, M., Föll, H.: Mat. Sci. Eng. B, **69**, 23 (2000)
5. Carstensen, J., Prange, R., Popkirov, G., Föll, H.: Appl. Phys. A **67**, 459 (1998)
6. Carstensen, J., Prange, R., Föll, H.: in: Proc ECS San Diego 1998, **98** 148
7. Carstensen, J., Prange, R., Föll, H.: J. Electrochem. Soc. 146(3), (1999) 1134
8. Propst, E. K., Kohl, P. A.: J. Electrochem. Soc., 141, (1994) 1006
9. Ponomarev, E. A., Levy-Clement, C.: J. Electrochem. Soc. Lett., 1, (1998) 1002
10. Föll, H., Carstensen, J., Christophersen, M., Hasse, G.: in: Proc. ECS 2001
11. Christophersen, M., Carstensen, J., Föll, H.: Phys. Stat. Sol. (a) 182, 45 (2000)

Evolution and Shapes of Dunes

Hans J. Herrmann and Gerd Sauermann

Institute for Computer Applications 1, University of Stuttgart Pfaffenwaldring 27, D-70569 Stuttgart, Germany

Abstract. We give a review on our field measurements on Barchan dunes. Then we present a model for dune formation that takes saturation transients in the saltation layer into account. We show that these transients introduce a characteristic length scale that is crucial for dune formation and the macroscopic shape of dunes, albeit determined by intrinsic properties of the saltation kinetics. Finally, we demonstrate that our model can explain the scale dependence of the shape of barchan dunes revealed in recent field

1 Introduction

We have all seen the majestic wave-like shapes of dunes in the desert. Sand dunes develop wherever loose sand is driven by a fluid (air, water ...) that lifts grains from the ground and entrains them into a surface flow. The diverse conditions of wind and sand supply in different regions on Earth give rise to a large variety of different shapes of aeolian dunes [1–3]. Moreover, dunes have been found on the sea–bottom and even on Mars. Despite the long history of the subject, the underlying physical mechanisms of dune formation are still not very well understood. How are aerodynamics (hydrodynamics) and the particular properties of granular matter acting together to create dunes? How is the shape of a dune maintained when it grows and moves? What determines the size of dunes? Due to the fact that neither now nor in the near future we will be able to simulate dunes on the grain scale (an average barchan comprises 10^{15} grains), we concentrate in the following on an effective continuum model that can be applied to sand dunes or other geomorphological problems on a large scale. Due to the highly complicated physical processes involved (saltation, turbulent wind) and the wide range of length and time scales to be covered — from the dynamics of single sand grains, the formation of ripples, to the genensis and migration of dune fields, the time and length scales span over more than seven orders of magnitude — the derivation of the model will not be given here. We will instead after reviewing our experimental measurements present the basic elements of the model and finally compare the two.

The simplest and best known type of dune is the barchan dune shown in Fig. 1, shaped in a crescent, which occurs if the wind comes steadily from the same direction throughout the year and if there is not enough sand to cover the entire surface. Barchan dunes move proportionally to the wind velocity and inversely proportionally to their height. They are encountered for instance in Peru [4–7], in Namibia [8] and Morocco [9]. On these dune fields, hundreds of barchans can be found, generally all of the same size. The dunes have heights between 1.5 and 10 m, while their bases are typically 40 to 150 m long and 30 to 100 m wide. The windward or stoss-side of the dune has typical slopes between 8^0 and 20^0 and is limited by a sharp edge, called the brink. The brink coincides in many cases with the crest of the dune and separates the slip face from the dune's windward side. Roughly speaking, its section is a parabola-like curve reaching from the tip of one horn to the point of maximum slip face height and back to the tip of the other horn. Despite the fact that for more than 50 years geologists and geographers have been measuring dunes in the field and have obtained data on height, width, length, volume and dune velocity, very little is yet known [10,11] about the exact quantitative shape of barchans. From a mathematical point of view, the barchan dune is a symmetrical object in the wind direction, but in nature there are many factors, like non-steady winds or inclined ground surfaces leading to asymmetrical shapes. Numerical simulations to predict the evolution of barchan dunes and their exact shape have been performed by [12,13].

Fig. 1. Barchan dunes near Laâyoune, Morocco. The dune in the front, on the left side was measured in detail during our field trip in May 1999, see [14].

2 Experimental Measurements of a Barchan Dune

2.1 Dune Morphology

Qualitatively the crescent-like shape of the barchan is well known. The first measurements concerning barchan dunes and their morphologic relationships were performed by Coursin (1964) [15] in Mauritania and Finkel (1959) [4] in the Pampa de La Joya in southern Peru. Hastenrath (1967,1987) [5,6] analysed barchans in the same area and revisited the site 20 years later. Additional investigations in the same field were undertaken by Lettau and Lettau (1969). [7] Slattery (1990) [8] measured barchan dunes in Namibia.

However, they did not measure the entire shape of the dune, but only the typical lengths. Our own field measurements [14] were performed in a dune field in the Sahara desert, located in southern Morocco (former Spanish Sahara) near the city of Laâyoune.

We define for each horn, the lengths L_a, L_b and the widths W_a, W_b independently, as did Finkel (1959) [4]. The orientation of the measuring axis is chosen according to the wind direction, which coincides with the symmetry line for a totally symmetric dune. Furthermore, we introduce the length of the slip face L_s and the length L_0 from the dune's toe on the windward side to the brink. Finally, the height of the slip face H is defined at the highest point of the brink, which is the intersection of the brink and the longitudinal centerline of the dune.

The relationship between the width of the horns $W = W_a + W_b$ and the height H of the slip face has been studied many times. An overview can be found by Hesp and Hastings (1998). [10] A linear relationship was found between the height H and the width of the horns W.

$$W = a_W H + b_W \tag{1}$$

For the management of dune movement the volume V of a dune is one of the most interesting features, apart from the rate of movement v_d. Together they determine the bulk flux Q_b of sand transported by the dunes.

$$Q_b = \rho_{sand} v_d V \tag{2}$$

To obtain the total flux Q_t, in addition to the bulk flux the inter-dune-flux Q_i, has to be taken into account.

$$Q_t = Q_b + Q_i \tag{3}$$

A study of these fluxes has been performed by Sarnthein and Walger (1974) [16]. In the following we concentrate on the bulk flux and the dune volume.

According to Oulehri (1992) [9], the rate of movement of the dunes in the area of Laâyoune is 32 m yr^{-1} for a dune of 9 m height. The bulk density of the dune sand is 1670 kg m^{-1}, on average. Using these data and the calculated volume of 23 000 m^3 for dune 7, we obtain a flux of 1.2 million t m^{-1} yr^{-1} or 736 000 m^4 yr^{-1}. Lettau and Lettau (1969) [7] estimated a bulk flux of 20 550 m^4 yr^{-1} for an average barchan of the Pampa de La Joya whose height was 3 m. A large barchan with a height of 5.2 m gave a bulk flux of 60 000 m^4 yr^{-1}.

The overall length L of a barchan is the sum of the length L_0 from the windward foot of the dune to its crest, the length of the slip face L_s, and the average of the horn lengths $L_h = (L_a + L_b)/2$

$$L = L_0 + L_s + \frac{L_a + L_b}{2} \tag{4}$$

This definition uses four lengths, of which only the length of the slip face L_s has an obvious dependence on the height H,

$$L_s = \frac{H}{\tan}, \Theta \tag{5}$$

where Θ is the angle of repose with typical values between 31° and 35°. This angle is an intrinsic property of the sand and is therefore independent on the aeolian processes.

Finkel [4] reported for the horn length.

$$L_x = a_x H + b_x \tag{6}$$

Further the ratio of the horn length L_h to the stoss-side length L_0 also depends on the dune's height. This shows clearly that the relative position of the slip face within the whole dune varies from small to large dunes and that the ratio of horn length to total length increases with the height as shown in Fig. 3. This disproves scaling invariance of barchan dunes. The size scaling of Barchans can be used to put them all on top of each other by rescaling the axis for each dune and using dimensionless variables. This is done for the longitudinal cross section in Fig. 2. We see that on the windward part the shape is a parabolloid and the brink position b_o moves to the right for larger dunes. In fact b_o is a linear function of the height. Therefore we have a deviation from perfect size scaling as shown in Fig. 3. Looking from the top, the brink line has also the shape of a parabola.

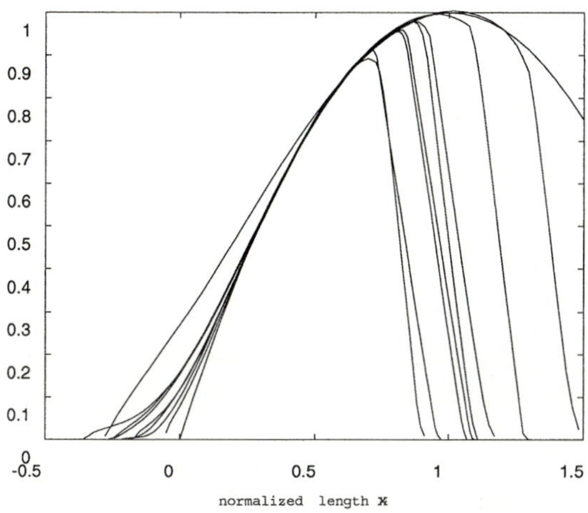

Fig. 2. Profile along the symmetry plane of the dunes (thin lines) using normalized variables and the standard parabola (thick line) from [14].

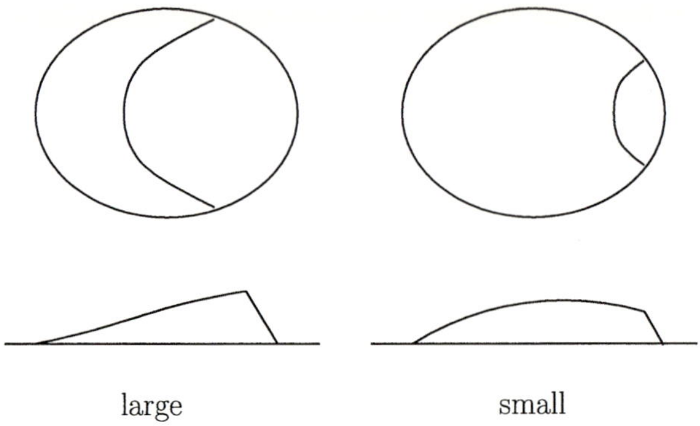

Fig. 3. Sketch of a small and a large dune.

2.2 Wind Velocity and Sand Flux

Correlated measurements of the wind velocity v and the sand flux q have been performed [17] on a large barchan dune near the beach of Jericoacoara (see Fig. 4) during the first week of December 2000, at the end of the dry season. The dune has approximately 34 m in height, a width of 600 m, and the length of its windward side is 200 m. When a fully turbulent atmospheric boundary

Fig. 4. Aerial photography of the coastal dune field near Jericoacoara.

layer develops over a flat surface it gives rise to a logarithmic velocity profile $v(z)$ [18],

$$v(z) = \frac{u_*}{\kappa} \ln \frac{z}{z_0}, \tag{7}$$

where u_* denotes the shear velocity, z_0 the roughness length of the surface, and $\kappa = 0.4$ the von Kármán constant. The shear velocity u_* has dimensions of velocity but is defined in terms of the shear stress $\tau = \rho_{\text{air}} u_*^2$ and density ρ_{air} of the air. According to Hunt et al. [19] the height l of the shear stress layer can be obtained implicitely through

$$l = \frac{2\kappa^2 L}{\ln l/z_0}, \tag{8}$$

where L is the characteristic length of the dune which is measured from the half height of the windward side to the crest, according to the definition of [19]. From (8), we obtain for $L = 100\,\text{m}$ and a roughness length $z_0 = 10^{-4}\,\text{m}$, a height of this layer of $l \approx 3.2\,\text{m}$ ($l \approx 4\,\text{m}$ for $z_0 = 10^{-3}\,\text{m}$). Hence, we placed the anemometers at a height of 1 m that is well inside the shear stress layer. One should note that it is very difficult to measure the wind velocity with standard anemometers within this layer for small dunes. This is the main reason for choosing a large dune. A reference anemometer has been placed approximately 300 m upwind of the dune's foot and has been kept there during all the measurements. With a second anemometer, we measured the average velocity v_i every 24 m on the central profile during 10 minutes. Finally, we normalized the average velocity $\langle v_i \rangle$ by the average velocity in the same period obtained from the reference anemometer, $\langle v_{r,i} \rangle$. By doing so, we could get rid of the long term (> 10 minutes) variations in the wind speed time series throughout the day and obtain the shear velocity $u_{*,i}$ through

$$1 + \hat{u}_{*,i} = \frac{u_{*,i}}{u_{*0}} = \frac{\langle v_i \rangle}{\langle v_{r,i} \rangle}, \tag{9}$$

where $\hat{u}_{*,i}$ is the dimensionless shear velocity perturbation of the air caused by the dune and u_{*0} is the undisturbed shear velocity far upwind of the dune. u_{*0} can be calculated assuming a typical logarithmic profile from turbulent flow (7). During our measurements, the wind was blowing quite constantly and had an average value of $7.5\,\text{m}\,\text{s}^{-1}$ at the reference station. Assuming a roughness length $z_0 = 2.5 \times 10^{-4}\,\text{m}$, we obtain from (7) the undisturbed shear velocity over the plane $u_{*0} = 0.36\,\text{m}\,\text{s}^{-1}$. The averaged and normalized measured shear velocities are plotted in Fig. 5.

We also measured the sand flux on the central slice of the dune using cylindrical traps with a diameter of 5 cm and an opening of 1 cm at the front. The back–side of the traps have an opening of 2 cm covered by a fabric with pores smaller than the grain diameter. The traps were placed at the same positions where the wind speed has been measured. From the mass m of the collected sand, the collection time T, and the width w of the opening, we calculated the sand flux $q = m/(T\,w)$. The measured sand fluxes are shown in Fig. 6.

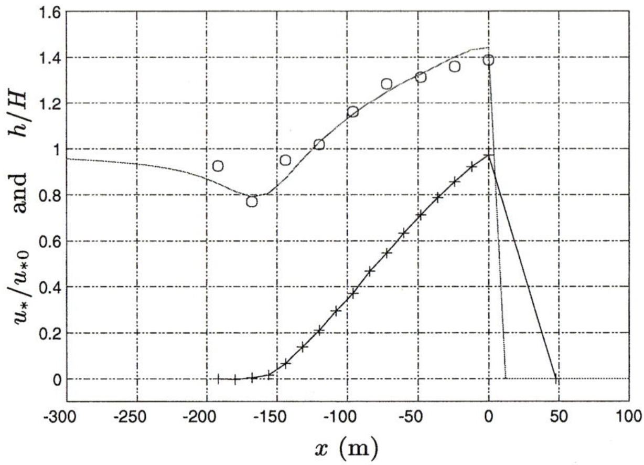

Fig. 5. The circles show the measured shear velocity u_* and wind speed $v(z = 1\text{m})$ normalized by their reference values u_{*0} and $v_0(z = 1\text{m})$, respectively. The solid line depicts the prediction of (11) using the measured height profile $h(x)$ shown in the bottom curve (crosses). The depression at the dune's foot is about 0.8 and the maximum speed–up at the brink is approximately 1.4 from [17].

3 The Model

Since we are interested in the formation and movement of dunes, the important time scale of our problem is defined by the erosion and deposition processes that changes the surface. A significant change of the surface happens within some hours or even days. In contrast to this, the time scale of the wind and the saltation process is of the order of seconds and therefore several orders of magnitude faster. Hence, we will use in the following stationary solutions for the wind field and the sand flux. Similarly we neglect the finite life time of avalanches (a few seconds) and consider them as instantaneous compared to the movement of the dunes. The separation of the different time scales and the resulting approximations lead to an enormous simplification, because it decouples the different physical processes. The entire model can be thought of as four (almost) independent parts: the stationary wind field over a complex terrain, the stationary aeolian sand transport, the time evolution of the surface due to erosion, and avalanches.

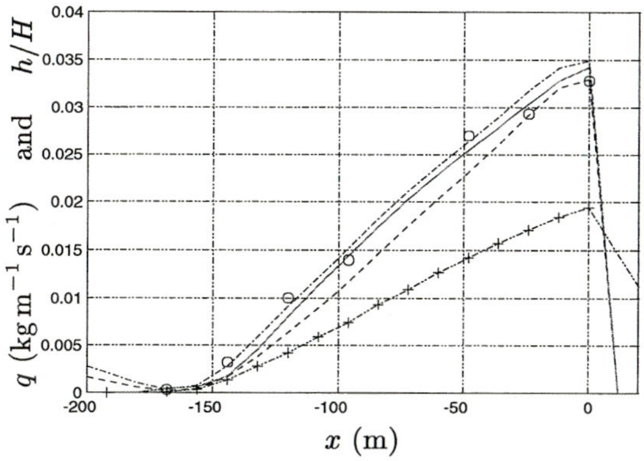

Fig. 6. Measured and calculated sand flux onto the central slice of a barchan dune. The circles denote measurements and the solid line the prediction of the non–equilibrium sand flux model, (13) (solid line). The two dashed lines correspond to the classical empirical relation for saturated flux: ref. [19] and ref. [27] is depicted. Yet, these saturated sand flux relations cannot correctly predict the sand flux near the dune's foot and show clearly that the assumption of saturation breaks down from ref. [17].

3.1 The Wind Shear Stress

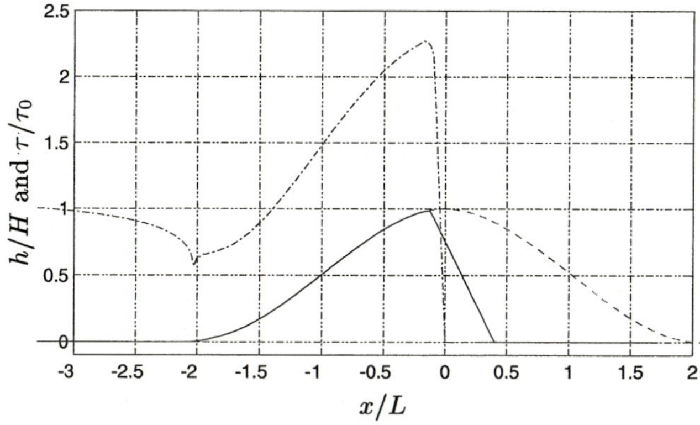

Fig. 7. The envelope $\tilde{h}(x)$ of the windward profile of a dune $h(x)$ (solid line) and the separating streamline $s(x)$ (dashed line) form together a smooth object which is used to calculate the air shear stress $\tau(x)$ (dash-dotted line) on the windward side. In the region of recirculation the air shear stress τ is set to zero.

The fully turbulent atmospheric boundary layer develops over a flat surface the logarithmic velocity profile $v(z)$ of 7 [18]. A perturbation of the ground $h(x)$ such as a dune or hill gives rise to a non–local perturbation $\hat{\tau}(x)$ of the undisturbed air shear stress τ_0,

$$\tau(x) = \tau_0 \left[1 + \hat{\tau}(x)\right] . \tag{10}$$

The functional dependence of the air shear stress perturbation is crucial for the understanding of the stability of dunes and to predict the sand flux onto the windward side of a dune. Analytical calculations of the flow over a gentle hill yield an analytical expression for the shear stress perturbation $\hat{\tau}(x)$ [19–21]. We performed further simplifications in order to obtain a minimal expression that captures the crucial features (and only those) and is applicable for sand dunes [22]. The resulting formula for the air shear stress perturbation $\hat{\tau}$ is,

$$\hat{\tau}(x) = A\left(\frac{1}{\pi}\int_{-\infty}^{\infty}\frac{h'}{x-\xi}d\xi + Bh'\right), \tag{11}$$

where h' denotes the spatial derivative of the dune's profile $h(x)$ in wind direction. The coefficients $A(L/z_0)$ and $B(L/z_0)$ depend only logarithmically on the ratio between the characteristic length L of the dune and the roughness length z_0 of the surface. For a dune with a length and width ratio $W/L \approx 1$ and $L/z_0 = 4.0\,10^5$ we obtain $A \approx 3.2$ and $B \approx 0.3$ from Ref. [14,22]. Equation (11) has several features that are important for dune formation. First, the air shear stress is completely scale invariant and leads to the same speed–up for small and large dunes. This is expected in the fully turbulent regime where no characteristic length exists. Secondly, the shear stress perturbation $\hat{\tau}(x)$, Equation (11), scales with the height H and inversely with the characteristic length L of the dune and thus with the average slope of the dune's windward side, $\hat{\tau} \propto H/L$. Thirdly, a depression of $\tau(x)$ in front of the hill occurs as a consequence of the strongly non–local contribution in Equation (11). Finally, the shear stress perturbation $\hat{\tau}(x)$ for the windward side of the dune is calculated using (11), the profile $h(x)$ on the windward side, and the separating streamline $s(x)$ on the lee side. The result is shown in Fig. 5 together with the measured mean values (averages over 10 minutes intervals) normalized according to 9. The agreement between model results and measurements is good. From this, we can conclude that the heuristic model of the separation bubble combined with the analytic expression, (11), provide a reasonable approximation for the wind field above the dune. This strategy enormously reduces the computational effort, compared to the numerical solution of turbulence models and the averaged three dimensional Navier–Stokes equation.

Equation (11) is based on a perturbation theory and can only be applied to smooth hills. Jackson and Hunt [20] assumed $H/L < 0.05$, whereas Carruthers et al. [23] showed that mean slopes up to $H/L \approx 0.3$ give reasonable results. The windward side of a barchan dune is always below the latter value and the formula should be applicable. However, flow separation occurs at the brink, which is out of the scope of the linear perturbation theory. A heuristic solution to solve this problem has been suggested by Zeman and Jensen [24]. They introduced a separation bubble that comprises the recirculating flow (the large eddy in the wake of the dune), which reaches from the brink (the point of detachment) to the bottom (to the point of reattachment) see fig 3.1. We model the separating streamline by a third order polynomial that is a smooth continuation of the profile $h(x)$ at the brink x_{brink} and at the reattachment point $x_{brink} + L_r$, i.e. $h(x_{brink}) = s(0)$, $h'(x_{brink}) = s'(0)$, $s(L_r) = 0$, $s'(L_r) = 0$, where $L_r \approx 6H$ is the downwind distance of the reattachment point from the brink. The shear stress perturbation $\hat{\tau}(x)$ for the windward side of the dune is finally calculated using equation (10), the profile $h(x)$ on the windward side, and the separating streamline $s(x)$ on the lee side. An example is depicted in fig. 3.1.

3.2 The Sand Flux

Sand transport has been studied already by Bagnold [25] and it was also him who proposed the first phenomenological law that predicted the sand transport from the shear stress of the air. Improved laws have been proposed by several authors in the meantime [26,7,27]. However, all these relations assume that the sand flux q is in equilibrium and can be written as a function of the shear stress τ, $q\left(\tau(x)\right)$. Temporal or spatial transients are completely neglected. In the following we will call such a relation *saturated*, because it predicts the amount of sand that can be maintained in the saltation layer at a certain air shear stress τ.

This condition is hardly fulfilled at the windward foot of an isolated dune [21], e.g. a barchan, where the bed changes rapidly from bedrock or vegetation to sand. Besides the particular conditions at the dune's foot, the sand flux may never reach saturation [28] on the entire windward side, where the shear velocity increases gradually from the foot to the crest. Wind tunnel measurements indicate that the typical time to reach saturation in saltation is approximately two seconds [29], which corresponds to a saturation length of the order of 10 m. This length is of the order of the dune size and can not be neglected if the sand flux on the entire windward side is significant. Furthermore, it has been observed that the time to reach saturation increases for shear velocities close to the threshold [29]. In this situation, the sand flux may never reach saturation on the entire windward side and should increase exponentially with distance from the dune's foot [28]. In recent years, several models to calculate the wind field have been developed, from analytic boundary layer approximations to numerical solutions of the Navier–Stokes equation with an enormous computational effort. Although some previous studies have discussed the limits of the saturation approximation in detail [21], much less effort has been dedicated to the development of sand flux relations that effectively incorporate non–saturation effects [30].

For the saturated flux many different functional forms of these sand transport laws exist and have been used in the past. For high shear stresses, however, they all converge to the simple relation proposed by [1],

$$q_s \propto \tau^{3/2}. \tag{12}$$

All other more elaborate relations add higher order corrections to the Bagnold formula that become important close to the air shear stress threshold. To overcome the limitation of saturation and to obtain information about the dynamics of the saltation process, numerical simulations on the grain scale have been performed in the last years [31–33]. Still, concerning the modeling of dune formation, both approaches had to be discarded. The microscopic models are computationally too expensive and the equilibrium assumption that is inherent in the simple flux relations does not hold on the entire windward side of a dune [11,12,14,21,34]. Since both known approaches cannot be used to model dune formation we developed a new phenomenological continuum saltation model that is computationally very efficient on the one side and on the other side incorporates the dynamics of the saltation layer and thus allows for saturation transients [34]. In this model the sand flux is defined by a differential equation of the form

$$\frac{\partial}{\partial x} q = \frac{1}{l_s} q \left(1 - \frac{q}{q_s}\right), \tag{13}$$

where $q_s(\tau)$ is the saturated sand flux and $l_s(\tau)$ the characteristic length of the saturation transients, called saturation length. The saturation length $l_s(\tau)$ depends on the air shear stress, but converges towards a constant value for $\tau \gg \tau_t$ [34]. A comparison between the saturated sand flux, our model, Equation (13), and the field measurements from Jericoacuara can be seen in Figure 6. One observes that Equation (13) gives the right behaviour at the foot of the Barchan while the neglect of the transient gives an unphysical dip.

3.3 The Surface Evolution

A spatial change in sand flux implies that erosion or deposition takes place and the surface changes in height. The time evolution of the surface can be calculated from the conservation of mass,

$$\frac{\partial h}{\partial t} = \frac{1}{\rho_{\text{sand}}} \frac{\partial q}{\partial x}, \tag{14}$$

where ρ_{sand} is the bulk density of dune sand. Finally, we note that Equation (14) is the only remaining time dependent equation and thus defines the time scale of the model.

The full dune model can be sketched as follows. An initial surface h is used to start the time evolution. If flow separation has to be modeled the

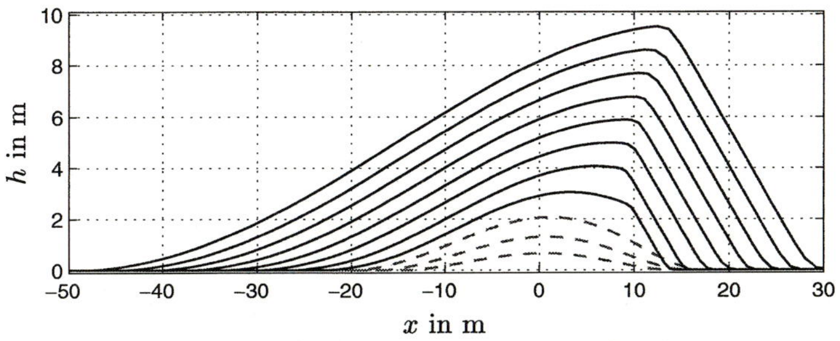

Fig. 8. The solutions for large volumes — above a critical heigth — are dunes including a slip face, whereas for small volumes heaps develop. An important fact is that the steepest lee side of a heap (dashed lines) is approximately 15°, which is well below the angle of repose of 34°.

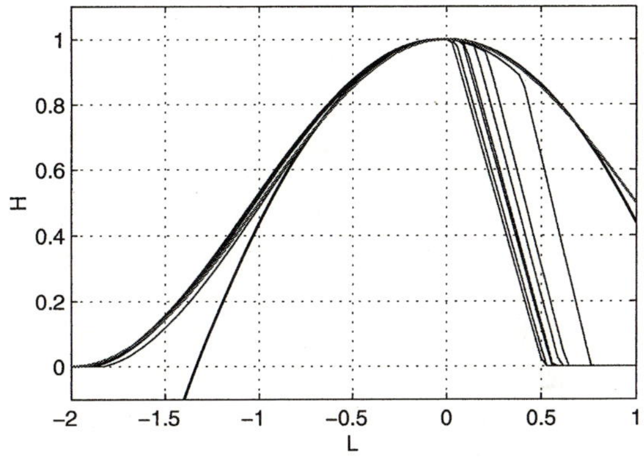

Fig. 9. Profiles along the symmetry plane of 3d Barchans of different size normalized on a single height as obtained from the model (from [35]).

separating streamline $s(x)$ is calculated. Next, the air shear stress $\tau(x)$ onto the given surface h (or h and s) is calculated using Equation (10). From the air shear stress $\tau(x)$ the sand flux can be determined using Equation (13). Then, the integration forward in time of the surface is calculated from the mass conservation, Equation (14). Finally, sand is eroded and transported downhill if the local angle $\partial_x h$ exceeds the angle of repose. This redistribution of mass (avalanches) is performed until the surface slope has relaxed below the critical angle. The time integration is calculated until the final shape-invariantly moving solution is obtained.

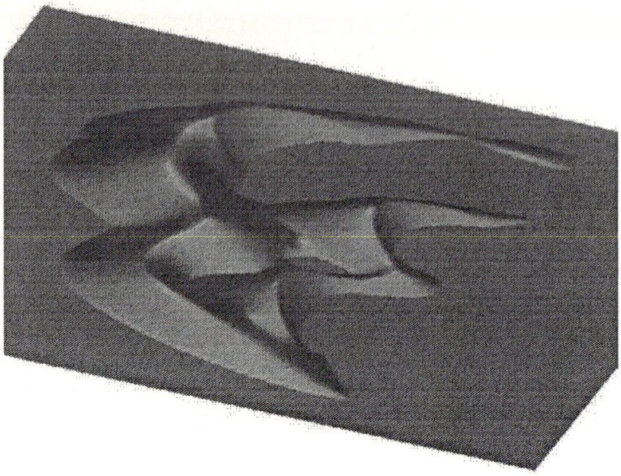

Fig. 10. Complex dune pattern, calculated with the full three dimensional model. Wind is blowing from the left to the right. When Barchan dunes are too close they interact, get eventually connected, and form complex dune structures. The large dunes are shielding the small dunes from the arriving sand flux which then constantly loose volume [35].

4 The Shape of the Dune and Outlook

In order to analyze the properties of the shape-invariantly moving solution of our model, we performed a series of calculations varying the volumes of the Gaussian hills that have been used as initial configuration. The final shape invariantly moving solutions are displayed in Fig. 8. For small volumes we obtained heaps without a slip face, whereas for large volumes dunes with a slip face developed. Hence, there is a minimal height for dune formation or, more precisely, a minimal height for the formation of a slip face. Empirically, this was observed many times in nature.

The simulation also showed that the Barchan shape is a steady state solution. Starting from different initial configurations having the same volume, one always obtains after a certain transient the same crescent shape dune. This dune moves with constant velocity inversely proportional to its height and quantitatively agrees with the ones measured in the field for corresponding volume.

Simulating 3d Barchans by starting from an initial Gaussian heap can be done under essentially two boundary conditions: an absorbing bedrock $[q = 0]$ for coastal dunes or a finite influx and $q \neq 0$ on the bedrock for desert dune fields. In Fig. 9 we see longitudinal profiles along the symmetry plane of desert Barchans for different initial volumes of dunes, normalized such that their heights fall on top of each other. One observes that the measured profiles from Fig. 2 agree very well with this prediction. The numerical study of Fig. 9 also shows that the tail on the windward side is not parabolic but rather logarithmic.

One can also using our programme construct virtual dunes and produce virtual desert landscapes. One example is shown in Fig. 10 where one sees the resulting coalescence of Barchans after they have developed from a set of several Gaussian heaps. In this way one can, starting from measured topographies predict the future evolution and therefore planify in advance to protect cities and fields against moving sand masses in the Sahara. Another perspective of the use of our equations of motion is the possibility to study techniques used to stop or destroy dunes, like Bofix, as introduced by Meunier in Nouakchott.

Acknowledgement. We want to thank our collaborators P. Rognon, J.S. de Andrade, A. Poliakov, L.P. Maia and K. Kroy.

References

1. Bagnold, R. A.: *The physics of blown sand and desert dunes.* Methuen, London, 1941.
2. Pye, K. and Tsoar, H.: *Aeolian sand and sand dunes.* Unwin Hyman, London, 1990.
3. Lancaster, N.: *Geomorphology of desert dunes.* Routledge, London, 1995.
4. Finkel, H. J.: The Barchans of Southern Peru. Journal of Geology 67, 614–647 (1959)
5. Hastenrath, S.: The Barchans of the Arequipa Region, Southern Peru, (1967).
6. Hastenrath, S.: The barchan dunes of Southern Peru revisited. Zeitschrift für Geomorphologie 31-2, 167–178 (1987)
7. Lettau, K., Lettau, H.: Bulk transport of sand by the barchans of the Pampa de La Joya in Southern Peru. Zeitschrift für Geomorphologie N.F. 13-2, 182–195 (1969)
8. Slattery, M. C.: Barchan migration on the Kuiseb river delta, Namibia. South African Geographical Journal 72, 5–10 (1990).
9. Oulehri, T.: Etude géodynamique des migrations de sables éoliens dans la région de Laâyoune (Nord du Sahara marocain). Ph.D. thesis, l'Universite Paris 6, Paris. no. 92-12 (1992)
10. Hesp, P. A., Hastings, K.: Width, height and slope relationships and aerodynamic maintenance of barchans. Geomorphology 22, 193–204 (1998).
11. Wiggs, G. F. S., Livingstone, I., Warren, A.: The role of streamline curvature in sand dune dynamics: evidence from field and wind tunnel measurements. Geomorphology 17, 29–46 (1996).
12. Wippermann, F. K., Gross, G.: The wind-induced shaping and migration of an isolated dune: A numerical experiment. Boundary Layer Meteorology 36, 319–334 (1986).
13. Howard, A. D., Morton, J. B.: Sand transport model of barchan dune equilibrium. Sedimentology 25, 307–338 (1978).
14. Sauermann G., Rognon P., Poliakov A., Herrmann H.J.: The shape of the barchan dunes of Southern Morocco. Geomorphology 36, 47–62 (2001).
15. Coursin, A.: Observations et expériences faites en avril et mai 1956 sur les barkhans du Souehel el Abiodh (région est de Port-Étienne). Bulletin de l' I. F. A. N. 22A, no. 3, 989–1022 (1964).

16. Sarnthein, M., Walger, E.: Der äolische Sandstrom aus der W-Sahara zur Atlantikküste. Geologische Rundschau 63, 1065–1087 (1974).
17. Sauermann, G., Andrade Jr., J.S., Maia, L.P., Costa, U.M.S., Araújo, A.D., Herrmann, H.J.: Wind Velocity and Sand Transport on a Barchan Dune (2001).
18. Landau and Lifshitz.: *Fluid Mechanics*, volume 6 of *Course of Theoretical Physics*. Pergamon Press, London, 1963.
19. Hunt, J. C. R., Leibovich S., Richards K. J.: Turbulent wind flow over smooth hills Q.J.R. Meteorol. Soc. 114, 1435–1470 (1988).
20. Jackson, P. S., Hunt J. C. R.: Turbulent wind flow over a low hill. Q.J.R. Meteorol. Soc. 101, 929 (1975)
21. Weng, W. S., Hunt J. C. R., Carruthers, D. J., Warren A., Wiggs, G. F. S., Livingstone, I., Castro, I.: An experimental study of Froud number effect on wind-tunnel saltation. Acta Mechanica Supppl. 1, 145-157 (1991).
22. Kroy, K., Sauermann, G., Herrmann, H. J.: A minimal model for sand dunes, *Phys. Rev. Lett.*, 88:054301, 2002. Preprint cond-mat/0203040.
23. Carruthers, D. J., Hunt J. C. R.: Atmospheric Processes over Complex Terrain. Vol. 23, Chapter Fluiid Mechanics of Airflow over Hills: Turbulence, Fluxes, and Waves in the Boundary Layer. Am. Meteorological Soc.
24. Zeman, O., Jensen N. O.: Progress report on modeling permanent form sand dunes. RisøNatinal Laboratory M-2738 (1988).
25. Bagnold R. A.: The movement of desert sand. *Proc. R. Soc. London, Ser. A*, 157:594–620, 1936.
26. Owen P.R.: Saltation of uniformed sand grains in air. *J. Fluid. Mech.*, 20:225–242, 1964.
27. Sørensen M.: An analytic model of wind-blown sand transport. Acta Mechanica (Suppl.) 1, 67-81 (1991).
28. Lancaster, N., Nickling W. G., McKenna Neumann C. K., Wyatt V. E.: Sediment flux and airflow on the stoss slope of a barchan dune. Geomorphology 17, 55-62 (1996).
29. Butterfield, G. R.: Sand transport response to fluctuating wind velocity. In N.J. Clifford, J.R. French, and J. Hardisty (eds.) Turbulence: Perspectives on Flow and Sediment Transport, Chapter 13, 305-335, John Wiley & Sons Ltd. (1993).
30. van Dijk, P. M., Arens, S. M., van Boxel J. H.: Aeolian processes across transverse dunes ii: Modelling the sediment transport and profile development. Earth Surf. Process. Landforms 24, 319-333 (1999).
31. Robert S. Anderson and Peter K. Haff.: Simulation of eolian saltation. *Science*, 241:820, 1988.
32. Anderson, R. S.: Wind modification and bed response during saltation of sand in air. *Acta Mechanica (Suppl.)*, 1:21–51, 1991.
33. McEwan, I. K. and Willetts, B. B.: Numerical model of the saltation cloud. *Acta Mechanica (Suppl.)*, 1:53–66, 1991.
34. Sauermann, G., Kroy, K. and Herrmann, H. J. : A phenomenological dynamic saltation model for dune formation. *Phys.Rev.E*, 64:31305, 2001.
35. Gerd Sauermann: Modeling of wind blown sand and desert dunes. Universität Stuttgart, 2001. PhD Thesis.

Morphogenesis of Growing Amorphous Films

Stefan J. Linz, Martin Raible and Peter Hänggi

Theoretische Physik I, Institut für Physik,
Universität Augsburg, 86135 Augsburg, Germany

Abstract. In a first part of this paper, we survey the conceptional and methodological background of the stochastic field equations approach to model surface growth processes. In the second part, we focus on recent progress in the modeling of such equations for the specific case of vapor deposited amorphous thin films that allow for a quantitative validation with experimental data.

1 Introduction

The fundamental physical laws governing the macroscopic and the microscopic world are, in principle, rather simple. Nevertheless, nature is able to create a highly complicated and structured world on the basis of these laws. Unraveling the hidden rules of nature how to fabricate such complex systems, i.e. systems being built up of many interacting constituents and exhibiting a complicated overall behavior that is not at all evident from the known underlying interaction between the individual constituents, has developed into a central area of current research in physics.

The morphogenesis or structure formation of growing surfaces and interfaces due to deposition processes constitutes a specific and nonetheless paradigmatic example of such a complex system [1,2]. Focussing on atomic deposition processes, the interplay of two individual components, i.e. atoms or molecules of the typical size of 10^{-1}nm, is basically determined by their electromagnetic interaction. This, however, does not give us immediate insights into the deposition processes of particles when they arrive at a surface profile of already collectively condensed particles. It is not at all obvious how to theoretically understand the plethora of different, experimentally observed surface structures that can be created by the variety of deposition methods, deposition conditions and specific type of particles. In this context, vapor deposited amorphous films exhibit the spectacular fact of building up surface patterns with some intrinsic regularity or periodicity on scales of the order of 10nm if the film thickness reaches values of several hundred nanometers [3–5].

Even if one were able to simulate the corresponding full ab-initio many-particle problem on square areas of the corresponding size and for correspondingly long deposition times, this would not directly lead to physical insights into the underlying physical phenomena happening on the mesoscale. In the last fifteen years, an alternative approach to study surface growth phenomena [1,2,6,7] basically pioneered by the seminal work of Kardar, Parisi and Zhang

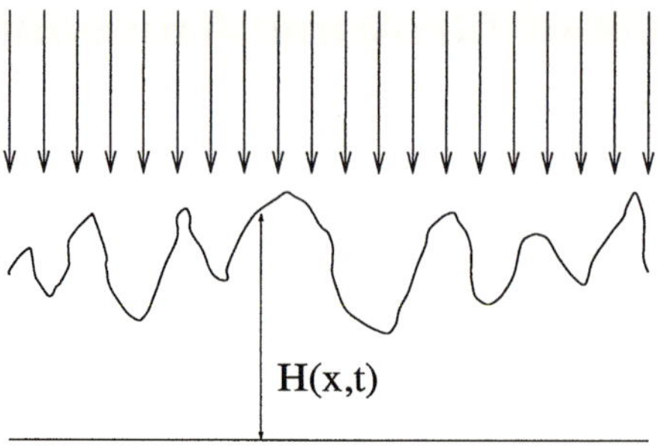

Fig. 1. Sketch of a representative deposition process: Particles from the incoming flux (denoted by the arrows) are deposited on the rugged surface and contribute to the spatio-temporal evolution of the surface.

[8] has attracted considerable interest and developed into a significant branch of surface physics. Here, the primary ideas are to consider the surface structure as a continuous two-dimensional manifold on a superatomic level as, for instance, represented by scanning tunneling microscopy images and then to model its spatio-temporal evolution by means of stochastic field equations (SFEs) that incorporate the relevant mesoscopic relaxation mechanisms of the deposited particles.

The focus of this work is twofold. In section 2, we critically review the basic strategies of the SFE approach for spatio-temporally evolving surface morphologies and put some emphasis on the pattern forming aspect of such equations. In section 3, we specifically apply this method to the problem of amorphous thin film growth by physical vapor deposition. We review and extend recent investigations that lead, by comparison with experimental data, to a rather complete picture of the significant relaxation phenomena and, most importantly, to an elementary description of the vapor deposition process in form of a minimal deposition equation.

2 Basic Concepts

General strategy. As starting point, we present here a synoptic account of the framework of stochastic field equations (SFEs) to model surface morphologies (for a thorough overview see Refs. [2,7]) including the clarification of some misconceptions in the literature. The general idea of the SFE approach is comparatively straightforward: A generally spatio-temporally varying flux of particles given by $I(\mathbf{x}, t)$ reaches the surface (cf. also Fig. 1), the particles from the beam are deposited at the surface and then undergo various surface diffusion processes until they arrive at their final position. The growing layer built up by the deposited particles forms a spatio-temporally evolving free surface that is characterized by its *height* or *morphology* $H(\mathbf{x}, t)$ at time t and the location $\mathbf{x} = (x, y)$ measured with respect to coordinates of the initially flat surface $H(\mathbf{x},0)=0$. The SFE approach disregards the microscopic details of the particle arrangement and interaction at the surface and considers the growth process on a slighty larger length scale, the meso- or nanoscale, where the (coarse-grained) surface morphology $H(\mathbf{x},t)$ can be regarded as a field variable evolving continuously in space and time. Then, the rate of change of the surface height $H(\mathbf{x},t)$ can be expressed in form a partial differential equation

$$\partial_t H(\mathbf{x},t) = G[\nabla H, ...] + I(\mathbf{x},t) \qquad (1)$$

where the functional $G[\nabla H, ...]$ depends only on spatial derivatives of H and their nonlinear combinations and comprises all physical mechanisms leading to growth and relaxational processes at the surface. Writing down (1), three fundamental symmetry requirements [2] for surface growth processes have been already incorporated: (i) no dependence of (1) on the specific choice of the origin of time implying *invariance under translation in time*, (ii) no dependence of (1) on the specific choice of the origin of the coordinate system implying *invariance under translation in the direction perpendicular to the growth direction*, and (iii) no dependence of (1) on the specific choice of the origin of the H-axis implying *invariance under translation in growth direction.* These symmetry requirements on the evolution equation (1) exclude any explicit dependence of the functional $G[..]$ on the time t, the spatial position $\mathbf{x}$, and the height H, respectively. Note, however, that these symmetry requirements do not necessarily apply to the solutions $H(\mathbf{x},t)$ of (1), too.

It is worthwhile noting that (1) contains several implicit assumptions: (i) the relaxation processes are local in space, (ii) no, in principle possible changes in bulk of the already built-up layer are taken into account, (iii) in order to guarantee single-valuedness of $H(\mathbf{x},t)$, no overhangs in the evolving surface structure are allowed, and (iv) to fulfill moderate existence and uniqueness requirements for $H(\mathbf{x},t)$, the (coarse-grained) surface profile can have at most some cusps with a *non*-zero opening angle.

How does stochasticity enter into (1)? In many physical applications such as vapor deposition experiments [3–5], the deposition flux is basically constant with some small superimposed spatio-temporal variations resulting from the particle source. As a consequence, the deposition flux can be split into a spatio-temporally constant mean deposition flux F and a fluctuating part $I(\mathbf{x},t) = F + \eta(\mathbf{x},t)$. As a simplest model for the generally not well-known fluctuations $\eta(\mathbf{x},t)$ one usually uses spatio-temporal Gaussian white noise determined by $\langle\eta(\mathbf{x},t)\rangle_\eta = 0$ and $\langle\eta(\mathbf{x},t)\,\eta(\mathbf{x}',t')\rangle_\eta = 2D\,\delta(\mathbf{x}-\mathbf{x}')\,\delta(t-t')$. Here, $\langle...\rangle_\eta$ denotes the ensemble average, and D the fluctuation strength. Since the mean deposition flux F is constant it also proofs useful to introduce the height profile $h(\mathbf{x},t) = H(\mathbf{x},t) - Ft$ in the frame comoving with the velocity F. Then, (1) simplifies to

$$\partial_t h = G[\nabla h] + \eta(\mathbf{x},t). \qquad (2)$$

As an aside, we note that the method leading to (2) also works if the mean deposition flux $F(t)$ contains an experimentally predetermined dependence on time t. Then one straightforwardly arrives at (2) if the definition $h(\mathbf{x},t) = H(\mathbf{x},t) - \int_0^t F(t')\,dt'$ is used.

How to specify the functional $G[..]$**?** As explained before, all physics of the growth process is hidden in the functional $G[..]$. So, the variety of different experimentally observable surface structures must be directly related to the specific functional form of $G[..]$. Basically two approaches are used in the literature. One way is to select known physical relaxation mechanisms (for collection of such mechanisms see e.g.[2]) that are considered to be relevant for the specific system, combines them and tries to compare the outcome with experimental data. This method has the drawback of being non-systematic. Another way is to start from the guiding principle of simplicity of the functional form of $G[..]$ combined with further symmetry requirements for the specifically considered system and a series expansion of $G[..]$ in small gradients of h up to some given order. In this case, one obtains a systematic skeleton of the functional form of the SFE; it is, however, not directly evident how to relate all the terms to underlying growth and relaxation processes.

In the remainder of this section, we focus on deposition processes that also possess *invariance under rotation and reflection* in the plane perpendicular to the growth direction. An example for such processes is amorphous growth where the isotropy of the amorphous phase implies such an invariance. This symmetry immediately excludes any odd derivatives of h in G and implies that ∇-operators entering the various contributions in G must be multiplied in couples by scalar multiplication. Assuming that all surface relaxation processes are local, we finally expand the functional G in a power series in all possible spatial derivatives of h and keep only the terms that are linear or quadratic in h and only possess a maximum of four ∇-operators. As a result of the afore-mentioned symmetries, the deterministic part of (2) can only consist of the terms $\nabla^2 h$, $(\nabla h)^2$, $\nabla^4 h$, $\nabla^2(\nabla h)^2$, $(\nabla^2 h)^2$, and $\nabla \cdot [(\nabla h)(\nabla^2 h)]$. The last term can be slightly rearranged in the form

$$\nabla \cdot [(\nabla h)(\nabla^2 h)] = \frac{1}{2}\nabla^2(\nabla h)^2 + 2M \tag{3}$$

with

$$M = \det \begin{pmatrix} \partial_x^2 h & \partial_y \partial_x h \\ \partial_x \partial_y h & \partial_y^2 h \end{pmatrix} . \tag{4}$$

Consequently, a *systematic* expansion of the functional form of the growth equation (1) that takes into account (i) all afore-mentioned symmetries and (ii) all admissible combinations of terms being linear or quadratic in $h(\mathbf{x},t)$ and containing up to a maximum of four ∇-operators is given explicitly by

[9]

$$\begin{aligned}\partial_t h = \;& a_1\nabla^2 h + a_2\nabla^4 h + a_3\nabla^2(\nabla h)^2 \\ & + a_4(\nabla h)^2 + a_5(\nabla^2 h)^2 + a_6 M + \eta.\end{aligned} \tag{5}$$

Eq.(5) constitutes the main result of this section.

Some general remarks. Equation (5) consists of two linear terms and four nonlinear terms in h. The term being proportional to a_6 becomes zero in the one-dimensional limit. This shows the principal problem that one-dimensionally motivated surface growth equations cannot be carried over to the two-dimensional case by simply replacing $\partial_x \to \nabla$. Moreover, it is interesting to note that Lai and Das Sarma [10] have also attempted to derive the leading order functional form of a growth equation using isotropy and the fact that the functional G in (2) is a scalar. Their result, however, significantly differs from (5) since the terms $a_5(\nabla^2 h)^2$ and $a_6 M$ are missing. Therefore, we conclude that Lai and Das Sarma's growth equation [10] represents an inconsistent systematic expansion since the terms $(\partial_x^2 h)(\partial_y^2 h) - (\partial_x\partial_y h)^2$ and $(\partial_x^2 h)^2 + (\partial_y^2 h)^2 + 2(\partial_x^2 h)(\partial_y^2 h)$ are not properly taken into account.

The growth equation (5) contains several known limiting cases. The limit $a_i = 0$ for $i = 1,..,6$, $\partial_t h = \eta$, is considered as an appropriate model for random deposition [2]. Setting $a_i = 0$ for $i = 2,..,6$, $\partial_t h = a_1\nabla^2 h + \eta$, yields the Edwards-Wilkinson (EW) equation originally motivated in the context of granular systems [11]. The limit $a_i = 0$ for $i = 2,3,5,6$, determines the Kardar-Parisi-Zhang (KPZ) equation, $\partial_t h = a_1\nabla^2 h + a_4(\nabla h)^2 + \eta$, being the paradigm for a stochastic roughening process [8]. Finally, the limit $a_i = 0$ for $i = 3,5,6$, $\partial_t h = a_1\nabla^2 h + a_2\nabla^4 h + a_4(\nabla h)^2 + \eta$, leads to the stochastic version of the Kuramoto-Sivashinsky (KS) equation [12].

Pattern forming aspects. From the viewpoint of nonlinear dynamics, the KPZ equation and the KS equation can be considered as antipodal paradigms although they only differ by the term $\nabla^4 h$. This can seen from the linearized version of (5) that dominates the initial stages of the growth process. Neglecting any stochasticity for the moment, inserting a solution or mode ansatz $h(\mathbf{x},t) = h_0 \exp(i\mathbf{k}\cdot\mathbf{x})\exp(\sigma t)$ into $\partial_t h = a_1\nabla^2 h + a_2\nabla^4 h$ directly leads to the dispersion relation for the growth rate, $\sigma(k) = -a_1 k^2 + a_2 k^4$, with $k = |\mathbf{k}|$ being the modulus of the wave vector $\mathbf{k}$. Depending on the sign of the coefficients a_1 and a_2, four main types of behavior can be distinguished. For $a_1 > 0$ and $a_2 \leq 0$, all modes are damped with time, whereas for $a_1 < 0$ and $a_2 \geq 0$ all modes grow with time and they grow the faster, the larger the wave number is. If $a_1 > 0$ and $a_2 > 0$, then all modes beyond the threshold $k_T = \sqrt{a_1/a_2}$ grow. The most important case, however, occurs if $a_1 < 0$ and $a_2 < 0$. Then, only wave numbers k in the range $0 \leq k \leq \sqrt{a_1/a_2}$ can grow with time and, moreover, there is a fastest growing mode $k_m = \sqrt{a_1/2a_2}$ that dominates the evolution of h. The latter result tells us that, at least for stages where the linearized version of the KS equation is sufficient, a pattern with an underlying periodicity dominated by k_m develops. This growth mechanism disappears in the KPZ-type limit when a_2 is set to zero. Physically interesting behavior of the KPZ equation appears when $a_1 > 0$. Then, the collaborate effect of the noise and the nonlinearity leads to a stochastic roughening of the surface with self-affine character [2]. The noisy KS equation for $a_1 > 0$ and $a_2 < 0$ leads to a similar behavior. If, however, $a_1 < 0$ and $a_2 < 0$, the pattern forming mechanism dominates (for small enough noise amplitudes) the early stages of the growth process. The role of the entering nonlinearity is to modify the pattern at later stages of the growth process.

Besides the invariances already invoked for its derivation, the growth equation (5) possesses an interesting additional symmetry: It remains invariant under the combined transformation

$$\{h, a_3, a_4, a_5, a_6\} \to \{-h, -a_3, -a_4, -a_5, -a_6\}. \tag{6}$$

As a consequence, one has to expect that a simultaneous change of the sign of the coefficients a_3, a_4, a_5, a_6 belonging to the nonlinear terms in (5) only leads to an inversion of the surface profile $h(\mathbf{x}, t)$ about $h = 0$. Note, however, that (5) does *not* possess mirror symmetry about $h = 0$, i.e. it does not fulfill the up/down invariance $h \to -h$ (without inversion of the signs of the nonlinear coefficients). This already implies some degree of asymmetry of the resulting surface profile $h(\mathbf{x}, t)$.

Conservative growth processes. A frequently invoked further requirement on surface growth equations [2,7] is that the functional G should be represented by the divergence of a surface current, $G = -\nabla \cdot \mathbf{j}_G(\nabla h)$ if no desorption of particles can occur. Such an assumption directly rules out the appearance of a KPZ term $(\nabla h)^2$ and the term $(\nabla^2 h)^2$ in G. It also implies that the spatial and ensemble averaged height is related to the deposition flux by $\langle\langle H(\mathbf{x}, t)\rangle_\eta\rangle_\mathbf{x} = Ft$ or, equivalently, that $\langle\langle h(\mathbf{x}, t)\rangle_\eta\rangle_\mathbf{x} = 0$ holds. The afore-mentioned assumption, however, also implicitly implies that no coarse-grained *density* variations can occur. In the presence of local density variations, a discussion starting from the condition that no incoming particles are lost (cf. the following section) can lead to a KPZ term $(\nabla h)^2$ and a term $(\nabla^2 h)^2$ and, therefore, also to a nonzero *excess velocity* $v = \langle\langle \partial_t h\rangle_\eta\rangle_\mathbf{x} = \langle\langle a_4(\nabla h)^2 + a_5(\nabla^2 h)^2\rangle_\eta\rangle_\mathbf{x}$ of the surface profile $h(\mathbf{x}, t)$.

How to make contact with experimental results? Modern experimental investigation tools such as scanning tunneling microscopy combined with image processing allow for a detailed resolution of the surface morphology and its spatio-temporal evolution [3–5]. Since the obtained data set is too immense and the data also contain some degree of stochasticity due to the small deposition noise resulting from the particle source, the *height-height-correlation function*

$$C(r, t) = \langle\langle [H(\mathbf{x} + \mathbf{r}, t) - \langle H\rangle_\mathbf{x}][H(\mathbf{x}, t) - \langle H\rangle_\mathbf{x}]\rangle_\eta\rangle_{\mathbf{x}, |\mathbf{r}|=r} \tag{7}$$

determines an appropriate quantitative statistical measure for the information on height variations and lateral correlations. In (7), $\langle ... \rangle_\eta$ represents an average over different samples (ensemble average), $\langle ... \rangle_{\mathbf{x}} = L^{-2} \int_0^L d^2x...$ the spatial average over a sample area of size L^2, and $\langle H \rangle_{\mathbf{x}} = \langle H \rangle_{\mathbf{x}}(t) = \langle H(\mathbf{x}, t) \rangle_{\mathbf{x}}$ the spatially averaged surface profile at time t. The height-height-correlation function $C(r, t)$ contains the two most important global quantities that characterize the surface morphology: (i) The *correlation length* $R_c(t)$ that is given by the first maximum of $C(r, t)$ for non-zero r, i.e. by $R_c(t) = \min\{r > 0 | \partial_r C(r, t) = 0, \partial_r^2 C(r, t) < 0\}$, and, therefore, determines the typical length scale over which height fluctuations are correlated, and (ii) the *surface roughness* $w(t)$ or root mean square deviation of the relative height fluctuations that is determined by the $r = 0$-limit of $C(r, t)$, $w^2(t) = C(0, t)$. Another often used statistical measure of the surface morphology is the *spectral power density* that is determined by

$$C(k, t) = C(|\mathbf{k}|, t) = \mathcal{F}[C(r, t)] = \mathcal{F}[C(|\mathbf{r}|, t)] \quad (8)$$

where $\mathcal{F}[...]$ represents the two-dimensional Fourier transform with respect to the wave vector $\mathbf{k}$. As a minimum requirement for a successful modeling attempt of the spatio-temporal evolution of $H(\mathbf{x}, t)$, the validation of the temporal evolution of $R_c(t)$ and $w(t)$ in comparison with the available experimental data needs to be achieved.

3 Deposition Equation for Thin Film Growth

In this section, we specifically focus on the growth of solid amorphous films generated by physical vapor deposition under normal incidence of the particle flux that is important e.g. in the context of coating and the manufacturing of thin glassy ZrAlCu films and has recently attracted interest [3–5,13] in materials science. We review and partly extend some major results obtained in recent works [13–16] on the development and detailed analysis of a minimal model in form of a SFE that (i) appropriately describes the spatio-temporal evolution of such amorphous surface growth processes and (ii) stands the test of a quantitative comparison with available experimental data [3–5,13]. From the theoretical point of view, amorphous film growth constitutes a particularly attractive testing ground for a *quantitative* comparison of experimental data and theoretical approaches since (i) there are not any long range ordering phenomena (as in epitaxial growth processes) to be expected, (ii) the effect of terrace formation and, therefore, the Ehrlich-Schwoebel effect being significant for epitaxial growth processes are absent, and (iii) the growing film should be spatially isotropic. As an appropriate starting point, we can directly use the general form (5) that contains all basic symmetries relevant for the vapor deposition process.

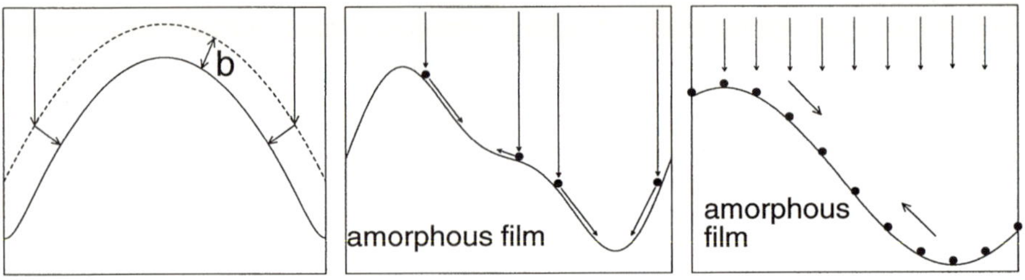

Fig. 2. Microscopic effects of amorphous surface growth. Left part: Inflection of particles due to interatomic interaction. Middle part: Surface diffusion of deposited particles due to surface relaxation. Right part: Equilibration of the inhomogeneous particle concentration due to the geometry of the surface.

Physics behind the growth equation. Guided by the principle that any mathematically admissible term might have some physical significance, we next relate all terms appearing in the growth equation (5) to the four competing microscopic mechanisms

- surface tension [18]
- concentration equilibration of deposited particles [19,20]
- steering of arriving particles [14]
- inhomogeneous density distribution [14,15]

that, as we shall see in the next section, seem to dominate physical vapor deposition and are all, at least at some stages of the growth process, important. Also the signs and the order of magnitude estimates of some coefficients in (5), as well as a physically motivated simplification of (5) are obtained.

The linear term proportional to a_2 in (5) can be interpreted as the result of a type of a microscopic surface tension effect as originally suggested by Mullins [18]. The basic idea behind this effect (cf. also the middle part of Fig. 2) is that the just deposited particles favoritely move to positions at the surface that have positive curvature $\nabla^2 h > 0$ since there, the already condensed surface particles form a local vicinity with higher binding energy. This gives rise to a diffusion current $\mathbf{j}_m \propto \nabla(\nabla^2 h)$ that, depending on the local curvature, can be uphill or downhill. The divergence of this current, $-\nabla \cdot \mathbf{j}_m = a_2 \nabla^4 h$, contributes to the surface evolution in (5) with a_2 being necessarily negative. This term basically tries to minimize the area of the surface and, as a consequence, to smooth the surface morphology.

The nonlinear term proportional to a_3 can be related to the tendency of equilibrating the non-homogeneous concentration c of the deposited particles just after arriving at the surface. This effect has originally been suggested by Villain [19] (cf. also [20]). The underlying reason is of purely geometric nature. Although the deposition flux is basically homogeneous, more particles per surface area arrive at positions with a small or zero modulus of slope ∇h

than at positions being strongly inclined with respect to the particle beam, cf. also the right part of Fig. 2. Therefore, the local concentration of the diffusing particles right after the deposition is not constant, but is weighted by the local slope of the surface, $c \propto 1/\sqrt{1+(\nabla h)^2}$, or in a small gradient expansion, $c \propto 1 - \frac{1}{2}(\nabla h)^2$. Then, the tendency to equilibrate the concentration is reflected by a diffusion current $\mathbf{j}_c \propto -\nabla c \propto \nabla(\nabla h)^2$, or, after taking the divergence, by the term $-\nabla \cdot \mathbf{j}_c = a_3 \nabla^2 (\nabla h)^2$ that contributes to the height changes in (5). Obviously, concentration equilibration requires that the coefficient a_3 is negative and also tries to smooth the surface morphology. A simple dimensional argument leads to an estimate for a_3. Equation (5) implies that the coefficient a_3 has the dimension of length^3/time. The magnitude of a_3 necessarily depends on the deposition flux F that possesses the dimension of length/time and the mean diffusion length l which is the only relevant length scale determining this process. The only combination of F and l leading to the correct dimension of a_3 is Fl^2. Therefore, one expects $a_3 \propto -Fl^2$. A thorough discussion of the concentration equilibration [14] supports this argument and yields the explicit relation $a_3 = -\frac{1}{8}Fl^2$. Moreover, one expects that the typical magnitude of l is of the order of several atom diameters.

The two terms in (5) that are proportional to a_1 and a_6 can microscopically be related to the steering of the arriving particles. Here, the basic idea [14] is that the particles from the beam experience close to the growing surface a deflection due to the interatomic attractive interaction with the already condensed surface particles. As a consequence, the particles do not hit the surface perpendicular to the substrate orientation, but perpendicular to the surface itself. This implies that more particles arrive at positions at the surface with negative curvature, $\nabla^2 h < 0$, than at positions with positive curvature $\nabla^2 h > 0$. Effectively, this leads to a tendency to roughen the surface morphology. We refer to Ref. [21] for experimental indications of the relevance of this effect. To model this scenario in a dynamical way [14], we use the idealization that the particles undergo a change of direction only after reaching a critical distance b, the effective range of the interaction, from the surface and are then attracted such that they arrive perpendicular to the surface, cf. the left part of Fig. 2. A detailed mathematical derivation [14] using a reparametrization in the coordinates of the imaginary surface where the interaction becomes effective (cf. the dotted line in the left part of Fig. 2) and a small gradient expansion in h in fact shows that this scenario gives simultaneously rise to the two contributions $a_1 \nabla^2 h$ and $a_6 M$ in (5). Moreover, the coefficients a_1 and a_6 can be related to the mean deposition flux F and the effective range b of the interatomic interaction yielding $a_1 = -Fb$ and $a_6 = Fb^2$ [14]. Although b cannot be directly measured its magnitude should be typically of the order of one atomic diameter and, therefore, much smaller than the radius of the surface curvature. This implies that the term proportional to a_6 is of minor relevance in comparison to the a_1-term and can be neglected. Moreover, the sign of a_1 is negative.

The physical origin of the nonlinear terms proportional to a_4 and a_5 is determined by the potential variations of the coarse-grained density [13,14]. These terms cannot result from particle desorption since the substrate is held at room temperature and the particle energy in the vapor beam is rather low (typically of the order 0.1eV). Therefore, *all* arriving particles finally contribute to the surface growth. As a consequence, any term that cannot be recast in form of the divergence of a current in (5) arises from changes of the coarse-grained density. Assuming for the moment that the deposition noise is zero ($\eta = 0$), particle conservation implies that the rate of change of the number of particles per substrate area above a given substrate location, C, is determined by a balance equation $\partial_t C = -\nabla \cdot \mathbf{j}_C + \rho_0 F$. Here the divergence of the current $\mathbf{j}_C$ is given by the combination of all surface relaxation processes (cf. the afore-mentioned arguments), i.e. by $-\nabla \cdot \mathbf{j}_C = \rho_0 [a_1 \nabla^2 H + a_2 \nabla^4 H + a_3 \nabla^2 (\nabla H)^2 + a_6 M]$, and ρ_0 represents the density of the growing film in the case of a horizontal surface. Allowing for density variations at the growing surface, the rate of change of C is related to the rate of change of the height H by $\partial_t C = \rho(\nabla H) \partial_t H$. Here $\rho(\nabla H)$ denotes the density at the surface. Without the incorporation of density changes ($\rho = \rho_0 = const.$), there is a direct proportionality $\partial_t C = \rho_0 \partial_t H$. If small density variations are taken into account, $\rho(\nabla H)$ can be expanded in the derivatives of H yielding $\rho(\nabla H) = \rho_0 [1 + q_1 (\nabla H)^2 + q_2 \nabla^2 H]$ in lowest order approximation. Therefore, $\partial_t H = \rho_0^{-1} [1 - q_1 (\nabla H)^2 - q_2 \nabla^2 H] \partial_t C$ holds. Inserting this in the balance equation from above, explains the presence of the two terms $-q_1 F (\nabla H)^2 = a_4 (\nabla h)^2$ and $-q_2 a_1 (\nabla^2 H)^2 = a_5 (\nabla^2 h)^2$ appearing in (5). From the physical point of view, however, density changes are primarily connected to the gradients of the surface profile reflecting the local arrangement of the particles at the surface and not so much to the surface curvature. Therefore, it is plausible to disregard the term $a_5 (\nabla^2 h)^2$ in a minimal description of the growth evolution. Since the density variations result from a widening of the mean inter-particle distances at the surface or an enlarged number of microscopic vacancies in the growing material one has to expect that they locally decrease the density implying that $a_4 > 0$ holds.

Taking into account the afore-mentioned physical arguments, the terms $a_5 (\nabla^2 h)^2$ and $a_6 M$ are negligible in leading order and, as a final result, we obtain the model equation for amorphous film growth [9,13–15],

$$\partial_t h = a_1 \nabla^2 h + a_2 \nabla^4 h + a_3 \nabla^2 (\nabla h)^2 + a_4 (\nabla h)^2 + \eta \tag{9}$$

with a_1, a_2, a_3 being negative and a_4 being positive.

Using stochastic numerical simulations of the surface growth equation (9) starting from a flat substrate (for details of the different methods see Ref. [17]), we investigate in the remainder of this contribution the evolution of the correlation length R_c and surface roughness w as a function of the experimentally measurable layer thickness $\overline{H}$. This quantity is determined by $\overline{H} = \langle \langle H(\mathbf{x}, t) \rangle_\eta \rangle_\mathbf{x} = Ft + \langle \langle h(\mathbf{x}, t) \rangle_\eta \rangle_\mathbf{x}$ and is, in general, implicitly connected to the time t via the solution of (9). The latter results from the fact that

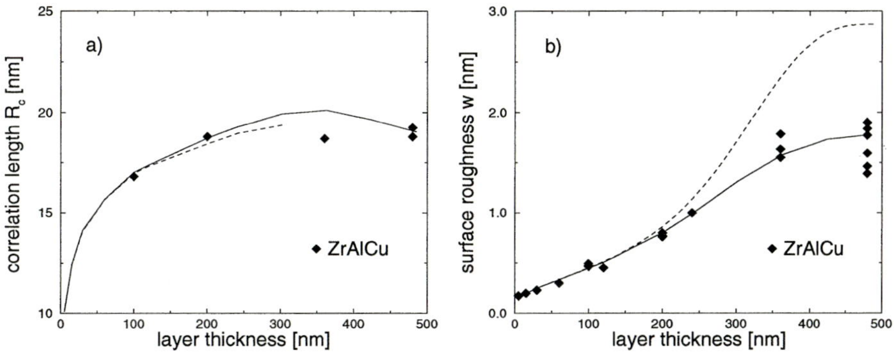

Fig. 3. Solid lines (dashed lines): Correlation length R_c and surface roughness w for the experimentally investigated thickness interval $0 \leq \overline{H} \leq 480\mathrm{nm}$ calculated from the nonlinear growth equation (9) using the parameters $a_1 = -0.0826\mathrm{nm}^2/\mathrm{s}$, $a_2 = -0.319\mathrm{nm}^4/\mathrm{s}$, $a_4 = 0.055\mathrm{nm/s}$, $D = 0.0174\mathrm{nm}^4/\mathrm{s}$ and $a_3 = -0.10\mathrm{nm}^3/\mathrm{s}$ ($a_3 = 0\mathrm{nm}^3/\mathrm{s}$). Diamonds represent the corresponding experimental results previously published in Ref.[3,22]. This figure is taken from Ref.[13].

the surface profile generated by (9) possesses a finite excess velocity, $v = \langle\langle \partial_t h \rangle_\eta\rangle_\mathbf{x} = \langle\langle a_4 (\nabla h)^2 \rangle_\eta\rangle_\mathbf{x}$. Since a_4 is positive the average of the surface morphology $H(\mathbf{x}, t) = Ft + h(\mathbf{x}, t)$ grows with a faster speed than F as a result of the inhomogeneous density distribution.

Selected results. Here, we show that the model equation (9) is indeed able to quantitatively reproduce experimental data on the correlation length R_c and surface roughness w if the coefficients a_1, a_2, a_3, a_4 and D are appropriately chosen. For the specific example of the growth of $Zr_{65}Al_{7.5}Cu_{27.5}$ films [3–5,13], a parameter estimation procedure discussed in detail in [13] yields for the coefficients in (9) $a_1 = -0.0826\mathrm{nm}^2/\mathrm{s}$, $a_2 = -0.319\mathrm{nm}^4/\mathrm{s}$, $a_3 = -0.10\mathrm{nm}^3/\mathrm{s}$, and $a_4 = 0.055\mathrm{nm/s}$ and for the strength of the deposition noise $D = 0.0174\mathrm{nm}^4/\mathrm{s}$. The experimentally determined mean deposition flux is given by $F = 0.79\mathrm{nm/s}$. For this set of parameter values, we show the dependence of the correlation length R_c and surface roughness w (solid lines) on the thickness of the amorphous film in Fig. 3 and infer a very good agreement with the corresponding experimental data. For comparison, the corresponding results of the Kuramoto-Sivashinsky limit (a_3=0, dashed lines) are given. Since the correlation length ceases to exist at a film thickness of about 300nm in this limit we also conclude that both nonlinear terms proportional to a_3 and a_4 are necessary to reproduce the experimental data. Leaving off the term that describes the effect of density inhomogeneities, $a_4 = 0$, the surface roughness increases strongly with time and does not show the cross-over to a saturation at layer thicknesses of about 480nm (for more details of this limit cf. Ref.[14,15]). Moreover, both linear terms proportional to a_1 and a_2 are necessary to excite the growth instability at the initial stages of the growth process [14]. Consequently, (9) must be considered as a minimal model for the growth of amorphous $Zr_{65}Al_{7.5}Cu_{27.5}$ films.

The extrapolated parameters a_1, a_2, a_3, a_4, and D also allow for microscopic estimates [13]. (i) Since $a_1 = -Fb$, the typical range b of the interaction between the surface atoms and the particles to be deposited is about 0.1nm, i.e. of the size of the radii (0.2nm) of the surface atoms. (ii) Since $a_3 = -Fl^2/8$, the diffusion length l is about 1.0nm. Consequently, the deposited particles experience a surface diffusion on a nanometer scale and do not just stick at the places where they hit the surface. (iii) If the particles arrive independently on the surface, the deposition noise is related to the particle volume Ω and the mean deposition rate F by $2D = F\Omega$ [14], yielding $\Omega = 0.04nm^3$. This is up to a factor of two the averaged particle volume of ZrAlCu. (iv) The local density of the growing film varies with the surface slope: On an inclined surface area the local density is decreased by $\rho(\nabla h) = \rho_0/\gamma$ with $\gamma = 1 + (a_4/F)(\nabla h)^2$ (where a_4/F is about 0.07). These finite density variations are physically compatible with the small diffusion length l of two to three atom diameters. At the layer thickness 480nm, this local density reduction γ (averaged over the surface) possesses a mean 1.021 and a standard deviation 0.017.

Finally, we present in Fig. 4 the dependence of the Fourier transform of the height-height correlation function, $C(k,t)$ (see (8)), on the modulus of the wave vector for five subsequent deposition times or layer thicknesses obtained from the numerical solution of (9). Also here, our numerical simulations show a striking agreement with previously published experimental data (cf. Fig. 4 [right part] in Ref.[22] and also Fig. 3 in Ref.[23]). As in these experiments, one recovers (i) the characteristic decay proportional to k^{-4} that gives clear evidence of the importance of the Mullins term $a_2\nabla^4 h$ in (9) and (ii) the slight buckling of $C(k)$ for the layer thickness of 480nm and wave numbers k somewhat larger than 10^{-1}nm^{-1} where in a narrow range of k a decay proportional to k^{-1} can be fitted. A detailed numerical comparison with and without the density variation term, $a_4(\nabla h)^2$, in (9) shows that this effect is directly related to the inclusion of density variations. Based on their experimental data (cf. Fig. 3 in Ref.[23]), Mayr and Samwer [23] have recently suggested an alternative explanation for this effect. Using the old idea tracing back to Mullins [18] that a *linear* term leading to a decay proportional to k^{-1} might be related to viscous flow in the bulk, these authors have designed a qualitative model of viscous hill coalescence to explain this feature. Since the decay of the power spectral density $C(k,t)$ proportional to k^{-1} in a small range of k is in our analysis a direct consequence of the *nonlinear* KPZ term, it does not seem to substantiate the presence of viscous flow in the bulk. Moreover, the remark in [23] that the KPZ term, $a_4(\nabla h)^2$, in (9) might be interpreted as the lowest order mathematical representation of viscous hill coalescence is not convincing. If no desorption takes place and a KPZ term,

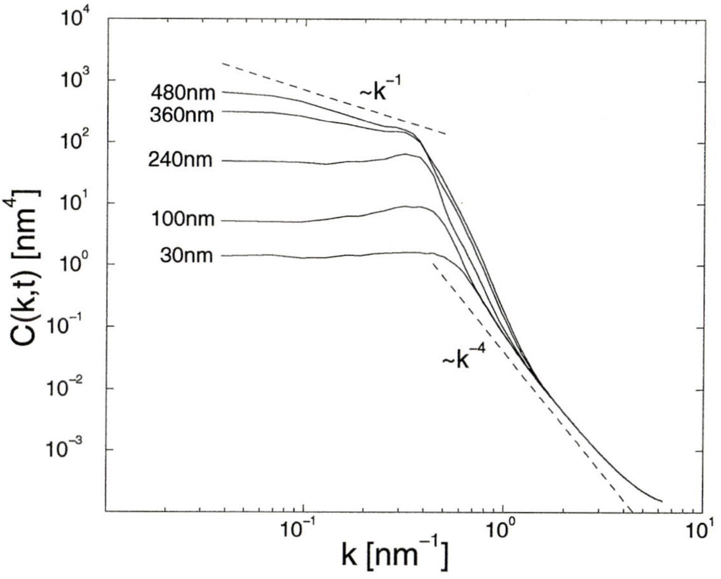

Fig. 4. Dependence of the Fourier transform of the height-height correlation function or power spectral density, $C(k,t)$, on the modulus of the wave vector for five subsequent deposition times or layer thicknesses calculated from the nonlinear growth equation (9) using the parameters $a_1 = -0.0826\text{nm}^2/\text{s}$, $a_2 = -0.319\text{nm}^4/\text{s}$, $a_3 = -0.10\text{nm}^3/\text{s}$, $a_4 = 0.055\text{nm/s}$ and $D = 0.0174\text{nm}^4/\text{s}$.

$a_4(\nabla h)^2$, is invoked to explain experimental data, then it must be related to density variations. All other effects that are based on transport mechanisms should be expressible in terms of divergences of currents.

Next, we explore some properties of the growth process in a layer thickness range up to 2000nm that has so far not yet experimentally investigated. The results in Fig. 3 up to a layer thickness of 480nm suggest that the growth process has not yet reached a final, not necessarily stationary state. Using again the afore-mentioned parameter values, the dependence of the correlation length R_c and surface roughness w (solid lines) on the thickness of the amorphous film is shown in Fig. 5 (the dashed lines refer to the special case a_3=0). Obviously, the surface roughness has reached an almost constant value for a layer thickness larger than 600nm that increases only very weakly as the growth process proceeds. In contrast to that, the correlation length steeply decays after reaching a maximum and then saturates in an almost constant value for a layer thickness larger than 600nm as the growth process proceeds. For further results, in particular the properties of the correlation function and the related height difference correlation function as well as visualizations of the surface morphology and a theoretical interpretation of the various stages of the growth process, we refer to Ref.[15].

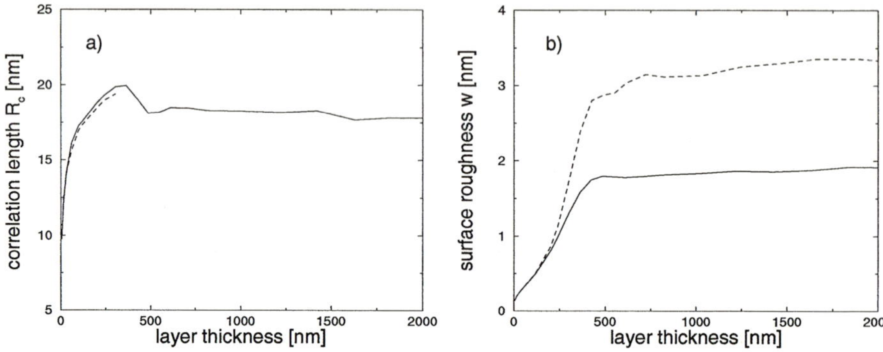

Fig. 5. Solid lines (dashed lines): Correlation length R_c and surface roughness w for the thickness interval $0 \leq \overline{H} \leq 2000$nm calculated from the nonlinear growth equation (9) using the parameters $a_1 = -0.0826\text{nm}^2/\text{s}$, $a_2 = -0.319\text{nm}^4/\text{s}$, $a_3 = -0.10\text{nm}^3/\text{s}$ ($a_3 = 0\text{nm}^3/\text{s}$), $a_4 = 0.055\text{nm/s}$, and $D = 0.0174\text{nm}^4/\text{s}$.

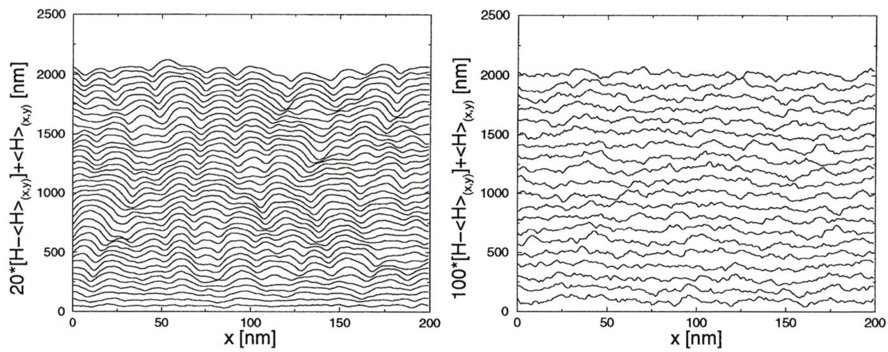

Fig. 6. Visualization of the cross-section of the spatio-temporal evolution of the growing surface of the film calculated from the nonlinear growth equation (9). Left part: realistic parameters for ZrAlCu $a_1 = -0.0826\text{nm}^2/\text{s}$, $a_2 = -0.319\text{nm}^4/\text{s}$, $a_3 = -0.10\text{nm}^3/\text{s}$, $a_4 = 0.055\text{nm/s}$, and $D = 0.0174\text{nm}^4/\text{s}$; right part: same parameters except of a sign change of a_1, i.e. $a_1 = 0.0826\text{nm}^2/\text{s}$.

To obtain further insight into the spatio-temporal evolution of the surface morphology, we present in the left part of Fig. 6 a representative one-dimensional cross-section of the growth of the surface profile (for $y = 0$) with increasing time or layer thickness. For demonstration purposes, the relative height fluctuations have been weighted by a factor of 20 relative to the mean thickness $\langle H \rangle_{\mathbf{x}} = \langle H \rangle_{(x,y)}$. From the left part of Fig. 6, three remarkable features can be read off. First, as the time proceeds and the layer builds up, the surface morphology develops into a predominantly almost periodic structure with an averaged periodicity length given by the correlation length R_c and some superimposed stochastic variations. Second, the evolving mound and dip structure is asymmetric in the sense that the dips are comparatively narrow in contrast to the wide mounds. Third and most remarkably, the surface morphology does not approach a stationary profile in the thickness range $1000\text{nm} \leq \langle H \rangle_{(x,y)} \leq 2000\text{nm}$. Despite the fact that statistical quantities such as the correlation length and the surface roughness are almost constant in this thickness interval, the surface profile still varies significantly with time. The right part of Fig. 6 depicts (up to some scaling of the amplitudes) the evolution of the surface morphology for the same parameters *except* that the sign of the coefficient a_1 has been inverted. Although this parameter set is not really physical, it shows the remarkable fact that, as a consequence of the inversion of the sign of a_1, the surface structure changes from a pattern with intrinsic regularity (depicted on the left panel of Fig. 6 with negative a_1) to a stochastically varying profile without any regularity when a_1 changes its sign (see right panel of Fig. 6). Beyond that, the roughness of the surface itself is largely reduced in comparison to the case $a_1 < 0$. The distinct surface evolution for positive and negative values of a_1 supports our previous statements about pattern forming aspects in section 2.

4 Conclusions and Perspectives

Based on (i) a systematically derived minimal functional form of a growth equation being appropriate for the understanding of amorphous thin film growth, cf. equation (5), and (ii) relations of the terms occuring in the functional form (5) to underlying microscopic surface relaxation mechanisms, a quantitative agreement of the temporal evolution of the correlation length and the surface roughness of the surface morphology with experimental data can be achieved. We expect that similar quantitative agreement of experimental data and appropriately modeled SFEs for the spatio-temporal evolution of surface morphologies can also be obtained for different systems such as for crystalline growth processes or sputter deposition.

Acknowledgement: This work has been supported by Sonderforschungsbereich 438 (TU München/Univ. Augsburg), Project A1.

References

1. Tong, W. M., Williams, R. S.: Kinetics of surface growth. Annu. Rev. Phys. Chem. **45** (1994) 401-438
2. Barabasi A.-L., Stanley, H. E. : *Fractal concepts in surface growth* (Cambridge University Press, Cambridge, 1995)
3. Reinker, B., Moske, M., Samwer, K.: Kinetic roughening of amorphous ZrAlCu films investigated in situ with scanning tunneling microscopy. Phys. Rev. **B 56** (1997) 9887-9893

4. Mayr, S. G., Moske, M., Samwer, K.: Early stages in amorphous $Zr_{65}Al_{7.5}Cu_{27.5}$ film growth on HOPG. Europhys. Lett. **44** (1998) 465-470
5. Mayr, S. G., Moske, M., Samwer, K.: Identification of key parameters by comparing experimental and simulated growth of vapor deposited amorphous $Zr_{65}Al_{7.5}Cu_{27.5}$ films. Phys. Rev. **B 60** (1999) 16950-16955
6. Krug, J.: Origins of scale invariances in growth processes. Adv. Phys. **46** (1997) 139-282
7. Marsili, M., Maritan, A., Toigo, F., Banavar, J. R.: Stochastic growth equations and reparametrization invariance. Rev. Mod. Phys. **68** (1996) 963-983
8. Kardar, M., Parisi, G., Zhang, Y. C.: Dynamic scaling of growing interfaces. Phys. Rev. Lett. **56** (1986) 889-892
9. Linz, S. J., Raible, M., Hänggi, P.: Stochastic field equation for amorphous surface growth. Lecture Notes in Physics **557** (2000) 473-483
10. Lai Z.-W., Das Sarma S.: Kinetic growth with surface relaxation: continuum versus atomistic models. Phys. Rev. Lett. **66** (1991) 2348-2351
11. Edwards S., Wilkinson D.R.: The surface statistics of a granular aggregate. Proc. Roy. Soc. London **A 381** (1982) 17-31
12. Drotar, J. T., Zhao, Y.-P., Lu, T.-M., Wang, G.-C.: Numerical analysis of the noisy Kuramoto-Sivashinsky equation in 2 + 1 dimensions. Phys. Rev. E **59** (1999) 177-185
13. Raible, M., Mayr, S. G., Linz, S. J., Moske, M., Hänggi, P., Samwer, K.: Amorphous thin film growth: theory compared with experiment. Europhys. Lett. **50** (2000) 61-67
14. Raible, M., Linz, S. J., Hänggi, P.: Amorphous thin film growth: minimal deposition equation. Phys. Rev. E **62** (2000) 1691-1705
15. Raible, M., Linz, S. J., Hänggi, P.: Amorphous thin film growth: effects of density inhomogeneities. Phys. Rev. E **64** (2001) 031506 1-11
16. Linz, S. J., Raible, M., Hänggi, P.: Amorphous thin film growth: modeling and pattern formation. Adv. Solid State Phys. **41** (2001) 391-403
17. Raible, M., Linz, S. J., Hänggi, P.: Amorphous thin film growth: simulation methods for stochastic deposition equations. Acta Phys. Pol. **B 33** (2002) 1049-1061
18. Mullins, W. W.: Theory of thermal grooving. J. Appl. Phys. **28** (1957) 333-339; Flattening of a nearly planar solid surface due to capillarity. J. Appl. Phys. **30** (1959) 77-83
19. Villain, J.: Continuum models of crystal growth from atomic beams with and without desorption. J. Physique **I 1** (1991) 19-42
20. Moske, M.: *Mechanische Spannungen als Sonde für Schichtwachstum und Schichtreaktionen* (Habilitation thesis, Universität Augsburg, 1997)
21. van Dijken, S., Jorritsma, L. C., Poelsema, B.: Steering-enhanced roughening during metal deposition at grazing incidence. Phys. Rev. Lett. **82** (1999) 4038-4041
22. Mayr, S. G., Moske, M., Samwer, K.:Identification of key surface processes for vapor deposited amorphous metallic film growth. Mater. Sci. Forum **343-346** (2000) 221-230
23. Mayr, S. G., Samwer, K.: Model for intrinsic stress formation in amorphous thin films. Phys. Rev. Lett. **87** (2001) 036105 1-4

Part II

INTERFACE DYNAMICS: Modelling and Simulation

Density Effects and Fluid Flow in Phase-field Models

Massimo Conti

Dipartimento di Matematica e Fisica, Universita' di Camerino, and Istituto Nazionale di Fisica della Materia, Via Madonna delle Carceri,I-62032 , Camerino, Italy

Abstract. In this paper we discuss a method to incorporate hydrodynamic effects into phase-field models for the solidification of both pure substances and binary alloys. We start from a generalised thermodynamic potential with squared gradient terms for the associated fields; the condition of local positive entropy production is then utilised to derive a set of equations which drive the system towards equilibrium. The models are numerically solved in one dimension; the effects of the flow field on the growth dynamics are presented and discussed.

1 Introduction

The change of density upon freezing is a common feature of both pure materials and alloys. The solid is generally denser than the liquid; the change ranges from a few percent for simple metals to more than 20% for some eutectic mixtures. The shrinking (or, in some cases, the dilatation) of the system, causes a fluid flow in the liquid phase towards the interface (or away from it). Then, even in absence of natural convection, the purely diffusive picture for the heat (and solute, for alloy solidification) released at the growing interface should be extended, to incorporate an accurate analysis of the flow field. Recently, this subject attracted an increased attention for practical reasons. The reduction in volume in metal solidification is a source of stress at the growing interface, and may induce cavitation, with formation of defects or micropores in cast products [1,2]. A further effect results from the dynamic pressure drop across the melt: at large growth rates the change of the melting temperature with pressure may significantly alter the undercooling of the liquid phase.

A stimulating approach to the density effects in solidification is based on the diffuse interface picture, through an extension of the phase-field model. In the classic formulation of the model, a non-conserved order parameter $\phi(\mathbf{x}, t)$ characterises the phase of the system at each point. A suitable free energy (or entropy) functional is then constructed, that depends on the order parameter as well as on the associate (conserved) fields and their gradients. The extremization of the functional with respect to these variables results in the dynamic equations for the evolution of the process. Studies conducted on solidification of both pure substances and binary alloys [3–11] pointed out that the model incorporates in a natural fashion the effects of the interface curvature and non-equilibrium phenomena as the trapping of solute into the solid phase and the kinetic undercooling of the solid-liquid interface. An extension of the phase-field model to incorporate flow effects was first proposed by Caginalp and Jones [12]. They derived a system of differential equations for the variables temperature, order parameter, fluid velocity, density and pressure. In the momentum equation capillary and viscosity effects were neglected. Anderson et al. [13] developed a phase-field model with convection in the melt, deriving an expression for the stress tensor within the arguments of the extended irreversible thermodynamics. The problem was treated in the quasi-incompressibility approximation, neglecting the pressure dependence of the local density, which is assumed uniform in the two bulk phases. The condition of positive entropy production was employed to derive the governing equations of the model. In a subsequent study the same authors [14] examined the sharp interface limit of the model. Their analysis recovered the standard interfacial conditions, including the Young-Laplace and Clausius-Clapeyron equations for the mechanical and chemical equilibrium at the interface. Conti [15] followed a different approach, starting from a grand canonical potential which includes the local density as a dynamic variable. Square gradient terms were allowed for both the structural order parameter and the density field. The stress tensor was derived assuming that any dissipation of tensorial order should be ascribed to the fluid viscosity. In this formulation, the pressure field is consistently related to the local density via an equation of state. The model equations for the order parameter and the local density, momentum and energy allow to describe the solidification process and the flow field, including the propagation of elastic waves. The model was numerically solved in one dimension to assess the effects of the fluid convection on the growth dynamics.

In the present paper we present a model which incorporates flow effects due to density change. The solid phase is modeled as an isotropic fluid with a viscosity much larger than that of the liquid. The entropy production equation, coupled with the balance of mass, momentum and energy, is used to derive governing equations that drive the system towards equilibrium. The model, derived for a pure substance, is then extended to describe alloy solidification. The equations of the model have been solved numerically in one dimension. As solidification starts, the sudden contraction of the liquid in front of the interface originates a pressure (and density) wave that propagates into the sample. The results of our simulations show that this phenomenon, as well as the other coupled effects of the chemical and mechanical relaxation, is properly described. In alloy solidification, trapping of solute is observed at large growth rates. Due to the flow effects on the melting point, the border between steady and diffusive growth, in the dynamical phase diagram, is slightly shifted.

2 Phase-field Model for a Pure Material

2.1 The Entropy Balance

Let us first consider a pure substance which undergoes a solid-liquid transition. The local state of the system is characterised by a coarse grained density $\rho(\mathbf{x}, t)$, the local temperature T and a nonconserved order parameter $\phi(\mathbf{x}, t)$, which is assumed to take the values $\phi = 0$ in the solid and $\phi = 1$ in the liquid. Let us denote the velocity field by $\mathbf{v}$, the specific energy by e' and the specific entropy by s'. The two latter quantities are determined by the specific free energy $f'(\rho, \phi, T, \nabla\rho, \nabla\phi)$ and involve gradient contributions. The stress tensor will be denoted by $\mathbf{P}$; $\mathbf{J}_{\mathrm{E}}$, $\mathbf{J}_{\mathrm{S}}$ stand for the energy and entropy flux vectors respectively, and σ is the entropy production rate. Finally, $\mathbf{g}$ stands for a specific body force field. In terms of these variables the classical balance laws read:

$$\frac{d\rho}{dt} = -\rho\nabla \cdot \mathbf{v} \tag{1}$$

$$\rho\frac{d\mathbf{v}}{dt} = \rho\,\mathbf{g} - \nabla \cdot \mathbf{P} \tag{2}$$

$$\rho\frac{de'}{dt} = -\nabla \cdot \mathbf{J}_{\mathrm{E}} - \mathbf{P} : \nabla\mathbf{v} \tag{3}$$

$$\rho\frac{ds'}{dt} = -\nabla \cdot \mathbf{J}_{\mathrm{S}} + \sigma \tag{4}$$

The constitutive relations and the explicit form of the fluxes will follow from the Courie principle and from the local form of the second law of thermodynamics, which implies $\sigma \geq 0$. We postulate a generalised specific Helmoltz free energy of the form

$$f'(\rho, \phi, T, \nabla\rho, \nabla\phi) = f(\rho, \phi, T) + \frac{1}{2\rho}\left[\delta_{\mathrm{F}}^2(\nabla\rho)^2 + \epsilon_{\mathrm{F}}^2(\nabla\phi)^2\right] \tag{5}$$

where $f(\rho, \phi, T)$ is the specific bulk free energy, and the the gradient terms account for nonlocal contributions in the interfacial region. We assume that ϵ_{F} and δ_{F} depend only on temperature. The non-gradient part of the specific energy and entropy are defined by

$$s(\rho, \phi, e) = -\frac{\partial f}{\partial T} \quad ; \qquad e(\rho, \phi, s) = f(\rho, \phi, T) + T\,s(\rho, \phi, e) \tag{6}$$

Similar relations are postulated for the corresponding quantities incorporating gradient terms. Denoting

$$\delta_{\rm S}^2 \equiv \frac{d\,\delta_{\rm F}^2}{dT} \quad ; \quad \epsilon_{\rm S}^2 \equiv \frac{d\,\epsilon_{\rm F}^2}{dT} \quad ; \quad \delta_{\rm E}^2 = \delta_{\rm F}^2 - T\,\delta_{\rm S}^2 \quad ; \quad \epsilon_{\rm E}^2 = \epsilon_{\rm F}^2 - T\,\epsilon_{\rm S}^2 \tag{7}$$

we obtain from (5)

$$s'(\rho,\phi,e,\nabla\rho,\nabla\phi) = s(\rho,\phi,e) - \frac{1}{2\rho}\left[\delta_{\rm S}^2(\nabla\rho)^2 + \epsilon_{\rm S}^2(\nabla\phi)^2\right] \tag{8}$$

$$e'(\rho,\phi,s,\nabla\rho,\nabla\phi) = e(\rho,\phi,s) + \frac{1}{2\rho}\left[\delta_{\rm E}^2(\nabla\rho)^2 + \epsilon_{\rm E}^2(\nabla\phi)^2\right] \tag{9}$$

Under the above assumptions the differential form of the second law of thermodynamics reads:

$$\begin{aligned} T\,ds' = de' - \frac{\partial f}{\partial\phi}d\phi - \frac{1}{\rho^2}\left[p - \frac{1}{2}\delta_{\rm F}^2(\nabla\rho)^2 - \frac{1}{2}\epsilon_{\rm F}^2(\nabla\phi)^2\right]d\rho \\ -\frac{1}{\rho}\left[\delta_{\rm F}^2(\nabla\rho)d(\nabla\rho) + \epsilon_{\rm F}^2(\nabla\phi)d(\nabla\phi)\right] \end{aligned} \tag{10}$$

Combining (10) with the balance equations (1,2,3) yields, after some manipulations:

$$\begin{aligned} \rho\frac{ds'}{dt} = -\frac{\nabla\cdot\mathbf{J}_{\rm E}}{T} - \frac{\delta_{\rm F}^2}{T}\nabla\cdot\left(\frac{d\rho}{dt}\nabla\rho\right) - \frac{\epsilon_{\rm F}^2}{T}\nabla\cdot\left(\frac{d\phi}{dt}\nabla\phi\right) \\ -\frac{1}{T}\frac{d\phi}{dt}\left(\rho\frac{\partial f}{\partial\phi} - \epsilon_{\rm F}^2\nabla^2\phi\right) - \frac{1}{T}\left(P_{ik}+T_{ik}\right)\partial_i v_k \end{aligned} \tag{11}$$

where we denote $\partial_i \equiv \partial/\partial x_i$, and T_{ik} is the capillary stress tensor, defined by

$$T_{ik} = \delta_{ik}\left[-p + \rho\delta_{\rm F}^2\nabla^2\rho + \frac{1}{2}\delta_{\rm F}^2(\nabla\rho)^2 + \frac{1}{2}\epsilon_{\rm F}^2(\nabla\phi)^2\right] - \delta_{\rm F}^2\rho_i\rho_k - \epsilon_{\rm F}^2\phi_i\phi_k \tag{12}$$

Notice that this form of the stress tensor extends to a solid-liquid phase transition with density change a well known result obtained either for fluid-fluid interfaces or for solidification without density change. We can rearrange (11) according to the entropy balance equation (4); the entropy production rate σ results as

$$\sigma = -\frac{1}{T}\frac{d\phi}{dt}\left(\rho\frac{\partial f}{\partial\phi} - \epsilon_{\mathrm{F}}^2\nabla^2\phi\right) + \mathbf{J}_{\mathrm{E}}\cdot\nabla\left(\frac{1}{T}\right) + \frac{d\rho}{dt}\nabla\rho\cdot\nabla\left(\frac{\delta_{\mathrm{F}}^2}{T}\right)$$
$$+\frac{d\phi}{dt}\nabla\phi\cdot\nabla\left(\frac{\epsilon_{\mathrm{F}}^2}{T}\right) - \frac{1}{T}\left(\mathbf{P}+\mathbf{T}\right):\nabla\mathbf{v} \qquad (13)$$

To simplify the discussion, extending the choice of Wang et al. in their model for solidification at constant density [6], we assume that $\epsilon_{\mathrm{E}}^2 = \delta_{\mathrm{E}}^2 = 0$; ϵ_{S}^2, $\delta_{\mathrm{S}}^2 = const$. Moreover in the sequel we shall neglect the spatial variation of both ϵ_{F}^2 and δ_{F}^2, which amounts to neglect the thermal gradient across the interface. Then, the constraint of local positive entropy production reduces to

$$\frac{d\phi}{dt} = -\Gamma\left(\rho\frac{\partial f}{\partial\phi} - \epsilon_{\mathrm{F}}^2\nabla^2\phi\right) \qquad (14)$$

$$\mathbf{J}_{\mathrm{E}} = -K\nabla T \qquad (15)$$

where Γ is a positive constant and K is the thermal conductivity. Moreover, assuming that the tensor contribution is only amenable to viscous dissipation, we obtain

$$\mathbf{P} = -\mathbf{T} - \mathbf{\Pi} \qquad (16)$$

with $\mathbf{\Pi}$ indicating the standard stress tensor for viscous fluids.

2.2 The Thermodynamic Potential

To proceed further we need an explicit expression for the specific free energy f. At equilibrium the latter should take the form of a double well over the ρ, ϕ plane, with the two minima centered at the bulk solid ($\rho = \rho_{\mathrm{s}0}, \phi = 0$) and liquid ($\rho = \rho_{\mathrm{l}0}, \phi = 1$), where we denoted as $\rho_{\mathrm{s}0}$, $\rho_{\mathrm{l}0}$ the equilibrium densities in the two phases at some reference temperature T_0 and pressure p_0. We start from a linearised equation of state which describes the bulk density shift out of equilibrium as

$$\rho - \rho_0 = -\beta\,\rho_0\,(T - T_0) + k\,\rho_0\,(p + \mathcal{P} - p_0) \qquad (17)$$

where β is the thermal expansion coefficient and k the isothermal compressibility; p is the thermodynamic pressure given by $p = \rho^2(\partial f/\partial\rho)$ and $\mathcal{P}$ is the excess pressure due to the capillary stress. In the case of a planar interface normal to z, the latter is written as $\mathcal{P} = (\epsilon_{\mathrm{F}}^2\,\phi_z^2 + \delta_{\mathrm{F}}^2\,\rho_z^2)/2 - \rho\,\delta_{\mathrm{F}}^2\,\rho_{zz}$

An expression of the free energy, consistent with (17) may be written as

$$f(\rho,\phi,T) = \frac{ag(\phi)}{\rho} + \frac{\beta}{k}(T-T_0)\left[\frac{\rho-\rho_0(\phi)}{\rho\rho_0(\phi)}\right] + (p_0-\mathcal{P})\left[\frac{\rho-\rho_0(\phi)}{\rho\rho_0(\phi)}\right] - \\ -CT\ln\frac{T}{T_0} + C(T-T_0) + p(\phi)L_0(1-\frac{T}{T_0}) - \\ -\frac{(p_0-\mathcal{P})}{\rho_0(\phi)} + \frac{\rho_0(\phi)}{2k}\left[\frac{\rho-\rho_0(\phi)}{\rho\rho_0(\phi)}\right]^2 \tag{18}$$

with $g(\phi) = (1/4)\phi^2(1-\phi)^2$. In Equation (18) C is the specific heat and L_0 the latent heat per unit mass in the reference state. We observe a Landau-Ginzburg contribution for the order parameter ϕ (first term in the r.h.s of the equation), and the elastic contribution of the last term. Along the way followed by [12] we shall neglect in the pressure field the contribution due to the density dependence of the well height. The function $p(\phi)$ is monotonic and increasing with ϕ, assuming the values $p(0) = 0$, $p(1) = 1$, and describes the transition of the free energy from the solid to the liquid phase. Choosing $p(\phi) = \phi^3(10-15\phi+6\phi^2)$ fixes the bulk solid and liquid to the values $\phi = 0$ and $\phi = 1$, respectively, for every value of temperature. The equilibrium density $\rho_0(\phi)$ is assumed to change in the interfacial region as $\rho_0 \equiv \rho_0(\phi) = \rho_{s0} + p(\phi)(\rho_{l0} - \rho_{s0})$. In the sequel we assume that all the bulk properties of the system change in the same fashion along the solid-liquid interface, for example we represent the local viscosity as $\eta = \eta_s + p(\phi)(\eta_l - \eta_s)$, with $\eta_s \gg \eta_l$. Then, Equations (1,2,3,14) along with specifications (15,16,18) represent the evolution equations for the system.

2.3 Nondimensional Equations in one Dimension

We assume equal values of the thermal diffusivity D and the specific heat (at constant pressure) C_p in both phases. A nondimensional form of the model equations is obtained adopting a reference length ξ and scaling time to $\tau = \xi^2/D$. Density is scaled as ρ/ρ_{l0} and a nondimensional temperature is introduced as $u = C_p(T-T_0)/L_0$, with T_0 being the melting temperature in the reference state. Specific energies will be scaled to v_0^2, where $v_0 = \xi/\tau$ is the natural reference for velocities, and the scale for the components of the stress tensor is $\rho_{l0}v_0^2$. The equilibrium density takes the form $\rho_0 = S + p(\phi)(1-S)$, with $S = \rho_{s0}/\rho_{l0}$. Notice that in the following we neglect thermal expansion effects and we assume equal compressibilities in both phases. Retaining for simplicity the same simbols for the scaled (nondimensional) quantities, and in absence of body forces, the model equations in one dimension read:

$$\frac{\partial\rho}{\partial t} + v\frac{\partial\rho}{\partial z} = -\rho\frac{\partial v}{\partial z} \tag{19}$$

$$\frac{\partial \phi}{\partial t} + v\frac{\partial \phi}{\partial z} = m\frac{\partial^2 \phi}{\partial z^2} - \frac{m}{\tilde{\epsilon}^2}\left[\frac{\partial g(\rho,\phi)}{\partial \phi} - p'(\phi)\,\rho\,\alpha\,\tilde{\epsilon}\,u\right] - \\ -\frac{1}{2}p'(\phi)\,\rho\,(1-S)\,R_1\left(\frac{\rho_0^2-\rho^2}{\rho^2\,\rho_0^2}\right) \tag{20}$$

$$\rho\frac{\partial v}{\partial t} + \rho v\frac{\partial v}{\partial z} = -\nabla\cdot\left[\frac{R_1\,R_3\,(\rho-\rho_0)}{m\qquad\rho}\right] + R_4\frac{\partial}{\partial z}\left(\lambda_1(\phi)\frac{\partial v}{\partial z}\right) \tag{21}$$

$$\frac{\partial u}{\partial t} + v\frac{\partial u}{\partial z} = -p'(\phi)\left(\frac{\partial \phi}{\partial t} + v\frac{\partial \phi}{\partial z}\right) + \frac{\partial^2 u}{\partial z^2} + \frac{R_4}{R_5}\left(\frac{\partial v}{\partial z}\right)^2 \tag{22}$$

with $\lambda_1(\phi) = 1 + [1 - p(\phi)][(\eta_s/\eta_l) - 1]$. The parameters appearing in the above equations are defined as:

$$m = \frac{\Gamma\epsilon_F^2\tau}{\xi^2};\quad \tilde{\epsilon}^2 = \frac{\epsilon_F^2}{a\,\xi^2};\quad \alpha = \frac{L_0}{C_p T_0}\frac{\rho_{10}L_0}{\epsilon_F}\frac{\xi}{\sqrt{a}};\quad R_1 = \frac{\Gamma\tau}{k};$$
$$R_2 = \frac{\delta_F^2\rho_{10}}{\xi^2 v_0^2};\quad R_3 = \frac{\epsilon_F^2}{\xi^2\rho_{10}v_0^2};\quad R_4 = \frac{\eta_l}{\tau\rho_{10}v_0^2};\quad R_5 = \frac{L_0}{v_0^2} \tag{23}$$

Notice that imposing $\rho_0 = S = 1$ the model collapses on the the classical phase-field description of the the solidification of a pure substance. In this case (see for example [6,16]) the model parameters $m, \tilde{\epsilon}, \alpha, \epsilon_F, a$ can be related to the material properties through

$$m = \frac{\mu\sigma T_0}{D\rho_0\,L};\qquad \tilde{\epsilon} = \frac{h}{\xi};\qquad \alpha = \frac{\xi}{6\sqrt{2}d_0};$$
$$\epsilon_F^2 = 6\sqrt{2}\,\sigma\,h;\qquad a = 6\sqrt{2}\,\frac{\sigma}{h} \tag{24}$$

where h is the interface thickness, and $d_0 = (\sigma\,C\,T_0)/(\rho_0 L_0{}^2)$ the capillary length (with σ being the surface tension); μ is the kinetic undercooling coefficient that relates the interface undercooling to the interface velocity v_I through $v_I = \mu(T_0 - T)$. We assume that the above equations still represent a reasonable estimation of the model parameters in terms of the thermophysical properties of the material.

3 Phase-field Model for a Binary Alloy

We are concerned with an ideal binary solution with components A (solvent) and B (solute). To simplify the discussion we neglect the difference of the molar mass $\bar{M}$ of the two components, assuming $\bar{M}_{\mathrm{A}} = \bar{M}_{\mathrm{B}} \equiv \bar{M}$, that means to identify the molar concentration with the mass fraction. A further balance equation, in addition to (1,2,3,4), must be supplemented for the solute concentration c. The latter reads

$$\rho \frac{dc}{dt} = -\nabla \cdot \mathbf{J}_{\mathrm{D}} \tag{25}$$

where $\mathbf{J}_{\mathrm{D}}$ is the diffusive solute flux. Internal energy, entropy and free energy, exhibit an additional gradient term for the local concentration, with coefficients $\eta_{\mathrm{E}}^2, \eta_{\mathrm{S}}^2, \eta_{\mathrm{F}}^2$, respectively. The entropy production rate reads, in this case

$$\begin{aligned}\sigma = -\frac{1}{T}\frac{d\phi}{dt}\left(\rho\frac{\partial f}{\partial \phi} - \epsilon_{\mathrm{F}}^2\nabla^2\phi\right) + \mathbf{J}_{\mathrm{E}} \cdot \nabla\left(\frac{1}{T}\right) + \\ +\frac{d\phi}{dt}\nabla\phi \cdot \nabla\left(\frac{\epsilon_{\mathrm{F}}^2}{T}\right) + \frac{d\rho}{dt}\nabla\rho \cdot \nabla\left(\frac{\delta_{\mathrm{F}}^2}{T}\right) + \frac{dc}{dt}\nabla c \cdot \nabla\left(\frac{\eta_{\mathrm{F}}^2}{T}\right) + \\ +\mathbf{J}_{\mathrm{D}} \cdot \nabla\left[\frac{(\mu_{\mathrm{A}} - \mu_{\mathrm{B}})}{T} + \frac{\eta_{\mathrm{F}}^2\nabla^2 c}{\rho T}\right] - \frac{1}{T}(\mathbf{P} + \mathbf{T}) : \nabla\mathbf{v}\end{aligned} \tag{26}$$

with the capillary stress tensor $\mathbf{T}$ given by

$$\begin{aligned}T_{ik} = \delta_{ik}\left[-p + \rho\delta_{\mathrm{F}}^2\nabla^2\rho + \frac{1}{2}\delta_{\mathrm{F}}^2(\nabla\rho)^2 + \frac{1}{2}\epsilon_{\mathrm{F}}^2(\nabla\phi)^2 + \frac{1}{2}\eta_{\mathrm{F}}^2(\nabla c)^2\right] - \\ -\delta_{\mathrm{F}}^2\rho_i\rho_k - \epsilon_{\mathrm{F}}^2\phi_i\phi_k - \eta_{\mathrm{F}}^2 c_i c_k\end{aligned} \tag{27}$$

Then, the constraint of positive local entropy production produces a further condition, in addition to (14,15,16), i.e.

$$\mathbf{J}_{\mathrm{D}} = M_{\mathrm{c}}\nabla\left[\frac{(\mu_{\mathrm{A}} - \mu_{\mathrm{B}})}{T} + \frac{\eta_{\mathrm{F}}^2\nabla^2 c}{\rho T}\right] \tag{28}$$

where M_{c} is a positive constant. The specific free energy is written as

$$f = (1-c)\left[f_{\mathrm{A}} + \frac{RT}{\bar{M}}\ln(1-c)\right] + c\left[f_{\mathrm{B}} + \frac{RT}{\bar{M}}\ln c\right] \tag{29}$$

where $f_{\mathrm{A}}, f_{\mathrm{B}}$ represent the free energy of pure A and B, obtained replacing in (18) the material parameters with the ones of the solvent and the solute, respectively, and the equilibrium density with

$$\rho_0(\phi) = \frac{\rho_{0\mathrm{A}}\,\rho_{0B}}{(1-c)\rho_{0B} + c\rho_{0\mathrm{A}}} \tag{30}$$

Then, the solute flux may be expressed as

$$\mathbf{J}_{\mathrm{D}} = -\frac{M_{\mathrm{c}}R}{\bar{M}c(1-c)}\nabla c + M_{\mathrm{c}}\nabla\left(\frac{\eta_{\mathrm{F}}^2}{\rho T}\nabla^2 c\right)$$
$$+M_{\mathrm{c}}\nabla\left\{\frac{(a_{\mathrm{A}}-a_{\mathrm{B}})}{\rho T}g(\phi) + \frac{p(\phi)}{T}\left[L_{0\mathrm{A}}\left(1-\frac{T}{T_{0\mathrm{A}}}\right) - L_{0B}\left(1-\frac{T}{T_{0B}}\right)\right]\right\}$$

We recover the standard definition of the solute diffusivity D_c taking

$$D_c = \frac{M_{\mathrm{c}}R}{\rho\bar{M}c(1-c)} \tag{31}$$

As the solute diffusivity is quite different in the solid and liquid phases,in the following D_c will be taken as $D_c = D_{\mathrm{S}} + p(\phi)(D_l - D_{\mathrm{S}})$, D_l and D_{S} being the diffusivities in the liquid and in the solid, respectively. We now adopt a reference scale for time as $\tau = \xi^2/D_l$; the nondimensional temperature is defined as $u = C_{\mathrm{p}}(T - T_{0\mathrm{A}})/L_{0\mathrm{A}}$ and density is scaled to ρ_{l0A}.

In nondimensional form the governing equations become

$$\frac{\partial\rho}{\partial t} + v\frac{\partial\rho}{\partial z} = -\rho\frac{\partial v}{\partial z} \tag{32}$$

$$\begin{aligned}\frac{\partial\phi}{\partial t} + v\frac{\partial\phi}{\partial z} &= m\frac{\partial^2\phi}{\partial z^2} - \frac{m}{\tilde{\epsilon}^2}g'(\phi)\left[(1-c)+c\omega_1\right]\\ &-\rho\, p'(\phi)\beta_1\left[(1-c)\left(1-\frac{T}{T_{0\mathrm{A}}}\right) + c\omega_2\left(1-\frac{T}{T_{0B}}\right)\right]\\ &-\frac{1}{2}\beta_2\rho\, p'(\phi)\left[(1-c)(1-S_{\mathrm{A}}) + c(E-S_{\mathrm{B}})\right]\left(\frac{\rho_0^2-\rho^2}{\rho^2\,\rho_0^2}\right)\end{aligned} \tag{33}$$

$$\rho\frac{\partial v}{\partial t} + \rho v\frac{\partial v}{\partial z} = -\nabla\cdot\left[\beta_3\frac{(\rho-\rho_0)}{\rho}\right] + \beta_4\frac{\partial}{\partial z}\left(\lambda_1(\phi)\frac{\partial v}{\partial z}\right) \tag{34}$$

$$\rho\frac{\partial c}{\partial t} + \rho v\frac{\partial c}{\partial z} = -\frac{\partial}{\partial z}\lambda_2(\phi)\rho c(1-c)\frac{\partial}{\partial z}\left\{\ln\frac{1-c}{c} + \frac{T_{0\mathrm{A}}}{T}\beta_5 g(\phi)(1-\omega_1)\right.$$
$$\left.+\frac{T_{0\mathrm{A}}}{T}p(\phi)\left(\beta_1\beta_5\frac{\tilde{\epsilon}^2}{m}\right)\left[\left(1-\frac{T}{T_{0\mathrm{A}}}\right) - \omega_2\left(1-\frac{T}{T_{0B}}\right)\right] + \frac{T_{0\mathrm{A}}}{T}\frac{\beta_6}{\rho}\frac{\partial^2 c}{\partial z^2}\right\}$$

$$\frac{\partial u}{\partial t}+v\frac{\partial u}{\partial z}=-p'(\phi)[(1-c)+c\omega_2]\left(\frac{\partial \phi}{\partial t}+v\frac{\partial \phi}{\partial z}\right)+\frac{D_T}{D_l}\frac{\partial^2 u}{\partial z^2}+\frac{\beta_2\beta_4}{\beta_1\beta_3}\left(\frac{\partial v}{\partial z}\right)^2 \tag{35}$$

with $\lambda_2(\phi)=(D_S/D_l)+p(\phi)[1-(D_S/D_l)]$, and the parameters defined as

$$m=\frac{\Gamma\epsilon_F^2\tau}{\xi^2};\quad \tilde{\epsilon}^2=\frac{\epsilon_F^2}{a_A\xi^2};\quad \beta_1=\Gamma\tau\rho_{l0A}L_{0A};\quad \beta_2=\frac{\Gamma\tau}{k};$$
$$\beta_3=\frac{1}{k\rho_{l0A}v_0^2};\quad \beta_4=\frac{\eta_l}{\tau\rho_{l0A}v_0^2};\quad \beta_5=\frac{a_A\bar{M}}{\rho_{l0A}RT_{0A}};$$
$$\beta_6=\frac{\eta_F^2\bar{M}}{\rho_{l0A}\xi^2RT_{0A}};\quad \beta_7=\frac{\delta_F^2\rho_{l0A}}{\xi^2v_0^2};\quad \omega_1=\frac{a_B}{a_A};\quad \omega_2=\frac{L_{0B}}{L_{0A}};$$
$$S_A=\frac{\rho_{s0A}}{\rho_{l0A}};\quad S_B=\frac{\rho_{s0B}}{\rho_{l0A}};\quad E=\frac{\rho_{l0B}}{\rho_{l0A}} \tag{36}$$

4 Numerical Solutions

4.1 Pure Substance

The equations of the model have been solved on the computational domain $0\le z\le z_m$. Initially a phase boundary at $z=z_0$ separates a solid region ($z<z_0,\phi=0,\rho=S$) from the liquid region ($z>z_0,\phi=1,\rho=1$). The system is initially at rest and the liquid is undercooled, i.e. $v(z,0)=0$, $u(z\le z_0,0)=0$, $u(z>z_0,0)<0$. For the phase, density and temperature fields we imposed Neumann boundary conditions; the velocity is fixed as $v(0,t)=0$ at the left end of the solid, while we chose $(\partial v/\partial z)=0$ at $z=z_m$. To advance the solution forward in time, we employed an explicit Euler integration scheme for the density, temperature and phase-field equations. To allow for the condition $\eta_s\gg\eta_l$, which originates a large diffusion term in the momentum equation, the latter was integrated with a Crank-Nicolson implicit scheme. Second order central differences were used to discretize the Laplace operator, and upwind differences for the convective terms. To ensure an accurate resolution of the solid-liquid interface the grid spacing was selected as $\Delta z=0.25\cdot\tilde{\epsilon}$; the time step required for numerical stability is $\Delta t=0.25\cdot10^{-7}$. We referred to the thermophysical properties of Nickel, fixing the parameters as $m=0.1$, $\alpha=265$, $\tilde{\epsilon}=0.001$, $S=1.05$, $R_1=5.29\cdot10^6$, $R_2=R_3=7.12\cdot10^{-3}$, $R_4=0.105$ and $R_5=4940$.

We first checked whether the model gives a consistent description of the mechanical effects due to the density change in solidification. To this aim we solved the model equations at constant temperature ($u=-0.005$), fixing our attention to the mechanical relaxation of the system. The contraction of the liquid in front of the interface originates a pressure wave which propagates both into the solid and into the liquid. This effect is illustrated in Fig. 1, where only the liquid portion of the system is shown. The solid-liquid interface is located near $z = 0.3$ (here and in the following all the numerical results will be presented in nondimensional form), and the pressure field is represented at different times. The initial pressure was initialised as $p(z, 0) = 0$, and we see the negative pressure front which propagates into the liquid. The velocity of the wave, estimated tracking the position of the wave front, results $v_S = 615.5$; this value is in excellent agreement with the theoretical one for the liquid in equilibrium, $(\phi = 1, \rho_l = 1)$, assumed as a pure elastic medium: the latter is $\sqrt{\partial p/\partial \rho} = \sqrt{R_1 \, R_3/m} = 613.7$.

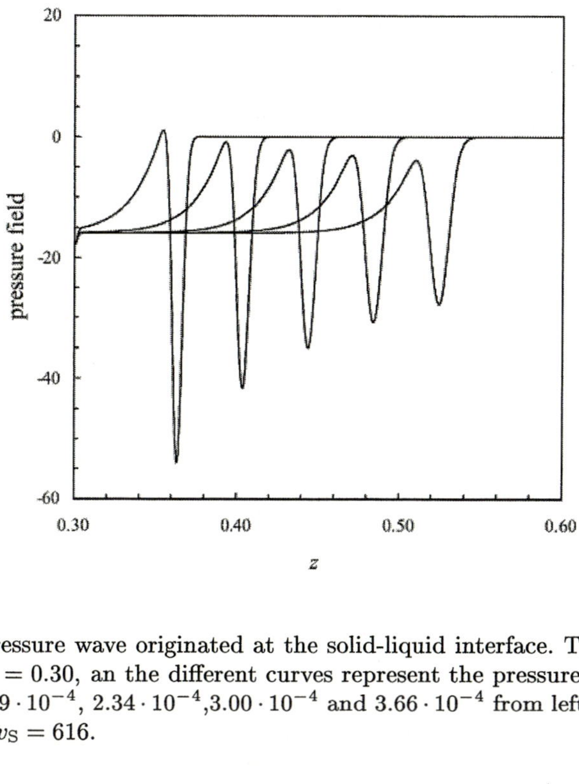

Fig. 1. The pressure wave originated at the solid-liquid interface. The interface is located near $z = 0.30$, an the different curves represent the pressure field at times $1.03 \cdot 10^{-4}$, $1.69 \cdot 10^{-4}$, $2.34 \cdot 10^{-4}$,$3.00 \cdot 10^{-4}$ and $3.66 \cdot 10^{-4}$ from left to right. The wave speed is $v_S = 616$.

After a short transient, growth at fixed temperature results in a steady advancement of the solid-liquid interface. In Figs. 2(a)- 2(c) we show the phase, density and velocity profiles obtained at different times, with $u = -0.004$. We observe that the solid is at rest, while the liquid is advected towards the interface with a velocity $v = -0.0356$. The interface velocity, as resulting from the numerical data, is $v_I = 0.7123$. Notice that this is the same value fixed by the mass conservation law through the relation $v = (1 - S)v_I$.

The conditions for mechanical equilibrium at a curved interface force a pressure jump across the phase boundary. This effect is shown in Fig. 3 for the growth of a circular germ (the initial undercooling is $\Delta \equiv u(z > z_0, 0) = -0.4$). The normal stress is represented versus the radial coordinate at different times. We observe the difference Δp of the pressure in the two phases, due to the Young-Laplace effect. Notice that the jump decreases with increasing the radius R of the germ. The product $R\,\Delta p$ remains constant during the process, and should correspond to the (nondimensional) surface tension. Indeed, the latter, given by $R_3/(\tilde{\epsilon}\sqrt{72})$ is in excellent agreement with the observed value $R\,\Delta p = 0.84$.

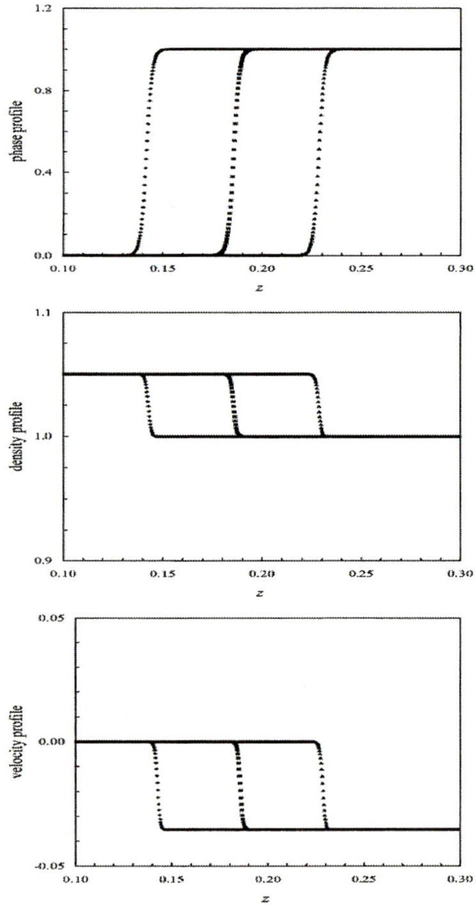

Fig. 2. The phase (a), density (b) and velocity (c) fields at different times. The curves are taken at $t = 0.06, t = 0.12, t = 0.18$ from left to right.

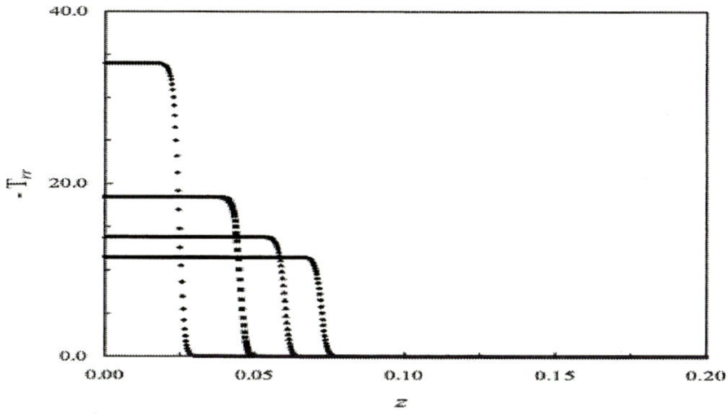

Fig. 3. Profile of the negative of the normal stress in circular growth. The curves are taken at times $0.875 \cdot 10^{-3}$, $2.250 \cdot 10^{-3}$, $3.625 \cdot 10^{-3}$,$5.000 \cdot 10^{-3}$ from left to right. We can observe the pressure jump across the interface due to the Young-Laplace effect.

4.2 Binary Alloy

In general we referred to phase diagram of an ideal solution of nickel (solvent) and copper (solute), fixing the model parameters as $m = 466$, $\beta_1 = 8.484 \cdot 10^8$, $\beta_2 = 6.681 \cdot 10^{10}$, $\beta_3 = 9.231 \cdot 10^{13}$, $\beta_4 = 1.282 \cdot 10^3$, $\beta_5 = 1.071$, $\beta_6 = 6.853 \cdot 10^{-7}$, $\omega_1 = 5.727 \cdot 10^{-1}$,$\omega_2 = 6.910 \cdot 10^{-1}$, $S_{\text{A}} = 1.051$, $S_{\text{B}} = 1.026$, $E = 0.983$. The interface width was chosen as $\tilde{\epsilon} = 8 \cdot 10^{-4}$, and the initial concentration of the solution is $c = 0.07214$, that belongs to the solidus line at $T = 1700$K. We recovered all the expected phenomenology. For example, we observed the transition from the kinetic to the diffusive regime in isothermal growth, as the temperature approaches the T_0 line from below, and solute trapping occurs at large growth rates. We want now to draw the attention on an interesting point. The flow effects result in a slower interface dynamics (see for example [1]). Then, we may expect that in the dynamic phase diagram (in the T, c plane) steady state solutions, due to density change, could be pushed into the region of diffusive regimes. This effect is illustrated in Fig. 4, where the growth rate is represented versus time, at $T = 1703$K. The upper set of data shows the kinetic (steady) growth obtained without density change (i.e. fixing $S_{\text{A}} = S_{\text{B}} = E = 1$). In the lower curve we accounted for the density effects, fixing $S_{\text{A}}y = S_{\text{B}} = 0.8$, $E = 1$. We see in this case the transition from the kinetic to the diffusive regime.

5 Conclusions

The classical phase-field model is a well established tool to describe solidification far from equilibrium. In this paper we derived the governing equations of the model in the framework of the extended irreversible thermodynamics, considering different densities of the solid and liquid phases and taking into account the effects of capillary stresses. The equations reduce to the classical formulation for equal solid and liquid densities. The numerical solutions of the model show that the sound wave propagation, the interfacial dynamics and the flow field are properly described. Then, the model has been extended to describe alloy solidification. In future work we plan to study the effects of the mechanical relaxation at the early stage of the growth. In a real process, as the nucleation starts, the homogeneity of the melt is rapidly lost and all the growing nuclei become the source of (primary or reflected) elastic waves. This phenomenon results in a new interaction mechanism between the growing germs, and could alter in a significant way the first stage of the crystal growth.

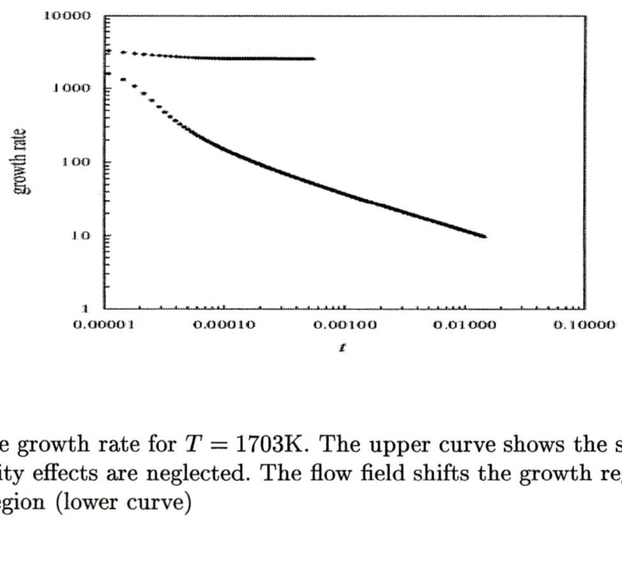

Fig. 4. The growth rate for $T = 1703\text{K}$. The upper curve shows the steady regime when density effects are neglected. The flow field shifts the growth regime into the diffusive region (lower curve)

References

1. Caroli, B., Caroli, C. and Roulet, B. in: *Solids far from equilibrium*, ed. Godreche, (Cambridge University Press, Cambridge, 1992).
2. Kurz, W. and Fisher, D. J.: *Fundamentals of solidification* (Trans Tech Publications, Aedermannsdorf, 1992).

3. Caginalp, G.: Arch. Ration. Mech. Anal. 92, 205 (1986).
4. Caginalp, G.: Phys. Rev. A 39, 5887 (1989).
5. Penrose, O. and Fife, P. C.: Physica D 43, 44 (1990).
6. Wang, S. L., Sekerka, R. F., Wheeler, A. A., Murray, B. T., Coriell, S. R., Braun, R. J. and McFadden, G. B.: Physica D 69, 189 (1993).
7. Caginalp, G. and Jones, J.: Annals of Physics 237, 66 (1995).
8. Wheeler, A. A., Boettinger, W. J. and McFadden, G. B.: Phys. Rev. E 47, 1893 (1993).
9. Conti, M.: Phys. Rev. E 56, 3197 (1997).
10. Ahmad, N. A., Wheeler, A. A., Boettinger, W. J. and McFadden, G. B.: Phys. Rev. E 58, 3436 (1998).
11. Zhiqiang Bi, Robert F. Sekerka, Physica A 261, 95 (1998).
12. Caginalp, G. and Jones, J.: in *IMA Volumes on Mathematics and its applications*, edited by Gurtin, M. E. and McFadden, G. B. 29-50, (1991).
13. Anderson, D. M., McFadden, G. B. and Wheeler, A. A.: Physica D 135, 175 (2000).
14. Anderson, D. M., McFadden, G. B. and Wheeler, A. A.: Physica D 151, 305 (2001).
15. Conti, M.: Phys. Rev. E 64, 051601 (2001).
16. Warren, J. A. and Boettinger, W. J.: Acta Metall. Mater. 43, 689 (1995).

A Lattice Boltzmann Method for the Mesoscopic Calculation of Anisotropic Crystal Growth

Wolfram Miller[1] and Igor Rasin[1,2]

[1] Institut für Kristallzüchtung (IKZ), Max-Born-Str.2, 12489 Berlin, Germany
[2] Brandburgische Technische Universität Cottbus

Abstract. We developed a new kind of a phase-field model in the framework of the lattice Boltzmann method. The latter is used to calculate the convection und heat transport. In the continuum limit of second order the new lattice phase-field model converges towards the ordinary continuum phase-field model. The calculation of dendritic growth in two dimensions shows the amenibility of the new method.

1 Introduction

In industry as well as in nature, crystal growth processes are often strongly influenced by convection. In addition, the convection can exhibit patterns on rather small scale, sometimes convection is even turbulent. During the last years the lattice Boltzmann method has been evolved as a powerful numerical tool for flow calculations [12,16], especially regarding its efficiency on parallel platforms [8,11]. This was the motivation to develop a new kind of lattice phase-field model for the lattice Boltzmann method [3,10]. In this paper we will focus on the description of the lattice phase-field model and its application to dendritic growth. Details of the lattice Boltzmann method can be found e.g. in [9,12,7].

2 Lattice Phase-field Model

In analogy to the continuum models for the phase transition we define a continous order parameter $-1 \leq \phi \leq +1$, which takes the value +1 for melt and -1 for solid. The order parameter is defined on a lattice point $\mathbf{r}_*$ at a discrete time t_*. Its evolution is given by a simple rate equation:

$$\phi(\mathbf{r}_*, t_* + 1) = \phi(\mathbf{r}_*, t_*) + \mathcal{R}(\mathbf{r}_*, t_*). \tag{1}$$

The rate $\mathcal{R}$ is given by the probability for melting and solidifcation, K^+ and K^-, respectively. The timescale for the phase transition is defined by a frequency factor f.

$$\mathcal{R}(\mathbf{r}_*, t_*) = f\{K^+(\mathbf{r}_*, t_*)[1 - \phi(\mathbf{r}_*, t_*)] - K^-(\mathbf{r}_*, t_*)[1 + \phi(\mathbf{r}_*, t_*)]\}. \tag{2}$$

For the probabilities we make the following approachs:

$$K^{\pm} = \frac{1}{2}\left(1 \pm \tanh\left\{\frac{(T' \mp T_s)}{T_w}\right\}\right)\sum_i \left(1 \pm \phi(\mathbf{r}_* + \mathbf{c}_i')\right)^2 \tag{3}$$

for model I and

$$K^{\pm} = \frac{1}{2}\left(1 \pm \tanh\left\{\frac{(T' \mp T_s)}{T_w}\right\}\right)\sum_i \left(1 \pm \phi(\mathbf{r}_*)\right)^2 \tag{4}$$

for model II. In model II we have an additional non-local term in (2):

$$f\delta^2 \sum_i \phi(\mathbf{r}_* + \mathbf{c}_i') - \phi(\mathbf{r}_*) \tag{5}$$

It can be shown that in the continuum limit of second order modell II converges towards the continuum phase-field equation

$$\frac{\partial \phi_s}{\partial t} = \mu_k \Gamma \left[\nabla^2 \phi_s - \frac{\phi_s(1-\phi_s)(1-2\phi_s)}{\delta^2}\right] + \mu_k(T_c - T)\frac{\phi_s(1-\phi_s)}{\delta}, \tag{6}$$

whereas model I includes additional first order derivatives in the phase-field ϕ [9]. $\phi_s = 1/2(1-\phi)$ is the solid fraction, μ_k the kinetic coefficient, Γ the Gibbs-Thomson coefficient, and 6δ is the width of the transition region $(-0.9 \leq \phi \leq 0.9)$.

The release of latent heat is included by a temperature change at the nodes:

$$\delta T_{\mathcal{L}}(\mathbf{r}_*) = -(\Delta T_*/\mathrm{St})(\delta\phi(\mathbf{r}_*)/2), \tag{7}$$

where St is the Stefan number or undercooling $\mathrm{St} = c_p \Delta T/L$, ΔT_* and ΔT are the characteristic temperature differences of the system in lattice and physical units, c_p and L are the specific and latent heat, respectively.
We want to mention that all entities are measured in lattice units (l.u.). The length scale is measured in units of lattice distances, the time scale in number of iterations, the temperature scale can be chosen arbitrarily with the restriction $|T| < 0.2$. The relevant energy scale is defined via 7.
Often, the surface energy of a crystal differs in the various spatial directions. This anisotropy can be considered in the phasefield model by introducing a dependency on the direction into the length and time scale, δ and f, respectively. The scales are modified in the following way:

$$\delta = \delta_0(1 + \epsilon \cos n\Theta), \qquad f = f_0(1 + \epsilon \cos n\Theta)^2.$$

Θ is the angle between the crystal axes and the growth direction and n defines the symmetry of the crystal. In this article we consider a fourfold symmetry, i.e. $n = 4$. Some additional modifications we have to made to our models in order to compute the growth direction. For model I these modifications are described in [9], for model II they will presented in a forthcoming article.

3 Results

We have studied the growth of a small circular seed into an undercooled melt. The problem is two-dimensional with a fixed temperature at the boundaries. The problem is characterized by a length, time and energy scale, represented by the capillary length number Ca, the Damköhler number Da and the Stefan number St. The definitions are $\mathrm{Ca} := d_0/(\sqrt{2}\delta)$, $\mathrm{Da} := \Gamma\mu_k/D$, and $\mathrm{St} = \Delta T c_p/L$. The relations in terms of the lattice parameters are: $\mathrm{Ca} = T_w \mathrm{St}/\sqrt{2}\Delta T_*$ and $\mathrm{Da} = \delta^2 f/2D_*$. The width of the transition region δ is an input parameter in model II but in model I it can be only measured after a run. In the case of anisotropic growth, the strength of anisotropy ϵ enters the list of characteristic parameters. Firstly, we present a result obtained with model I. The capillary length, the Damköhler, and the Stefan number were $\mathrm{Ca} = 3.5 \times 10^{-3}$, $\mathrm{Da} = 0.2$, and $\mathrm{St} = 0.1$, respectively. Choosing an anisotropy of $\epsilon = 0.22$ we obtained a dendritic growth. In the first run the crystal axes coincided with the lattice directions, in the second run they were tilted by 18°. The resulting dendrite was the same in both cases (see. Fig. 1) [10]. If we switch on thermal convection and define the gravity in y-direction,

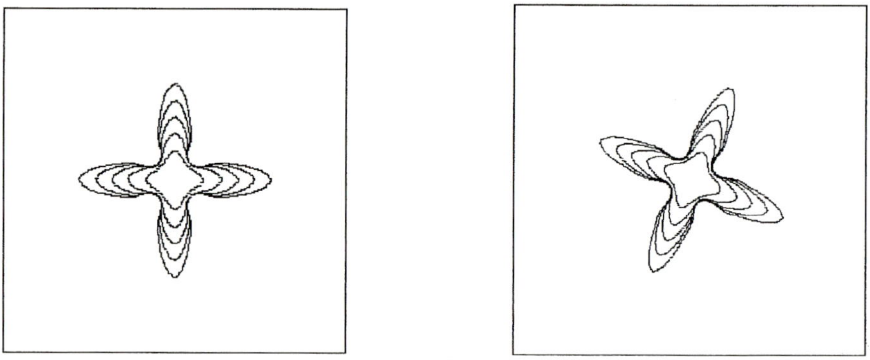

Fig. 1. Dendritic growth computed with model I. Interfaces ($\phi = 0$) are shown for different timesteps (4000, 8000, 12000, 16000, and 20000 iterations). In the figure on the right the crystal axes were tilted by 18° according to the lattice.

the symmetry is broken and the resulting crystals will not be the same for the tilted and the untilted case. In Fig. 2 the shape of the dendrite and the convection field is shown for the case of a tilted crystal. The runs were performed on a 200×200 lattice. The melt had a Prandtl number of $\mathrm{Pr} = 0.1$ and the Rayleigh number was $\mathrm{Ra} = 1 \times 10^5$. The Prandtl number is the ratio of viscous to thermal diffusivity. Due to the better heat transport, the dendrite is growing is faster in the direction of upstreaming flow. Our observations are consistent with those of Bänsch and Schmidt [1]. Other authors have considered cases of dendritic growth in a forced flow, either with a single dendrite [14] or an array of dendrites on a surface [15]. With model II we

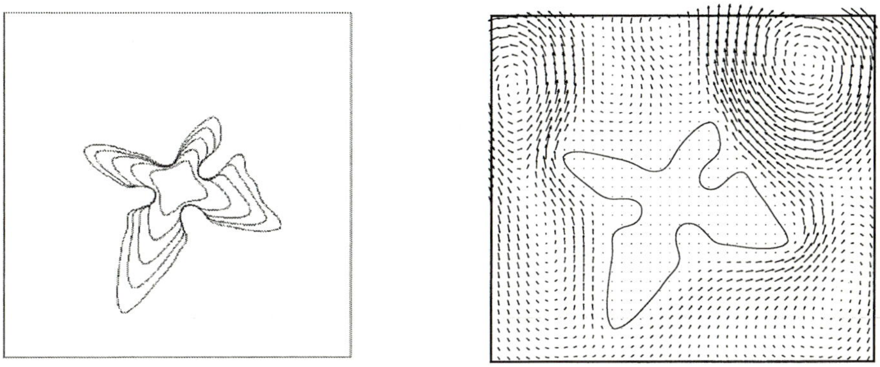

Fig. 2. Dendritic growth computed with model I. The thermal convection is induced by the heated crystal. On the lefthand side the interface ($\phi = 0$) is shown for different timesteps (4000, 8000, 12000, 16000, and 20000 iterations) and righthand side the velocity field in the vicinity of the crystal (iteration 15000) is represented by arrows.

have calculated the dendritic with a parameter set of the same order as in the paper of Tong et al. [13]. In particular, we have chosen Ca = 0.1, Da = 0.063, St = 0.5, and $\epsilon = 0.05$. The resulting dendrite at iteration 5000 is shown in Fig. 3. The computational domain was 400×400. If we increase the Damköhler number by increasing the frequency for the phase transition the dendrite is growing faster with a smaller tip radius (Fig. 4). The tip radius is evaluated numerically by interpolation with a parabola. On the lefthand side of Fig. 4 the accuracy of this approach is chown in comparison with the real shape of the tip. We measured a tip radius of ρ_{tip}=2.2 l.u. and a tip velocity of v_{tip}= 0.033 l.u., which yield a tip Peclét number of Pe=0.145. The tip Peclét number is defined via Pe $= \rho_{\text{tip}} v_{\text{tip}}/2D_*$ and according to the linear solvability theory of Ivantsov the Peclét number depends on the undercooling only [6,4]. Nevertherless, more eleborate solvability theories and numerical calculations show an decreasing Peclét number with increasing isotropy [5]. The Peclét numbers of our computations are in good agreement with the oberservations of Karma and Rappel [5]. Another point of discussion is the shape of the tip. In an experiment with xenon Bisang and Bilgram observed a tip shape, which deviates significantly from a parabola. Instead of an exponent of 2 the shape was fitted well by a power law with an exponent of 1.67 [2]. We did not find such an deviation in our calculations. The parabola fits the shape of the tip quite well as it can be seen in Fig. 4.

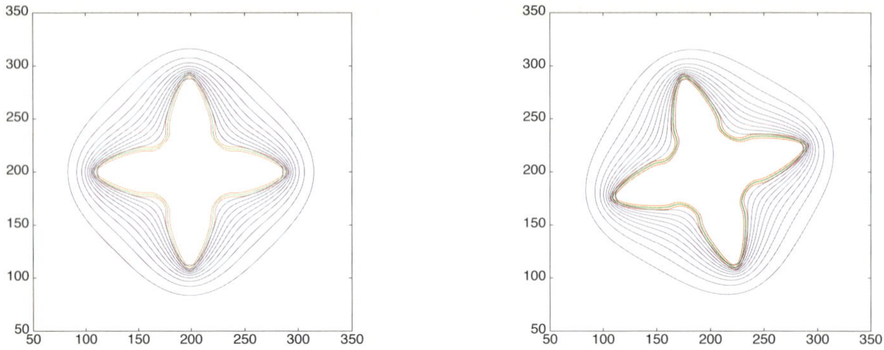

Fig. 3. Dendritic growth computed with model II. Isolines of phase-field ($\phi = -0.9, 0.0, 0.9$) and temperature isolines for iteration 5000. The righthand figure presents the result, when crystal and lattice axes are tilted by 17.1^o.

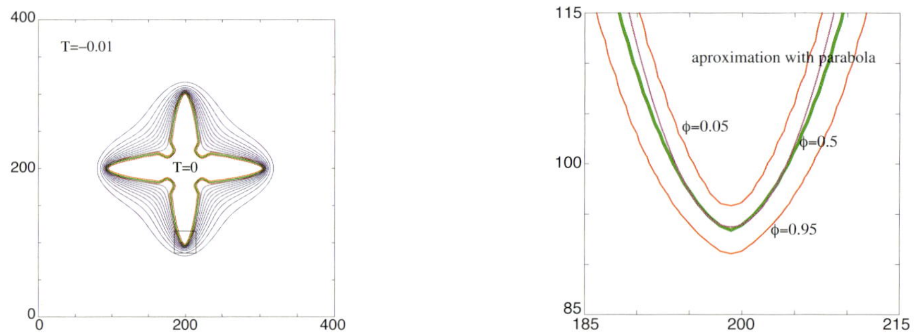

Fig. 4. Dendritic growth computed with model II. Isolines of phase-field ($\phi = -0.9, 0.0, 0.9$) and temperature isolines for iteration 3000 (left). Magnification of the lower tip region and comparison of the tip interface with a theoretical parabola curve (right).

4 Conclusions

We have presented a lattice version of the phasefield model, which is embedded in the lattice Boltzmann method for the calculation of convection and heat transport. We showed its amenability to calculate dendritic growth in two dimensions without exhausting much computer time. This result is encouraging to develope an efficient parallel version of the programm for large scale simulations in the future. This will also include calculations of growth and flow in three dimensions.

Acknowlegement

Part of the model has been jointly developed with Sauro Succi. Financial support from ESF-PESC Scientific Programme "Challenges in molecular simulations: bridging the length and time scale gap" is kindly acknowledged.

References

1. Bänsch, E. and Schmidt, A.: Simulation of dendritic crystal growth with thermal convection, Interfaces and Free Boundaries **2** (2000), 95–115.
2. Bisang, U. and Bilgram, J. H.:, Dendritic solidification of xenon, J. Crystal Growth **166** (1996), 212–216.
3. de Fabritiis, G., Mancini, A., Mansutti, D. and Succi, S.: Mesoscopic models of liquid/solid phase transitions, Int. J. Mod. Phys. C **9** (1998), 1405–1415.
4. Ivantsov, G. P.: Dokl. Akad. Nauk SSSR **58** (1947).
5. Karma, A. and Rappel, W. -J.: Quantitative phase-field modeling of dendritic growth in two and three dimensions, Phys. Rev. E **57** (1998), 4323–4349.
6. Langer, J. S.: Instabilities and pattern formation in crystal growth, Rev. Mod. Phys. **52** (1980), 1–28.
7. Massaioli, F., Succi, S. and Benzi, R.: Exponential tails in Rayleigh-Bénard convection, Europhys. Lett. **21** (1993), 305.
8. Miller, W. and Succi, S.: Parallel three-dimensional lattice Boltzmann hydrodynamics on the IBM 9076-SP1 and SP2 scalable parallel computers, in Computational Fluid Dynamics '96, Désidéri, J. -A., Hirsch, C., Tallec, P., Pandolfi, M. and Périaux, J.: eds., ECCOMAS, John Wiley & Sons, 1996, 1052–1058.
9. ——, A lattice Boltzmann model for anisotropic crystal growth from melt, J. Stat. Phys. **107** (2002), 173–186.
10. Miller, W., Succi, S. and Mansutti, D.:, A lattice Boltzmann model for anisotropic liquid/solid phase transition, Phys. Rev. Lett. **86** (2001), 3578–3581.
11. Punzo, G., Massaioli, F. and Succi, S.: High-resolution lattice-Boltzmann computing: on the SP1 scalable parallel computer, Comput. Phys. **5** (1994), 1–7.
12. S. Succi The lattice Boltzmann Equation, Numerical Mathematics and Scientific Computation, Clarendon Press, Oxford, 2001.
13. Tong, X., Beckermann, C. and Karma, A.: Velocity and shape selection of dendritic crystals in a forced flow, Phys. Rev. E **61** (2000), R49–R52.
14. Tong, X., Beckermann, C., Karma, A. and Li, Q.:, Phase-field simulations of dendritic growth in a forced flow, Phys. Rev. E **63** (2001), p. 061601.
15. Tönhardt, R. and Amberg, G.:, Dendritic growth of randomly oriented nuclei in a shear flow, J. Crystal Growth **213** (2000), 161–187.
16. Wolf-Gladrow, D. A.:, Lattice-Gas cellular automata and lattice Boltzmann models, vol. 1725 of Lecture Notes in Mathematics, Springer-Verlag, Berlin Heidelberg New York, 2000.

A Phase-field Model for the Solidification Process in Multicomponent Alloys

H. Garcke[1], B. Nestler[2] and B. Stinner[1]

[1] Naturwissenschaftliche Fakultät I – Mathematik, Universiät Regensburg
[2] Fachbereich Informatik, FH Karlsruhe

Abstract. Our aim is to describe phase transitions in a system of an arbitrary number of components and phases. Based on a gradient flow for the entropy (including surface anisotropy) we propose a phase field model that can be regarded as an extension of the Penrose-Fife model and that is thermodynamically consistent. By formal asymptotic expansions we see that the considered domain splits into several phases. We define the surface entropies on the phase boundaries and then we can show that in the limit the model satisfies the Gibbs–Thomson relation and other conditions known from classical sharp interface models. Finally, some possibilities to linearize the equations are outlined.

1 Solidification Effects and Length Scales

The solidification process in multicomponent alloys is charactarized by effects that occur on several length and time scales and that influence each other. On a large length scale castings reveal a separation into several grains involving various kinds of grain boundaries. This separation into grains is combined with a complex microstructure. Examples are given in Fig. 1.

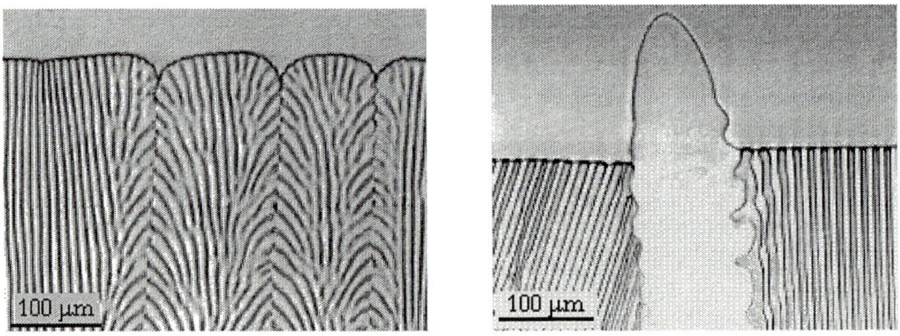

Fig. 1. Examples for microstructures occuring during the solidification process of alloys

On the first image we can see a dendrite combined with lamellar eutectic structures growing into a liquid melt. The second image shows growing eutectic grains/cells, so called eutectic colonies. The effective growth laws of the solid–liquid front which is influenced by "macroscopic" variables (as e.g. the temperature) are of interest as well as the influence of the microstructure formation on such macroscopic variables. Modelling a solidification process means therefore that one has to take into consideration the different scales as well as their interactions.

Mathematically, phase field models can be used to avoid topological difficulties with the motion of free boundaries in sharp interface models; by the technique of matched asymptotic expansions such a diffuse interface model can be related to sharp interface models. Our goal is to develop a phase field model for alloy systems with multiple components and phases with the following properties:

- consistency with the second law of thermodynamics, i.e. there is an entropy inequality and the total entropy grows,
- the model is related to classical sharp interface models for systems with multiple components and phases,
- only the knowledge of the bulk free energies of the occuring phases, the surface free energies of the phase transitions and some mobility and diffusion coefficients is necessary to derive the governing equations,
- earlier results obtained for phase field models are generalized, e.g.
  - by Caginalp ([2] and [3]) for the limiting free boundary problem,
  - by Penrose, Fife [14], Alt, Pawlow [1] and Wang, Sekerka, Wheeler, Murray, Coriell, Braun, McFadden [17] for thermodynamically consistent models,
  - by Wheeler, Boettinger, McFadden [18] and Caginalp, Xie [4] for binary alloys,
  - by Steinbach, Pezolla, Nestler, Seeßelberg, Prieler, Schmitz, Rezende [15] and Garcke, Nestler, Stoth [7] for systems with multiple phases,
  - by Wheeler, Boettinger, McFadden [19] for eutectic systems, by Lo, Karma, Plapp [10] for peritectic systems and by Nestler, Wheeler [11] for both.

In the following section we present the model in its general form. The corresponding sharp interface model is described in the third section. In the last section we will present some linearized models.

2 Description of the Model

The model is based on an entropy functional of the form

$$S(e,c,\phi) = \int_\Omega \left(s(e,c,\phi) - (\varepsilon a(\phi,\nabla\phi) + \frac{1}{\varepsilon} w(\phi)) \right) dx. \tag{1}$$

The unknown variables are the internal energy density e, the concentrations of the N components c_i, $1 \leq i \leq N$, and the phase field variables $\phi = (\phi_\alpha)_{\alpha=1}^M$

(ϕ_α gives the fraction of phase α). For the concentrations and the phase field variables the following constraints have to be fulfilled:

$$\sum_{i=1}^{N} c_i = 1, \quad \sum_{\alpha=1}^{M} \phi_\alpha = 1, \tag{2}$$

i.e. $c \in \Sigma^N$ and $\phi \in \Sigma^M$ where $\Sigma^K := \{d \in \mathbb{R}^K : \sum_{k=1}^{K} d_k = 1\}$.

The entropy density and the internal energy density are given in terms of the free energy density f by the fundamental thermodynamic relations $s = -f_{,T}$ and $e = f + Ts$ respectively where T is the temperature. Knowing the free energies $f^\alpha(T, c)$ of the pure phases we obtain the total free energy as a suitable interpolation, i.e. $f(T, c, \mathbf{e}_\alpha) = f^\alpha(T, c)$ where $\mathbf{e}_\alpha$ is the α'th unit coordinate vector. As an example we obtain the ideal solution free energy density

$$\begin{aligned} f(T, c, \phi) &:= \sum_{\alpha=1}^{M} \sum_{i=1}^{N} c_i L_i^\alpha \left(\frac{T - T_i^\alpha}{T_i^\alpha} \right) h(\phi_\alpha) \\ &\quad + \sum_{\alpha=1}^{M} \left(kT \sum_{i=1}^{N} (c_i \ln(c_i)) - c_v T(\ln(T) - 1) \right) \phi_\alpha . \end{aligned}$$

The terms a and w are introduced to model surface contribution to the entropy; possible choices are:

$$a(\phi, \nabla\phi) = \sum_{\alpha<\beta} \frac{\tilde{\gamma}_{\alpha,\beta}}{\tilde{m}_{\alpha,\beta}} |\phi_\alpha \nabla\phi_\beta - \phi_\beta \nabla\phi_\alpha|^2, \tag{3}$$

$$w_{st}(\phi) = 9 \sum_{\alpha<\beta} \tilde{m}_{\alpha,\beta} \tilde{\gamma}_{\alpha,\beta}\, \phi_\alpha^2 \phi_\beta^2 . \tag{4}$$

In [8], an obstacle potential was introduced which allows for a simple calibration between coefficients in the phase field model on the one hand and the surface free energies and mobility coefficients of the sharp interface model on the other hand.

The governing set of equations consists of energy balance, mass balance and a diffuse Gibbs–Thomson equation:

$$\partial_t e = -\nabla \cdot J_0, \tag{5}$$

$$\partial_t c_i = -\nabla \cdot J_i, \quad 1 \le i \le N, \tag{6}$$

$$\omega \, \varepsilon \, \partial_t \phi_\alpha = \frac{\delta S}{\delta \phi_\alpha} - \lambda, \quad 1 \le \alpha \le M, \tag{7}$$

where ω is a constant and λ is an appropriate Lagrange multiplier such that the constraint (2) for the phase field variable is satisfied. The variation of the entropy with respect to ϕ is

$$\frac{\delta S}{\delta \phi} = \varepsilon \left(\nabla \cdot a_{,X}(\phi, \nabla\phi) - a_{,\phi}(\phi, \nabla\phi) \right) - \frac{1}{\varepsilon} w_{,\phi}(\phi) - \frac{f_{,\phi}(T, c, \phi)}{T} \tag{8}$$

and the fluxes $J_0, \ldots, J_N$ for the conserved quantities are given by the phenomenological equations (see [9])

$$J_0 = L_{00}(T, c, \phi) \nabla \frac{1}{T} + \sum_{j=1}^{N} L_{0j}(T, c, \phi) \nabla \frac{-\mu_j}{T}, \tag{9}$$

$$J_i = L_{i0}(T, c, \phi) \nabla \frac{1}{T} + \sum_{j=1}^{N} L_{ij}(T, c, \phi) \nabla \frac{-\mu_j}{T}. \tag{10}$$

Observe that there is the thermodynamic relation

$$ds = \frac{1}{T} de + \sum_i \frac{-\mu_i}{T} dc_i + \sum_\alpha \frac{-f_{,\phi_\alpha}}{T} d\phi_\alpha$$

where $\mu_j = f_{,c_j}$ is the chemical potential of the component $j \in \{1, \ldots, N\}$. By Onsager's postulate (see [12] and [13]) every thermodynamic flux is given by a linear combination of the thermodynamic forces which are here given by $\nabla \frac{1}{T}, \nabla \frac{-\mu_1}{T}, \ldots, \nabla \frac{-\mu_N}{T}$.

The matrix of mobility coefficients $L = (L_{ij})_{i,j=0}^{N}$ is assumed to positiv semi–definite and symmetric (Onsager relations). For the first constraint of (2) to be satisfied we assume

$$\sum_{i=1}^{N} L_{ij} = 0, \quad 0 \leq j \leq N. \tag{11}$$

In [5] we derive an entropy inequality for this model, hence it is consistent with the second law of thermodynamics. For closed systems we show that the total entropy cannot decrease in time.

3 The Related Sharp Interface Model

By formally matched asymptotic expansions it is shown in [5] that the above set of equations is related to the following sharp interface model:

In every phase $q \in \{1, \ldots, M\}$ there are the evolution equations for energy and mass:

$$\partial_t e^q = -\nabla \cdot \left(L^q_{00}(T^q, c^q) \nabla \frac{1}{T} + \sum_{j=1}^{N} L^q_{0j}(T^q, c^q) \nabla \frac{-\mu^q_j}{T^q} \right), \tag{12}$$

$$\partial_t c^q_i = -\nabla \cdot \left(L^q_{i0}(T^q, c^q) \nabla \frac{1}{T} + \sum_{j=1}^{N} L^q_{ij}(T^q, c^q) \nabla \frac{-\mu^q_j}{T^q} \right) \quad \forall i; \tag{13}$$

we obtain the terms with the upper index q by setting $\phi = \mathbf{e}_q$, the q'th prinzipal coordinate vector, in the correspondings terms of the phase field model.

On a (smooth) boundary Γ between two phases α and β the temperature and the generalized chemical potential differences are continuous, there are jump conditions for the energy and the mass, and the Gibbs–Thomson relation must be satisfied:

$$T^\alpha = T^\beta =: T, \tag{14}$$

$$\bar{\mu}_i^\alpha = \bar{\mu}_i^\beta =: \bar{\mu}_i \quad \forall i, \tag{15}$$

$$[e]_\alpha^\beta \, v = [J_0]_\alpha^\beta \cdot \nu, \tag{16}$$

$$[c_i]_\alpha^\beta \, v = [J_i]_\alpha^\beta \cdot \nu \quad \forall i, \tag{17}$$

$$m_{\alpha,\beta} v = \gamma_{\alpha,\beta}\kappa + \frac{[f]_\alpha^\beta - \sum_i \bar{\mu}_i \, [c_i]_\alpha^\beta}{T}. \tag{18}$$

Here, $\nu = \nu_{\alpha,\beta}$ is the unit normal pointing into β, v is the speed of Γ in this direction, κ the mean curvature and

$$\bar{\mu}_i^q = \mu_i^q - \frac{1}{N}\sum_{j=1}^N \mu_j^q = \frac{1}{N}\sum_{j=1}^N (\mu_i^q - \mu_j^q) \text{ where } \mu_i^q = f^q_{,c_i}(T,c) \tag{19}$$

are the generalized chemical potential differences. $\gamma_{\alpha,\beta}$ is the surface entropy; as observed in [16] in the isotripic case and later more generally in [7] the relation between the surface entropy and the phase field model is given by

$$\gamma_{\alpha,\beta} = \inf_p \left\{ 2\int_{-1}^1 \sqrt{w(p)}\sqrt{a(p, p' \otimes \nu)} \right\} \tag{20}$$

where the infimum is taken over all Lipschitz continuous functions p connecting the mimima of w which correspond to the phases adjacent to the interface, i.e. $p(-1) = \mathbf{e}_\alpha$ and $p(1) = \mathbf{e}_\beta$. A function $\tilde{p} : \mathrm{I\!R} \to \Sigma^M$ is a solution of the (suitably reparametrized) Euler–Lagrange equation of (20) if and only if

$$0 = -P^M \left(\partial_z (a_{,X}(\tilde{p}, \partial_z \tilde{p} \otimes \nu))\nu + a_{,\phi}(\tilde{p}, \partial_z \tilde{p} \otimes \nu) + w_{,\phi}(\tilde{p}) \right). \tag{21}$$

Here, P^M is the projection of $\mathrm{I\!R}^M$ onto the tangential space $T\Sigma^M = \{d' \in \mathrm{I\!R}^M : \sum_{k=1}^M d'_k = 0\}$ of Σ^M. The boundary conditions for $\tilde{p}$ are

$$\tilde{p}(z) \stackrel{z \to \infty}{\longrightarrow} e_\beta, \quad \tilde{p}(z) \stackrel{z \to -\infty}{\longrightarrow} e_\alpha. \tag{22}$$

The above Euler–Lagrange equation is exactly the equation to lowest order obtained from the ϕ-equation of the phase field system after expanding in the (diffuse) phase transition, and the boundary conditions are the matching conditions with the outer expansion in the adjacent bulk phases. Another representation of the surface entropy is

$$\gamma_{\alpha,\beta} = \int_{-\infty}^{\infty} (a(\tilde{p}, \partial_z \tilde{p} \otimes \nu) + w(\tilde{p}))dz. \tag{23}$$

Finally, $m_{\alpha,\beta}$ is a mobility coefficient which is also given in terms of that mimimizer p respectivly $\tilde{p}$:

$$m_{\alpha,\beta} = \int_{-\infty}^{\infty} |\partial_z \tilde{p}|^2 dz. \tag{24}$$

4 Examples

It can easily be verified that the above model reduces to the Penrose–Fife model [14] if there is only a pure substance (i.e. $N = 1$) and if there are two phases (i.e. $M = 2$). In this case the system for $(T, \phi) := (T, \phi_2 - \phi_1)$ consists of the equations derived by Penrose and Fife.

Various simplifications of the model are possible by partially linearizing the full set of equations. We will give an example for a binary alloy with three phases (i.e. $M = 3$, $N = 2$). We set $c = c_1$, then $c_2 = 1 - c$ holds.

By a suitable choice of the diffusion coefficients it is possible to linearize the balance equations. Set

$$L_{00} := (e_{,c})^2 D \frac{T}{f_{,cc}} + kT^2, \quad L_{01} = L_{10} := e_{,c} D \frac{T}{f_{,cc}}, \quad L_{11} := D \frac{T}{f_{,cc}}.$$

This yields the equations

$$\partial_t e = \nabla \cdot \left(k \nabla T + e_{,c} D \nabla c + e_{,c} D \frac{f_{,c\phi}}{f_{,cc}} \nabla \phi \right), \tag{25}$$

$$\partial_t c = \nabla \cdot \left(D \nabla c + D \frac{f_{,c\phi}}{f_{,cc}} \nabla \phi \right). \tag{26}$$

If $e_{,c}$ and $c_v = e_{,T}$ are constant then in a pure phase (i.e. $\nabla \phi = 0$) we obtain the simple equations

$$\begin{aligned} c_v \partial_t T &= k \Delta T \quad \text{(Fourier's law)}, \\ \partial_t c &= D \Delta c \quad \text{(Fick's law)}. \end{aligned}$$

We remark that (26) becomes linear even across a phase transition (where $\nabla \phi \neq 0$) if we choose f such that $f_{,c\phi}/f_{,cc}$ is constant.

Finally let us consider a case in which the internal energy does not depend on c. More precisely we set

$$\begin{aligned} f(T, c, \phi) := \sum_{\alpha=1}^{3} L^\alpha \left(c \frac{T - T_1^\alpha}{T_1^\alpha} + (1 - c) \frac{T - T_2^\alpha}{T_2^\alpha} \right) \phi_\alpha \\ + kT(c \ln(c) + (1 - c) \ln(1 - c)) - c_v T(\ln(T) - 1). \end{aligned}$$

In addition we take the gradient term $a(\phi, \nabla \phi) := \frac{1}{2} \sum_\alpha |\nabla \phi_\alpha|^2$, we choose w to be the standard multi well potential, and let $\omega = 1$, $L_{00} := T^2$, $L_{10} = L_{01} := 0$, $L_{11} := D\, c(1 - c)$. Approximating the nonlinear term $\frac{1}{T}$ by

$$\frac{1}{T} \approx \frac{1}{T_2^\alpha} - \left(\frac{1}{T_2^\alpha} \right)^2 (T - T_2^\alpha).$$

the governing equations become

$$\partial_t e = \partial_t (c_v T - \sum_\alpha L^\alpha \phi_\alpha) = \Delta T,$$

$$\partial_t c = kD\,\Delta c + \nabla \cdot D\, c(1-c)\nabla \left(\sum_\alpha \left(\frac{1}{T_1^\alpha} - \frac{1}{T_2^\alpha} \right) L^\alpha \phi_\alpha \right),$$

$$\varepsilon \partial_t \phi_\alpha = \varepsilon \Delta \phi_\alpha - \frac{1}{\varepsilon} w_{,\phi_\alpha}(\phi) \\ - L^\alpha c \left(\frac{1}{T_1^\alpha} - \frac{1}{T_2^\alpha} \right) - L^\alpha \left(\frac{1}{T_2^\alpha} \right)^2 (T - T_2^\alpha) - \lambda.$$

This set of equations can be interpreted as a generalization of the phase model studied by Caginalp in [3] which can be obtained from the Penrose–Fife model by a similar linearization as already done.

5 Further Generalizations

Observe that in this paper the internal energy contains no constributions by surface terms. The surface entropy density of an interface is independent of the temperature, hence its surface free energy density is linear in T so that the internal energy density of the interface vanishes. In a forthcoming paper we will study more general dependences of the surface free energy density on the temperature.

Furthermore, the influence of the components will be taken into consideration. This will be done via a dependence of the surface free energy density on the generalized chemical potential differences $\bar{\mu}_i$ which are continuous and hence well defined on an interface.

References

1. H. W. Alt and I. Pawlow: Models of non–isothermal phase transitions in multicomponent systems. Part I: Theory, SFB256 University Bonn, Preprint 222 (1992)
2. G. Caginalp: Stefan and Hele–Shaw type models as asymptotic limits of the phase–field equations, Phys. Rev. A **39**, 5997 (1989)
3. G. Caginalp: An analysis of a phase field model of a free boundary, Arch. Rational Mech. Anal., **92**, 205-245 (1986)
4. G. Caginalp and W. Xie: Phase–field and sharp–interface alloy models, Phys. Rev. E **48**, 1897 (1993)
5. H. Garcke, B. Nestler and B. Stinner: A diffuse interface model for alloys with multiple components and phases, in preparation for submission to Phys. Rev. Lett.
6. H. Garcke, B. Nestler and B. Stinner: ..., in preparation for submission to ...
7. H. Garcke, B. Nestler and B. Stoth: On anisotropic order parameter models for multi–phase systems and their sharp interface limit, Physica D **115**, 87-108 (1998)

8. H. Garcke, B. Nestler and B. Stoth: A multi phase field concept: numerical simulations of moving phase boundaries and multiple junctions, SIAM J. Appl. Math. **60**, No.1, 295-315 (1999)
9. J. S. Kirkaldy and D. J. Young: Diffusion in condensed state, The Institute of Metals, London (1987)
10. T. Lo, A. Karma, M. Plapp: Phase-field modeling of microstructural pattern formation during directional solidification of peritectic alloys without morphological instability, Phys. Rev. E. **63**, 031504 (2001)
11. B. Nestler and A. A. Wheeler: A multi phase field model of eutectic and peritectic alloys: numerical simulation of growth structures, Physica D **138**, 114 (2000)
12. L. Onsager: Reciprocal relations in irreversible processes I, Phys. Rev. **37**, 405 (1931)
13. L. Onsager: Reciprocal relations in irreversible processes II, Phys. Rev. **38**, 2265 (1931)
14. O. Penrose and P. C. Fife: Thermodynamically consistent models of phase–field type for the kinetics of phase transitions, Physica D **43**, 44 (1990)
15. I. Steinbach, F. Pezolla, B. Nestler, M. Seeßelberg, R. Prieler, G. Schmitz, J. Rezende: A phase field concept for multiphase systems, Physica D **94**, 135-147 (1996)
16. P. Sternberg: Vector–valued local minimizers of nonconvex variational problems, Rocky Mt. J. Math. **21** No. 2, 799-807 (1991)
17. S-L. Wang, R. F. Sekerka, A. A. Wheeler, B. T. Murray, S. R. Coriell, R. J. Braun, G. B. McFadden: Thermodynamically–consistent phase–field models for solidification, Physica D **69**, 189-200 (1993)
18. A. A. Wheeler, W. J. Boettinger and G. B. McFadden: Phase–field model for isothermal phase transitions in binary alloys, Phys. Rev. A **45**, 7424 (1992)
19. A. A. Wheeler, W. J. Boettinger and G. B. McFadden: Phase–field model for solidification of a eutectic alloy, Proc. R. Soc. Lond. A **452**, 495-525 (1996)

Planar Solidification from Undercooled Melt: An Approximation of a Dilute Binary Alloy for a Phase-field Model

Denis Danilov

Laboratory of Condensed Matter Physics, Physical Faculty, Udmurt State University, 1 Universitetskaya Street, Izhevsk 426034, Russia
Email: ddanilov@uni.udm.ru

Abstract. Planar solidification from an undercooled melt has been considered using the phase-field model. The solute and the phase fields have been found in the limit of small impurity concentration. These solutions in the limit of vanishing velocity of the interface motion give the equilibrium partition coefficient and the liquidus slope. Asymptotic expansions for the solute and for the phase fields and the relation between the diffusive speed and the parameters of the phase field model have been found at high growth velocity. A comparison with numerical calculations is presented.

1 Introduction

Classical macroscopic models of solidification are based on the assumption that the interface between regions of different phases have zero thickness [1]. Sharp-interface descriptions require the introduction of separately derived nonequilibrium models for the behavior of the interfacial temperature and of the solute concentrations. In contrast to the sharp-interface models, the phase-field models [2,3] describe the bulk phases as well as the interface, i.e. these models treat the system as a whole and eliminate the need to specify the interfacial conditions separately. In phase-fields models, equilibrium behavior is recovered at low growth velocities and nonequilibrium effects naturally emerge at high growth velocities. From a computational point of view, the advantage of the phase-field formulation is that the interface is not tracked but is given implicitly by the value of the phase-field variable. However, theoretical analysis solutions were obtained by matching of separately derived "inner" (near the interface) and "outer" (far from the interface) solutions in an intermediate region [4,5]. In the work [6], phase-fields and solute profiles have been obtained without using the multiply-variable expansion, which leads to "inner" and "outer" solutions, but the profiles were obtained from a simplified set of equations in which the interface kinetics was eliminated.

In this work an analytical solution of the phase-filed model for planar solidification from an undercooled melt as an expansion in terms of the solute concentration far from the interface is presented. This solution is valid for the bulk phases as well as for the interface and includes the interface kinetics. To verify the expansion, numerical simulations of the problem were carried out.

2 The Model

Let us consider the motion of a planar solid-liquid interface during the solidification of a binary alloy. The governing equations for the phase-field $\phi(x,t)$ and for the solute concentration $c(x,t)$ are given by the following set of equations [2]

$$\frac{\partial \phi}{\partial t} = M_1 \left(\epsilon^2 \frac{\partial^2 \phi}{\partial x^2} - \frac{\partial f}{\partial \phi} \right), \qquad \frac{\partial c}{\partial t} = \frac{\partial}{\partial x} M_2 c(1-c) \frac{\partial}{\partial x} \frac{\partial f}{\partial c}, \tag{1}$$

which satisfy the conservation of solute and which have been derived under the assumption that the total free-energy decreases monotonically in time. The parameters M_1, M_2, and ϵ are related to the growth kinetics, to the diffusion coefficient, and to the surface energy respectively.

In the approach of ideal solution, the free-energy density $f(\phi, c, T)$ at a temperature T is given by [2]

$$f(\phi, c, T) = c f_B(\phi, T) + (1-c) f_A(\phi, T) + \frac{RT}{v_m} [c \ln c + (1-c) \ln(1-c)], \tag{2}$$

where R is the universal gas constant, v_m is the molar volume, which is assumed to be constant. Free-energy densities of the pure materials A and B are assumed to be of the form [2]

$$f_i(\phi, T) = W_i \int_0^\phi p(p-1) \left(p - \frac{1}{2} - \beta_i(T) \right) dp = \frac{W_i}{4} g(\phi) + \frac{W_i \beta_i(T)}{6} p(\phi) \tag{3}$$

where

$$g(\phi) = \phi^4 - 2\phi^3 + \phi^2, \quad p(\phi) = \phi^2(3 - 2\phi), \qquad \frac{W_i \beta_i(T)}{6} = L_i \frac{T - T_i}{T_i}. \tag{4}$$

Here W_i is a constant, L_i is the latent heat per unit volume, T_i is the melting point of material $i = A, B$.

To investigate steady-state interface motion, the boundary conditions will be taken in the form

$$c|_{x \to \infty} = c_\infty, \quad \phi|_{x \to \infty} = 0, \qquad \frac{\partial c}{\partial x}|_{x \to -\infty} = 0, \quad \phi|_{x \to -\infty} = 1, \tag{5}$$

and the moving frame z will be adopted according to the equation

$$z = x - Vt, \tag{6}$$

so that the interface (given by $\phi = 1/2$) corresponds to $z = 0$ at this frame. In the moving frame, equations (1) may be written by using (2) as

$$\epsilon^2 \frac{\partial^2 \phi}{\partial z^2} + \frac{V}{M_1} \frac{\partial \phi}{\partial z} - \left[\frac{\partial f_A}{\partial \phi} + c \left(\frac{\partial f_B}{\partial \phi} - \frac{\partial f_A}{\partial \phi} \right) \right] = 0, \tag{7}$$

$$\frac{M_2RT}{v_m}\frac{\partial^2 c}{\partial z^2}+V\frac{\partial c}{\partial z}+M_2\frac{\partial}{\partial z}\left[c(1-c)\left(\frac{\partial f_B}{\partial \phi}-\frac{\partial f_A}{\partial \phi}\right)\frac{\partial \phi}{\partial z}\right]=0. \tag{8}$$

For further analysis we introduce the notation for the diffusion coefficient $D = M_2RT/v_m$ and integrate equation (8) with respect to z over an interval $[z, \infty)$. Employing the boundary conditions (5) gives

$$D\frac{\partial c}{\partial z}+V(c-c_\infty)+M_2c(1-c)\left(\frac{\partial f_B}{\partial \phi}-\frac{\partial f_A}{\partial \phi}\right)\frac{\partial \phi}{\partial z}=0. \tag{9}$$

3 An Approximation of a Small Impurity Concentration

Let us examine a dilute binary alloy with initial impurity concentration $c_\infty \ll 1$. For this purpose we shall consider the expansion in terms c_∞ for the phase-field, the solute concentration, and the temperature

$$\phi(z)=\phi_0(z)+c_\infty\phi_1(z)+O(c_\infty^2), \tag{10}$$
$$c(z)=c_\infty c_1(z)+O(c_\infty^2), \tag{11}$$
$$T=T_0+c_\infty T_1+O(c_\infty^2), \tag{12}$$

As a result, in the zeroth-order approximation, from (7) by using (3) we obtain the equation for $\phi_0(z)$

$$\epsilon^2\frac{\partial^2\phi_0}{\partial z^2}+\frac{V}{M_1}\frac{\partial\phi_0}{\partial z}-\left(\frac{W_A}{4}\left.\frac{\partial g}{\partial\phi}\right|_{\phi_0}+\frac{W_A\beta_A}{6}\left.\frac{\partial p}{\partial\phi}\right|_{\phi_0}\right)=0, \tag{13}$$

The solution of (13), which satisfies the boundary conditions (5), describes the phase field during the solidification of a pure material A and has the form

$$\phi_0(z)=\frac{1}{2}\left[1-\tanh\left(\frac{z}{2l_A}\right)\right], \tag{14}$$

where l_A is the interface thickness and the interface velocity V is related to the temperature T_0 by the relationship

$$V=-M_1l_AW_A\beta_A(T_0)=\frac{6M_1l_AL_A}{T_A}(T_A-T_0), \tag{15}$$

which determines the interface kinetics for a pure material. The relationship (15) corresponds to the model of collision limited growth [7] with a kinetic coefficient

$$\mu=\frac{6M_1l_AL_A}{T_A},\quad T_0=T_A-\frac{V}{\mu}. \tag{16}$$

In the first-order approximation, the equations (7) and (9) give

$$\epsilon^2 \frac{\partial^2 \phi_1}{\partial z^2} + \frac{V}{M_1}\frac{\partial \phi_1}{\partial z} - \left[\phi_1 \left.\frac{\partial^2 f_A}{\partial \phi^2}\right|_{\phi_0,T_0} + T_1 \left.\frac{\partial^2 f_A}{\partial T \partial \phi}\right|_{\phi_0,T_0} + c_1 \left.\left(\frac{\partial f_B}{\partial \phi} - \frac{\partial f_A}{\partial \phi}\right)\right|_{\phi_0,T_0} \right] = 0, \quad (17)$$

$$\frac{\partial c_1}{\partial z} + \frac{V}{D}(c_1 - 1) + c_1 \frac{\partial \Delta F}{\partial z} = 0, \quad (18)$$

where function $\Delta F(z)$ is defined by expression

$$\Delta F(z) = \frac{v_m}{RT_0}\left[f_B(\phi_0(z), T_0) - f_A(\phi_0(z), T_0)\right]. \quad (19)$$

Taking into account the boundary conditions (5), the solution of (18) is

$$c_1(z) = \frac{V}{D}\int_{-\infty}^{z} e^{V(z'-z)/D + \Delta F(z') - \Delta F(z)} dz'. \quad (20)$$

The point $z = 0$ corresponds to the solid-liquid interface, i. e. $\phi(0) = \phi_0(0) = 1/2$, and to the inflection of the phase field, therefore

$$\phi_1|_{z=0} = 0, \qquad \left.\frac{\partial^2 \phi_1}{\partial z^2}\right|_{z=0} = 0. \quad (21)$$

As for metallic systems $M_1 \sim 10^8\,\mathrm{cm}^3/\mathrm{J{\cdot}s}$ [2], then the relation $V/M_1 \ll 1$ will be satisfied. Taking into account the conditions (21) at $z = 0$, one can find from (17) the relationship between T_1 and c_1

$$T_1 = -\frac{\left.\left(\frac{\partial f_B}{\partial \phi} - \frac{\partial f_A}{\partial \phi}\right)\right|_{\phi_0,T_0}}{\left.\frac{\partial^2 f_A}{\partial T \partial \phi}\right|_{\phi_0,T_0}} c_1|_{z=0}. \quad (22)$$

Equations (11), (12), (16) and (22) give the expression for the temperature

$$T = T_A - \frac{V}{\mu} - m(V)c|_{z=0}, \quad (23)$$

where the function $m(V)$ is defined by

$$m(V) = \frac{\left.\left(\frac{\partial f_B}{\partial \phi} - \frac{\partial f_A}{\partial \phi}\right)\right|_{\phi_0,T_0}}{\left.\frac{\partial^2 f_A}{\partial T \partial \phi}\right|_{\phi_0,T_0}} = \left.\frac{W_B \beta_B(T) - W_A \beta_A(T)}{W_A \frac{\partial \beta_A}{\partial T}}\right|_{T = T_A - V/\mu} \quad (24)$$

4 Stationary Interface

Now we consider the solute field due to the stationary solid-liquid interface, at $V = 0$, which corresponds to the equilibrium state. Let us introduce the definitions for the bulk concentrations in the liquid $c_L = c(z \to \infty)$ and in the solid $c_S = c(z \to -\infty)$. For a stationary interface, first it follows from (16) that $T_0 = T_A$, and second the equation (18) gives the solute field

$$\frac{c(z)}{c_L} = e^{-\Delta F(z)}, \tag{25}$$

which leads to the equilibrium partition coefficient

$$k_e = \frac{c_S}{c_L} = \exp[-\Delta F(z \to -\infty)] = \exp\left(-\frac{v_m L_B (T_A - T_B)}{R T_A T_B}\right). \tag{26}$$

This expression has the same form as in [6], where the equilibrium partition coefficient has been obtained from a simplified set of equations in which the effects of interface attachment kinetics are eliminated.

Using (4), from equation (24) at $T_0 = T_A$ we obtain

$$m_0 = m(0) = \frac{L_B}{L_A}\frac{T_A}{T_B}(T_A - T_B). \tag{27}$$

The solute concentration at the interface $z = 0$, in accordance with (25), is

$$c|_{z=0} = c_L \exp\left(-\frac{v_m}{R T_A}\left[\frac{W_B - W_A}{64} + \frac{1}{2}\frac{L_B (T_A - T_B)}{T_B}\right]\right). \tag{28}$$

Thus equations (23), (27), and (28) give the temperature at the equilibrium solid-liquid interface

$$T = T_A - m_0 c|_{z=0} = T_A - m_0 \exp\left(-\frac{v_m}{R T_A}\left[\frac{W_B - W_A}{64} + \frac{1}{2}\frac{L_B (T_A - T_B)}{T_B}\right]\right) c_L = T_A - m_e c_L, \tag{29}$$

which yields the following equation for equilibrium liquidus slope

$$m_e = \frac{L_B}{L_A}\frac{T_A}{T_B}(T_A - T_B) \exp\left(-\frac{v_m}{R T_A}\left[\frac{W_B - W_A}{64} + \frac{L_B (T_A - T_B)}{2 T_B}\right]\right). \tag{30}$$

5 Large-Velocity Asymptotics

To investigate the behavior of the solute concentration under rapid solidification conditions, we consider an asymptotic analysis in the limit $V \gg 1$. The solution (20) can be written in the form

$$c_1(z) = \frac{V}{D} \int_0^{\infty} e^{-Vz'/D + \Delta F(z-z') - \Delta F(z)} dz'. \tag{31}$$

Applying the Laplace's theorem about asymptotic expansions [8] to this integral, we obtain the solute field with an accuracy to second order

$$c_1(z) = \frac{c(z)}{c_\infty} = 1 - \frac{D}{V}\frac{\partial \Delta F}{\partial z}. \tag{32}$$

According to (32), a maximum value of solute is reached at $z = 0$. Therefore the segregation coefficient at large velocities is given by

$$\begin{aligned} k = \frac{c_\infty}{c|_{z=0}} = \frac{1}{1 - \frac{D}{V}\left.\frac{\partial \Delta F}{\partial z}\right|_{z=0}} &= 1 + \frac{D}{V}\left.\frac{\partial \Delta F}{\partial z}\right|_{z=0} \\ &= 1 - \frac{D}{4 l_A V}\frac{v_m}{RT}\left.\left(\frac{\partial f_B}{\partial \phi} - \frac{\partial f_A}{\partial \phi}\right)\right|_{\phi=1/2}. \end{aligned} \tag{33}$$

For the sharp-interface model the nonequilibrium segregation coefficient has been derived by Aziz [9] and in the limit of dilute alloy is

$$k^{(a)} = \frac{k_e + V/V_D}{1 + V/V_D}, \tag{34}$$

where diffusive speed V_D is a characteristic trapping velocity. For $V/V_D \gg 1$ the segregation coefficient $k^{(a)}$ can be approximated by

$$k^{(a)} = 1 - (1 - k_e)\frac{V_D}{V}. \tag{35}$$

A comparison of the equations (33) and (35) gives

$$V_D = \frac{D}{4 l_A (1 - k_e)}\frac{v_m}{RT}\left.\left(\frac{\partial f_B}{\partial \phi} - \frac{\partial f_A}{\partial \phi}\right)\right|_{\phi=1/2, T_0} \tag{36}$$

According to (36), the diffusive speed V_D depends on the diffusion coefficient, on the temperature, and on the difference of free energy densities of the pure materials.

6 Numerical Calculations

To validate the obtained solution, we will compare it with numerical solutions of the steady-state governing equations (7) and (8). The material parameters used in the numerical calculations are given in Table 1. These parameters are similar to parameters employed in [6] and correspond to the alloy Ni–Cu [2].

Table 1. Thermophysical properties used in calculations

Parameter	Value		Reference
L_A	2350	J/cm^2	[6]
L_B	1725	J/cm^2	[6]
T_A	1728	K	[6]
T_B	1358	K	[6]
D	10^{-5}	cm^2/ñ	[6]
σ	2.8×10^{-5}	J/cm^2	[6]
l_A	6.48×10^{-8}	cm	[6]
v_m	7.4	cm^3/mol	[2]
μ	24	cm/(s·K)	[10]

The value of kinetic coefficient was set to $\mu = 24$ cm/(s·K) according to the work [10] in which good agreements between the model for dendritic growth and experimental data have been achieved. The far-field concentration was set to $c_\infty = 0.1$. The parameter ϵ is related to the surface energy σ and to the interface thickness l_A by equation $\epsilon^2 = 6\sigma l_A$ [2,6]. The parameters W_A and W_B were chosen to be equal and are given as $W_A = W_B = W = 12\sigma/l_A$ [6].

Employing (3) and (4), a finite-difference approximation of the equations (7) and (8) on a uniform grid gives the following set of equations

$$\frac{\phi_i^{k+1} - \phi_i^k}{\Delta t} = \frac{V}{M_1 \epsilon} \frac{\phi_{i+1}^k - \phi_{i-1}^k}{2\Delta x} + \frac{\phi_{i+1}^k - 2\phi_i^k + \phi_{i-1}^k}{\Delta x^2} - \frac{W}{4}\left(\frac{\partial g}{\partial \phi}\right)_i^k - \frac{W(c\beta_B + (1-c)\beta_A)}{6}\left(\frac{\partial p}{\partial \phi}\right)_i^k, \quad (37)$$

$$\frac{c_i^{k+1} - c_i^k}{\Delta t} = \frac{V}{M_1 \epsilon} \frac{c_{i+1}^k - c_{i-1}^k}{2\Delta x} + \frac{D}{M_1 \epsilon^2} \frac{c_{i+1}^k - 2c_i^k + c_{i-1}^k}{\Delta x^2} + \frac{M_2}{M_1 \epsilon^2} \frac{q_{i+1}^k - q_{i-1}^k}{2\Delta x}, \quad (38)$$

$$q_i^k = c_i^k(1 - c_i^k)\left[\left(\frac{\partial f_B}{\partial \phi}\right)_i^k - \left(\frac{\partial f_A}{\partial \phi}\right)_i^k\right] \frac{\phi_{i+1}^k - \phi_{i-1}^k}{2\Delta x}, \quad (39)$$

which determine the discrete functions of the solute and of the phase field at a point $z = i\Delta x$ on a $(k+1)$th step of iteration with a time step Δt. To equalize the velocity V of the moving frame with the velocity of the interface, we introduce the equation

$$\tau_f \frac{dV(\tau)}{d\tau} = V_s(\tau) = M_1 \epsilon \frac{dz_f}{d\tau}, \quad (40)$$

where $V_s(\tau)$ is the shift velocity of the interface relative to the moving frame, τ_f is an accessory parameter, τ is a time variable corresponding to the iteration process, and the interface coordinate z_f is given by the condition $\phi(z_f) = 1/2$. A finite-difference approximation of equation (40) gives

$$\frac{V^{k+1} - V^k}{\Delta t} = \frac{\epsilon M_1}{\tau_f} \frac{z_f^{k+1} - z_f^k}{\Delta t}. \tag{41}$$

Results of numerical calculations of the equations (37)–(41) are shown in the following figures. Fig. 1 shows a good agreement between the analytical solution [equations (11), (20), (23), and (24)] and the numerical results in the wide range of undercoolings.

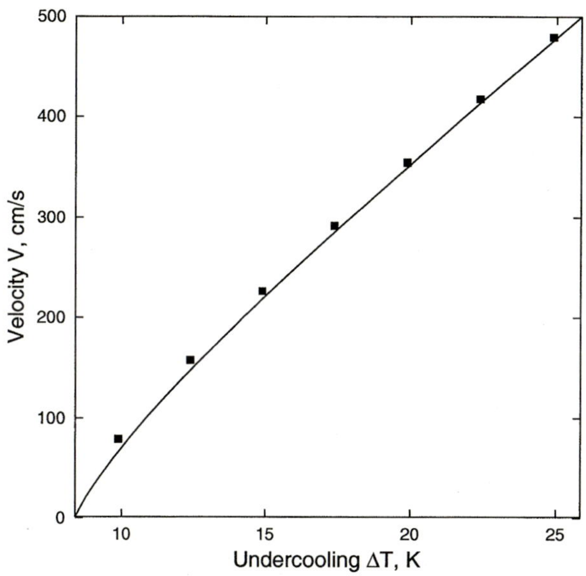

Fig. 1. Kinetic curve "interface velocity V — initial undercooling ΔT". The solid curve shows the approximated solution, data points denote the result of the numerical calculations.

Fig. 2 shows the solute and the phase fields. At the interface, the solute field is smooth as against sharp-interface models, which gives a jump in concentration. At low undercoolings, the analytical solution gives some excessive value of solute concentration in the vicinities of the interface. The series of numerical calculations given in Fig. 2 demonstrate the degeneration of the solute to uniform field with increasing interface velocity and undercooling.

Such behavior of the solute field is in accordance with the equation (32), in which the deviation from the far-field concentration is proportional to $1/V$.

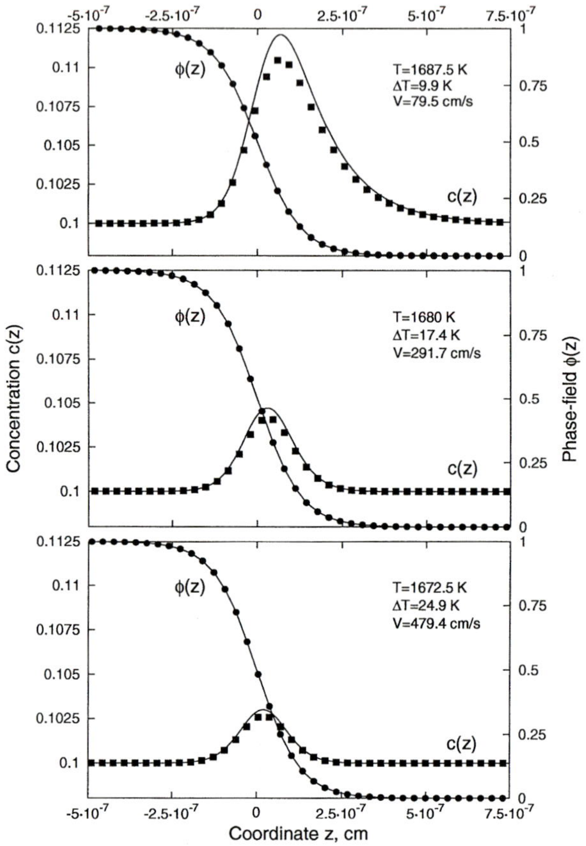

Fig. 2. Solute profiles and the phase field for different values of the initial undercooling. The solid curve shows the approximated solution, data points denote the result of numerical calculations.

7 Conclusions

Planar solidification from undercooled melt was investigated by numerical and analytical methods in the limit of small impurity concentration. Comparing the obtained results with those of [6] we note that while equilibrium partition coefficient (26) has the same form as in the work [6], the expression (30) for the liquidus slope differs from the one given in the work [6]. However the evaluated equilibrium liquidus slope has very close values, $m_e = 306.2$ K from equation (30) and $m_e = 306.9$ K from equation (A13) in [6]. The diffusive speed has a weak dependence on the temperature, Eq. (36), and can be accepted to be constant as well as the diffusive speed obtained in [6]. The numerical calculations presented here show a good agreement with the analytical solution that indicate a sufficient accuracy of the expansion (10)–(12).

The obtained solutions simultaneously are valid for the bulk phases and the interface. Therefore they can be suitable for the purpose of numerical simulations: for instance to construct initial splitting of adaptive finite-element grid and as a first approximation for the numerical solution of transcendental equations.

References

1. Kurz W., Fisher D. J.: Fundamentals of Solidification, 3rd ed. Aedermannsdorf: Trans Tech Publication, (1992) 305.
2. Wheeler A. A., Boettinger W. J., McFadden G. B.: Phase-field model for isothermal phase transitions in binary alloys. Phys. Rev. A **45** (1992) 7424–7438.
3. Bi Z., Sekerka R. F.: Phase-field model of solidification of a binary alloy. Physica A **261** (1998) 95–106.
4. Wheeler A. A., Boettinger W. J., McFadden G. B.: Phase-field model of solute trapping during solidification. Phys. Rev. E **47** (1993) 1893–1909.
5. Elder K.R., Grant M., Provatas N., Kosterlitz J. M.: Sharp interface limits of phase-field models. Phys. Rev. E **64** (2001) 021604.
6. Ahmad N. A., Wheeler A. A., Boettinger W. J., McFadden G. B.: Solute trapping and solute drag in a phase-filed model of rapid solidification. Phys. Rev. E **58** (1998) 3436–3450.
7. Turnbull D.: Metastable structures in metallurgy. Metall. Trans. A. **12** (1981) 695–708.
8. Olver F. W. J.: Asymptotics and Special Functions. New York: Academic Press, 1974. 584 p.
9. Aziz M. J.: Model for solute redistribution during rapid solidification. J. Appl. Phys. **53** (1982) 1158–1168.
10. Galenko P. K., Danilov D. A.: Model for free dendritic alloy growth under interfacial and bulk phase nonequilibrium conditions. J. Crystal Growth **197** (1999) 992–1002.

Initial Transients in the Symmetric Model for Directional Solidification

Raúl Benítez and Laureano Ramírez-Piscina

Departament de Física Aplicada. Universitat Politècnica de Catalunya. Av. Dr. Marañón 44, E-08028 Barcelona. Spain.

Abstract. We study the initial transient during the directional solidification of a dilute mixture in the symmetrical constant-gap approximation. We perform phase-field simulations of the transient recoil stages and compare the results with predictions obtained from the sharp-interface model. In particular, we focus in the evolution of the front position and of the transient dispersion relation, obtaining quantitative agreement between theory and simulations. Results are applied to the destabilization of the front by fluctuations.

1 Introduction

The selection of a dendritic pattern during the directional solidification of a dilute binary alloy is a complex problem which depends on initial conditions[1], and in particular on the first wavelengths that appear in the destabilization of the planar front induced by fluctuations[2,3]. Recent work has focused in the importance of internal fluctuations in solidification patterns [4,5]. Quantitative agreement with experiments has only been obtained for the solidification of pure substances, whereas in solutal growth the origin itself of the fluctuations is still an open problem [6,7]. The aim of this work is to use phase-field techniques to study the effects of transients and fluctuations in the selection problem of directional solidification. For this objective we have performed a quantitative comparison between phase-field simulations and predictions for the sharp-interface model.

2 The Sharp Interface Model

In a directional solidification experiment, a thermal gradient $G\hat{\mathbf{z}}$ is moved in the $\tilde{z}$ direction along the sample at constant pulling velocity $V\hat{\mathbf{z}}$. Provided the sample is thin, we take the system as 2D and describe the interface position in the moving frame of the gradient by $\tilde{z} = \tilde{\xi}(\tilde{\boldsymbol{\rho}}, \tilde{t})$, where $\tilde{\boldsymbol{\rho}} = \tilde{x}\hat{\mathbf{x}} + \tilde{y}\hat{\mathbf{y}}$. The solid phase is located in the region where $\tilde{z} < \tilde{\xi}$, and the liquid where $\tilde{z} > \tilde{\xi}$. We will consider the particular case of symmetric directional solidification, which assumes the same solute diffusivity D in both phases. We introduce diffusion length $l = D/V$ and time $\tau = D/V^2$ to scale variables as $\mathbf{r} = \tilde{\mathbf{r}}/l$ and $t = \tilde{t}/\tau$. We also introduce a diffusive field in each phase $u_i(\mathbf{r}, t) = \frac{C_i - C_\infty}{\Delta C_0}$ (i=1 solid, i=2 liquid), where C_i is the solute concentration, C_∞ the concentration far away from the solid-liquid interface, and $\Delta C_0 = [C_2 - C_1]_{int}$ the concentration jump across the interface. In the moving frame and in reduced variables, the fields u_i evolve according to the diffusion equation

$$(\frac{\partial}{\partial t} - \frac{\partial}{\partial z} - \nabla^2) u_i(\mathbf{r}, t) = 0. \tag{1}$$

The diffusion fields u_i must satisfy some moving boundary conditions at the interface position which can be written as

$$[u_1 - u_2]_{int} = -1 \tag{2}$$
$$u_2|_{int} = 1 - \tfrac{l}{l_T}\xi + \tfrac{d_0}{l}\kappa \tag{3}$$
$$\hat{\mathbf{n}} \cdot \left[\boldsymbol{\nabla} u_1 - \boldsymbol{\nabla} u_2\right]_{int} = -n_z(1 + \dot{\xi}) \tag{4}$$

Equations (1-4) define the so called sharp-interface description of the symmetric directional solidification problem. Equation (2) relates solute concentrations at both sides of the interface. We have assumed the additional approximation of having a constant concentration jump across the interface. This particular assumption is equivalent to suppose that the mixture has parallel solid and liquid branches in the $T(C)$ coexistence diagram, which is valid only for liquid crystals and alloys with a partition coefficient close to 1. Eq. (3) is the Gibbs-Thompson equation (local equilibrium at the interface). In this relation $l_T = |m_L|\Delta C_0/G$ is a thermal length imposed by the temperature gradient, and d_0 is the capillary length. Eq. (4) describes the solute conservation across the interface, and $\hat{\mathbf{n}}$ represents a normal unitary vector pointing to the liquid. Using Green's function techniques [8], it is possible to derive a closed integral expression for u_i at each side of the interface. Introducing the notation $p = (\boldsymbol{\rho}, z, t)$, $p_S = (\boldsymbol{\rho}, \xi, t)$, the concentration at the interface verifies

$$\begin{aligned} \tfrac{1}{2}u_i(p_S) = (-)^i \Big(& \textstyle\int_{\xi(\boldsymbol{\rho},t_0)}^{(-)^i\infty} d\mathbf{r}' \cdot G_i(p_S, \mathbf{r}', t_0) \cdot u_i(\mathbf{r}', t_0) \\ & - \textstyle\int_{t_0}^{t} dt' \int d\boldsymbol{\rho}' \cdot [1 + \dot{\xi}(\boldsymbol{\rho}', t')] \cdot u_i(p'_S) \cdot G(p_S, p'_S) \\ - \textstyle\int_{t_0}^{t} dt' \int dS' \hat{\mathbf{n}}' \cdot & [G_i(p_S, p'_S)\boldsymbol{\nabla}' u_i(p'_S) - u_i(p'_S)\boldsymbol{\nabla}' G_i(p_S, p_{S'})] \Big) \end{aligned} \tag{5}$$

where $G_i(p, p')$ is the retarded Green's function of the diffusion problem

$$G(p, p') = \frac{\theta(t - t')}{[4\pi(t - t')]^{\frac{3}{2}}} \exp\left\{ -\frac{(\boldsymbol{\rho} - \boldsymbol{\rho}')^2 + (z - z' + t - t')^2}{4(t - t')} \right\}. \tag{6}$$

Summing Eq. (5) over $i = 1, 2$, and using the moving boundary conditions Eqs. (2-4), an integro-differential equation can be obtained for the interface position during the transient

$$\frac{l}{l_T}\xi(\rho,t_0) = -\int_{-\infty}^{\xi(\rho,t_0)} d\mathbf{r}' G(p_s;\mathbf{r}',t_0)u_1(\mathbf{r}',t_0) -$$
$$\int_{\xi(\rho,t_0)}^{\infty} d\mathbf{r}' G(p_s;\mathbf{r}',t_0)u_2(\mathbf{r}',t_0) + \int_{t_0}^{t} dt' \int d\rho'(1+\dot{\xi}(\rho',t))G(p_s,p_s')$$
$$+\frac{1}{2}(1+erf(\frac{\xi(\rho,t)-\xi(\rho,t_0)+t}{2\sqrt{t}})) \tag{7}$$

where $\xi = 0$ stands for the steady position of an unperturbed planar interface. Note that Eq. (7) includes transients from the initial condition at t_0.

The next step is to perform a linear stability analysis of the problem to obtain a transient dispersion relation describing the time evolution of a sinusoidal modulation with wavevector kl. Within an adiabatic approximation [2,3] we derive the growth ratio of the mode as

$$\omega(kl,t) = \frac{l}{l_T}\dot{\xi} - (1+\dot{\xi})^2 + q_2(1+\dot{\xi}) - (q_1+q_2)\left[\frac{l}{l_T} + \frac{d_0}{l}(kl)^2\right], \tag{8}$$

where

$$q_\alpha = \tfrac{(-1)^\alpha}{2}(1+\dot{\xi}) + \sqrt{\tfrac{1}{4}(1+\dot{\xi})^2 + (kl)^2 + \omega(kl,t)}. \tag{9}$$

Therefore the prediction of the transient dispersion relation consists of two steps. Firstly the numerical resolution of Eq. (7) for a planar interface is performed by a Newton-Raphson method, yielding the transient front position $\xi(t)$. This function is then introduced into Eq. (8) to obtain $\omega(kl,t)$ in this adiabatic approximation.

3 The Phase-field Model

In the last few years, phase-field models have become a quantitative technique to numerically simulate solidification phenomena [9]. This method introduce a continue variable (the phase-field) $\phi(\mathbf{r},t)$ which takes different constant values in the solid and liquid phases and which localizes the solid-liquid interface in a diffuse region of thickness W. This field is coupled with the solute diffusion field, and the parameters are chosen in such a way that in the limit $\varepsilon \doteq W/l \to 0$ the interface dynamics recovers the behavior described by the sharp-interface model. We will use a phase-field model introduced by Karma et. al. [10] for symmetric directional alloy solidification with constant concentration gap. For isotropic solidification with no kinetic dynamics, the equations in reduced variables take the form

$$\tfrac{a_1 a_2}{d_0/l}\varepsilon^3\partial_t\phi = \varepsilon^2\nabla^2\phi + (1-\phi^2)(\phi - \varepsilon\tfrac{a_1}{d_0/l}(1-\phi^2)(u + \tfrac{z-t}{l_T/l})) \tag{10}$$
$$\partial_t u = \nabla^2 u + \tfrac{1}{2}\partial_t\phi. \tag{11}$$

In this model, $a_1 = \frac{5\sqrt{2}}{8}$, $a_2 = 0.6267$ are integral constants obtained when performing a thin-interface limit [9]. Note that the equation for the phase-field evolution contains three parameters: two main control parameters coming from the sharp-interface model (l_T/l and d_0/l), and the model-specific parameter ε.

4 Results and Discussion

We will consider the transient from the rest, i.e. for an initial condition at $t_0 = 0$ consisting of an equilibrium solid-liquid planar interface located at $\xi(t = 0) = l_T/l$. In this case $C_1(\mathbf{r}, 0) = C_\infty - \Delta C_0$ $(u_1(\mathbf{r}, 0) = -1)$, and $C_2(\mathbf{r}, 0) = C_\infty$ $(u_2(\mathbf{r}, 0) = 0)$.This corresponds to taking $u(\mathbf{r}, 0) = -1$ as initial condition for the phase-field model. Trying to mimic a real experiment, we consider here that at $t = 0$ the pulling velocity suddenly takes the final value V (1 in scaled variables).

We perform numerical integration of the phase-field Eqs. (10,11) with an explicit finite-difference scheme with $\Delta x = 0.8$ and $\Delta t = 0.08$. We first study the 1D dynamical evolution during the transient, comparing the results with the sharp-interface predictions. Fig. 1 presents the front position for two different values of the control parameter $l_T/l = 12.5$ and 2.5. For each case, we compare simulations for three different values of ε ($\varepsilon = 0.5$, $\varepsilon = 0.25$ and $\varepsilon = 0.125$) with the front position obtained with direct resolution of the integral equation (7). Convergence to the sharp-interface limit can be observed as ε decreases, and good agreement is found for a value of $\varepsilon = 0.125$. We now estimate the transient dispersion relation $\omega(kl, t)$ from phase-field

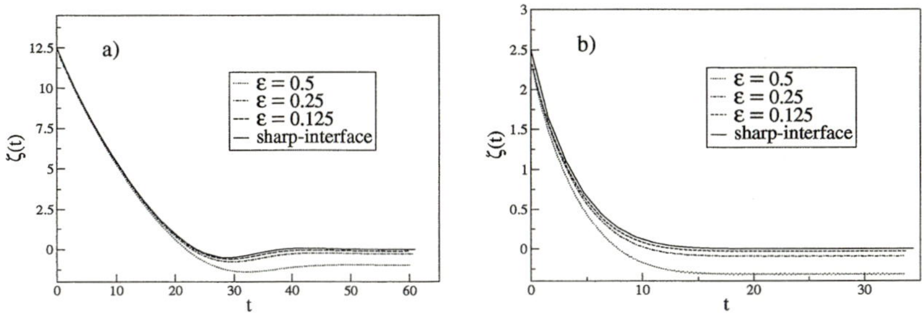

Fig. 1. Evolution of the interface position during the transient for (a) $l_T/l = 12.5$ and (b) $l_T/l = 2.5$, and convergence to the sharp-interface as $\varepsilon \to 0$.

simulations and compare it with the sharp interface prediction of Eqs. (8, 9). To this end we simulate for each desired value of t a (planar) 1D interface evolving from $t_0 = 0$ to t. At that moment we introduce a sinusoidal interface perturbation with wavevector kl, and continue the simulation in 2D. The spectral analysis of the front allows us to locate the regime where the mode evolution is linear. This is represented in Fig. 2a, where three definite regions can be observed. From the linear region the value of the transient growth rate $\omega(kl, t)$ is calculated. Fig. 2b shows the growth rate at different times obtained for two different modes $kl = 1.25$ and $kl = 2.5$ in the case of $l_T/l = 12.5$ and $d_0/l = 0.06923$. Quantitative agreement is observed for all times. In the last

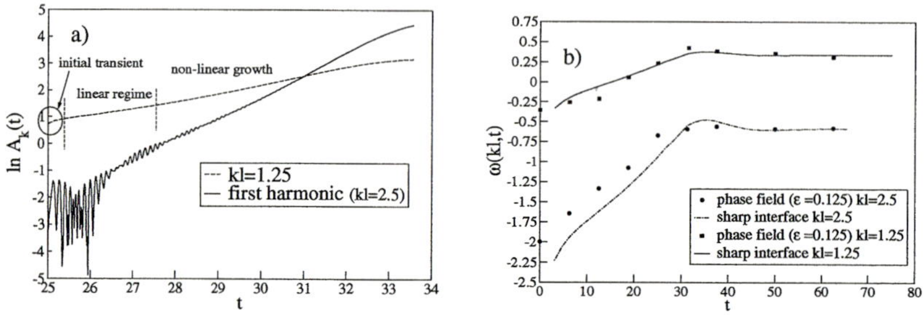

Fig. 2. a) Localization of the linear growth regime of the mode $kl = 1.25$. b) Time evolution of two different modes $kl = 1.25$ and $kl = 2.5$ for $l_T/l = 12.5$.

part of this work we have introduced internal fluctuations in the diffusion equation of the phase-field by following the formalism introduced in Ref. [11]. In simulations the growth of a range of wavelengths can be observed until a selected mode dominates. The spectral analysis of the front reveals a good agreement during the linear regime between the growth of each mode and the transient dispersion relation predicted by the sharp interface model. In Fig. 3 we present the amplitude of one of the modes ($kl = 2.5$, $l_T/l = 12.5$) averaged for 15 different noise realizations. The agreement is quite good in the regime with positive ω (i.e. where an increasing $A_k(t)$ is predicted).

5 Concluding Remarks

In this paper we have studied the initial stages of a directional solidification experiment in the context of the symmetric model. We have presented predictions for the transient recoil of the sharp interface model and for the transient dispersion relation by using an adiabatic approximation. We have performed phase field simulations of this initial regime. Simulations quantitatively agree with theoretical calculations. In particular the analysis of the evolution of single modes and of fronts with fluctuations, both from theory and from simulations, should be of relevance in the dendritic selection problem. Work in progress [12] will extend these results to the one-side model and to the Langevin formalism for the modes of Ref. [2].

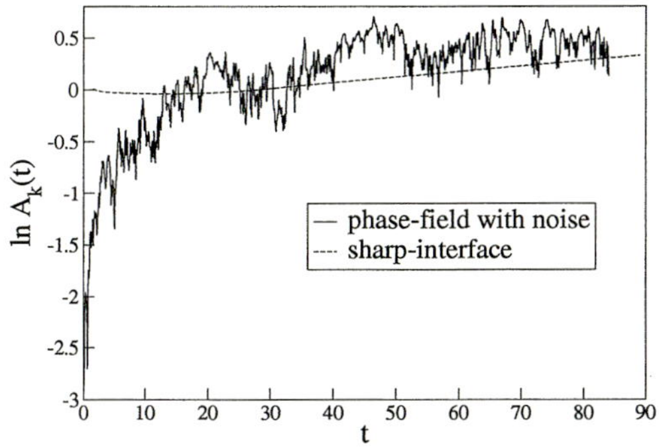

Fig. 3. Time evolution of the mode $kl = 2.5$ ($l_T/l = 12.5$) with the phase-field with noise, and comparison with the sharp-interface prediction.

We acknowledge financial support from Dirección General de Investigación Científica y Técnica (Spain) (Project BFM2000-0624-C03-02) and Comissionat per a Universitats i Recerca (Spain) (Project 2001SGR 00221).

References

1. J. A. Warren, J. S. Langer, Stability of dendritic spacings. Phys. Rev. E **42** (1990) 3518–3525.
2. J. A. Warren, J. S. Langer, Prediction of dendritic spacings in a directional-solidification experiment. Phys. Rev. E **47** (1993) 2701–2712.
3. B. Caroli, C. Caroli, L. Ramírez-Piscina, Initial front transients in directional solidification of thin samples of dilute alloys. J. Cryst. Growth **132** (1993) 377.
4. E. Brener, D. Temkin, Noise-induced sidebranching in the three-dimensional nonaxissymmetric dendritic growth. Phys. Rev. E **51** (1995) 351–359.
5. A. Pocheau, M. Georgelin, M.: Validation criterion for noise-induced mechanism of sidebranching in directional solidification. Eur. Phys. J. B. **21** (2001) 229–240.
6. W. Losert, B. Q. Shi, H. Z. Cummins, Proc. Natl. Acad. Sci. USA **95** (1998) 439–442.
7. R. González-Cinca, L. Ramírez-Piscina, J. Casademunt, A. Hernández-Machado, Noise-induced sidebranching in a phase-field model for solidification. Phys. Rev. E **63** (2001) 051602.
8. B. Caroli, C. Caroli, R. Roulet, On the linear stability of needle crystals: evolution of a Zel'dovich localized front deformation. J. Phys. France **48** (1987) 1423–1437.
9. A. Karma, W. J. Rappel, Quantitative phase-field modeling of dendritic growth in two and three dimensions. Phys. Rev. E **57** (1998) 4323–4349.
10. W. Losert, D. A. Stillman, H. Z. Cummins, P. Kopczynski, W. J. Rappel, A. Karma, Selection of doublet cellular patterns in directional solidification through spatially periodic perturbations. Phys. Rev. E **58** (1998) 7492–7506.
11. A. Karma, J. W. Rappel, Phase-field model of dendritic sidebranching with thermal noise. Phys. Rev. E **60** (1999) 3614–3625.
12. R. Benítez, L. Ramírez-Piscina, In preparation.

Dynamics of a Faceted Nematic-Smectic B Front in Thin-Sample Directional Solidification

Tamás Börzsönyi[1,2], Silvère Akamatsu[1], and Gabriel Faivre[1]

[1] Groupe de Physique des Solides, CNRS UMR 75-88, Universités Paris VI et VII, Tour 23, 2 place Jussieu, 75251 Paris Cedex 05, France
[2] Research Institute for Solid State Physics and Optics, Hungarian Academy of Sciences, H-1525 Budapest, P.O.B.49, Hungary

Abstract. We study the dynamics of directional-solidification fronts presenting facets in one direction. We focus on the phenomena occurring immediately above the Mullins-Sekerka instability threshold when the facet is tilted with respect to the growth direction. The experimental system we use is a nematic - smectic B front in the planar configuration. We observe 1.) drifting shallow cells, which we explain by means of a linear stability analysis including diffusive and kinetic anisotropies; 2.) a new type of oscillatory solitary wave, called "faceton". For a numerical approach, we use a phase field model with sharp capillary and kinetic anisotropies.

1 Introduction

Extensive studies have been devoted to the directional solidification (a thin sample is pulled with a constant velocity V in a fixed temperature gradient G) patterns of fully non-faceted systems, in which the surface tension γ and the kinetic coefficient β (defined as the ratio of the kinetic undercooling to the growth velocity) are non-singular functions of the orientation of the interface with respect to the crystal lattice. In contrast, little is yet known about the directional-solidification dynamics of faceted crystals. Here we study the dynamics of solidification fronts that are rough in all but one direction, and have no forbidden directions. In such a system, when the facet is tilted with respect to the isotherms, the dynamics of the front must obviously be that of a non-faceted crystal as long as the deformation of the front remains small, that is, below the cellular threshold V_c and in a small range of V above V_c. Facets only appear at higher V when the deformation of the interface is large [1,2].

The nematic-smectic B front of the liquid crystal CCH4 (MERCK) fulfils such conditions [3,4]. A long-range order exists in the smectic B phase perpendicularly to the molecular layers, so that this phase actually is a lamellar crystal. We use 12-μm thick samples in the planar configuration (director parallel to the plane of the sample), so that the front –including facets, if any– remains perpendicular to the sample plane. The observations are recorded with a polarizing microscope equipped with a CCD camera. The video signal is analyzed with digital image processing. Further details about our setup are given elsewhere [5,6].

2 Results

2.1 Experiments

The sequence of morphologies observed as a function of V for all (except 0° and 90°) values of the tilt angle θ_0 of the facet is illustrated in Fig. 1. Drifting shallow cells are observed from V_c ($\approx 2\ \mu ms^{-1}$) to about 8 μms^{-1}. In the same range of V, solitary waves called "facetons" consisting essentially of a drifting single facet are also observed. At higher V, arrays of narrow tilted faceted fingers showing various types of oscillatory behaviors are observed. We focus here on the morphologies observed near the threshold (drifting cells and facetons).

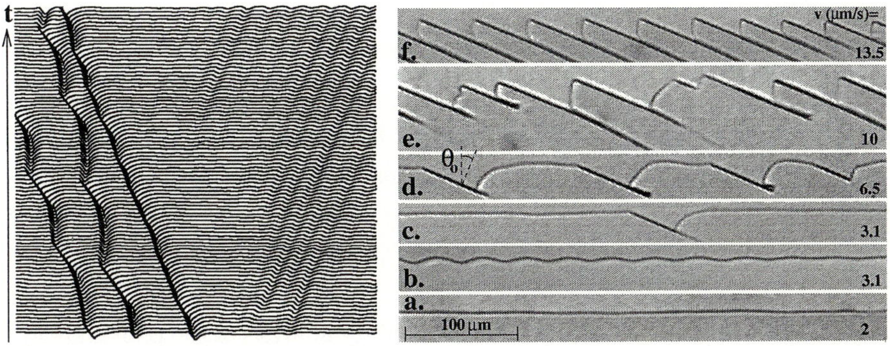

Fig. 1. Left: Spatiotemporal diagram showing the trajectories of three oscillatory facetons (on the left) and of shallow cells (on the right). $V = 6.5\mu ms^{-1}$. $\theta_0 = 55^o$. Right : Growth morphologies observed at the indicated values of V for $\theta_0 = 25^o$. (a) Planar front; (b) drifting shallow cells; (c, d) facetons; (e,f) arrays of faceted fingers. $G \approx 50\ Kcm^{-1}$.

The observed drifting cells certainly are Mullins-Sekerka cells. The fact that they drift indicates that the system has a strong diffusive or kinetic anisotropy [1,7]. The value of their drift velocity V_d depends on θ_o, as it should (see Fig. 2 below), but not on the cell amplitude. Facetons appear when the amplitude of the cells is large enough for their sides to be tilted to an angle larger than θ_0. This does not occur spontaneously in the considered range of V, but only under the effect of strong external perturbations (see ref. [6]). Once created, facetons are stable. They propagate through the whole sample erasing the shallow cells that exist in front of them, and leaving behind a planar front on which shallow cells slowly reappear. The total width of these localized objects is of about $200\mu m$. Their drift direction is globally opposite to that of the shallow cells, but, most frequently, their drift velocity is not constant. A close examination of the spatiotemporal diagrams shows that facetons oscillate between two different quasi-stationary states. The times of residence in these two states shows a great variability (the reason of this is not known), but the drift velocity of the facet in these states is well defined. In the so-called "blocked" state, the drift velocity is such that the normal growth rate of the facet is much lower than V, as revealed by the fact that its trajectory merges into its image. On the contrary, in the "unblocked" state, the dynamical behavior of the facet is essentially the same as that of the near-by rough regions of the front. Their drift velocity is then of an opposite sign, and of a much smaller absolute value, than in the blocked state.

Knowing that the growth rate of a facet is governed by the motion of molecular steps flowing out of specific sources [8,9], these observations strongly suggest that the transition between blocked and unblocked states corresponds to the making, or the break, of a contact with step sources. Observations give no direct information about the nature of the involved step sources. A possibility is that the steps flow from the rough region of the front contiguous to the colder edge of the facet when this region is of a concave shape, and stop to flow when it is of a convex shape. The blocked-to-unblocked transition would then be of a purely morphological nature (the lattice defects of the crystal would play no essential part in it). If this is true, it should be possible to reproduce the observed transition numerically with a phase field model.

2.2 Linear Stability Analysis

We have performed numerically a linear stability analysis of the planar steady state including simultaneously the anisotropy of β, and those of the diffusion coefficients in the nematic and in the smectic phase. The calculation is standard in its principle [2,7,10] but lengthy, and cannot be reproduced here.

The aim is to fit the calculated dependence of V_d on θ_o to the measured one, and hence to determine (or, at least, fix bounds to) the values of the anisotropy parameters in our system. Let us review the involved parameters. In the nematic (smectic) phase, the diffusion coefficient $D_{\parallel}$ ($D_{\parallel}^S$) parallel to the director is different from the diffusion coefficient $D_{\perp}$ ($D_{\perp}^S$) normal to the director. The anisotropy of the interfacial properties is characterized by the functions $\gamma(\theta - \theta_o)$ and $\beta(\theta - \theta_o)$, where θ is the angle between the normal to the interface and the growth axis $\mathbf{z}$. The stability analysis is performed at fixed value of θ_o. It involves the values of $D_{zz} = cos^2\theta_o D_{\parallel} + sin^2\theta_o D_{\perp}$, $D_{xx} = sin^2\theta_o D_{\parallel} + cos^2\theta_o D_{\perp}$, $D_{xz} = sin\theta_o cos\theta_o (D_{\parallel} - D_{\perp})$, and those of $\gamma + \gamma$" (surface stiffness), β and β' , where γ" $= d^2\gamma/d\theta^2$ and $\beta' = d\beta/d\theta$, at $\theta = 0$. The angular dependence of γ is known from previous experiments [4], but its actual value is not. We therefore took $\gamma(\theta_{ref})$, where θ_{ref} is an orientation of reference to be defined later, as an adjustable parameter. We have advanced previously qualitative arguments showing that diffusion anisotropy cannot be the main cause of the drift of the cells in our system [6]. In a first approach, we therefore ignored diffusion anisotropy. Two adjustable parameters (β and β') are then left for a given value of θ_o. These parameters of course cease to be independent when one imposes that the function $\beta(\theta)$ is smooth. Given the value of $\beta(\theta_{ref})$, the best-fit value of $\beta(\theta)$ is then uniquely determined. Calculations performed with various values of the diffusion anisotropy in the two bulk phases delivered values of $\beta(\theta)$ that were not substantially different from those obtained by setting this anisotropy to zero.

The measured values of $V_d(\theta_o)$ and the function $\beta(\theta)$ calculated numerically for the indicated values of the parameters are shown in Fig. 2. The fact that V_d goes to zero at $\theta_o = 0^o$ and 90° is in agreement with the symmetry of the system. Another zero of $\theta_o = 0^o$ that is not prescribed by any a priori argument, seems to exist at $\theta_o \approx 12^o$. We have chosen $\theta_{ref} = 12^o$, and imposed that $\beta'(\theta)_{ref} = 0$. Note that the calculated function $\beta(\theta)$ is an increasing function of θ for $\theta > 12^o$, and that the ratio of its maximum to its minimum value is large (about 3). The meaning of this as concerns the structure of the interface on a molecular scale is not obvious. The range $0^o < \theta < 12^o$, in which $\beta(\theta)$ is a (slightly) decreasing function is perhaps the vicinal range, but this is not certain. For consistency, it must be assumed that β has a discontinuity in $\theta = 0$ (facet devoid of steps), with $\beta(0) \gg \beta(0_\pm)$.

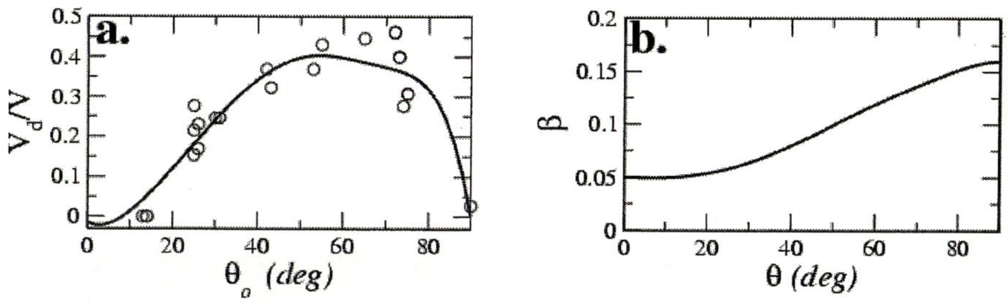

Fig. 2. (a) Symbols: Measured values of V_d/V for $V = 6.5\mu ms^{-1}$. Curve: calculated values of V_d/V_c for the values of β given in (b). (b) Calculated $\beta(\theta)$ curve for $\beta(12^o) = 0.05$, $d_o(12^o) = 0.001$, $K = 0.25$. $d_o = a_o G/\Delta T_o^2$ (a_o: Gibbs-Thomson capillary coefficient; ΔT_o: thermal gap of the alloy) is the reduced capillary length. a_o is proportional to $\gamma + \gamma$".

2.3 Phase-field Simulations

In this Section, we discuss the possibility of modeling the growth of our faceted system with the help of the phase field model proposed in [11,12]. The equations we use are:

$$\tau(\theta)(\partial_t - V\partial_z)\phi = (1-\phi^2)\left[\phi + \alpha(\phi^2-1)\left(u + \frac{z-z_0}{l_T}\right)\right] \quad (1)$$

$$+\nabla(W(\theta)^2\nabla\phi) + \partial_x\left(|\nabla\phi|^2 W(\theta)\frac{\partial W(\theta)}{\partial\phi_x}\right) + \partial_z\left(|\nabla\phi|^2 W(\theta)\frac{\partial W(\theta)}{\partial\phi_z}\right)$$

$$(\partial_t - V\partial_z)u = D\nabla^2 u \frac{1}{2}(\partial_t - V\partial_z)\phi \tag{2}$$

For the angular dependence of $W(\theta)$, we take $W(\theta) = W_0\ (1 + (W_a - 1)|sin(\theta)|$ where W_a denotes the anisotropy (W_{max}/W_{min}) of the surface tension. This function is close to the one previously determined experimentally [4]. As for the angular dependence of $\tau(\theta)$, we assumed that interfacial kinetics has very sharp minima (slow kinetics) corresponding to the directions of faceting, as explained above. This has been realized by using $\tau(\theta) = \tau_0(1 + c_1(1 - |sin(\theta)|)^{c_2})$

The $W(\theta)$ and $1/\tau(\theta)$ functions are shown in Fig. 3 for the parameters $W_a = 3$, $c_1 = 5 \cdot 10^6$ and $c_2 = 1000$. Numerical simulations have been carried out by using these functions in order to trace the dynamics of the front in the presence of a purely capillary anisotropy, and in the presence of a purely kinetic anisotropy. The initial state was a stationary planar front (at $V = 0.06$) which we perturbed by adding a spatially and temporally uncorrelated noise with the amplitude of 0.1 (zero mean value) to ϕ near the front (at $-0.9 < \phi < 0.9$). Due to the perturbation the front became unstable and converged to the pattern of stationary faceted fingers. By analyzing the near-threshold behavior using a diffusional anisotropy in both phases and by tuning $\tau(\theta)$ the results of the linear stability analysis have been verified. The final morphologies are shown in Fig. 3. They bear some similarity with facetons in their "unblocked" states. A morphology with facets in a "blocked" state has not yet been obtained.

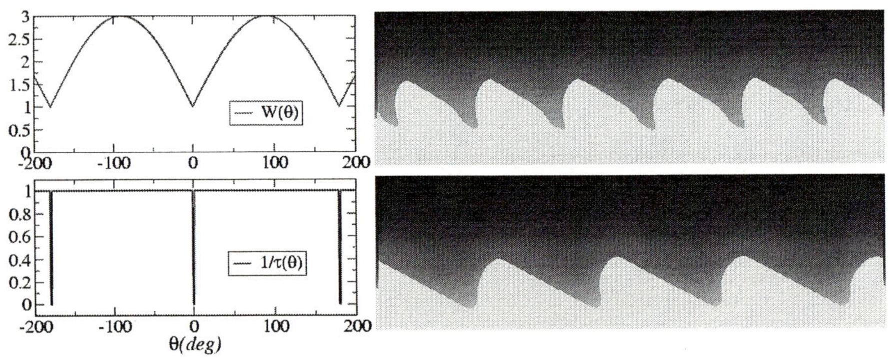

Fig. 3. The functions $W(\theta)$ and $1/\tau(\theta)$ and the corresponding solution of stationary faceted fingers at $V = 0.06$, $D = 2$, $dx = 0.8$, $dt = 0.03$ and $l_T = 250$.

As a conclusion, we wish to underline some general characteristics of the modeling of systems with large anisotropies using phase field models.

1. Since $\frac{\partial W(\theta)}{\partial \theta}$ has a jump at $\theta = \pi/2$ in the vicinity of the singularity a special treatment is needed. We found that by replacing the singularity by a

parabolic function in the case when $\pi/2 - \epsilon < \theta < \pi/2 + \epsilon$ using $\epsilon = \pi/50$, gave a smoother dynamics of the front.

2. By introducing a very slow ($1/\tau \to 0$) kinetics in a narrow range of θ (see Fig. 3), we found that the propagation of the facet was not stationary but small steps are running along the facet, as it is seen in Fig. 4, where three snapshots of the front position with respect to a perfect (rectilinear) facet (which would be a horizontal line in this graph) and the normal velocity of the facet is plotted.

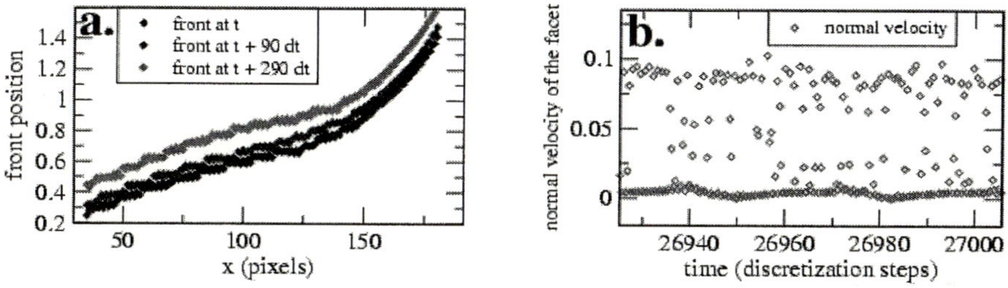

Fig. 4. a.) The deviation of the "flat part" of the front from a perfect (rectilinear) facet, b.) The normal velocity of the "facet" as a function of time.

References

1. F. Melo and P. Oswald, Phys. Rev. Lett., **64**, 1381 (1990).
2. F. Melo and P. Oswald, J. Phys. II, **1**, 353 (1991). Also see F. Melo, Thése de Doctorat, Université Lyon-I (1991).
3. Á. Buka, T. Tóth-Katona and L. Kramer, Phys. Rev. E **51**, 571 (1995).
4. T. Tóth-Katona, T. Börzsönyi, Z. Váradi, J. Szabon, Á. Buka, R. González-Cinca, L. Ramirez-Piscina, J. Casademunt, and A. Hernández-Machado, Phys. Rev. E **54**, 1574 (1996).
5. S. Akamatsu, G. Faivre and T. Ihle, Phys. Rev. E **51**, 4751 (1995).
6. T. Börzsönyi, S. Akamatsu and G. Faivre, Phys. Rev. E **65** 11702 (2001).
7. S. R. Coriell and R. F. Sekerka J. Cryst. Growth **34**, 157 (1976).
8. W. K. Burton, N. Cabrera and F. C. Frank, Phil. Roy. Soc. London **243**, 299 (1951).
9. A. A. Chernov, Contemporary Physics, **30** 251 (1989).
10. A. Classen, C. Misbah, H. Müller-Krumbhaar and Y. Saito Phys. Rev. A **43**, 6920 (1991).
11. W. Losert, D.A. Stillman, H.Z. Cummins, P. Kopczyński, W.-J. Rappel and A. Karma. Phys. Rev. E **58**, 7492 (1998).
12. W.-J. Rappel and A. Karma. Phys. Rev. E **53**, R3017 (1996).

Last Stage Solidification of Alloys: A Theoretical Study of Dendrite Arm and Grain Coalescence

M. Rappaz, A. Jacot and W. J. Boettinger

- Laboratoire de Métallurgie Physique, Ecole Polytechnique Fédérale de Lausanne, MX-G, CH-1015 Lausanne, Switzerland (M. Rappaz, A. Jacot)
- Calcom SA, PSE-EPFL, CH-1015 Lausanne, Switzerland (A. Jacot)
- Metallurgy Division, National Institute of Standards and Technology, Gaithersburg MD 20899 USA (W.J. Boettinger)

This extended abstract is based on a paper submitted to Metallurgical and Materials Transactions (January 2002)[1]

1 Introduction and Theoretical Aspects

Solidification of metallic alloys has been extensively studied (dendrite tip kinetics, microsegregation, coarsening of dendrite arms, etc.) but surprisingly, in the absence of eutectic reactions (i.e., low concentration alloys), little is known about the last stage of solidification when the primary phase regions impinge. Yet, the details of the final stages of solidification of multigrain and/or dendritic materials have significant impact on casting defects such as hot tearing. Indeed, models of hot tearing [2,3] require input of a value of fraction solid or temperature, sometimes called the coherency point, that defines the situation where the two-phase liquid plus solid material is sufficiently strong to withstand thermal contraction stresses without rupture.

For pure substances, coalescence or bridging of two planar solid-liquid interfaces can be modeled by considering that the excess free energy of the planar solid-liquid interfaces deviates from $2\gamma_{sl}$, as the thickness of liquid between the two interfaces approaches atomic dimensions. Here, we define the interface energies as those for isolated interfaces, i.e., for interfaces of two semi-infinite media. Fig. 1a shows a schematic representation of the excess surface energy, $\gamma(h)$, as a function of the liquid width, h, for a pure material for the two most common cases encountered in solidification : (i) for two solids originated from the same nucleus, there is no grain boundary energy when the liquid film disappears (i.e., $\gamma_{gb} = 0$) ; (ii) most grain boundaries are "repulsive", i.e., $\gamma_{gb} > 2\gamma_{sl}$[4,5]. The first case is referred to as "intragranular coalescence", whereas the second one corresponds to "intergranular coalescence". The thickness-dependent interfacial energy of these two interfaces can be described, somewhat arbitrarily as :

$$\gamma(h) = 2\gamma_{sl} + (\gamma_{gb} - 2\gamma_{sl})exp\left(\frac{-h}{\delta}\right), \tag{1}$$

where δ is a measure of the interface thickness. For temperatures T different from the melting temperature T_m, the excess free energy, $G(h)$, of the two interface structure per unit area is given by :

$$G(h) = (G_l - G_s)h + \gamma(h) = L\frac{T_m - T}{T_m}h + \gamma(h) = \Delta s_f h \Delta T + \gamma(h), \quad (2)$$

where G_l and G_s are the bulk free energy per unit volume of the liquid and solid, respectively, L is the heat of fusion per unit volume, Δs_f is the entropy of fusion, $\Delta T = (T_m - T)$ is the undercooling. Fig. 2 shows $G(h)$ for the intragranular and intergranular cases of Fig. 1, when the temperature is above and below the melting point, respectively. In the first case, a local maximum exists and the problem is equivalent to nucleating a liquid layer in between two superheated grains : a thin liquid layer will not remain unless its thickness h becomes larger than the critical value h^*, above which the whole domain melts. During solidification (i.e., undercooled liquid present), two such close interfaces within a single grain are "attractive" and the liquid film morphology is unstable (i.e., a decrease of the liquid layer thickness further decreases the excess interfacial energy). For intergranular coalescence with $\gamma_{gb} > 2\gamma_{sl}$, a minimum exists, i.e., a stable liquid film of thickness h^* can remain in between two repulsive grains even below the melting point. It is straightforward to show that such a situation occurs for an undercooling ΔT^* given by :

$$\Delta T^* = \frac{\gamma_{gb} - 2\gamma_{sl}}{\Delta s_f}\frac{1}{\delta}\, exp\left(\frac{-h^*}{\delta}\right). \quad (3)$$

Therefore, the film will disappear ($h* = 0$) at the coalescence or bridging undercooling, ΔT_b, given by :

$$\Delta T_b = \frac{\gamma_{gb} - 2\gamma_{sl}}{\Delta s_f}\frac{1}{\delta} = \frac{\Delta\Gamma_b}{\delta}, \text{ with } \Delta\Gamma_b = \frac{\gamma_{gb} - 2\gamma_{sl}}{\Delta s_f}, \quad (4)$$

where $\Delta\Gamma_b$ is somewhat like a differential Gibbs-Thomson coefficient associated with the closing grain boundary.

In alloys, this undercooling will be clearly influenced by solute rejection into, and diffusion parallel to, the liquid film located in between the two impinging arms, as well as by diffusion of solute in the two solids perpendicularly to the interfaces (back-diffusion).

Two numerical approaches have been undertaken in the present study. For 1D situations, a sharp-interface front-tracking method [6], which incorporates the $\gamma(h)$ expression (1), has been employed. This model allows to consider the influence of back-diffusion on coalescence. In order to handle 2D situations, a multi-phase field approach based upon a Lagrangian multiplier method [7] has been used. In this case, the free energy functional describing the system is given by :

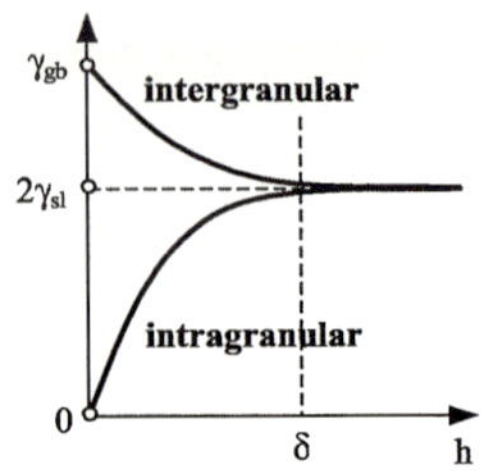

Fig. 1. Thickness dependence of the interfacial energy for intra- and inter-granular situations.

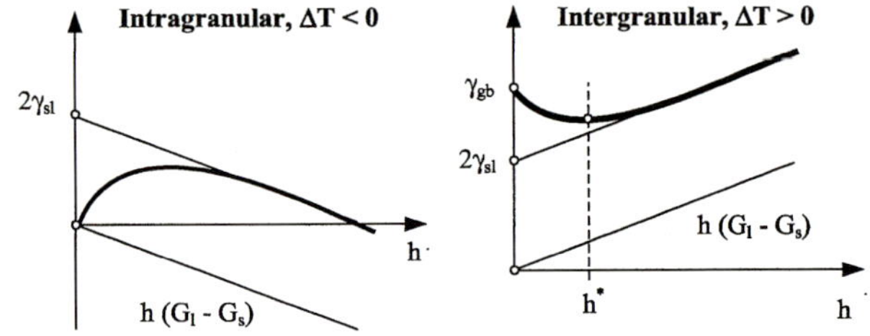

Fig. 2. Evolution of the Gibbs free energy for a pure substance, as a function of the gap between two interfaces for two situations : intra- and inter-granular.

$$\mathcal{R} = \int_{\Omega} [g(\Phi_1, \Phi_2, \Phi_3, T, c) \\ - \sum_{i,k<i} 0.5\epsilon_{ik}^2 \left|\Phi_i \mathbf{grad}\Phi_k - \Phi_k \mathbf{grad}\Phi_i\right|^2 - \Lambda(\Phi_1 + \Phi_2 + \Phi_3 - 1)] \, dV, \tag{5}$$

where the Φ_i's are the phase fields corresponding to the two grains ($i = 1, 2$) and to the liquid ($i = 3$), g is the bulk free energy, including double-well potential barriers W_{ik} between two phases i and k, ϵ_{ik} is the other interfacial energy contribution between these two phases, which multiplies the square of the gradients of these phase fields, Λ is a Lagrangian multiplier associated with the constraint $\Phi_1 + \Phi_2 + \Phi_3 = 1$. More specifically, one has :

$$g(\Phi_1, \Phi_2, \Phi_3, T) = \Delta s_f \Delta T \Phi_3^3 \left[10 - 15\Phi_3 + 6\Phi_3^2\right] + \sum_{i,k<i} W_{ik} \Phi_i^2 \Phi_k^2. \tag{6}$$

Taking the functional derivatives of Eq. (5) with respect to all the variables, Φ_1, Φ_2 and Φ_3, the phase evolution equations are then given by :

$$\frac{1}{M_\Phi} \frac{\partial \Phi_i}{\partial t} = -\frac{\delta \mathcal{R}}{\delta \Phi_i}, \tag{7}$$

where the phase field mobility, M_{Φ}ÿ as well as W_{ik} and ϵ_{ik} were chosen to match the sharp interface model parameters (i.e., γ_i and kinetic coefficient μ), viz. :

$$W_{ij} = \frac{3\gamma_{ij}}{\delta}, \quad \epsilon_{ij} = \sqrt{6\gamma_{ij}\delta}, \quad M_{\Phi} = \frac{\mu T_m}{6\delta L}. \tag{8}$$

For binary alloys, the solute diffusion equation was appended to the system of phase field equations [8]:

$$\frac{\partial c}{\partial t} = \nabla \left\{ D \left[\nabla c - \frac{(1-k)c}{\Phi_3 + k(1-\Phi_3)} \nabla \Phi_3 \right] \right\} = \nabla \left\{ D \left[\nabla c - (c_1 - c_s) \nabla \Phi_3 \right] \right\}, \tag{9}$$

where D is a diffusion coefficient averaged over the solid and liquid phases.

2 Results and Discussion

The time independent 1-D solutions of the phase field equations for a pure material are first shown in Fig. 3 for an intergranular case ($\gamma_{gb} = 0.4Jm^{-2}, \gamma_{sl} = 0.1Jm^{-2}$) and an undercooling $\Delta T = 4K$. Using the present form of the free energy, the identity of the phase at any position is the one with the largest value of Φ_i. Since $\Phi_1 = \Phi_2$ at the center of the grain boundary, a value of $\Phi_3 > 1/3$ indicates that a thin liquid layer is present between the two grains. As can be seen in Fig. 4, $\Phi_3 = 0.5$ at the center of the grain boundary, thus indicating that a thin layer of liquid still exists at 4 K undercooling for this repulsive case. It has been checked that the undercooling for which $\Phi_3 = 1/3$ at the grain boundary follows essentially Eq. (4). In particular, the multi-phase approach predicts that coalescence at a neutral boundary (i.e., $\gamma_{gb} = 2\gamma_{sl}$) precisely occurs at zero undercooling.

Turning now to the dynamic evolution of two planar interfaces of an alloy, Fig. 4 shows the evolutions of the average concentrations of the liquid and solid phases, as predicted with the phase-field and sharp interface models. These evolutions are superimposed on the phase diagram. The maximum liquid phase field value at the centerline of the boundary calculated with the multi-phase field model is also represented as a function of temperature, with a scale that can be read on the top horizontal axis. All the parameters were the same in both models, except for the parameter δ which was adjusted in the sharp-interface model in order to match the coalescence point calculated with the phase field method. As can be seen, the overall behavior of the curve calculated with the multi-phase field approach is qualitatively similar to the prediction of the sharp interface model. At the start of solidification, one can notice in both models a small deviation from the liquidus line which is due to the use of the average liquid concentration instead of the interfacial value (i.e., solutal undercooling). As solidification proceeds, the amount of liquid

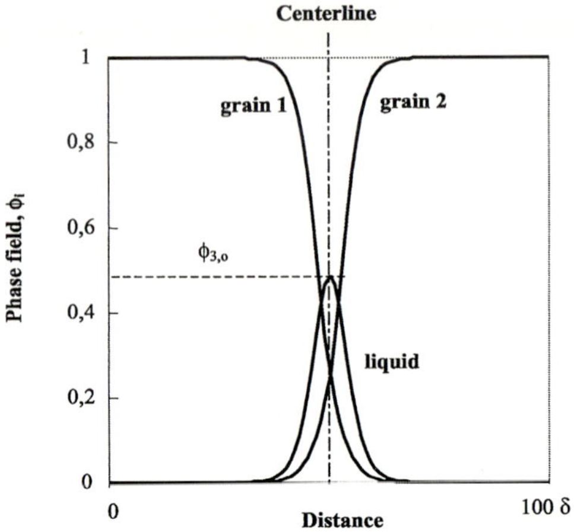

Fig. 3. 1-dimensional profile of the phase variables at $\Delta T = 4K$, as calculated with the multi-phase field simulation for a pure system. $\gamma_{gb} = 0.4Jm^{-2}, \gamma_{sl} = 0.1Jm^{-2}, \Delta s_f = 10^6 Jm^{-3}K^{-1}, \delta = 2nm$.

decreases and a state of complete mixing of solute in this phase is reached. Therefore, one follows the liquidus line until the two interfaces start to interact. In the multi-phase field approach, the start of coalescence can be clearly identified by a very rapid decrease of the liquid phase field at the centerline of the grain boundary. Surprisingly, during this initial rapid decrease, the average concentration in the liquid slightly increases, even though the two interfaces are repulsive ($\gamma_{gb} = 3\gamma_{sl}$). Then, the liquid phase field in the centerline decreases at a much slower rate and the average composition of the liquid follows the trend of the curve calculated with the sharp interface model. The coalescence point ($\Phi_3 = 1/3$ at the grain boundary centerline) calculated with the multi-phase field approach was used to adjust the value δ of the sharp interface approach and consequently the coalescence line (dashed line parallel to and 200 K below the liquidus in Fig. 4).

Finally, Fig. 5 shows 2D results obtained with the multi-phase field model for a square domain $10 \times 10\mu m^2$. As initial conditions, a seed of grain 1 was placed in the upper right corner and the remaining part of the domain was set as liquid. The initial fraction of grain 2 was set to zero everywhere. The concentrations of the seed and of the liquid were set to the equilibrium values given by the phase diagram for a very small undercooling (i.e, kc_o and c_o, respectively). The boundary condition for the concentration on the limits of the domain was set to that of a closed system. For the liquid phase, a similar condition was applied on all the boundaries. For the solid phase, the following symmetric conditions were applied : $\mathbf{grad}\Phi_1 \cdot n = -\mathbf{grad}\Phi_2 \cdot n$. Such a boundary condition, which is equivalent to that of a closed system for

$\Phi_s = (\Phi_1 + \Phi_2)$, allows to simulate the growth of two other mirror dendrite arms of grain 2 in a left and bottom symmetric square domain. The domain was cooled with a constant heat extraction rate corresponding to a relatively low cooling rate of $-10K/s$ (higher rates lead to dendritic morphologies). The domain was enmeshed with fairly coarse volume elements, except along the left and bottom boundaries where coalescence is expected (mesh size in these regions on the order of δ).

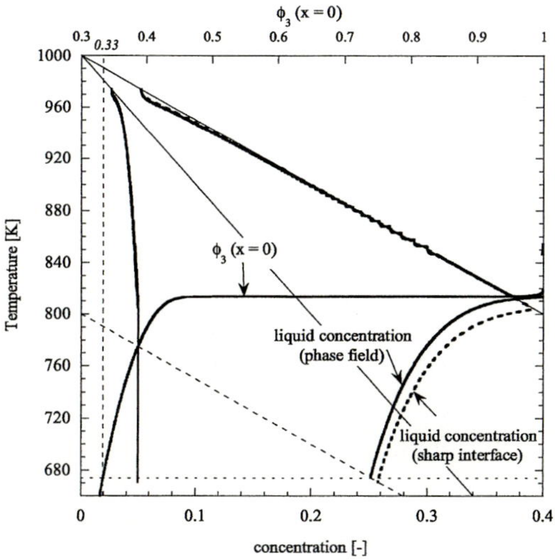

Fig. 4. Solidification paths of a repulsive grain boundary ($\gamma_{gb} = 3\gamma_{sl}$) predicted with the sharp interface and multi-phase field models for the same set of parameters, except for the value of δ (sharp interface: $\delta = 4.5\ 10^{-9}m$, phase field: $\delta = 2.0\ 10^{-9}m$). The liquid phase field at the centerline of the grain boundary calculated with the phase field model is also shown.

In Fig. 5 (left), the shape of grain 1 is shown at 3 different times (or temperatures) for an attractive case with $\gamma_{gb} = \gamma_{sl}$. After an initial circular growth stage (not shown here), the solute layers of both grains interact on the left and bottom boundaries and the grain becomes more square-shape, with an arc of a circle at the quadruple junction (top left, $T = 922.4K$). The bottom interface is still fully liquid ($\Phi_3 = 1$), but the first instabilities of the liquid film, induced in the present case by the noise of the numerical calculation, can be observed. Just $0.75s$ later ($T = 877.55K$), the first bridges can be visualized directly on the grain pictures (middle left of Fig. 5) and also in Fig. 6, where the liquid phase field profile along the bottom grain boundary has been plotted (continuous curves). In this attractive case, the grain boundary quickly closes after these first initial bridges and at $t = 49s, T = 877.55K$

(bottom left of Fig. 5), most of the boundary is solid. As can be seen in Fig. 6, a sharp transition from $\Phi_3 < 1/3$ to $\Phi_3 > 1/3$ can be observed in the last profile. The remaining liquid region has changed its morphology, from a concave morphology in the two upper pictures, to a convex droplet at the bottom. In this last case, one can notice the formation of a well-defined triple point with a dihedral angle of 120 deg. corresponding to $\gamma_{gb} = \gamma_{sl}$. The multi-phase field method is definitely capable of reproducing the observed coalescence of dendrite arms within a grain (change from concave to convex pockets of liquid), as well as the equilibrium condition at triple junctions.

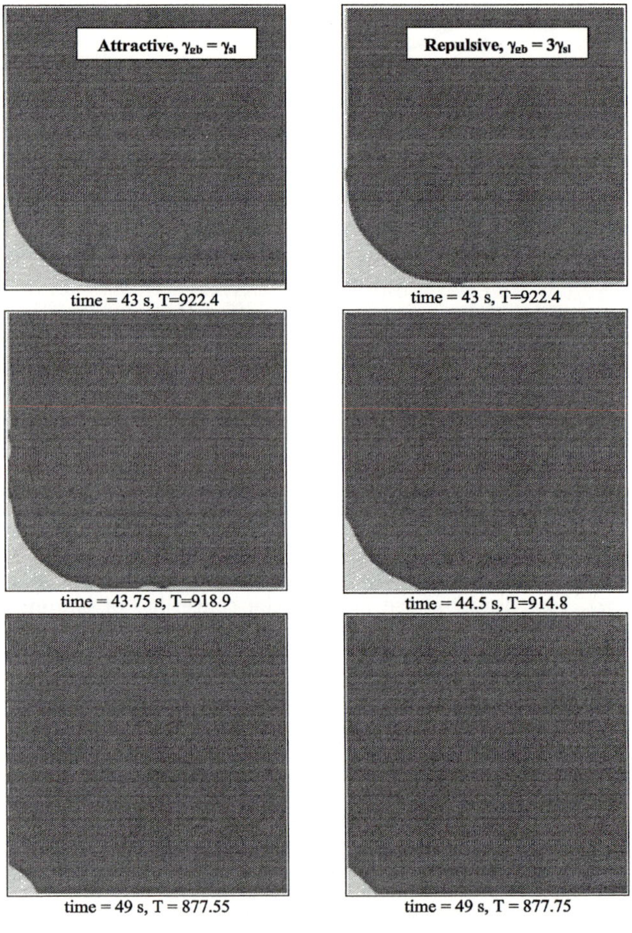

Fig. 5. Maps of the liquid phase field, Φ_3 (white region), as calculated with the multi-phase field model for two situations : attractive grains ($\gamma_{gb} = \gamma_{sl}$, left) and repulsive grains ($\gamma_{gb} = 3\gamma_{sl}$, right). The time and temperature at which these maps and profiles are shown during cooling are also indicated.

More interesting is the case shown in Fig. 5 (right) for repulsive grain boundaries with $\gamma_{gb} = 3\gamma_{sl}$. Solidification starts exactly in the same way and at the corresponding time $t = 43s$, the temperature ($T = 922.43K$) and the shape of the grain are essentially the same. However, one can notice that the initial attractive behavior of the interfaces predicted by the multi-phase field approach when $\Phi_3 > 0.4 - 0.5$ is indeed reproduced here. The instability is already visible in the top figure, but does not develop into bridges ($\Phi_3 < 1/3$). Indeed, after this initial attractive stage is over, the liquid phase field along the bottom boundary remains between 0.4 and 0.5 (central picture in Fig. 5, right), as can be checked in Fig. 6 (dashed curves). This means that the grain boundary is still wet. At that time, the shape of the solid-liquid interface has not a constant-sign curvature near the bottom left corner : the remaining liquid droplet has the shape of a cross. The convex shape of the liquid droplet occurs near the grain boundaries, in order to ensure continuity of the concentration field, since c_l at the grain boundary is larger in the initial attractive stage. At $t = 49s$ ($T = 877.75K$), the remaining pocket of liquid has about the same volume as in the attractive case of Fig. 5 (left), but has the surprising shape of a diamond. At that time, Fig. 6 indicates that the boundary has coalesced on a fairly long distance ($\Phi_3 < 1/3$ for nearly 80% of the boundary), but there is no real triple junction in this case. Instead, the slope of $d\Phi_3/dx$ for $\Phi_3 = 1/3$ along the boundary is greatly reduced and about $7 - 8\%$ of the boundary is characterized by $1/3 < \Phi_3 < 1$. In other words, the remaining intergranular liquid has a diamond shape, with very thin wedges of liquid (smaller than δ) extending from the corners of the diamond along the boundaries.

3 Conclusion

The last stage solidification of a binary alloy, which does not exhibit eutectic solidification, has been investigated from a fundamental point of view. Using first a sharp interface model, it has been shown that dendrite arms belonging to the same grain ($\gamma_{gb} = 0$) are attracted as the thickness of the interdendritic liquid film is on the order of the interaction distance, δ. This locally increases the liquid concentration above the liquidus line. Neutral liquid-solid interfaces, characterized by $\gamma_{gb} = 2\gamma_{sl}$, end their solidification on the liquidus line. Finally, for most grain boundaries, γ_{gb} is thought to be larger than $2\gamma_{sl}$ (repulsive case) and accordingly a coalescence line with undercooling $\Delta T_b = \Delta \Gamma_b \delta^{-1}$, where $\Delta \Gamma_b = (\gamma_{gb} - 2\gamma_{sl}) \Delta s_f^{-1}$, can be drawn below the liquidus. Solidification ends when the combination of temperature and liquid concentration of the thin liquid film between the grains reach this line. The trajectory of the concentration-temperature path on the phase diagram is strongly influenced by back-diffusion [1]. A multi-phase field approach to the problem with a solution method based on a Lagrange multiplier technique has also been developed. For pure substances, the equilibrium condition where a

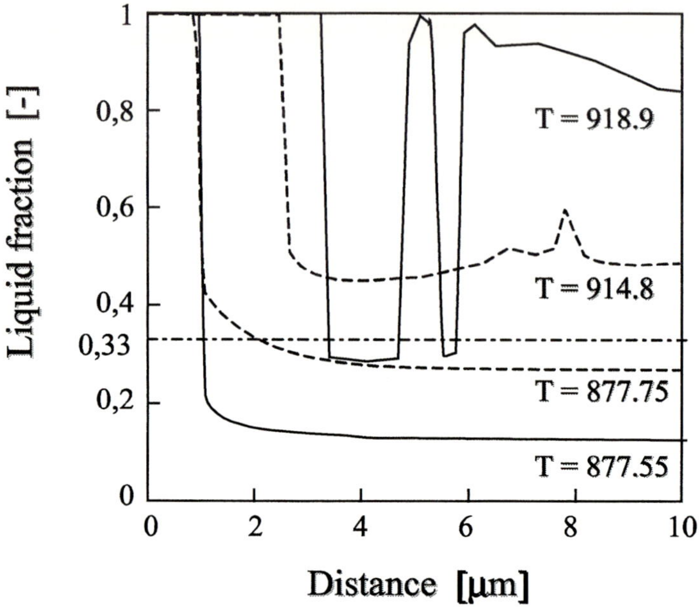

Fig. 6. Liquid phase field profile along the bottom grain boundary of Fig. 5 for the attractive case ($\gamma_{gb} = \gamma_{sl}$, continuous lines) and repulsive case ($\gamma_{gb} = 3\gamma_{sl}$, dashed lines). The two curves for each case with the indicated temperatures correspond to the middle and bottom pictures of Fig. 5.

thin liquid layer remains for repulsive interfaces below the melting point is retrieved. For the dynamic simulation of alloys, the overall features of attractive and repulsive interfaces are reproduced, but some details require further examination. This work is a first step in understanding the coalescence behavior of dendritic alloys during the last stage solidification and ultimately modeling hot-cracking tendency.

4 Acknowledgements

The authors would like to thank the Office Fédéral de l'Education et de la Science, Bern, Switzerland, for the partial financial support of this research within the framework of the Brite-Euram project VIRCAST (OFES contract # 99-00644-2)

References

1. M. Rappaz, A. Jacot, W. J. Boettinger, Last Stage Solidification of Alloys : a Theoretical Study of Dendrite Arm and Grain Coalescence, submitted to Met. Mater. Trans. (2002).

2. T. W. Clyne and G. J. Davies, Br. Foundryman, 68 (1975) 238.
3. M. Rappaz, J. -M. Drezet and M. Gremaud, Metall. Matls. Trans. A, 30 (1999) 449-455.
4. N. Eustathopoulos, Int. Met. Rev, 28 (1983) 189-210.
5. D. A. Porter and K. E. Easterling, Phase Transformations in Metals and Alloys, 1981 Van Nostrand Reinhold, UK, p. 122.
6. H. Combeau, A. Mo, J.-M. Drezet and M. Rappaz, Met. Mater. Trans., 27A (1996) 2314-27.
7. B. Nestler and A. A. Wheeler, Physica D, 138 (2000) 114.
8. J. Tiaden, B. Nestler, H. J. Diepers, I. Steinbach, Physica D, 115 (1998) 73-86.

Phase-field Modeling of Eutectic Solidification: From Oscillations to Invasion

Roger Folch* and Mathis Plapp

Laboratoire de Physique de la Matière Condensée, CNRS/École Polytechnique, 91128 Palaiseau, France

Abstract. We develop a phase-field model of eutectic growth that uses three phase fields, admits strictly binary interfaces as stable solutions, and has a smooth free energy functional. We use this model to simulate oscillatory limit cycles in two-dimensional lamellar growth, and find a continuous evolution from low-amplitude oscillations to successive invasions of one solid phase by the other when the lamellar spacing is varied.

1 Introduction

Solidification is both a fascinating example of pattern formation out of equilibrium and a phenomenon of practical importance in metallurgy. The most common solidification microstructures found in industrial alloys are dendrites and eutectic composite structures (rods or lamellae). Eutectic growth occurs when two solid phases of different composition can solidify from the same melt. The interplay between the redistribution of chemical components in front of the moving phase boundary and the effects of capillarity along the curved interfaces gives rise to a wealth of different patterns and nonlinear phenomena such as bifurcations, limit cycles, solitary waves, and even chaotic states [1].

Phase-field modeling has become the method of choice for simulating solidification fronts. Its main advantage is that it avoids explicit tracking of the solidification front by introducing "phase fields" that vary continuously between constant values corresponding to the bulk phases, thus replacing the sharp fronts by diffuse interfaces with a finite thickness. The connection to the traditional free-boundary formulation of solidification is established by the technique of matched asymptotic expansions around the equilibrium front in the limit where the thickness of the diffuse interfaces is much smaller than all other relevant length scales. Recently, the computational efficiency of phase-field models for single-phase solidification of pure substances and binary alloys has been drastically enhanced by pushing this perturbation analysis to second order [2]. This, combined with a random walk algorithm for an efficient simulation of the diffusion equation, has made it possible to simulate quantitatively the dendritic growth of a pure substance in three dimensions for experimentally relevant parameters [3].

* rf@pmc.polytechnique.fr

Therefore, it would be highly desirable to extend this second-order asymptotic analysis to phase-field models for multi-phase solidification; we will specifically address the case of eutectic growth, where two solid phases denoted by α and β grow from the liquid. The first phase-field models for eutectic growth used the standard phase field to distinguish between solid and liquid, and a concentration field [4,5] or a second phase field [6] to distinguish between the two solids. However, a solid-liquid interface then involves variations of the two variables and the asymptotic analysis becomes extremely cumbersome. The more recent multi-phase-field approach [7] assigns one phase field to each phase and interprets these fields as local volume fractions. Across an interface, in principle only the volume fractions of the two bulk phases limiting the interface need to vary, and one of them can be eliminated in terms of the other. Hence, this method allows for equilibrium interface solutions depending on a single variable that are equivalent to the usual "binary" interfaces of the standard phase-field model. However, to achieve this the model has to be carefully designed, and so far strictly binary interfaces have only been obtained by using a singular free energy (double obstacle potential) [8], which might complicate the asymptotic analysis.

We present a phase-field model for eutectic growth that uses three phase fields, yields exactly binary interfaces at equilibrium, and has a smooth free energy functional. Furthermore, the free energy is designed to keep the interfaces binary to first order in the usual non-equilibrium asymptotic expansion, which makes it a promising starting point for a second-order analysis.

To illustrate the capabilities of the model in its present form, we simulate lamellar eutectic growth of a simple model alloy with a symmetric phase diagram. We show that small-amplitude oscillations of the lamellar width, "giant oscillations", and successive invasions of one solid phase by the other, each of which is seen in experiments, all belong to the same branch of oscillatory limit cycles that is parametrized by the lamellar spacing.

2 Model

We use three phase fields $p_i \in [0,1]$ representing the local volume fractions of each phase (α, β or liquid), and thus $\sum_i p_i = 1$. Their dynamics is derived from a free energy functional $\mathcal{F}$,

$$\frac{\partial p_i}{\partial t} = -\frac{1}{\tau}\frac{\delta \mathcal{F}}{\delta p_i} , \tag{1}$$

where τ is a relaxation time and the variational derivative has to take into account the constraint $\sum_i p_i = 1$. Furthermore, the concentration field c obeys the conserved dynamics

$$\partial c/\partial t = \boldsymbol{\nabla}\left[M(p_\alpha, p_\beta, p_l)\boldsymbol{\nabla}\mu\right] , \tag{2}$$

where $\mu \equiv \delta\mathcal{F}/\delta c$ is the chemical potential and M is a phase-dependent mobility.

The free energy functional $\mathcal{F} = \int_V f dV$ is the volume integral of a free energy density f that is conveniently decomposed into $f = f_{\rm grad} + f_{\rm TW} + f_c$, where $f_{\rm TW}$ is a triple well potential with minima on each pure phase — the equivalent of the double well potential of the standard phase-field model, $f_{\rm grad}$ contains the gradients of the phase fields and ensures a finite interface thickness and surface tension, and f_c couples the phase fields and the concentration and drives the system out of equilibrium.

Due to the constraint $\sum_i p_i = 1$, the free energy density and the interface solutions can be conveniently represented on a *Gibbs simplex*, that is an equilateral triangle where the distance to each side of the triangle from a given point represents the value of one of the phase fields. Thus, each vertex corresponds to a pure phase, and each side to a purely binary interface. Our goal is to obtain strictly binary interfaces, i.e., the equilibrium solutions of (1) for an interface connecting two phases should exactly run along the edges of the simplex. One way to achieve this is to choose

$$f_{\rm grad} = \frac{W^2}{2} \sum_i |\nabla p_i|^2 \ , \tag{3}$$

and to design the function f_{TW} in such a way that the three minima are connected by “saddles” that run along the edges and that have vanishing derivatives in the direction normal to the edges. Such functions can be constructed using the geometry of the Gibbs simplex, as will be detailed elsewhere. The simplest choice,

$$f_{\rm TW} = \sum_i p_i^2 (1 - p_i)^2 \ , \tag{4}$$

is plotted in Fig. 1a and reduces to the standard fourth-degree double well potential of the binary phase-field model on the edges. With these choices, the free energy is completely symmetric with respect to the interchange of any two phases, and as a consequence all three surface tensions are equal. However, the general case of unequal surface tensions can also be treated by adding other terms in the free energy density that modify the height of the saddle points.

Out of equilibrium, the triple well is “tilted” by the term f_c in order to favor one phase over the others. Since we want to keep the interfaces binary, this term needs to be carefully constructed. The elementary building blocks are tilt functions g_i that satisfy $g_i = 1$ for $p_i = 1$, $g_i = 0$ for $p_j = 1$, $j \neq i$, and that have vanishing derivatives in the direction normal to the edges of the simplex. One possible choice is

$$\begin{aligned} g_i = 1 - \big[& (1 + p_j - p_i)^2 (2 + p_i - p_j)^2 + (1 + p_k - p_i)^2 (2 + p_i - p_k)^2 \\ & + (p_k - p_j)^2 (3 - |p_k - p_j|)^2 \big] /12 \ . \end{aligned} \tag{5}$$

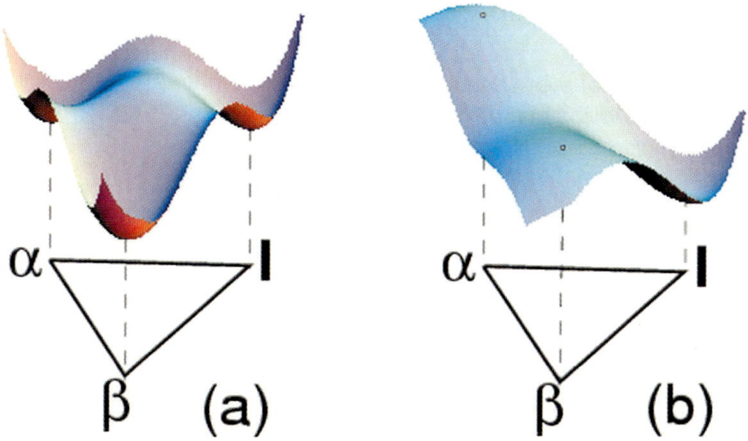

Fig. 1. Triple-well potential f_{TW} (a) and elementary tilt function $-g_l$ (b) drawn as "landscapes" over the Gibbs simplex.

In Fig. 1b, we plot $-g_l$ that lowers the well for the liquid with respect to the two others. With the help of these functions,

$$f_c = \frac{1}{2}\left[c - \sum_i A_i g_i\right]^2 + \sum_i B_i g_i \tag{6}$$

interpolates between the concentration-dependent free energies $f_i = (c - A_i)^2/2 + B_i$ of each phase i. Here, $c = (C - C_E)/\Delta C$ is a scaled concentration variable, where $\Delta C = C_\beta - C_\alpha$ is the width of the eutectic plateau, and C_α, C_β, and C_E are the equilibrium compositions of the two solids and the liquid at the eutectic temperature. An arbitrary phase diagram can be constructed by performing the well-known double tangent construction, which yields $c_i^{ij} = A_i + (B_j - B_i)/(A_j - A_i)$ for the equilibrium concentration c_i^{ij} of phase i coexisting with phase j, and by choosing appropriate $A_i(T)$, $B_i(T)$, where T is the temperature. Note that by the same procedure, a peritectic phase diagram can be constructed [9].

Finally, we need to specify the mobility function in (2). Since $\partial^2 f_i/\partial c^2 \equiv 1$, the choice $M(p_\alpha, p_\beta, p_l) = D p_l$ yields a constant diffusivity D in the liquid, and no diffusion in the solid (one-sided model).

3 Simulations

In thin-sample directional solidification of eutectic alloys with global composition in the eutectic range, the basic pattern is a periodic array of alternating lamellae of each solid phase, growing perpendicular to the large-scale solidification front and parallel to the imposed temperature gradient. This basic state exhibits various instabilities that have been carefully studied both experimentally in the transparent organic alloy CBr_4-C_2Cl_6 [1] and by numerical simulations using the boundary integral method [10]. For fixed experimental parameters, the stability of the basic state is controlled by the width of one lamella pair, or *lamellar spacing.* For sample compositions close to the eutectic point, the first instability that is encountered for increasing spacing is a period-preserving oscillatory instability, that is, the lamellae start to oscillate, with all the lamellae of the same phase oscillating in phase, and the global spacing left unchanged. Beyond the onset of this instability, stable limit cycles are found, with an oscillation amplitude that increases with spacing. In the boundary integral simulations, this branch of solutions terminates when the amplitude becomes too large because the thinner lamellae pinch off. In contrast, in experiments "giant oscillations" with very large amplitudes can be obtained [11]. Furthermore, when the experiments are started from a single solid phase growing into the liquid with a few widely spaced nuclei of the other solid phase on the solidification front, the nuclei grow and spread along the front forming "invasion tongues" [12]. When two tongues growing in opposite directions collide, they do not coalesce; instead, a narrow channel of the other phase remains and the whole process starts over with the role of the two phases reversed. The interface dynamics is hence still oscillatory, even if the patterns look completely different. In this regime, the boundary integral method becomes inapplicable because it uses the quasistationary approximation of the diffusion equation that is valid only if the interface motion is slow compared to the diffusive solute redistribution. Here, we investigate oscillatory limit cycles with our phase-field model, that does not have this restriction.

The sample is solidified with pulling speed V in a constant temperature gradient G directed along the z direction. For simplicity, we use a symmetric phase diagram that has constant concentration jumps across the interfaces (parallel liquidus and solidus lines). This is implemented by choosing $A_\alpha = -0.5$, $A_\beta = 0.5$, $A_l = 0$, $B_\alpha = B_\beta = 0$ and $B_l = -(z - Vt)/l_T$, where $l_T = (m\Delta C)/G$ is the thermal length, with m being the liquidus slope (equal for both phases). The other relevant physical length scales are the diffusion length $l_D = D/V$ and the capillary length $d_0 = \Gamma/(m\Delta C)$, where Γ is the Gibbs-Thomson constant. In terms of the model parameters, $d_0 = (2\sqrt{2}/3)W$. All simulations discussed here were performed with $l_T/d_0 = 530$ and $l_D/d_0 = 212$ and at exactly eutectic composition (that is, $c = 0$ far ahead of the front).

The equations are integrated using a standard explicit finite-difference scheme on a regular grid of spacing 0.8 W for which we have checked that the discretization has converged. Since the diffusivity vanishes in the solid, the diffusion equation only needs to be solved in interfacial and liquid regions. Moreover, far ahead of the solidification front, it is solved using a variant of the random walker algorithm of Ref. [3] that will be detailed elsewhere. Two types of initial conditions are used. To obtain lamellar steady states, simulations are started from two flat lamellae in contact with the liquid. If the lamellar state is unstable, the numerical noise generated by the random walkers triggers the instability. Alternatively, simulations are started from a flat single solid phase at equilibrium with the liquid. Upon pulling, the interface recoils towards colder temperatures as a diffusion layer of the rejected component is gradually formed. When the undercooling reaches a predetermined value, nucleation is mimicked by placing a large enough semicircular nucleus of the other phase on top of the interface.

We restrict our attention here to period-preserving oscillations without lateral drift. Therefore, it is sufficient to simulate two adjacent half lamellae or half a nucleus of one solid phase on top of the other with reflecting (no-flux) boundary conditions on the sides of the simulation box that are parallel to the lamellae. The lamellar spacing λ is varied by changing the lateral size of the simulation box. In the figures below, for clarity we have reconstructed a whole lamella pair by reflecting the system with respect to one side of the box.

We start by constructing the branch of stable lamellar steady-state solutions and find the minimal undercooling spacing at $\lambda_{\mathrm{min}} \approx 85d_0$, in good agreement with the prediction of the Jackson-Hunt theory [13], $\lambda_{\mathrm{min}} \approx 79d_0$. Next, we start from a single nucleus and monitor the dynamics for increasing spacings. We find a bifurcation toward oscillatory limit cycles for λ close to $2\lambda_{\mathrm{min}}$. As shown in Fig. 2a, for $\lambda = 2\lambda_{\mathrm{min}}$ the oscillation already has an amplitude of about a quarter of the lamellar spacing, but its shape is still close to a sinusoidal wave. At $\lambda = 4\lambda_{\mathrm{min}}$ (Fig.2 2b), the oscillation is highly asymmetric, and has an amplitude close to half the lamellar spacing, i.e., only a thin channel of the other solid phase is left. Finally, at $8.48\lambda_{\mathrm{min}}$ (Fig. 2c), three different regimes become apparent: (i) initially, the nucleus grows slowly and spreads along the interface to form an invading finger with a well-defined shape, (ii) the finger speeds up with an approximately constant acceleration, leaving a straight border with the other solid phase behind, and (iii) when it approaches a finger growing in the opposite direction (in our simulations, its mirror image generated by the reflecting boundary conditions), it "feels" the diffusion field generated by the other and slows down very rapidly, allowing for the other phase to emerge through a narrow channel left between the two fingers and thus restart the process with the role of the two solids reversed.

Remarkably, our largest simulations reproduce many characteristic features of the "successive invasions" observed in the experiments [12], in particular the approximately constant acceleration of invasion tongues and the leftover of narrow channels between "colliding" fingers. From our simulations, it hence appears that this regime can be understood as period-preserving oscillations with very large amplitudes. Indeed, the morphology at $4\lambda_{\mathrm{min}}$ already presents stages (i) and (iii), but not (ii). Apparently, to observe this intermediate, constant acceleration regime, it is sufficient to have enough "free space" left in front of the propagating finger (i.e., large lamellar spacings). It

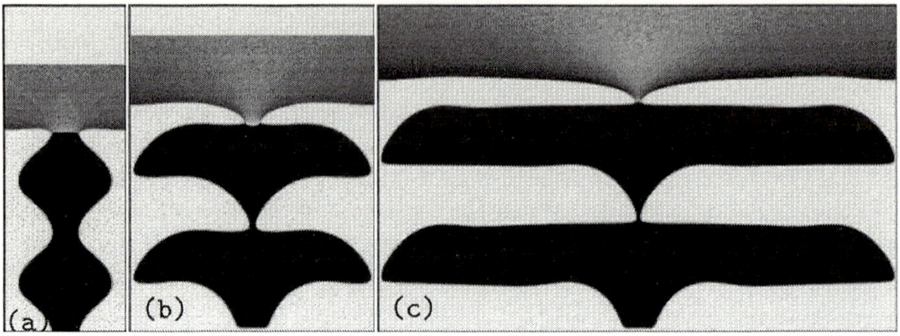

Fig. 2. Snapshot pictures of typical oscillatory structures. Growth is upwards, the two solids are black and white, and the greyscale in the liquid is proportional to the solute concentration. (a) low-amplitude oscillations ($\lambda/\lambda_{\min} = 2$), (b) giant oscillations ($\lambda/\lambda_{\min} = 4$), (c) successive invasions ($\lambda/\lambda_{\min} = 8.48$). The scale is the same on all three figures.

is also interesting to note that in peritectic alloys, simulations of spreading in the same geometry produce stages (i) and (ii), but then colliding fingers coalesce, and to obtain an oscillatory dynamics explicit nucleation events have to be introduced [9,14].

The fact that we can switch from gentle oscillations to long invading fingers just by changing the lamellar spacing demonstrates that both belong to the same branch of oscillatory limit cycles. This branch of solutions is indeed an attractor, since simulations at fixed spacing and started from very different initial conditions always converge to the same cycle.

4 Conclusion

We have developed a phase-field model of eutectic solidification that has a smooth free energy landscape and yields exactly binary interfaces away from the trijunction points. These properties make it a promising starting point for a second-order asymptotic analysis. Furthermore, we have used the model to simulate oscillatory limit cycles in a model alloy that has a symmetric phase diagram, and we have found that low-amplitude oscillations, "giant oscillations", and successive invasions, all observed in experiments, lie on a single branch of solutions that is parametrized by the lamellar spacing and that bifurcates from the steady-state branch at about twice the minimum undercooling spacing.

Here, we have focused on period-preserving oscillatory modes; many other instability modes exist, for example the tilt instability or period-doubling oscillations. These instabilities can be studied using less restrictive boundary conditions. Preliminary simulations show that for the case studied here, namely an alloy with symmetric phase diagram at its eutectic composition, period-preserving oscillations remain the only stable limit cycles. In contrast, if we use off-eutectic sample compositions or an asymmetric phase diagram, and hence break the complete symmetry between the two solid phases, other modes become active. Therefore, our model can be used to obtain a more complete bifurcation diagram, which is the subject of ongoing work.

We thank Silvère Akamatsu and Gabriel Faivre for many stimulating discussions and Jean-François Gouyet for his help with Fig. 1. R. F. acknowledges financial support from project FMRX-CT96-0085 of the European Commission and from Centre National d'Études Spatiales (France).

References

1. Ginibre, M., Akamatsu, S., Faivre, G.: Experimental determination of the stability diagram of a lamellar eutectic growth front. Phys. Rev. E **56** (1997) 780–796
2. Karma, A., Rappel, W.-J.: Quantitative phase-field modeling of dendritic growth in two and three dimensions. Phys. Rev. E **57** (1998) 4323–4349; Karma, A.: Phase-Field Formulation for Quantitative Modeling of Alloy Solidification. Phys. Rev. Lett. **87** (2001) 115701
3. Plapp, M., Karma, A.: Multiscale random-walk algorithm for simulating interfacial pattern formation. Phys. Rev. Lett. **84** (2000) 1740–1743; Plapp, M., Karma, A.: Multiscale Finite-Difference-Diffusion-Monte-Carlo method for simulating dendritic solidification. J. Comp. Phys. **165** (2000) 592–619
4. Karma, A.: Phase-field model of eutectic growth. Phys. Rev. E **49** (1994) 2245–2250
5. Elder, K. R., Drolet, F., Kosterlitz, J. M., Grant, M.: Stochastic eutectic growth. Phys. Rev. Lett. **72** (1994) 677–680
6. Wheeler, A. A., McFadden, G. B., Boettinger, W. J.: Phase-field model for solidification of a eutectic alloy. Proc. R. Soc. Lond. A **452** (1996) 495–525
7. Steinbach, I., Pezzola, F., Nestler, B., Seeßelberg, M., Prieler, R., Schmitz, G. J.: A phase field concept for multiphase systems. Physica D **94** (1996) 135–147
8. Garcke, H., Nestler, B., Stoth, B.: A multi phase field concept: numerical simulations of moving phase boundaries and multiple junctions. SIAM J. Appl. Math **60** (1999) 295–315
9. Lo, T. S., Karma, A., Plapp, M.: Phase-field modeling of microstructural pattern formation during directional solidification of peritectic alloys without morphological instability. Phys. Rev. E **63** (2001) 031504
10. Karma, A., Sarkissian, A.: Morphological instabilities of lamellar eutectics. Metall. Mater. Trans **27A** (1996) 635–656
11. Akamatsu, S., Faivre, G.: Private communication.
12. Akamatsu, S., Moulinet, S., Faivre, G.: The Formation of Lamellar-Eutectic Grains in Thin Samples. Metall. Mater. Trans. **32A** (2001) 2039–2047
13. Jackson, K. A., Hunt, J. D.: Lamellar and rod eutectic growth. Trans. Metall. Soc. AIME **236** (1966) 843–852
14. Nestler, B., Wheeler, A. A.: A multi-phase-field model of eutectic and peritectic alloys: numerical simulation of growth structures. Physica D **138** (2000) 114–133

Phase-field Theory of Nucleation and Growth in Binary Alloys

László Gránásy, Tamás Börzsönyi* and Tamás Pusztai

Research Institute for Solid State Physics and Optics, POB 49, H-1525 Budapest, Hungary; *Universités Paris VI at VII, Tour 23, 2 place Jussieu, 75251 Paris Cedex 05, France

Abstract. We present a phase field theory for binary crystal nucleation. Using the physical interface thickness, we achieve quantitative agreement with computer simulations and experiments for unary and binary substances. Large-scale numerical simulations are performed for multi-particle freezing in alloys. We deduce the Kolmogorov exponents for dendritic solidification and for the "soft-impingement" of crystallites interacting via diffusion fields.

1 Introduction: Diffuse Interface and Nucleation

The solid-liquid interface is known to extend to several molecular layers, as indicated by experiment [1], computer simulations [2], and statistical mechanical treatments based on the density functional theory [3]. This plays an essential role in modeling nucleation, as the typical size of critical fluctuations is comparable to the physical thickness of the interface. Modern nucleation theories, which consider the molecular scale diffuseness of the interface, removed the many orders of magnitude difference seen between nucleation rates from the classical sharp interface approach and experiment [4].

The phase field theory (PFT) predicts a diffuse interface. In conventional applications for solidification at small undercoolings, the diffuse interface is both an advantage and a disadvantage: An advantage, since all mesh points can be handled uniformly (no need for front tracking), and a disadvantage, as the physical interface thickness is by orders of magnitude smaller than the size of the modeled particles, thus due to computational limitations, the true interface thickness cannot be realized. Therefore, one is compelled to use an artificially broad interface, which is accompanied with artifacts. Imaginative work by Karma and coworkers (see [5] and references therein) established operations, such as the introduction of interface currents and specific choice of the phase field parameters, which remove these artifacts and ensure the convergence of the diffuse interface results to the sharp interface solution.

In modeling nucleation, we face a different situation. Here, we are able to choose the model parameters so that the true interface thickness is recovered. We demonstrate that under such conditions the PFT is quantitative for nucleation in unary and binary systems. Note that convergence to the sharp interface model is not a requirement, as the sharp interface model is inaccurate for nucleation [4].

2 Phase-field Theory of Nucleation

We adopt the standard binary phase field theory (for review see [6,7]). Our starting point is the free energy functional

$$F = \int d\mathbf{r} \left\{ \frac{\epsilon^2 T}{2} (\nabla\phi)^2 + f(\phi, c) \right\} , \tag{1}$$

where ϕ and c are the phase and concentration fields, $f(\phi, c) = WTg(\phi) + [1 - P(\phi)]f_S + P(\phi)f_L$ is the local free energy density, $W = (1-c)W_A + cW_B$ the free energy scale, the quartic function $g(\phi) = \phi^2(1-\phi)^2/4$ ensures the double-well form of f, while the function $P(\phi) = \phi^3(10 - 15\phi + 6\phi^2)$ switches on and off the solid and liquid contributions $f_{S,L}$, taken from the ideal solution model. (A and B indicate the constituents.)

For binaries the model contains three parameters ϵ, W_A and W_B that reduce to two (ϵ and W) in the one-component limit. They can be fixed if the respective interface free energy γ, melting point T_f, and interface thickness δ are known. Such information is available for the Lennard-Jones, ice-water, and Cu–Ni systems [8–10].

2.1 Nucleation in 3D

The critical fluctuation (nucleus) is a non-trivial time-independent solution of the governing equations:

$$\frac{\partial \phi}{\partial t} = -M_\phi \frac{\delta F}{\delta \phi} , \tag{2}$$

$$\frac{\partial c}{\partial t} = \nabla M_c \nabla \frac{\delta F}{\delta c} , \tag{3}$$

where $(\delta F/\delta x)$ stands for the functional derivatives ($x = \phi, c$), M_x are mobilities. For spherical symmetry (a reasonable assumption), the phase field equation reduces to $\nabla^2\phi = \Delta\mu(\phi, c)/(\epsilon^2 T)$. Here $\Delta\mu(\phi, c) = WTg'(\phi) + [(1 - c)\Delta f_A + c\Delta f_B]P'(\phi)$ is the local chemical potential difference relative to the initial liquid, prime stands for differentiation with respect to the argument, the local concentration is related to the phase-field as $c(\phi) = c_\infty e^{-y}/(1 - c_\infty + c_\infty e^{-y})$, where $y = v(W_B - W_A)g(\phi)/R + v(\Delta f_B - \Delta f_A)[P(\phi) - 1]/RT$, while Δf_i are the volumetric free energy differences between the pure liquid and solid phases. Solving these equations numerically under boundary conditions that prescribe bulk liquid properties far from the fluctuations ($\phi \to 1$, and $c \to c_\infty$ for $r \to \infty$), and zero field-gradients at the center, one obtains the free energy of critical fluctuation as $W^* = F - F_0$. Here F is obtained by numerically evaluating (1) after having the time-independent solutions inserted, while F_0 is the free energy of the initial liquid. This is compared with $W^* = (16\pi/3)\gamma_f^3/\Delta f^2$ from the sharp interface "droplet" model of the classical nucleation theory, where $\gamma_f = \gamma(T_f)$.

The homogeneous nucleation rate is calculated as $J = J_0 exp\{-W^*/kT\}$, where the nucleation prefactor J_0 of the classical kinetic approach is used, which proved consistent with experiments [11].

2.2 Multi Particle Solidification in 2D

To treat nucleation and growth of particles on equal footing, we extend the standard binary phase field theory [6,7] via incorporating a non-conservative orientational field θ, which assumes a constant value between 0 and 1 in the crystal that determines crystal orientation in the laboratory frame as done recently by Kobayashi, Warren, and Carter [12]. We assume, however, that θ is defined in the liquid as well, where it fluctuates. Via this, we capture the feature that the short-range order in the solid and liquid are usually similar, with the obvious difference that the building units have a uniform orientation in the crystal, while their orientation fluctuates in the liquid. Following [12], we assume that the grain boundary energy acts in the solid and is proportional to $|\nabla\theta|$. We realize this by adding $f_{ori} = \mathcal{M}|\nabla\theta|$ to f_S, where $\mathcal{M}$ is taken composition independent. The governing equation is

$$\frac{\partial\theta}{\partial t} = -M_\theta \frac{\delta F}{\delta\theta}. \tag{4}$$

For $\phi < 1$, equation (4) switches in orientational ordering, and chooses the value of θ that survives as the orientation of the particle relative to which the anisotropy of γ is measured. Unlike previous work, in our approach the orientation field θ is coupled to the phase field, and extends to the liquid phase where crystallographic orientation develops from orientational fluctuations. While our model incorporates grain boundary dynamics, M_θ is set so in the solid that grain rotation is negligible on the time scale of freezing.

Since PFT is a mean field type approach, nucleation is introduced as follows. Method I.: By including noise into (2)–(4). Method II.: The simulation area is divided into domains according to the local composition. The time-independent solution is found for these compositions. Critical fluctuations of statistically correct numbers following Poisson distribution are placed into these areas in every time step. The added small-amplitude noise makes these fluctuations either grow or shrink. In nucleation-growth processes the transformed fraction often follows the Kolmogorov scaling, $X(t) = 1 - exp\{-(t/t_0)^p\}$, where the "Kolmogorov exponent" p is representative of the mechanism of the phase transformation, and is evaluated from the slope of the plot $ln[-ln(1-X)]$ vs. lnt. We define the crystalline fraction via $\phi < 0.5$. Simultaneous modeling of nucleation and growth raises the question of interface thickness: When using the nanometer size physical interface thickness, growth of large objects cannot be treated. If, in turn, we use a broad interface for large particles, relying on methods for quantitative description of dendritic solidification [5], we face with difficulties when addressing nucleation and particle impingement. Therefore, following previous work [10], we perform the calculations for a hypothetical system of a broader interface ($\delta = 41.6$ nm), a reduced interfacial free energy ($\gamma/6$), and an increased diffusion coefficient ($\sim 100 \times D_L$). The consistent treatment of nucleation and growth in our model, in which both processes emerge from the same governing equations, allows us to describe the life of particles from birth to impingement on other particles, a feature essential for understanding the variation of Kolmogorov exponent with transformed fraction during "soft-impingement".

Equations (2)–(4) were solved numerically on 2000 × 2000 and 7000 × 7000 grids, using parallel processing (MPI protocol) on a computer cluster consisting of 36 PCs of 1.0 – 1.6 GHz Athlon processors.

3 Results and Discussion

First we compare the phase-field predictions for the nucleation rate with computer simulations and experimental results for unary systems. In the Lennard–Jones system all relevant properties are known from molecular dynamics simulations [8]. The phase field profiles indicate that for typical undercoolings the critical fluctuations are ramified, and do not show bulk crystal properties. While the phase-field predictions for the nucleation rate agree with results from computer simulations [13], those from the classical sharp interface theory are too low by eight to ten orders of magnitude (Fig. 1a). Similar relationships are observed between predictions and experiment [14] for the ice-water system (Fig. 1b). While the input data are less reliable for the Cu–Ni system, the critical undercoolings computed for the realistic range of nucleation rates ($J = 10^{-4}$ to 1 drop^{-1}s^{-1} for electromagnetically levitated droplets of 6 mm diameter) are also in a reasonable agreement with experiment [15] (Fig. 1c). Note that without adjustable parameters, a quantitative

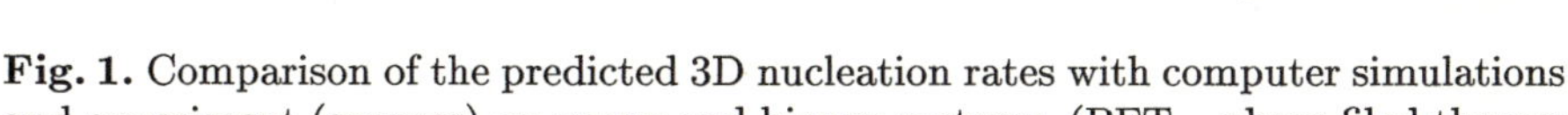

Fig. 1. Comparison of the predicted 3D nucleation rates with computer simulations and experiment (squares) on unary and binary systems. (PFT – phase filed theory; CNT – classical sharp interface theory; dotted lines – limits due to the error of γ_f.)

agreement has been achieved with simulations and experiment, indicating that the phase-field theory is a powerful tool to attack nucleation problems, provided that the physical interface thickness is used.

Multi-particle freezing has been studied in 2D using the thermodynamic properties of Cu–Ni, and a large anisotropy for the interface free energy ($\gamma_0 \{1 + s_0 cos[4\vartheta]\}$ with $s_0 = 0.25$). With appropriate choices of M_θ and $\mathcal{M}$ single orientation nucleates per particle. The freezing morphologies and the variation of exponent p with crystalline fraction are shown as a function of liquid composition in Fig. 2. Here the particles were nucleated by Method II. (Method I yields less regular shapes.) At small dimensionless nucleation rates $\tilde{J} = 0.49$ dendritic particles evolve, yielding $p \approx 3$, that obeys the relationship $p = 1+d$ (constant nucleation and growth rates in d-dimensions), which follows from the steady-state traveling tip solution of the diffusion equation. At large $\tilde{J}$ pillow-shaped particles form. The variation of p with transformed fraction depends on the liquid composition. Similar behavior has been observed during the formation of nanocrystalline alloys.

At reduced M_θ, different orientations nucleate at the perimeter, of which those with their fastest growth direction perpendicular to the interface survive, yielding spherulitic growth morphology (Fig. 3).

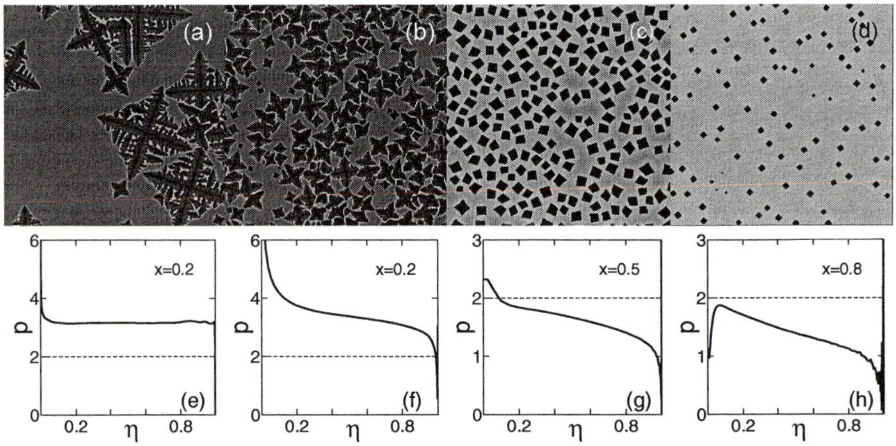

Fig. 2. Transformation kinetics in 2D vs. composition in the Cu–Ni system at 1574 K ($x = (C - C_l)/(C_s - C_l)$, where C_l and C_s are the solidus and liquidus compositions). (a)–(d) Snapshots of concentration distributions (black – solidus; light grey – liquidus); (e)–(f) Kolmogorov exponent vs. normalized transformed fraction $\eta = X/X_{max}$, where X_{max} is the maximum crystalline fraction.

This work has been supported by the ESA Prodex Contract No. 14613/00 /NL/SFe, by the Hungarian Academy of Sciences under contract Nos. OTKA-T-025139 and T-037323, by the EU grant HPMF-CT-1999-00132, and forms part of the ESA MAP Project No. AO-99-101.

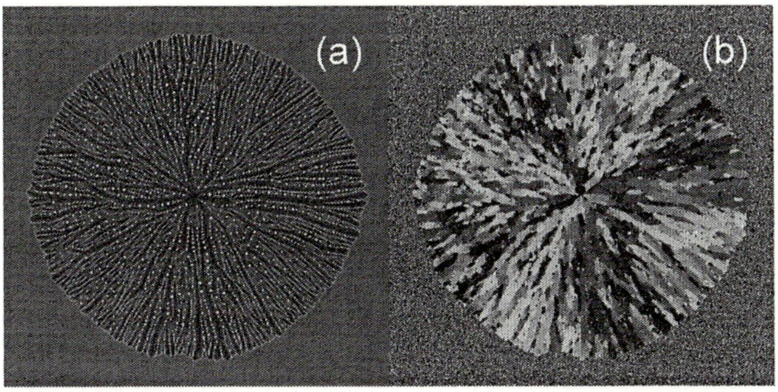

Fig. 3. Spherulitic growth at $x = 0.23$: (a) Composition and (b) orientation maps. Note the self-organized radial grain structure.

References

1. Huisman, W. J., Peters, J. F., Zwanenburg, M. J., de Vries, S. A., Derry, T. E., Albernathy, D., van der Veen, J. F.: Layering of a liquid metal in contact with a hard wall. Nature **390** (1997) 379–381
2. Davidchack, R. L., Laird, B. B.: Simulation of the hard-sphere crystal-melt interface. J. Chem. Phys. **108** (1998) 9452–9462
3. Ohnesorge, R., Löwen, H., Wagner, H.: Density functional theory of crystal-fluid interfaces and surface melting. Phys. Rev. E **50** (1994) 4801–4809
4. Gránásy, L., Iglói, F.: Comparison of experiments and modern theories of crystal nucleation. J. Chem. Phys. **107** (1997) 3634–3644
5. Karma, A.: Phase-field formulation for quantitative modeling of alloy solidification. Phys. Rev. Lett. **87** (2001) 115701-1–115701-4
6. Warren, J. A., Boettinger, W. J.: Prediction of dendritic growth and microsegregation patterns in a binary alloy using the phase-field method. Acta Metall. Mater. **43** (1995) 689–703
7. Boettinger, W. J., Warren, J. A.: The phase-field method: Simulation of alloy dendritic solidification during recalescence. Metall. Mater. Trans. A **27** (1996) 657–669
8. Broughton, J. Q., Gilmer G. H.: Molecular dynamics investigation of the crystal–fluid interface. I-VI. J. Chem. Phys. **79** (1983) 5090–5127; **84** (1986) 5741–5768
9. Gránásy, L.: Cahn–Hilliard-type density functional calculations for homogeneous ice nucleation in undercooled water. J. Mol. Struct. **485–486** (1999) 523–536
10. Conti, M.: Growth of a needle crystal from an undercooled alloy melt. Phys. Rev. E **56** (1997) 3197–3202
11. Kelton, K. F.: Crystal nucleation in liquids and glasses. Solid St. Phys. **45** (1991) 75–177
12. Kobayashi, R., Warren, J. A., Carter, W. C.: Vector-valued phase field model for crystallization and grain boundary formation. Physica D **119** (1998) 415–423
13. Báez, L. A., Clancy, P.: The kinetics of crystal growth and dissolution from the melt in Lennard-Jones systems. J. Chem. Phys. **102** (1995) 8138–8148
14. Taborek, P: Nucleation in emulsified supercooled water. Phys. Rev. B **32** (1985) 5902–5906
15. Willnecker, R., Herlach, D. M., Feuerbacher, B.: Nucleation in bulk undercooled nickel-base alloys. Mater. Sci. Eng. **98** (1988) 85–88

Modelling of Phase Transformations in Titanium Alloys with a Phase-field Model

Benoît Appolaire and Elisabeth Gautier

LSG2M, Ecole des Mines, Parc de Saurupt F-54042 Nancy Cedex, France

Abstract. We have used a phase field model devoted initially to solidification for studying some phase transformations in metastable β titanium alloys. First calculations have been carried out to determine if we can rely on the model to get quantitative information on transformation kinetics. For a simple 1D problem, the phase field model, a sharp interface model and an analytical solution have been compared. The agreement was good even for quite coarse meshes. Then, 2D calculations have been done starting from micrographs of real microstructures. For a moderate departure from equilibrium, we obtained kinetics of the order of magnitude of some experimental measurements. For a larger departure and with the same parameters as before, the calculations failed because of the propagation of a spurious intermediate state, which could be avoided in decreasing drastically the mesh size.

1 Introduction

Titanium alloys have more and more applications due to their desirable combination of properties. First, titanium alloys have low densities which provide very attractive strength to weight ratios allowing lighter structures. Secondly, they show superior corrosion and erosion resistance in many environments. Finally, they have a high temperature capability. Although titanium alloys are still considered as expensive materials, their cost can be justified on the basis of their versatile properties for many applications. Their good mechanical properties are achieved thanks to proper thermo- mechanical treatments, optimised to design the complex microstructures hierarchy.

From a metallurgical point of view, the microstructure formation obtained by solid/solid phase transformations is easy to study by microscopy when considering β metastable alloys. Indeed, for these alloys it is possible to freeze the microstructures during a transformation without subsequent occurence of any displacive transformation at low temperatures.

During their thermo-mechanical treatments, most of the titanium alloys are deformed in the high temperature range where the most stable phase is β (bcc) and further cooled to room temperature where the material is a mixture of β and α (hcp) phases. Different morphologies of the α phase are observed depending on the cooling rate and the previous deformation conditions:

- for small departures from equilibrium conditions, α allotriomorph grains at the β grain boundaries are first observed (Fig. 1a).

- For greater departures from equilibrium conditions, α plates of widmanstätten have grown from or near the allotriomorph grains (Fig. 1b).
- And finally, for the greatest departures from equilibrium conditions, acicular α grains nucleated on defects are observed inside the β grains (Fig. 1c).

Furthermore, deforming the material prior to the cooling promotes the apparition of the allotriomorphs together with the widmanstätten plates.

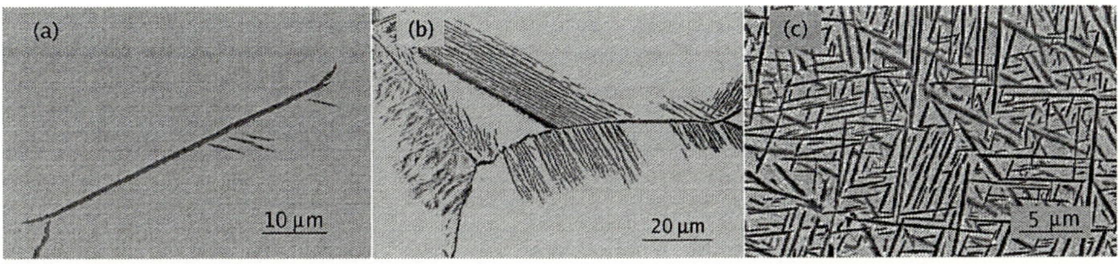

Fig. 1. Scanning electron microscopy (SEM) images of microstructures of a β titanium alloy (a) isothermally treated at 790 °C during 5 min.; (b) isothermally treated at 790 °C during 10 min.; (c) isothermally treated at 700 °C during 6 min. .

To explain these different morphologies, some mechanisms have been proposed ([1] and references therein) mainly from post-mortem observations with microscopy. However some features are unclear which could be explained thanks to 2D/3D numerical modelling.

2 Which Model to Choose?

Among the different modelling techniques which can tackle this kind of problem, the phase field models have proved to be the best suited for the moment. This is in great extent because the interfaces of complex morphologies have not to be tracked explicitely. Another advantage of this kind of models, which is of great importance for solid state transformations, is the possibility to incorporate easily phenomena which can be cast into the thermodynamical formalism, such as elasticity [2].

Two methodologies distinguish between phase field models: on the one hand those devoted to solid state transformations with full coherency at the interfaces [2]; and on the other hand those devoted to solidification. We have chosen the second one because of the nature of the α/β interfaces that observations with Transmission Electronic Microscope have shown to be at least semi-coherent [1].

We have used the model of Kim et al. [3] developed initially for the solidification of binary alloys because of its following features:

- first, it is thermodynamically consistent, i.e. the evolution equations are gradient flows for the free energy functional of the system. Hence, in principle, it should be quite easy to incorporate elasticity.
- Secondly, it is not linked to a particular phase diagram. The homogeneous free energy is defined with a kind of rule of mixture between the energies of both phases, augmented with the 'double well potential' accounting for the energy penalty of the interface.
- Thirdly, it is based on a special definition of the interface as a mixture of the two phases with different compositions but with equal chemical potentials, which prevents a chemical contribution to be added to the interfacial energy.
- Finally, a thin interface asymptotic analysis has been performed, in a simplified form compared to the original one of Karma et al. [4], which may allow the use of coarser grids than the classical sharp interface analysis, so as to enhance computational efficiency.

We have adapted this model to the case of our multicomponent titanium alloy with the composition reported in Tab. 1. The α phase can be assumed stoechiometric as suggested by experimental measurements. The β phase is described as an ideal solid solution for simplicity at a first step, because the agreement between our experimental measurements and a more sophisticated description based on the CALPHAD method was not satisfactory.
With the previous hypotheses, the evolution equations are straightforward to obtain and are not reported here for brevity.
Lastly, the necessary thermo-chemical data necessary for the calculations (i.e. equilibrium data and diffusion coefficients) have been obtained by suitable and careful experiments.

Table 1. Composition of the β CEZ alloy.

Molar %	Ti	Al	Mo	Cr	Fe	Zr	Sn
Nominal	balance	8.83	2.07	2.01	0.78	2.10	0.84
α phase	balance	10.90	0.19	0.27	0.04	2.10	0.84

3 Can we Rely on the Model to Get Quantitative Predictions?

In order to check if the model of Kim et al. provides quantitative information (i.e. kinetics) together with the morphological evolutions, we have first performed comparisons for a simple 1D planar growth problem between the

phase field model, a numerical sharp interface model and the analytical solution for a system of semi-infinite extent.
A nucleus of the growing α phase has been put at one side of the system, small as compared to the size of the system in order to avoid boundary effects on the growth rate, at least at the begining of the transformation. Moreover, zero flux boundary conditions were applied on both sides of the system. The initial composition in the β matrix has been set at the nominal one (Tab. 1). Then the system was quenched down to 830 ^{o}C (corresponding to an undercooling of 60 ^{o}C) and held at this temperature.
The main calculation parameters relative to this test case are reported in Tab. 2. Because the interface has been assumed to be at local equilibrium, the phase field mobility was calculated in such a way that the interfacial kinetics vanishes, thanks to a thin interface asymptotic analysis [3].

Even for a large interface width (i.e. 3 μm), a very good agreement between the different methods has been achieved on the growth velocity as shown in Fig. 2a. The agreement is also good for the time evolutions of the interfacial concentrations of all the chemical species except for Molybdenum (Fig. 2b). The moderate discrepancy for Molybdenum which is the slowest diffusing species ($D_{\mathrm{Mo}} = 5.8 \cdot 10^{-15}\, m^2/s$ at 830 ^{o}C) is due to solute trapping. Indeed, this effect is minimized by choosing smaller mesh sizes because the interfacial Péclet number for Molybdenum is not small enough for large mesh sizes (it is $\mathcal{O}\left(10^{-1}\right)$). Nonetheless, at this stage, we could hope to perform calculations on real systems with a moderate computational power (i.e. a single processor workstation). Thus, a series of calculations has been carried

Table 2. Parameters of the calculations.

	System size (μm or μm^2)	mesh size (μm or μm^2)	interface width (μm)	interfacial energy (J/m^2)	interf. kinetic coeff. (W/m^4)
1D	100	0.5	3	1	0
2D	20×20	0.15×0.15	0.6	0.3	0

out in 2D, starting from real SEM microstructures, with the main calculation parameters reported in Tab. 2. As a compromise between accuracy and moderate computational power, we picked up square zones of 20×20 μm^2 in the SEM images, that have been recasted with a pixel resolution of 128×128 corresponding to the computational grid.

The first calculation was performed starting from equilibrium microstructures at 870 ^{o}C. The initial composition was set at the equilibrium one in the β phase for this temperature. Then, the system was quenched down to 830 ^{o}C and held at this temperature for 1000 s. Figs. 3a, b and c show snapshots of the microstructures respectively at 0 s, 100 s and 1000 s.

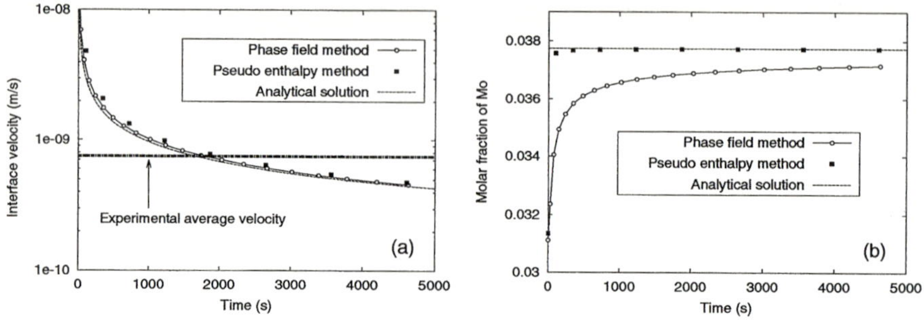

Fig. 2. Comparison of the phase field model with a pseudo-enthalpy model and an analytical solution for the growth of a planar interface: (a) growth rate; (b) molar fraction of Mo at the interface in the β matrix.

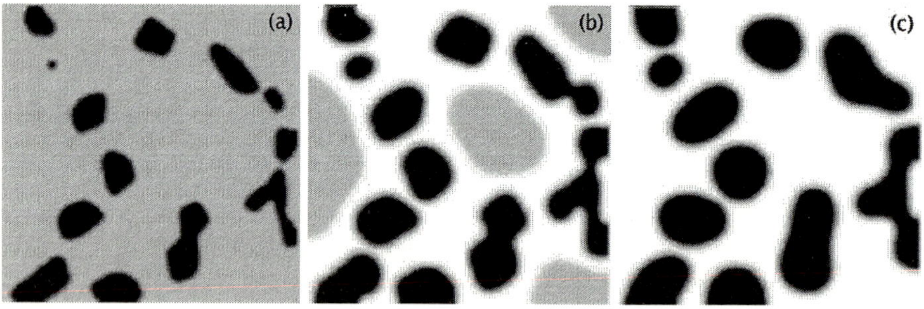

Fig. 3. Snapshots of the 2D microstructures (α precipitates are in black and the grey levels visualize the molar fraction field of Al) when quenched from 870 °C to 830 °C and held at this temperature for (a) 0 s; (b) 100 s; (c) 1000 s.

The α phase amount and mean size increase as expected. However, because we did not consider any anisotropy (interfacial energy or interfacial kinetics), the α precipitates become rounded as the transformation proceeds, contrary to what is observed experimentally. Indeed, in experiments, growth is slightly favoured along β grain boundaries. Nonetheless, a qualitative comparison between SEM images of equilibrium microstructures at 830 °C and the snapshot at 1000 s reveals a quite good agreement for the mean size of the α precipitates. Concerning the kinetics, the transformation is nearly complete after 1000 s in agreement with resistivity measurements of transformations in similar conditions.

We repeat the same kind of calculation but quenching microstructures obtained at 820 °C down to 750 °C (so involving a driving force for the transformation greater than in the preceding case). In Fig. 4 snapshots of the phase field at successive times show the appearance of a spurious intermediate state (in grey) which is destabilized as the transformation progresses. This spurious state gives rise to alternating bands of α and β phases which finally split into strings of small α spherical precipitates. Following Glasner et al. [5], this phenomenon can be attributed to the existence of an internal minimum of the free energy density with respect to the phase field when a parameter proportional to the driving force and to the interface width is greater than a critical value. This may be a serious limitation of the model when applied to solid state transformations because the driving forces involved are usually large, and should be offset against a very small interface width to get rid of the spurious state.

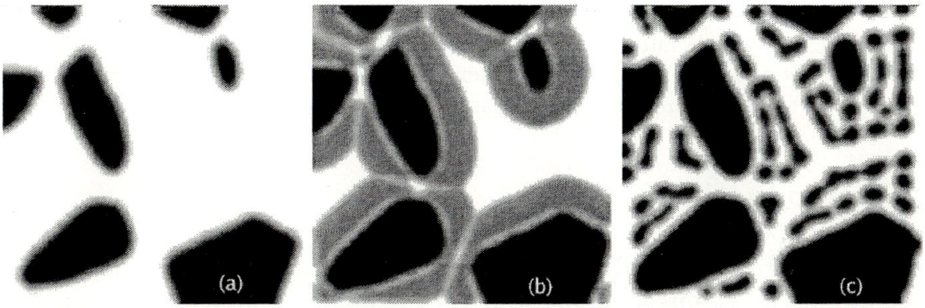

Fig. 4. Snapshots of the phase field (black and white correspond to the α and β phases, and grey represents the spurious intermediate state) when quenched from 820 °C to 750 °C and held at this temperature for (a) 0 s; (b) 70 s; (c) 500 s.

References

1. Aaronson, H.I.: Atomic mechanisms of diffusional nucleation and growth and comparisons with their counterparts in shear transformations. Met Trans. A **24** (1993) 241–276
2. Jin, Y.M., Artemev A., Khatchaturyan, A.G.: Three-dimensional phase field model of low-symmetry martensitic transformation in polycristal: simulation of ζ_2' martensite in AuCd alloys. Acta Mater. **49** (2001) 2309–2320
3. Kim, S.G., Kim, W.T., Suzuki, T.: Phase-field model for binary alloys. Phys. Rev. E **60** (1999) 7186–7197
4. Karma, A., Rappel, W.-J.: Quantitative phase-field modeling of dendritic growth in two and three dimensions. Phys. Rev. E **57** (1998) 4323–4349
5. Glasner, K., Almgren, R.: Dual fronts in a phase field model. Physica D **146** (2000) 328–340

Spreading of Liquid Monolayers: From Kinetic Monte Carlo Simulations to Continuum Limit

M. N. Popescu and S. Dietrich

Max-Planck-Institut für Metallforschung, Heisenbergstraße 1, D-70569 Stuttgart and Institut für Theoretische und Angewandte Physik, Universität Stuttgart, Pfaffenwaldring 57, D-70569 Stuttgart, Germany

Abstract. We study with kinetic Monte Carlo simulations the two-dimensional spreading of a fluid monolayer that is extracted from a reservoir. The asymptotic spreading behavior is analyzed within a continuum limit and compared with previous predictions. The influence of a chemically patterned substrate (longitudinal stripes) on the spreading behavior of such fluid monolayers is briefly discussed.

1 Introduction

Recent experiments on dynamics of wetting [1] performed with nano-liter volumes of liquid spreading on atomically smooth surfaces have shown one or few layers of a precursor film with *molecular-thickness* and *macroscopic extent* advancing in front of the macroscopic wedge of the droplet. For both spreading of nano-droplets and capillary rise geometries the linear extent $X(t)$ of the extracted film grows in time according to $X(t) = \tilde{A}t^{1/2}$ as long as the reservoir (droplet) is not emptied. Several models have been proposed [2–4] to explain these observations. For the case of a single monolayer precursor, a microscopic approach based on the dynamics of a lattice gas of interacting particles has been developed by Burlatsky et al. [2] and recently generalized by Oshanin et al. [3]. Several mean-field assumptions allowed them to derive analytically the $t^{1/2}$ law for the spreading and the prefactor $\tilde{A}$.

In this paper we present results of kinetic Monte Carlo (KMC) simulations of a model closely related to the one in Refs. [2] and [3]. The asymptotic behavior of the transversally averaged density profile is analyzed within a continuum limit and compared with the previous mean-field predictions [2,3]. Finally, we briefly discuss the influence of a chemically patterned substrate with longitudinal stripes on the spreading behavior of such liquid monolayers.

2 Model and KMC Simulations

The following rules describe the model.
(i) The spreading geometry is rectangular ($x-y$ plane). The half-plane $x < 0$ is occupied by a reservoir which maintains a fixed *average* density C_0 on the line $x = 0$. C_0 is defined as the average of the local occupation number

$\eta(\mathbf{r};t) \in \{0,1\}$ on the lattice sites $(x=0,y)$ (see (ii) below). At time $t=0$ the half-plane $x>0$ is empty; thus, the liquid spreads in the x-direction.

(ii) The substrate is homogeneous, and the substrate-fluid interaction is modeled as a periodic potential (square lattice) in which the particle motion proceeds by activated jumps to nearest-neighbor wells. Evaporation from the substrate is not allowed. The activation barrier U_A defines the hopping rate $\Omega = \nu_0 \exp[-U_A/k_B T]$, where ν_0 is an attempt frequency which defines the time unit $1/\nu_0$, k_B is the Boltzmann constant, and T is the temperature. This hopping rate is absorbed into the one-particle diffusion coefficient $D_0 = \Omega a^2/z$ on the bare substrate, where z is the coordination number of the underlaying lattice, and a is the lattice constant.

(iii) The pair interaction between fluid particles is assumed to be strongly repulsive at short range, preventing double occupancy of the wells, i.e., hard-core repulsion, with a long-ranged attractive interaction $-U_0/r^6$ for $r/a \geq 1$, mimicking dispersion forces.

(iv) The probability of a jump from lattice site $\mathbf{r}$ to $\mathbf{r}'$ is biased by the fluid-fluid energy landscape and is given by

$$p(\mathbf{r} \to \mathbf{r}';t) = Z^{-1}(\mathbf{r};t)\exp\{\beta[\tilde{U}(\mathbf{r};t) - \tilde{U}(\mathbf{r}';t)]/2\}, \tag{1}$$

where $1/\beta = k_B T$, $Z(\mathbf{r},t) = \sum_{\mathbf{r}'} \exp\{\beta[\tilde{U}(\mathbf{r};t) - \tilde{U}(\mathbf{r}';t)]/2\}$ is the normalization factor, $\tilde{U}(\mathbf{r};t) = -U_0 \sum_{|\mathbf{r}-\mathbf{r}''| \leq 3a} \frac{\eta(\mathbf{r}'';t)}{|\mathbf{r}-\mathbf{r}''|^6}$, and the summation has been restricted to three lattice units for computational convenience. This is similar to the cut-off used in molecular dynamics simulations for the attractive part of Lennard-Jones interactions.

(v) The advancing edge Γ_t of the monolayer at time t is defined as the union of the most advanced particles from each line along the spreading direction. In order to control the rate of *two*-dimensional "evaporation", an additional constraint is imposed by rejecting the moves from Γ_t toward sites $\mathbf{r}'$ ahead of Γ_t for which $|\tilde{U}(\mathbf{r}';t) + U_0| < U_c$, where $U_c \geq 0$ is a fixed threshold value. We have used $U_c = U_0/3^6$, a value for which the evaporation is negligible.

The simulations have been carried using periodic boundary conditions along the transversal, i.e., y-direction (appropriate for simulating an infinitely wide substrate), a hopping barrier $\beta U_A = 3.5$, and a time increment Δt which is a random variable distributed according to $P(\Delta t) = N\Omega \exp(-N\Omega\Delta t)$, N being the number of particles in the film at time t [5]. The density $\rho(\mathbf{r};t)$ is defined as $\rho(\mathbf{r};t) = \langle \eta(\mathbf{r};t) \rangle$, where $\langle \cdot \rangle$ means average over different KMC runs. The transversally averaged density $C(x,t)$ and the average position of the advancing edge of the monolayer are defined as $C(x,t) = \langle L^{-1} \sum_{y=1}^{L} \eta(x,y;t) \rangle$ and $X(t) = \langle L^{-1} \sum_{\mathbf{r} \in \Gamma_t} x \rangle$, respectively, where L is the width of the lattice.

3 Results and Discussion

A. Time dependence of spreading

Figure 1(a) shows the scaled time dependence of the front line, $X(t)/\sqrt{t}$, for several values of $W_0 = \beta U_0$ and C_0. The results confirm that the spreading follows $X(t) = \tilde{A}t^{1/2}$, and the time independent dimensionless prefactor $A \equiv \tilde{A}/\sqrt{D_0}$ depends on both W_0 and C_0 as shown in Fig. 1(b). For

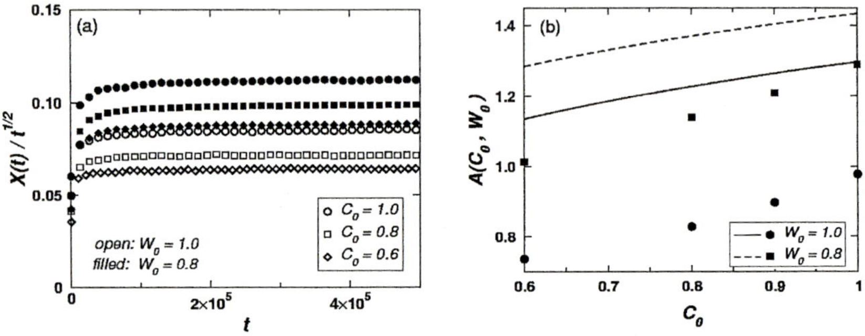

Fig. 1. (a) Position of the front line as a function of time for $W_0 = 0.8$ (filled symbols) and $W_0 = 1.0$ (open symbols), respectively. (b) Dependence of the prefactor A on the density of the reservoir, C_0, for $W_0 = 1.0$ (circles) and $W_0 = 0.8$ (squares), respectively. Symbols: KMC results, lines: mean-field prediction from Ref.[2].

comparison, we also show (lines) in Fig. 1(b) the theoretical predictions [1] from Ref. [2]. The results of the simulations are smaller, and the decrease with increasing W_0 seems to be more pronounced than predicted. These discrepancies are most probably due to the fact that the simulations capture, even far away from the advancing edge, the existence of a very small asymmetry between the rates of jumping toward the reservoir and the advancing edge, respectively, which has been neglected in the mean-field description [2,3].

B. Continuum limit

Assuming $< \eta(\mathbf{r};t)\eta(\mathbf{r}';t') >=< \eta(\mathbf{r};t) >< \eta(\mathbf{r}';t') >$, i.e., neglecting correlations, the following master equation for the local density $\rho(\mathbf{r};t)$ is obtained:

$$\frac{d\rho(\mathbf{r};t)}{d(\Omega t)} = \sum_{\mathbf{r}'}\{-\rho(\mathbf{r};t)p(\mathbf{r}\to\mathbf{r}')[1-\rho(\mathbf{r}';t)]+[1-\rho(\mathbf{r};t)]p(\mathbf{r}'\to\mathbf{r})\rho(\mathbf{r}';t)\}.$$

[1] In order to use Eq. (8) in Ref. [2], the value μ for the ratio of the probabilities for jumping ahead and back from Γ_t is needed. We have determined this ratio from our KMC data and obtained $\mu \simeq 0.95$ for $W_0 = 0.8$ and $\mu \simeq 0.93$ for $W_0 = 1.0$.

In the continuum space and time limit ($a \to 0$, $\Omega^{-1} \to 0$, D_0 finite), by taking Taylor expansions for $p(\mathbf{r} \to \mathbf{r}')$ and $\rho(\mathbf{r}'; t)$ around $\mathbf{r}$ and keeping terms up to second order in the spatial derivative of $\rho(\mathbf{r}; t)$, one obtains [3]

$$\partial_t \rho = D_0 \{ \nabla^2 \rho + \beta \nabla \left[\rho (1-\rho) \nabla U \right] \}. \tag{3}$$

Using [2] $U(\mathbf{r}; t) \simeq -g U_0 \rho(\mathbf{r}; t)$, where $g = \sum_{a \le |\mathbf{r}| \le r_c} |\mathbf{r}/a|^{-6}$ is a geometrical factor depending on the lattice type and on the cut-off range r_c, assuming that the monolayer is homogeneous along y, which is supported by the simulation results, and changing variables to $\lambda = x/\sqrt{D_0 t}$, one obtains that asymptotically ($t \gg 1$) the density $C(\lambda)$ should satisfy [7]

$$\frac{d^2 C}{d\lambda^2} + \frac{\lambda}{2} \frac{dC}{d\lambda} - g W_0 \frac{d}{d\lambda} \left[C(1-C) \frac{d\rho}{d\lambda} \right] = 0. \tag{4}$$

An important consequence of Eq. (4) is the existence of a critical value $W_{0,c}$ [6] given by $W_{0,c} = 4/g$. For the present case of a square lattice with a cut-off at $r_c = 3$ one has $g \simeq 4.64$, and thus $W_{0,c} \simeq 0.86$. Here we will discuss only results for values $W_0 < W_{0,c}$ not too close to the critical value, in particular $W_0 = 0.6$. A detailed analysis for $W_0 \gtrless W_{0,c}$ will be presented elsewhere [7].

In Fig. 2(a) we show the density profiles as a function of the scaled variable λ (and as a function of x in the inset) for $C_0 = 1.0$. The data collapse for

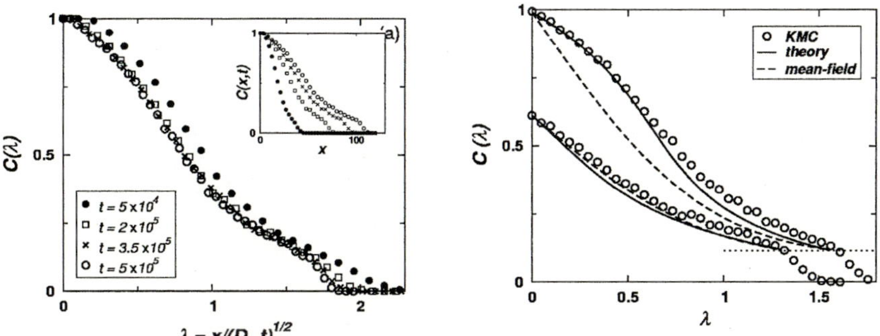

Fig. 2. (a) Density profiles $C(\lambda)$ as a function of the scaling variable λ (respectively x in the inset). (b) Density profiles $C(\lambda)$ for $C_0 = 1.0$ and $C_0 = 0.6$ obtained from Eq. (4) (full lines), KMC simulations (symbols), and the mean-field approach (dashed lines) from Ref. [2]. The horizontal dotted line shows $C_1 = 0.12$.

times $t > 1 \times 10^5$ is good and thus validates the scaling assumption leading to Eq. (4). We have obtained similar results for all the values W_0 and C_0 which have been investigated. In order to numerically integrate Eq. (4) for a

[2] The first order term, $\sim \nabla \rho$, is zero due to symmetry and summation over all $\mathbf{r}'$.

quantitative analysis of the predicted density profiles, it must be augmented by corresponding initial or boundary conditions. One boundary condition, i.e., $C(\lambda = 0) = C_0$ is obvious, and a second boundary condition can be obtained [7] at $\lambda = A$ based on the constraint (v) in Sec. 2 as $C_1 \equiv C(\lambda = A) \simeq 0.12$ (for $U_c = U_0/3^6$). In Fig. 2(b) we show the predicted density profiles (full lines) $C(\lambda)$ for $C_0 = 1.0$ and $C_0 = 0.6$, respectively, along with results of KMC simulations (symbols) at time $t = 5 \times 10^5$ (which can be regarded as being close to the asymptotic limit) and with predictions of the mean-field approach (dashed lines) from Ref. [2]. At low densities of the reservoir the mean-field results and the predictions of Eq. (4) are almost identical because the gas is dilute and thus the correlations are less effective. At high densities of the reservoir, however, there are significant differences in the qualitative behavior of the two theoretical predictions, and the simulation results clearly favor the behavior predicted by Eq. (4).

4 Patterned Substrates

As a simple example of spreading on heterogeneous substrates we are investigating the case of chemical patterns in the form of longitudinal (along the x-direction) wettable stripes on a non-wettable substrate. We choose the same lattice constant a for both. The difference between wettable and non-wettable regions is modeled, as shown in Fig. 3(a), through an additional energy barrier ΔU at the border between the two types of substrates, which acts like a spatially varying chemical potential.

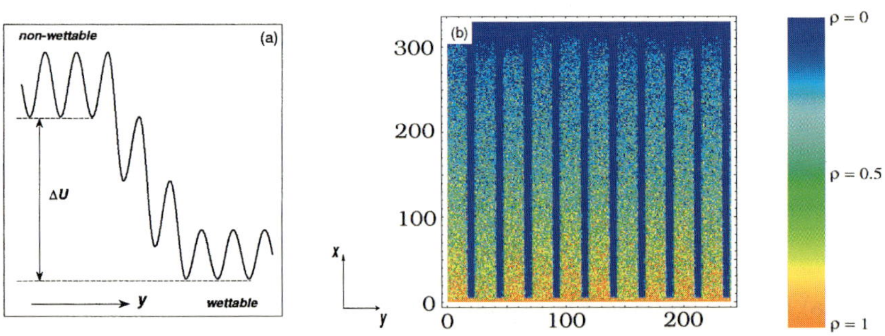

Fig. 3. (a) Potential landscape at the border between the two types of substrates (left: non-wetting, right: wetting). (b) Density profile $\rho(\mathbf{r};t)$ at time $t = 10^5$ for spreading (along the vertical direction) on a periodic distribution of longitudinal wettable stripes (width $= 16a$) separated by non-wettable stripes (width $= 6a$) for $C_0 = 0.9$ and $W_0 = 0.4$. The color coding (right) is a linear function of density.

This potential difference can be changed in real time by manipulating the substrate properties, e.g., by using photo responsive coatings as in Ref. [8]. Adjusting ΔU to a large enough value should facilitate a confining of the spreading. Such an example of control is shown in Fig. 3(b). For the parameters used, $\beta U_A = 0.8$ (wettable), $\beta U_A = 1.0$ (non-wettable), and $\beta \Delta U = 6.5$, respectively, the long-time confined-spreading is achieved even with very narrow ($6a$) separating, non-wettable stripes.

5 Summary and Conclusions

We have studied with KMC simulations a model for the spreading of a liquid monolayer being extracted from a reservoir, and we have briefly discussed the influence of a chemically patterned substrate on the spreading behavior of such liquid monolayers. Our main results are:
1) The model captures the dependence $X(t) = \tilde{A}t^{1/2}$ for the spreading edge, but the previous mean-field approach [2] overestimates the rate of spreading.
2) Mean-field theory correctly predicts the density profiles $C(\lambda)$ for small values C_0 and W_0, but breaks down at large values; an explicit inclusion of the particle attraction as in Eq. (4) leads to correct predictions for all C_0.
3) Patterning of the substrate facilitates a confining of the liquid.
Work in progress [7] addresses further questions like the effect of impurities and of external fields on the spreading behavior.

Acknowledgments

Research done within the priority program "Wetting & Structure Formation at Interfaces", Grant DI 315/7-3, of the Deutsche Forschungsgemeinschaft.

References

1. F. Heslot, A. M. Cazbat, and N. Fraysse: Diffusion-controlled Wetting Films. J. Phys.: Cond. Matt. **1** (1989) 5793-5798; F. Heslot, A. M. Cazbat, P. Levinson, and N. Fraysse: Dynamics of Wetting of Tiny Drops: Ellipsometric Study of the Late Stages of Spreading. Phys. Rev. Lett. **62** (1989) 1286-1289; N. Fraysse, M. P. Valignat, F. Heslot, A. M. Cazbat, and P. Levinson: The Spreading of Layered Microdroplets. J. Coll. Int. Sci. **158** (1993) 27-32.
2. S. F. Burlatsky, G. Oshanin, A. M. Cazbat, and M. Moreau: Microscopic Model of Upward Creep of an Ultrathin Wetting Film. Phys. Rev. Lett. **76** (1996) 86-89.
3. G. Oshanin, J. De Coninck, A. M. Cazabat, and M. Moreau: Microscopic Model for Spreading of a Two-dimensional Monolayer. J. Mol. Liquids **76** (1998) 195-219.
4. P. G. de Gennes and A. M. Cazbat: Films Diffusifs de Mouillage. C. R. Acad. Sci. **310 II** (1990) 107-111; D. B. Abraham, R. Cuerno, and E. Moro: A Microscopic Model for Thin Film Spreading. preprint, cond-mat/0201218.
5. E. Adam, L. Billard, and F. Lançon: Class of Monte Carlo Algorithms for Dynamic Problems Leads to an Adaptive Method. Phys. Rev. E **59** (1999) 1212-1216.
6. G. Giacomin: Van der Waals Limit and Phase Separation in a Particle Model with Kawasaki Dynamics. J. Stat. Phys. **65** (1991) 217-234.
7. M. N. Popescu and S. Dietrich: unpublished.
8. S. Abbott, J. Ralston, G. Reynolds, and R. Hayes: Reversible Wettability of Photoresponsive Pyrimidine-Coated Surfaces. Langmuir **15** (1999) 8923-8928.

A Multi-mesh Finite Element Method for Phase-field Simulations

Alfred Schmidt

Zentrum für Technomathematik, Universität Bremen, Germany
schmidt@math.uni-bremen.de

Abstract. In phase field models, different components of the solution (temperature and phase variable) exhibit a strongly different local behaviour. An optimal discretization should use a mesh for each component. We present a general framework for the adaptive solution of coupled systems and its application to phase field simulations.

Free boundary problems give rise to coupled systems of partial differential equations, which are prominent examples of systems where a careful numerical discretization is needed in order to resolve the solution's behaviour. Usually, a high resolution is essential near the free boundary, while coarser meshes are sufficient in the bulk. A reasonable numerical method should use locally adapted meshes to fulfil both the needs of accuracy and efficiency. For coupled systems like the phase field system, where components of the solution show strongly different local behaviour, an optimal discretization should use specially adapted meshes for each component.

1 Models for Solidification

An undercooling or oversaturation of a liquid leads to a rapid solidification of the material. Models for this behaviour include energy and/or mass transport by diffusion and/or convection as well as energy/mass conservation conditions across the interface, like the Stefan condition. On a certain (meso-) scale, additional surface effects at the phase boundary play an important role, depending on curvature C or velocity V of the moving interface Γ, usually described by a Gibbs-Thomson relation like

$$\varepsilon_c C_\Gamma + \varepsilon_v V_\Gamma + \theta = 0 \qquad \text{on } \Gamma.$$

Anisotropic surface effects lead to dendritic growth.

Mathematical models for solidification with surface effects differ mainly in the treatment of the free boundary, defining it as a sharp interface, a level set, or a diffuse interface (phase field). Numerical methods for solidification simulations are developed following the same lines.

1.1 Sharp Interface Models

In a sharp interface model, the phase boundary is a (smooth) hyper-surface of the underlying space (a surface in 3D or a curve in 2D). The advantage of the treatment of the interface as a smooth surface is its lower dimension and that curvature, which appears in the Gibbs-Thomson law, is well defined. Drawbacks are that the standard model allows no changes in interface topology, no nucleation. The notion of varifolds allows for some possibilities in this direction.

The free boundary problem leads to a degenerate parabolic equation for the interface motion, like an anisotropic mean curvature flow equation, coupled with heat or mass diffusion in solid and liquid phases. Additionally, convection can be added by coupling to Navier-Stokes' equations in the time dependent liquid phase.

Numerical Methods for Sharp Interface Simulations. The sharp interface model leads to numerical methods with separate discretizations of bulk (3D) and interface (2D) (or 2D & 1D).

In the bulk, a fine mesh is essential near the interface to sufficiently resolve the behaviour of solutions (temperature, concentration). Adaptively refined meshes in finite element methods are crucial to get enough resolution, especially in 3D. A numerical sharp interface finite element method with adaptive meshing in 2D and 3D was presented in [7,8], with additional convection in 2D in [2].

1.2 Diffuse Interface Models

Phase field models can be formulated either with a double obstacle potential or a smooth double well potential. We want to restrict ourself here to an obstacle formulation, as introduced by Blowey and Elliott [3].

The model describes the evolution of temperature θ and phase variable χ in a domain Ω and is a system of two (degenerate) parabolic equations:

$$\partial_t(\theta + \lambda\chi) - \kappa\Delta\theta = f$$
$$\varepsilon\,\partial_t\chi - \varepsilon\,\mathrm{div}(a(\nabla\chi)) + \Lambda(\chi) - \tfrac{1}{\varepsilon}\beta\chi \ni \gamma\theta$$

in $\Omega \times (0,T)$ plus boundary and initial conditions. The function a may include anisotropic solidification parameters, and Λ is a (set valued) maximal monotone graph, the subdifferential of the double obstacle potential,

$$\Lambda(s) = \begin{cases} (-\infty, 0] & \text{if } s = -1 \\ 0 & \text{if } s \in (-1,1) \\ [0, +\infty) & \text{if } s = +1 \end{cases}$$

with the effect that values of χ are in the interval $[-1, +1]$. f is a given heat source density, λ, κ, and β are non-negative, material-dependent coefficients, and $\varepsilon > 0$ a small parameter. The phase variable χ is equal to -1 (solid) or $+1$ (liquid) everywhere but in a narrow transition region of width $O(\varepsilon)$. For $\varepsilon \to 0$, the solution converges against a solution of the sharp interface model with a corresponding Gibbs-Thomson relation. Figure 1 shows the evolution of the phase boundary during the solidification of an undercooled liquid.

Fig. 1. Solid-liquid interface at different times

Both components of the solution, temperature and phase variable, show a strongly different local behaviour, see Fig. 2.

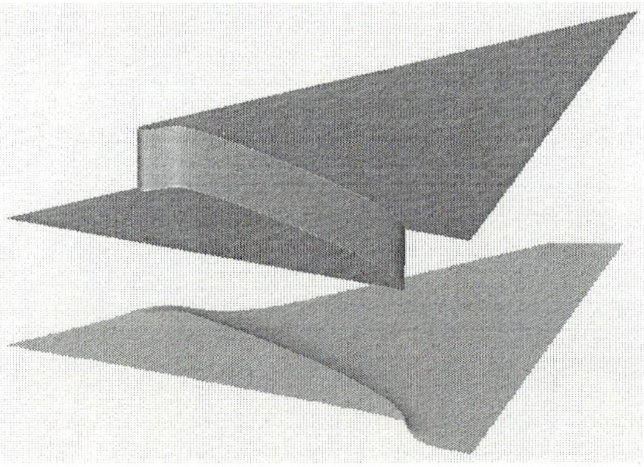

Fig. 2. Graphs of phase variable and temperature on 1/8 domain

- The phase variable is constant outside of a moving narrow strip of width $O(\varepsilon)$, where the phase transition occurs. Here, $|\nabla\chi| = O(\varepsilon^{-1})$.
- The temperature satisfies the heat equation outside this strip, thus it is smooth. Inside the moving strip, the gradient of temperature changes

rapidly. In the limit $\varepsilon \to 0$, the Stefan condition holds at the sharp interface,

$$(\kappa\nabla\theta_{\text{liquid}} - \kappa\nabla\theta_{\text{solid}}) \cdot n = -\lambda V.$$

In order to resolve this behaviour with a finite element approximation, different requirements to the discretization hold for both components.

- The mesh for discretization of χ must have a local mesh width of $h_S < c\varepsilon$. Outside this strip, where χ is constant, the mesh might be arbitrarily coarse. In order to be able to track nucleations, the phase field system should be solved in the whole domain, not only near the current interface.
- To resolve the temperature behaviour, a much coarser mesh is sufficient in the strip than is needed for the phase variable. On the other hand, the mesh must have a sufficient fineness also in the rest of the domain.

A separate discretization for both temperature and phase variable is needed in order to meet all requirements in an efficient numerical method.

2 Adaptive Finite Element Methods for Coupled Systems of PDE

Besides phase transitions, many physical problems lead to coupled systems of partial differential equations, too. Especially in case of nonlinear phenomena, the components u_i, $i = 1, \ldots, n$ may show a strongly different behaviour (smoothness of solutions etc.) in the common underlying domain $\Omega \subset \mathbb{R}^d$.

The usual adaptive discretization for systems of PDEs uses the same mesh for all components, locally refined based on a posteriori error indicators for the sum of error contributions on each mesh element.

An optimal adaptive discretization should use different locally refined meshes for different components of the solution, controled by separate error indicators for each component.

In [6] we present a general concept for adaptive finite element methods for stationary or time dependent coupled problems.

Solution components u_i are discretized in finite element spaces X_i, which are based on different locally refined simplicial grids $\mathcal{S}_i$, where all meshes $\mathcal{S}_i$ are refinements of the same macro triangulation $\mathcal{S}_0$ of Ω, see Fig. 3 for a simple example.

The advantages of this approach are

- separately adapted mesh for each component of the solution;
- altogether less degrees of freedom, thus more efficient;
- by the common macro mesh, a direct local hierarchy of the triangulations $\mathcal{S}_i$ is given. This enables an *exact* evaluation of each other component U_j on the elements of mesh $\mathcal{S}_i$.

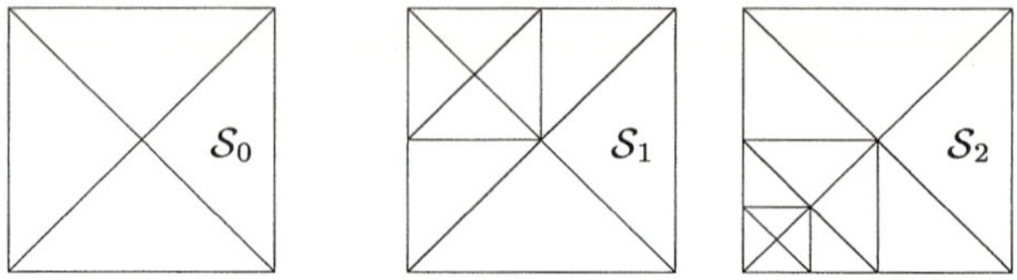

Fig. 3. Two different refinements of a macro triangulation $\mathcal{S}_0$

A somewhat similar idea for phase field simulations was proposed by Elliott and Gardiner [5], who use two separate meshes for temperature and phase variable. A fine mesh with mesh size $h \ll \varepsilon$ for the phase variable is used only near the transition region (selected by a 'mask'), a coarse mesh with mesh size $H = 4h$ for the temperature. Use of the 'mask' reduces computations for the phase variable significantly, but does not allow to track nucleations, e. g. Both meshes are aligned, but no error estimators or local mesh refinements are used.

2.1 Aspects of Implementation

The adaptive multi-mesh method is implemented in the finite element toolbox ALBERT, a joint development with K. G. Siebert [9,10]. The toolbox uses simplicial meshes in 2D (triangles) and 3D (tetrahedra) and local refinement by bisection of elements, which induces a hierarchical structure of meshes and finite element spaces.

The multi-mesh concept introduces the traversal of a common, *virtually refined* mesh $\mathcal{S}_M$, see Fig. 4, for calculation of coupling terms, which involve several components or functions from finite element spaces defined on different meshes, like $\int_\Omega U_i \Phi$ with $U_i \in X_i$, $\Phi \in X_j$.

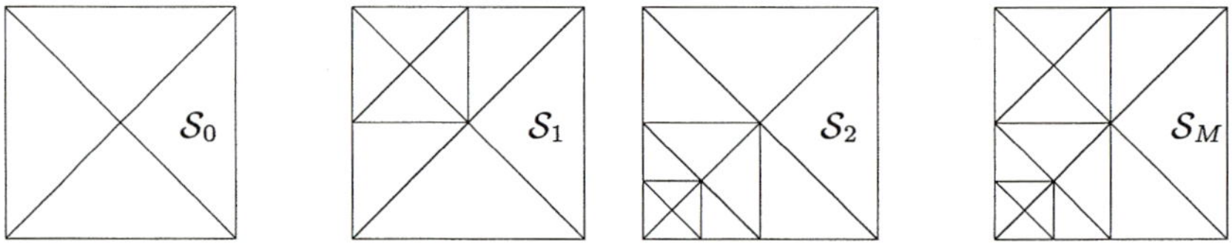

Fig. 4. Virtual mesh $\mathcal{S}_M$ with locally maximal refinement from $\mathcal{S}_1, \mathcal{S}_2$

2.2 Numerical Analysis and Adaptive Methods

We recall the standard a posteriori error estimates and adaptive finite element methods for a scalar elliptic problem. The error $||u-U||$ between the solution u

and a discrete approximation U is estimated by a sum of local error indicators, e. g.

$$\|u - U\|^2 \le \sum_{S \in \mathcal{S}} \eta_S^2,$$

where the indicators η_S must be computable from the discrete solution U and given data of the problem (*a posteriori*).

On an quasi-optimal mesh for a given error tolerance *tol*, the local error indicators are equally distributed over all mesh elements,

$$\eta_S^2 \approx \frac{tol^2}{\#\mathcal{S}} \qquad \text{for all } S \in \mathcal{S},$$

compare [1], e. g. A similar approach is used in case of coupled systems. The goal is an estimate of the errors $\|u_i - U_i\|$ by a sum of local error indicators on separate meshes, like

$$\sum_{i=1}^{n} \|u_i - U_i\|^2 \le \sum_{S \in \mathcal{S}_1} \eta_{1,S}^2 + \cdots + \sum_{S \in \mathcal{S}_n} \eta_{n,S}^2.$$

In order to describe a quasi-optimal set of meshes, one possibility is to split the total tolerance *tol* to the separate meshes

$$tol_1^2 + \cdots + tol_n^2 = tol^2$$

and optimize every single mesh. This leads to decoupled equidistribution conditions

$$\eta_{i,S}^2 \approx \frac{tol_i^2}{\#\mathcal{S}_i} \quad \text{for all } S \in \mathcal{S}_i, i = 1, \ldots, n.$$

3 Adaptive Method for Phase-field Models

In a joint paper with Z. Chen and R.H. Nochetto [4], we derive error estimates and adaptive methods for the double obstacle phase field system. Denoting by $u = \theta + \lambda\chi$ the energy density, and again discrete (finite element) functions by uppercase letters, the estimate looks like

$$\begin{aligned} &\|u - U\|_{L^\infty(0,T;H^{-1}(\Omega))} + \sqrt{\varepsilon}\|\chi - X\|_{L^\infty(0,T;L^2(\Omega))} \\ &+ \left(\int_0^T \|\theta - \Theta\|_{L^2(\Omega)}^2 + \alpha\varepsilon\, \|\nabla\chi - \nabla X\|_{L^2(\Omega)}^2 \, dt\right)^{1/2} \\ &\le \text{computable terms, localizable to mesh elements.} \end{aligned}$$

In [4], we use a common mesh for the discretization of temperature and phase variable. The local error indicators η_S contain contributions from both components,

$$\text{Estimated error } \le \eta_0 + \max_m \sum_{S \in \mathcal{S}^m} \eta_S^2,$$

where m is the time step index. A quasi-optimal common mesh fulfils the equidistribution condition for the local indicators

$$\eta_S^2 \approx \frac{tol^2}{\#\mathcal{S}^m} \quad \text{for all } S \in \mathcal{S}^m.$$

Here, the total mesh element count $\#\mathcal{S}^m$ is very large because of the fine resolution in the strip, thus the local temperature error must be very small also in elements far from the interface.

Adaptive FE Method with Separate Meshes

A separation of contributions from temperature and phase variable in the error indicator is possible:

$$\text{Estimated error } \leq \eta_0 + \max_m \Big(\sum_{S \in \mathcal{S}_\theta^m} \eta_{\theta,S}^2 + \sum_{S \in \mathcal{S}_\chi^m} \eta_{\chi,S}^2 \Big).$$

A split of the tolerance to both meshes, as described above, leads to a smaller element count $\#\mathcal{S}_\theta^m$, thus larger local error tolerances for the temperature error. It follows that the temperature mesh contains much less elements than a common mesh. Also the mesh for the phase variable contains less elements, as a fine resolution is needed only in the transition region. Fig. 5 shows element counts over time. A time-dependent tolerance was used for the error in each time step, which reflects the fact that the interface length grows during the simulation.

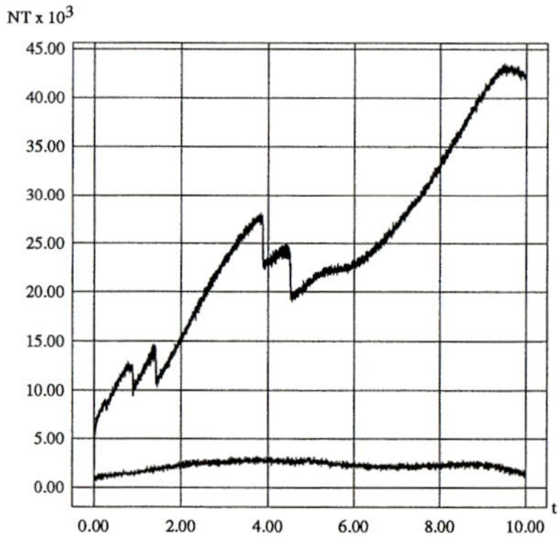

Fig. 5. Element counts for phase variable and temperature meshes over time

Figures 6 and 7 show the temperature and phase variable meshes with zooms to the interface region. Finally, we present in Figs. 8-10 a comparison of temperature and phase variable meshes from simulations with three different error tolerances *tol*.

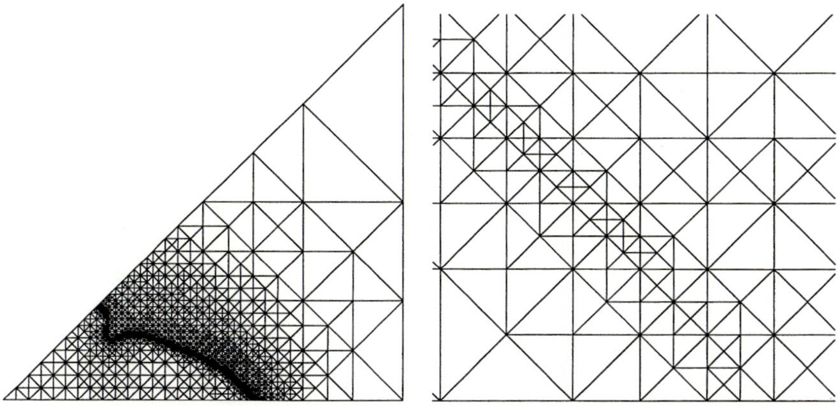

Fig. 6. Temperature mesh (2608 elements) and zoom to interface

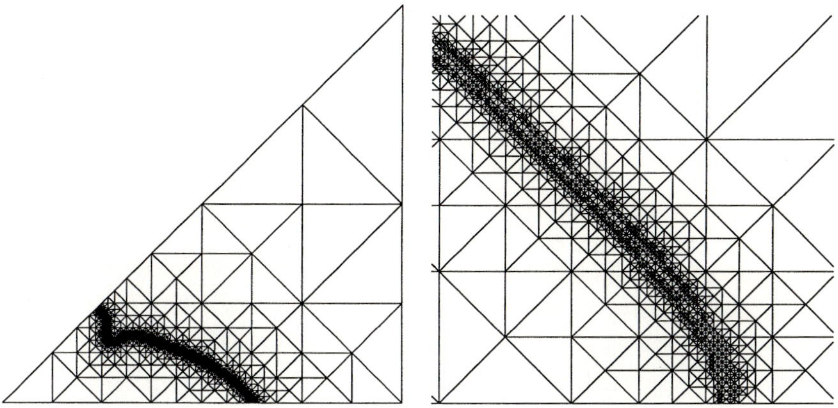

Fig. 7. Phase variable mesh (23015 elements) and zoom

Conclusion

We described an efficient adaptive finite element method for phase field calculations, together with some 2D simulation results. A more detailed description of the multi-mesh method and results from 3D simulations will be presented in forthcoming articles.

Acknowledgements

This article reports partly joint work with Z. Chen (Beijing), R. H. Nochetto (College Park), and K. G. Siebert (Freiburg).

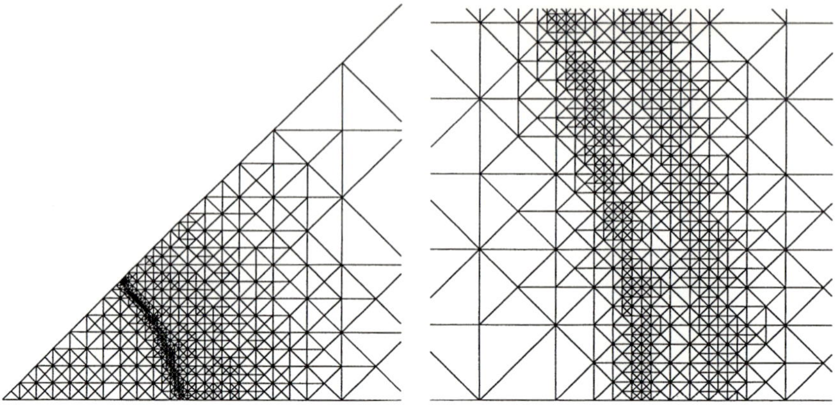

Fig. 8. *tol*=10: Meshes with 1122 resp. 7440 elements

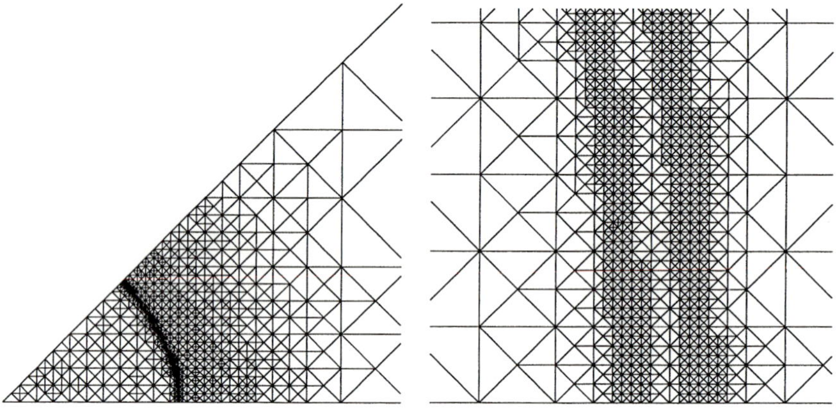

Fig. 9. *tol*=7: Meshes with 1568 resp. 11813 elements

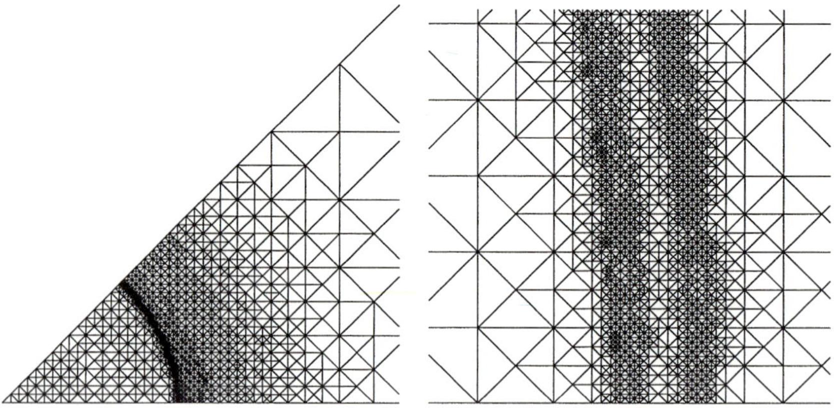

Fig. 10. *tol*=5: Meshes with 2245 resp. 20631 elements

References

1. Babuška, I. and Rheinboldt, W., *Error estimates for adaptive finite element computations*, SIAM J. Numer. Anal., 15 (1978), pp. 736–754.
2. Bänsch, E. and Schmidt, A., *Simulation of dendritic crystal growth with thermal convection*, Interfaces and Free Boundaries, 2 (2000), pp. 95–115.
3. Blowey, J. and Elliott, C., *Curvature dependent phase boundary motion and parabolic double obstacle problems.*, in Ni, Wei-Ming (ed.) et al., Degenerate diffusions. IMA Vol. Math. Appl. 47, 19-60 , 1993.
4. Chen, Z., Nochetto, R. H. and Schmidt, A., *Adaptive finite element methods for diffuse interface models.* In preparation.
5. Elliott, C. and Gardiner, A., *Double obstacle phase field computations of dendritic growth.* Report 96/19, University of Sussex.
6. Schmidt, A., *Adaptive methods for coupled systems.* In preparation.
7. ———, *Computation of three dimensional dendrites with finite elements*, J. Comput. Phys., 125 (1996), pp. 293–312.
8. ———, *Approximation of crystalline dendrite growth in two space dimensions*, Acta Math. Univ. Comenianae, 67 (1998), pp. 57–68.
9. Schmidt, A. and Siebert, K. G., *ALBERT: An adaptive hierarchical finite element toolbox.* Preprint 06/2000 Freiburg, 2000. Documentation.
10. ———, *ALBERT – Software for scientific computations and applications*, Acta Math. Univ. Comenianae., 70 (2001), pp. 105–122.

Transport of Point Defects in Growing Si Crystals

Axel Voigt,Christian Weichmann

research center caesar, Crystal Growth, Friedensplatz 16, 53111 Bonn, Germany

Abstract. Silicon crystal growth is one of the key processes that determine the yield and the profitability in semiconductor device manufacturing. The art of growing silicon crystals is today highly developed and the quality, purity and size of today's crystals have reached an outstanding level. However, the so-called grown-in defects, which are related to the pulling of crystals, can not be perfectly controlled. Such defects can deteriorate devices, and the ongoing increase of integration density requires wafer with even lower defect concentration. Hence the control and reduction of grown-in defects still remains an important challenge for crystal growers. The formation of grown-in defects is predominantly affected by intrinsic point defects, which diffuse, recombine and aggregate in the silicon lattice during growth and cooling of the ingot. We present a detailed defect model for silicon crystals which incorporates all relevant phenomena. Conservation equations for self-interstitials and vacancies are introduced and a finite element method is provided for predicting the concentration of point defects in growing Czochralski silicon crystals.

1 Introduction

Intrinsic point defects, i.e. self-interstitials and vacancies, impurity atoms and defect cluster are present in the crystal lattice at finite temperature due to energetic and entropic reasons. Vacancies and self-interstitials are formed during solidification at the solid/liquid interface. It is generally assumed that they are incorporated at their thermal equilibrium concentration. Once the silicon atoms have crystallized, additional mechanisms for the creation are known.

- The Schottky mechanism, where a bulk lattice atom jumps to an interstitial site and diffuses to the surface, where it is added.
- The Frenkel mechanism, where a silicon atom leaves the lattice site by simultaneously creating a vacancy and a self-interstitial.
- Recombination is the inverse process of the Frenkel mechanism. A vacancy and a self-interstitial diffuse towards each other and form a lattice atom.
- The free energy of a larger defect is generally smaller than that of a larger number of smaller defects. Therefore vacancies and self- interstitials can also be consumed by preexisting agglomerates. But is is also possible that they are released at the expense of the agglomerates.

Today it is generally assumed that vacancies and self-interstitials are simultaneously present in a silicon crystal under thermal equilibrium. The presence of point defects in the lattice is due to thermal vibrations of the silicon atoms at finite temperatures. For a detailed discussion of intrinsic point defect diffusion in silicon, we refer to [5]. In a continuum approach we describe diffusion fluxes of atomic species that are driven by the gradient of the chemical potential, which is a function of temperature, pressure and the concentration distribution of the diffusion species. Diffusion of point defect species due to concentration and temperature gradients are governed by Fickian and thermal diffusion laws. Pressure gradients occur in cooling crystals in the form of thermal stresses.

Self-interstitials and vacancies are incorporated into the crystal in their equilibrium concentrations and then diffuse and recombine rapidly in the high temperature portion of the crystal. Depending on the growth conditions, one of these defect species survives in excess and leads to the formation of the corresponding microdefects. The source of most observable microdefects is the condensation of supersaturated point defects as the crystal cools during growth. In order to describe the formation of microdefects nucleation and aggregation of defect clusters have to be considered. The model in [12] can explain qualitatively the transition between self-interstitial and vacancy related microdefects as a function of pull rate and thermal fields.

2 Point defect modeling

The conservation equations for intrinsic point defects include convective, Fickian diffusion, thermal diffusion and recombination terms. The equations can be derived from basic principles of thermodynamic, see [2]. More detailed models for the diffusion of point-defects in semiconductors can be found in [4] and [7]. The governing equations for the concentrations of interstitials and vacancies are

$$\frac{\partial C_i}{\partial t} + \mathbf{u}_c \cdot \nabla C_i - \nabla \cdot \left(D_i \nabla C_i - \frac{Q_i D_i C_i}{k T_c^2} \nabla T_c \right) = k_{rec} \left(C_i^{eq} C_v^{eq} - C_i C_v \right),$$

$$\frac{\partial C_v}{\partial t} + \mathbf{u}_c \cdot \nabla C_v - \nabla \cdot \left(D_v \nabla C_v - \frac{Q_v D_v C_v}{k T_c^2} \nabla T_c \right) = k_{rec} \left(C_i^{eq} C_v^{eq} - C_i C_v \right),$$

where C_i and C_v are the concentrations of interstitials and vacancies, D_i and D_v the diffusion coefficients, Q_i and Q_v the activation enthalpy for thermal diffusion, C_i^{eq} and C_v^{eq} the equilibrium concentrations, T temperature, $\mathbf{u_c}$ the velocity of the crystal, k the Boltzmann constant and k_{rec} the reaction coefficient for recombination.

Convection of self-interstitials and vacancies occurs as bulk transport in the axial direction with a given pull rate u_{pull} and in the tangential direction with a given rotation rate u_{rot}. Fickian diffusion is driven by concentration gradients. The flux of the diffusing species $J = -D\nabla C$ is in the direction of lower concentrations. The diffusion coefficient D depends on temperature and concentration. Due to the low concentrations of intrinsic point defects, $10^{16} cm^{-3}$, the concentration dependence is neglected. Thermal diffusion is driven by temperature gradients in the direction of higher temperature and the flux is proportional to its activation energy Q, the diffusion coefficient D and the temperature T. The effect of thermal diffusion was shown to be small compared with the other transport mechanisms [3], and will therefore be neglected. Recombination is identical for both species and describes the elimination of self-interstitials and vacancies by the creation of perfect lattice sites. Recombination is a three step process. First self-interstitials and vacancies diffuse as delocalized point defects in the silicon lattice by random motion until they approach each other to a capture radius a_r and become localized. Then they have to overcome the recombination free energy barrier and finally the self-interstitial will be trapped by the vacancy.

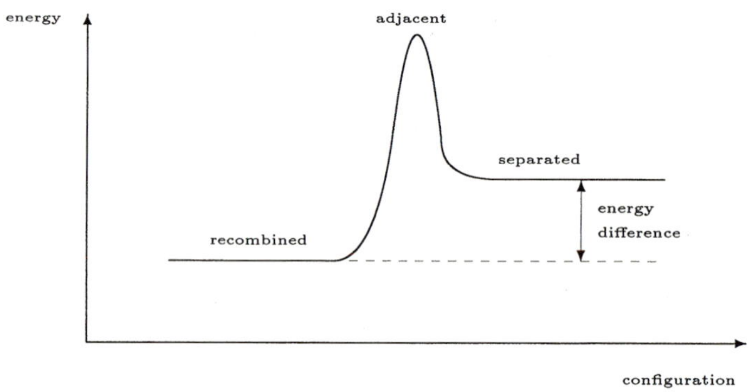

Fig. 1. Energy diagram for self-interstitial and vacancy recombination

The first part of the reaction coefficient describes the diffusion and the second the activation dependence

$$k_{rec} = \frac{4\pi a_r}{V_{Si} C_{Si}} (D_I + D_V) e^{-\frac{\triangle H_{IV}^r - T \triangle S_{IV}^r}{kT}},$$

with a_r the capture radius, V_{Si} the volume of a silicon unit cell, C_{Si} the atomic density, $\triangle H_{IV}^r$ the enthalpy barrier for recombination and $\triangle S_{IV}^r$ the entropic contribution.

$$k_{rec} = 1.2 \cdot 10^{-6} (D_I + D_V) e^{-\frac{0.61 + kT\left(-2.30 + 7.38 \cdot 10^{-3} T\right)}{kT}}.$$

The thermophysical properties in the equations are experimentally because of the high temperatures not directly measurable. One way to circumvent this problem is to calculate the microscopic properties with the help of statistical mechanics and molecular-dynamic simulations. The parameter set in this study has been taken from [8]. The data were obtained with the use of a simple Stillinger-Weber potential to approximate the microscopic parameters for point-defects in silicon crystals. To prevent finite size effects in the molecular-dynamic simulations it is necessary to use millions of particles to realize the needed length scales. Massive parallel algorithms are used to overcome the complexity [1]. In the last years more complex potentials have been developed to simulate the structural properties of silicon [9]. The use of Tersoff-potentials for the purpose of predicting material properties is in progress and will allow more precise point defect simulations. The estimates for the diffusion coefficients and equilibrium concentrations are given in Arrhenius form as

$$D = D^0 e^{-\frac{H^m}{kT}},$$
$$C_{eq} = \rho e^{-\frac{S^f(T)}{k}} e^{-\frac{H^f(T)}{kT}},$$

where ρ is the density of the lattice site, H^m the migration energy, H^f the formation enthalpy and S^f the formation entropy. The following parameter set is used [8].

$$D_I = 3.30 \cdot 10^{-1} \cdot e^{-\frac{0.937}{kT}} \left[\frac{cm^2}{s}\right]$$

$$D_I = 1.30 \cdot 10^{-3} \cdot e^{-\frac{0.456}{kT}} \left[\frac{cm^2}{s}\right]$$

$$C_I^{eq} = 2.90 \cdot 10^{23} \cdot e^{\frac{+1.4+3.85\cdot 10^{-3}T}{k}} \cdot e^{-\frac{3.46+3.08\cdot 10^{-4}T}{kT}} \left[\frac{1}{cm^3}\right]$$

$$C_V^{eq} = 4.97 \cdot 10^{22} \cdot e^{\frac{-3.7+3.53\cdot 10^{-3}T}{k}} \cdot e^{-\frac{2.48+2.33\cdot 10^{-4}T}{kT}} \left[\frac{1}{cm^3}\right]$$

By writing the transport equation in nondimensional form we define the Peclet number Pe, which compares the convective with the Fickian diffusion term, the Damköhler number Da, comparing recombination with Fickian diffusion as well as a nondimensional number for thermal diffusion compared with convection Td,

$$Pe = \frac{u_c L}{D}, \quad Da = \frac{k_{rec} C^{eq} L^2}{D}, \quad Td = \frac{u_c L k T^2}{Q D \nabla T},$$

with a characteristic length L. Self-interstitials and vacancies are set to equilibrium concentration at the solid/liquid interface. An essential boundary condition for the point-defects at the crystal rim and cone can physically described by the Schottky defect mechanism, which assumes that point-defects diffusing to the surface are added to the lattice and are therefore in equilibrium concentration,

$$C_i = C_i^{eq},$$
$$C_v = C_v^{eq}.$$

Initial conditions are specified for time t_0 assuming equilibrium concentration in the crystal for self-interstitials and vacancies.

3 Finite element discretisation

If we multiply the reaction-diffusion equations by a test function $\phi \in H_0^1(\Omega)$, integrate over the region Ω, use Green's theorem and apply the Dirichlet conditions we get

$$\int_\Omega \left(\frac{\partial C}{\partial t} + (\mathbf{u}_c \cdot \nabla C) \right) \phi d\Omega + \int_\Omega D \nabla C \cdot \nabla \phi d\Omega - \int_\Omega \frac{QDC}{kT^2} \nabla T \cdot \nabla \phi d\Omega = \int_\Omega k_{rec} \left(C_I^{eq} C_V^{eq} - C_I C_V \right) \phi d\Omega.$$

For existence of a unique weak solution see [6]. The discretisation combines piecewise linear finite elements in space and a time discretisation including the convection that is based on the method of characteristics. The characteristic finite difference method is based on the approximation

$$\frac{\partial C(\mathbf{x},t)}{\partial t} + \mathbf{u}_c \cdot \nabla C(\mathbf{x},t) \approx \frac{C(\mathbf{x},t) - C(\mathbf{x} - \tau \mathbf{u}_c, t - \tau)}{\tau},$$

with an appropriate time step size τ. For each time step (t_n, t_{n+1}) let $\mathcal{T}_{n+1}$ denote an adaptively constructed conforming mesh of the domain $\Omega \subset I\!R^d$. Let W^{n+1} denote the finite element space of continuous piecewise linear functions of $\mathcal{T}_{n+1}$,

$$W^{n+1} = \left\{ w \in C^0(\bar{\Omega}); w_{|T} \in I\!P_1(T), T \in \mathcal{T}^{n+1} \right\},$$
$$W_0^{n+1} = W^{n+1} \cap H_0^1.$$

We approximate the concentrations as a linear combination of the basis functions such that

$$C^{n+1} = \sum_{i=1}^{n} C_i^{n+1} \phi_i^{n+1},$$

where $C_i^{n+1} \in I\!R$ and $\phi_i^{n+1} \in W_0^{n+1}$. The discrete problem reads as follows: Given $C^n \in W^n$ find $C^{n+1} \in W^{n+1}$ such that for all $\phi^{n+1} \in W_0^{n+1}$

$$\frac{1}{\tau^n} \left(C^{n+1} - \bar{C}^n, \phi^{n+1} \right)_\Omega^{n+1} + \left(D \nabla C^{n+1}, \nabla \phi^{n+1} \right)_\Omega^{n+1} - \left(\frac{QDC^{n+1}}{kT^2} \nabla T, \nabla \phi^{n+1} \right)_\Omega^{n+1} = \left(k_{rec} \left(C_I^{eq} C_V^{eq} - C_I^{n+1} C_V^{n+1} \right), \phi^{n+1} \right)_\Omega,$$

with $\bar{C}^n(\mathbf{x}, t^n) = C^n(\mathbf{x} - \tau^n \mathbf{u}_c, t^n)$. In each time step the equation leads to the system of nonlinear algebraic equations

$$\mathbf{M}^C \mathbf{C}_I^{n+1} + \mathbf{A}^{C_I} \mathbf{C}_I^{n+1} + \mathbf{B}^{C_I} \mathbf{C}_I^{n+1} + \mathbf{C}^{C_V} \mathbf{C}_I^{n+1} = \mathbf{R}^{C_I},$$
$$\mathbf{M}^C \mathbf{C}_V^{n+1} + \mathbf{A}^{C_V} \mathbf{C}_V^{n+1} + \mathbf{B}^{C_V} \mathbf{C}_V^{n+1} + \mathbf{C}^{C_I} \mathbf{C}_V^{n+1} = \mathbf{R}^{C_V},$$

with the diagonal lumped mass matrix $\mathbf{M}^C = (M_{ii}^C)$,

$$M_{ii}^C = \frac{1}{\tau^n}(\phi_i^{n+1}, \phi_i^{n+1})_\Omega,$$

the stiffness matrices $\mathbf{A}^{C_I} = (A_{ij}^{C_I})$ and $\mathbf{A}^{C_V} = (A_{ij}^{C_V})$,

$$A_{ij}^{C_I} = (D_I \nabla \phi_i^{n+1}, \nabla \phi_j^{n+1})_\Omega,$$
$$A_{ij}^{C_V} = (D_V \nabla \phi_i^{n+1}, \nabla \phi_j^{n+1})_\Omega,$$

the gradient matrices $\mathbf{B}^{C_I} = (B_{ij}^{C_I})$ and $\mathbf{B}^{C_V} = (B_{ij}^{C_V})$,

$$B_{ij}^{C_I} = (\frac{Q_I D_I}{kT^2} \nabla T \phi_i^{n+1}, \nabla \phi_j^{n+1})_\Omega,$$
$$B_{ij}^{C_V} = (\frac{Q_V D_V}{kT^2} \nabla T \phi_i^{n+1}, \nabla \phi_j^{n+1})_\Omega,$$

the recombination matrices $\mathbf{C}^{C_I} = (C_{ij}^{C_I})$ and $\mathbf{C}^{C_V} = (C_{ij}^{C_V})$,

$$C_{ij}^{C_I} = (k_{rec} C_V \phi_i^{n+1}, \phi_j^{n+1})_\Omega,$$
$$C_{ij}^{C_V} = (k_{rec} C_I \phi_i^{n+1}, \phi_j^{n+1})_\Omega,$$

and the right hand sides $\mathbf{R}^{C_I} = (R_{ij}^{C_I})$ and $\mathbf{R}^{C_V} = (R_{ij}^{C_V})$,

$$R_i^{C_I} = \frac{1}{\tau^n}(\bar{C}_i^n, \phi_i^{n+1})_\Omega^{n+1} + (k_{rec} C_I^{eq} C_V^{eq}, \phi_i^{n+1})_\Omega$$
$$R_i^{C_V} = \frac{1}{\tau^n}(\bar{C}_v^n, \phi_i^{n+1})_\Omega^{n+1} + (k_{rec} C_I^{eq} C_V^{eq}, \phi_i^{n+1})_\Omega.$$

Because of the nonlinearities we use a Newton method to solve the nonlinear system. The generalized minimum residual method (GMRES) is used for solving the linear system in each iteration.

The ingot length enlarges proportional to the pulling velocity during the growth of the crystal, resulting in a deformation of the domain. Therefore the numerical mesh has to adapt itself to the growing crystal. The shape of the crystal and the temperature distribution in each timestep was calculated in a thermal simulation taking account of conduction, melt convection, heat radiation and a free solidification interface [10] and [13]. The meshes resulting from this simulations are used for the point defect calculation.

4 Results

Quantitative results of the coupled heat transfer and point-defect system were verified in [11], where an experimentally satisfied phenomenological law was used which qualitatively relates point defects to growth conditions [3]. Starting from this justification of the model, realistic Czochralski furnaces for the growth of 200mm silicon crystals with different pulling velocities where

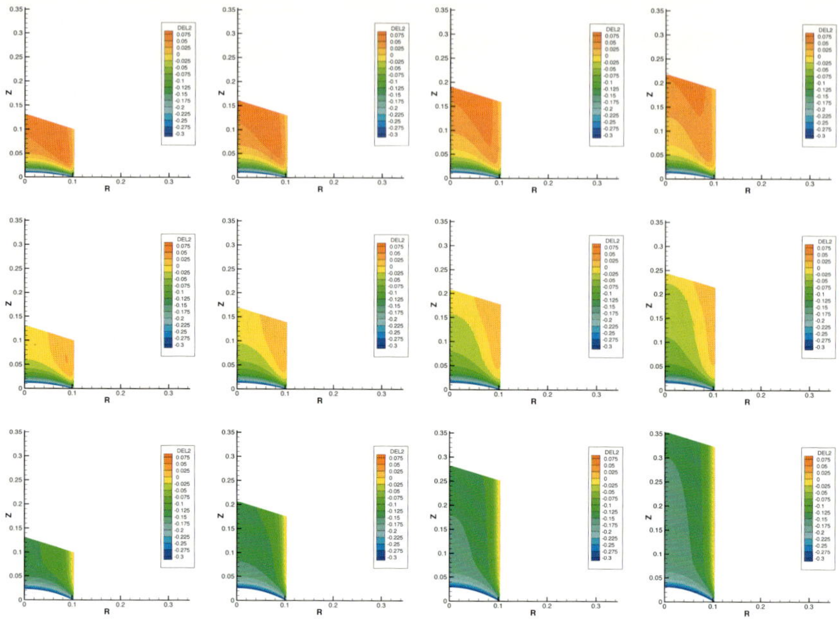

Fig. 2. Defect concentration at various time steps for different pulling velocities

simulated, in order to relate the point-defect distribution in the crystal to the pulling velocity. Due to the axisymmetry of the crystal the calculations are performed in 2D.

In the same hot zone three different pull rates where used, $v_{p1} = 0,348$ mm/min, $v_{p2} = 0,454$mm/min and $v_{p3} = 0,892$mm/min, to simulate the growing process over a time interval of 250min. All simulations start with an initial quasi steady state computation for a crystal length of 10cm. Highest self-interstitial concentrations occur in the crystal which is grown with the lowest pull rate v_{p1}. This crystal is fully self-interstitial-rich after cooling down. The crystal grown with pull rate v_{p2} shows vacancy-rich as well as self-interstitial-rich regions and the crystal grown with pull rate v_{p3} is completely vacancy-rich after cooling down. Fig. 2 shows normalized differences between the concentration of self-interstitials and vacancies in the crystal at various time steps.

5 Conclusions

Coupled transient heat transfer and transient point defect transport have been successfully simulated with constant pull rates for growing silicon crystals. The change in the phase boundary during the growth process influences the defect distribution. In order to predict the concentration of intrinsic point defect in the crystal the thermal history of the crystal should therefore no longer be neglected.

6 Acknowledgment

This work was partially supported by Wacker Siltronic Inc. and the DFG-Sonderforschungsbereich 611 Bonn, TP C5.

References

1. Caglar, A.: (2000) On the numerical simulation of fullerene nanotubes: $C_{100.000.000}$ and beyond! Proc. of NIC Workshop on Molecular Dynamics on Parallel Computers, World Scientific.
2. De Groot, S. R., Mazur, P.: (1984) Non-equilibrium Termodynamics. North-Holland, Amsterdam.
3. Dornberger, E.: (1997) Prediction of OSF Ring Dynamics and Grown-in Voids in Czochralski Silicon Crystals. PhD Thesis, Universite Catholique de Louvain.
4. Dunham, S. T.: (1992) A Quantitative Model for the Coupled Diffusion of Phosphorus and Point Defects in Silicon. J. Electrochem. Soc. **1** 2 - 35.
5. Fahey P., Griffin P., Plummer, J.: (1989) Point defects and dopant diffusion in silicon. Rev. Modern Physics **61** 291 - 306.
6. Lang, J.: (2001) Adaptive Multilevel Solution of Nonlinear Parabolic PDE Systems. Springer, Berlin.
7. Merz, M.: (1999) Analysis und Numerische Berechnung der Diffusion von Fremdatomen in Homogenen Strukturen. HabilThesis, Technische Universität München.
8. Sinno, T. R.: (1998) Defects in Crystalline Silicon: Intergated Atomistic and Continuum Modeling. PhD Thesis, MIT, Cambridge.
9. Tersoff: (1986) New Empirical Model for the Structural Properties of Silicon. Physical Review Letters **56(6)** 632 - 635.
10. Voigt, A.: (2001) Numerical Simualtion of Industrial Crystal Growth. PhD Thesis, Technische Universiät München.
11. Voigt, A., Nitschkowski, J., Weichmann, C., Hoffmann, K.-H. (2002) Controlling Point Defects in Single Silicon Crystals Grown by the Czochralski Method. In: Breuer, M., Durst, F., Zenger, C. (Ed.) High Performance Scientific and Engineering Computing, Springer, Berlin, 229 - 236.
12. Voronkow, V.: (1982) The mechanism of swirl defects formation in silicon. J. Cryst. Growth **59** 625 - 643.
13. Weichmann, C., Nitschkowski, J., Hoffmann, K.-H., Voigt, A.: (2002) Time dependent simulation of heat transfer in a Czochralski furnace. WIT Press. in press.

Atomistic Simulation of Transport Phenomena in Simple and Complex Fluids and Fluid Mixtures

Kurt Binder[1], Jürgen Horbach[1], Walter Kob[2], and Fathollah Varnik[1]

[1] Institut für Physik, Johannes Gutenberg–Universität, D–55099 Mainz, Staudinger Weg 7, Germany
[2] Laboratoire des Verres, Université Montpellier II, Place E. Bataillon, cc69, 34095 Montpellier, France

Abstract. Computer simulations of fluids in thermal equilibrium can yield information on transport coefficients such as self–diffusion and interdiffusion coefficients, viscosity, and thermal conductivity. While the estimation of self–diffusion coefficients from the mean square displacements of the respective particles is rather straightforward, the estimation of other transport coefficients is less straightforward, and can be based on either an analysis of time correlation functions of the appropriate collective variables, or on nonequilibrium techniques where the linear response to appropriate perturbations is measured.

To concretize ideas, we first discuss a very simple example, interdiffusion in a binary (AB) noninteracting lattice gas, based on Monte Carlo simulation of diffusion in the framework of the vacancy mechanism. Measuring fluxes induced by chemical potential gradients, Onsager coefficients are deduced, while the interdiffusion coefficient is measured from an instationary Nonequilibrium Monte Carlo (NEMC) method, studying the decay of concentration waves. Then Nonequilibrium Molecular Dynamics (NEMD) methods are described, where fluid flow between plates is used to extract both viscosity and thermal conductivity. As an alternative method, the use of the Kubo formula for the viscosity and the analysis of time–dependent specific heats (to estimate the thermal conductivity) are mentioned. Application examples range from molten SiO_2 to polymer melts.

1 Introduction

Since the early days of computer simulation [1,2] it has been a very important goal of such work to estimate transport coefficients and to elucidate the relation between structural properties of the simulated model systems and transport phenomena [3]. Both Molecular Dynamics methods [4–8], suitable for applications to simple and complex fluids and their mixtures, and Monte Carlo methods [5,7–11], suitable for the study of diffusion phenomena in solids [12,13] and viscous fluids such as polymer melts [14–17], are used. Understanding the dynamics of fluctuations and associated transport phenomena in equilibrium, one can proceed further to simulate the dynamics of phase changes such as phase separation in mixtures (via "spinodal decomposition" [18–21] or nucleation [22,23]), crystal growth from the vapor or melt [24,25], growth of wetting layers at surfaces [26,27], dynamics of adsorption (e.g. [28]) and related situations where interfaces are formed and grow [29,30]. All these latter phenomena far from equilibrium, however, are beyond the scope of the present article, which focuses on transport phenomena in systems containing a single phase in equilibrium, so transport across interfaces between coexisting phases will here not be considered. For the sake of simplicity, all the examples treated are taken from the own research of the authors, notwithstanding the fact that much excellent work from other groups does exist on related topics: however, the purpose of this article is not a comprehensive review, but rather to give a tutorial introduction to some of the many methodologies that have been proposed in the literature, and to describe the "state of the art" that one can reach with different simulational strategies with representative examples.

One of the simplest cases is the simulation of diffusion in crystalline solids. As a representative example, the simulation techniques to elucidate interdiffusion in binary alloys [31,32] will be described in Sec. 2. The related topic of interdiffusion in polymer mixtures [33–36] has been the subject of a longstanding controversy.

Sec. 3 then considers the estimation of selfdiffusion coefficients in fluids, treating SiO_2 and SiO_2–Na_2O mixtures [37–41] and glassforming polymer melts [42–45] as an example. Sec. 4 discusses the estimation of the viscosity, both via Nonequilibrium Molecular Dynamics (NEMD) methods [46–60] and via a Kubo formula [38], while Sec. 5 discusses the estimation of the thermal conductivity [59,61]. Sec. 6 finally summarizes a few conclusions.

2 Interdiffusion and Selfdiffusion in Binary Mixtures (A,B)

First we recall the phenomenological theory of interdiffusion. One writes down so–called constitutive equations for the current densities $\mathbf{j}_A, \mathbf{j}_B$ of A– and B–atoms in the mixture [62]. Note that this is just a postulate, relating the current densities and the corresponding driving forces linearly. These driving forces are the gradients of the chemical potential differences between A– or B–atoms and vacancies, respectively: $\mu_A - \mu_V$ and $\mu_B - \mu_V$. The coefficients in these linear relations ($\Lambda_{AA}, \Lambda_{BB}, \Lambda_{BA} = \Lambda_{AB}$) are the Onsager coefficients,

$$\mathbf{j}_A = -\frac{\Lambda_{AA}}{k_B T}\nabla(\mu_A - \mu_V) - \frac{\Lambda_{AB}}{k_B T}\nabla(\mu_B - \mu_V) \tag{1}$$

$$\mathbf{j}_B = -\frac{\Lambda_{BA}}{k_B T}\nabla(\mu_A - \mu_V) - \frac{\Lambda_{BB}}{k_B T}\nabla(\mu_B - \mu_V) \; . \tag{2}$$

Note that this (deterministic!) ansatz neglects statistical fluctuations, it is a kind of mean field theory. By invoking the continuity equations, which simply

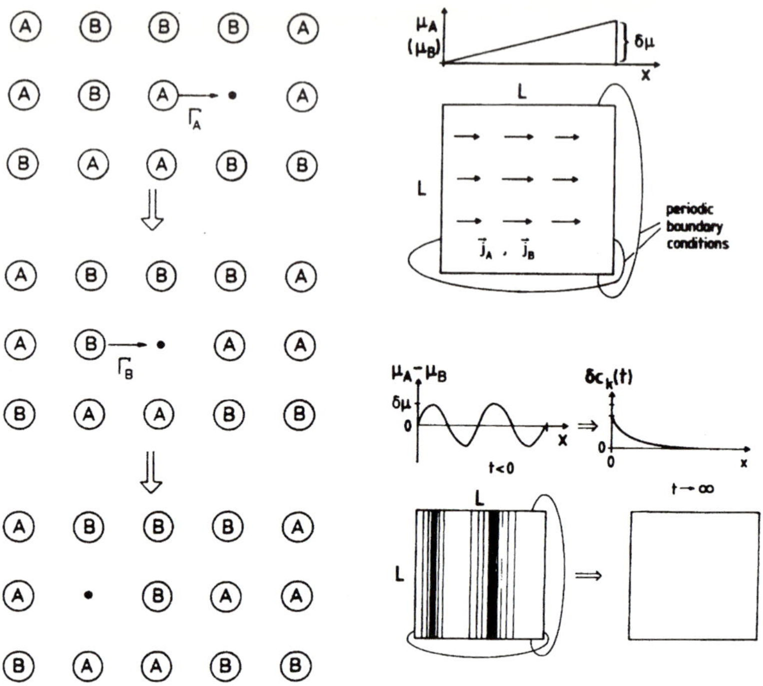

Fig. 1. (a) Schematic description of interdiffusion in a random mixed crystal, where two different kinds of atoms (A,B) may occupy the sites of a regular lattice. Occupied sites are indicated by circles, an empty site by a dot. Diffusion proceeds by the vacancy mechanism: A–atoms may jump to vacant sites with a jump rate $\Gamma_{\rm A}$, B–atoms with a jump rate $\Gamma_{\rm B}$. For simplicity, it is assumed that these jump rates do not depend on the occupation of neighboring lattice sites, implying that there are no interactions depending on $c_i^{\rm A}$, the occupation variables of the lattice sites, (= 1 if site i is taken by an A–atom, zero else) and $c_i^{\rm B}$. (b) Top: Schematic sketch to indicate how a chemical potential difference $\delta\mu = \mu_{\rm A} - \mu_{\rm V}$ (or $\delta\mu = \mu_{\rm B} - \mu_{\rm V}$) over a linear dimension L in an $L \times L$ system with periodic boundary conditions creates currents $\mathbf{j}_{\rm A}, \mathbf{j}_{\rm B}$. The lower part indicates a system prepared with a chemical potential difference $\mu_{\rm A} - \mu_{\rm B} = \delta\mu \sin(\lambda x/L)$, with L/λ integer, to create a concentration wave δc_k, that is switched off at time $t = 0$, to monitor the decay of the concentration wave with time t.

express the fact that the total concentrations are conserved, and hence are exact,

$$\frac{\partial c_{\rm A}}{\partial t} + \nabla \cdot \mathbf{j}_{\rm A} = 0, \quad \frac{\partial c_{\rm B}}{\partial t} + \nabla \cdot \mathbf{j}_{\rm B} = 0\,, \tag{3}$$

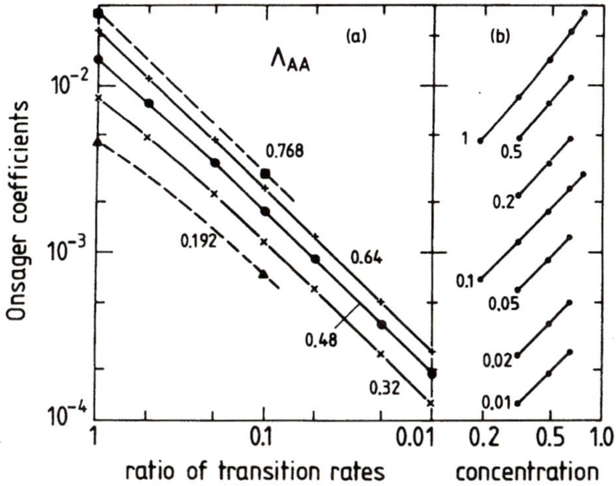

Fig. 2. Onsager coefficient $\Lambda_{\rm AA}$ plotted vs. the ratio of jump rates $\Gamma_{\rm A}/\Gamma_{\rm B}$ (a) or against the concentration $c_{\rm A}$ of A–atoms (b), for a two–dimensional square $L \times L$ lattice ($L = 80$) using a vacancy concentration $c_{\rm V} = 0.04$, for the random binary alloy ("ABV model") shown in Fig. 1. Parameter of the curves in part (a) is $c_{\rm A}$, in part (b) is $\Gamma_{\rm A}/\Gamma_{\rm B}$. From Kehr et al. [31].

one obtains a closed set of equations (note $c_{\rm A} + c_{\rm B} + c_{\rm V} = 1$ for a lattice model, Fig. 1a, where a lattice site is either taken by an A atom or a B atom or empty). In the non–interacting case, the free energy simply contains entropy of mixing terms,

$$F/k_BT = c_{\rm A} \ln c_{\rm A} + c_{\rm B} \ln c_{\rm B} + c_{\rm V} \ln c_{\rm V} \ , \tag{4}$$

and $\mu_\alpha = \partial F/\partial c_\alpha$. With some algebra, it is then possible to express the interdiffusion coefficient in terms of the Onsager coefficients [31]:

$$D_{\rm int} = \frac{\Lambda_{\rm AA}\Lambda_{\rm BB} - \Lambda_{\rm AB}^2}{\Lambda_{\rm AA} + 2\Lambda_{\rm AB} + \Lambda_{\rm AB}} \left(\frac{1}{c_{\rm A}} + \frac{1}{c_{\rm B}}\right), \ c_{\rm V} \to 0 \ . \tag{5}$$

However, there remain several problems: how are the Onsager coefficients related to the jump rates $\Gamma_{\rm A}, \Gamma_{\rm B}$ of the simple atomistic model (Fig. 1a)? Can the nondiagonal term be neglected? After all, there are only 2 independent jump rates ($\Gamma_{\rm A}, \Gamma_{\rm B}$) but three Onsager coefficients. In fact, analytical theories in most cases assume $\Lambda_{\rm AB} = 0$ [33–36]. Assuming then that $\Lambda_{\rm AA}$ is proportional to $\Gamma_{\rm A}$ and hence to the selfdiffusion coefficient $D_{\rm A}$ of the A–atoms, and likewise $\Lambda_{\rm BB} \propto D_{\rm B}$ one arrives at the so–called "slow mode theory" [33,35]. It then is the slower diffusing species that controls interdiffusion. On the other hand, assuming already in (1) and (2) that the vacancies always are in equilibrium, $\nabla\mu = 0$, then (5) does not hold, and one gets the "fast

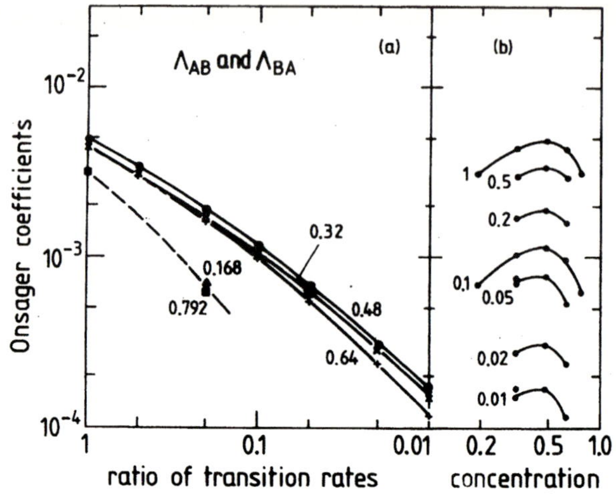

Fig. 3. Same as Fig. 2 but for $\Lambda_{AB} = \Lambda_{BA}$. From Kehr et al. [31].

mode theory" [34]: now it is the faster diffusing species that controls interdiffusion! Now one could say, let the experiment decide which theory is correct — however, different research groups have claimed experimental evidence for both theories [36]! Since these theories drastically contradict each other, and cannot both be correct in the same situation, there is clearly a problem in the comparison between experiment and theory. In fact, it is not at all clear whether the experiments really correspond to the simple model of Fig. 1a: they are all done on fluid polymer mixtures [36] rather than mixed crystals where single atoms diffuse. Even if one could do experiments studying interdiffusion for such mixed crystals, there would be interactions between the A and B atoms and thus deviations from random mixing would occur, invalidating (4). In addition, elastic distortions (due to the fact that the atomic radii of A and B are not precisely equal) play a role, as well as lattice defects (grain boundaries, dislocations etc. may greatly affect diffusion processes in crystals). So different experiments correspond to different physical situations that often are incompletely characterized, and the model may be (perhaps?) inadequate for a particular experiment. Thus the approximations made in the mathematical treatment of the model cannot be tested conclusively by a comparison with experiments! Here simulation has an advantage: precisely the same model (Fig. 1a) can be simulated on which the theory is based, all parameters that the theory uses are either given or can be independently estimated (Fig. 1b). So no uncontrolled approximations occur.

Now answers to all above questions can be gotten from "taylored" simulations, as indicated in Fig. 1b. In order to obtain the Onsager coefficients, one simply follows the definitions, (1), (2): one imposes a small gradient $\delta\mu/L$

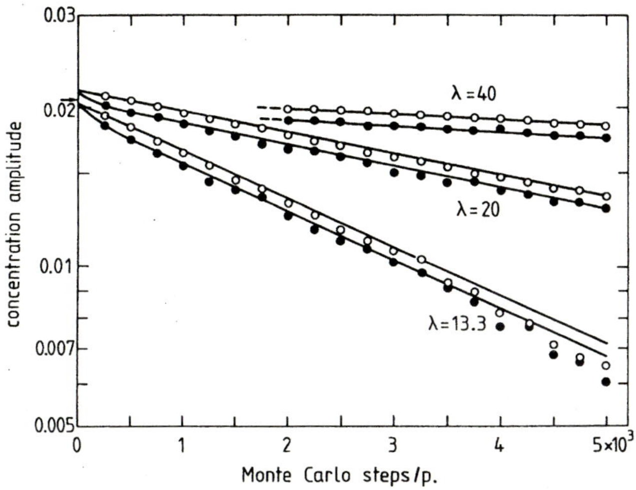

Fig. 4. Concentration amplitude $\delta c_k(t)$ plotted vs. time (in units of Monte Carlo steps per particle), for the same model as in Figs. 1–3, using $c_\mathrm{A} = c_\mathrm{B} = 0.48$, $\Gamma_\mathrm{A}/\Gamma_\mathrm{B} = 0.1$, $\delta c_\mathrm{A}(t=0) - \delta c_\mathrm{B}(t=0) = 0.02$. Note that a single exponential with D_int given by (5) results for $c_\mathrm{V} \to 0$ only, while for $c_\mathrm{V} \neq 0$ two exponentials with different decay constants superimpose [31], resulting in separate curves for A–particles (open dots) and B–particles (full dots). Three wavelengths λ commensurate with $L = 80$ are shown, to test that the hydrodynamic limit ($k \to 0$) actually is realized. Curves are the result of an analytic calculation, based on (1)–(4) and the "measured" Onsager coefficients. From Kehr et al. [31].

of the chemical potential in x–direction, acting on the A particles. One then measures the resulting steady state currents of A–atoms and B–atoms. Because of the periodic boundary conditions, atoms that leave the simulation box at the right boundary, re–enter it at the left boundary, and so a steady state current is established after a transition period. This procedure yields the Onsager coefficients Λ_AA, Λ_BA. Then the same is done with a chemical potential gradient acting on the B particles, to obtain Λ_BA and Λ_BB. So all Onsager coefficients are gotten (Figs. 2, 3). In fact, it is not true that the off–diagonal Onsager coefficient is always negligible in comparison with the diagonal ones: for small $\Gamma_\mathrm{A}/\Gamma_\mathrm{B}$ one sees that Λ_AB and Λ_AA are of the same order, and hence Λ_AB must not be neglected!

For the Monte Carlo estimation of the interdiffusion constant, one follows the recipe of linear response theory: one imposes a wave–like chemical potential difference $\delta\mu = \mu_\mathrm{A} - \mu_\mathrm{B}$ (Fig. 1b, lower part) and thus prepares a system with a superimposed "concentration wave" $\delta c_k(k = 2\pi/\lambda)$ as an initial state. At the time $t = 0$, the chemical potential difference is then switched off, and one watches the decay of the amplitude of the concentration wave with time, $\delta c_k(t) = \delta c_k(0) \exp(-D_\mathrm{int} k^2 t)$. This decay should be exponential

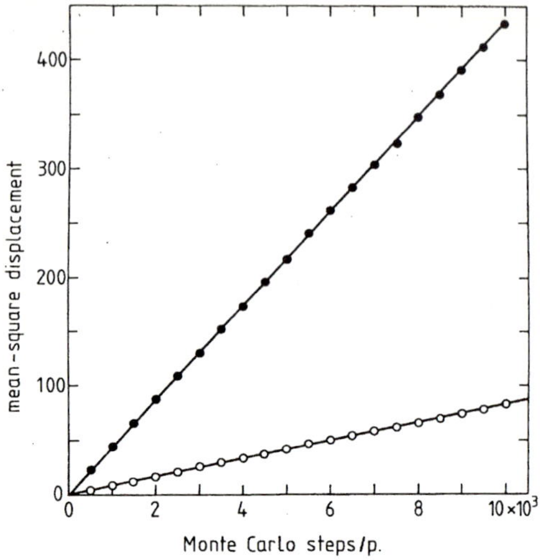

Fig. 5. Mean square displacements of tagged A (open circles) and B (solid circles) particles as a function of Monte Carlo steps per particle, for a 50^3 lattice with $c_A = c_B = 0.45$ and $\Gamma_A/\Gamma_B = 0.1$. From Kehr et al. [31].

(which needs to be checked in the simulation! [31]), and the decay constant (for $k \to 0$) yields the interdiffusion constant D_{int}. Fig. 4 shows the outcome of such a linear response "computer experiment". Choosing a logarithmic ordinate scale and a linear scale for the time t, one verifies by observing straight lines on this plot that the decay indeed is exponential with time, proportional to $\exp(-\Gamma_k t)$. Using three different wavelengths ($\lambda = L/2, L/4$ and $L/6$), it is verified that the decay constant Γ_k scales with k as $\Gamma_k = D_{\text{int}} k^2$. Actually if the amplitude $\delta\mu$ is small but not infinitesimal, there are two different curves for A and B atoms with slightly different amplitudes. The curves in Fig. 4 were calculated [31] from the above phenomenological theory, (1)–(4), using the Onsager coefficients as estimated from the simulations, such as shown in Figs. 2, 3. Thus, there are no adjustable parameters whatsoever, and the agreement between the phenomenological theory and the simulations hence is quite perfect! However, neither the slow mode theory [33,35] (in its traditional form, assuming $\Lambda_{AB} = 0$, relating Λ_{AA}, Λ_{BB} to selfdiffusion constants D_A, D_B) nor the fast mode theory [34] work for the present simple model.

One can obtain the selfdiffusion coefficients very simply from the Einstein relations, considering the mean square displacements of "tagged particles" with time t

$$\langle [\mathbf{r}_i(t) - \mathbf{r}_i(0)]^2 \rangle = 2dD_\alpha t \quad , \quad t \to \infty, \tag{6}$$

where d is the dimensionality of the system, and we average over all particles (with label i) of type α to improve the statistics, in order to estimate D_α ($\alpha =$ A or B). Fig. 5 shows that for our simple model (6) actually holds already for rather early times. Of course, Figs. 1–5 refer to a simplistic example: the random hopping processes of Fig. 1a are straightforward to simulate by Monte Carlo, and there are no problems of equilibration whatsoever. "Real life"–problems, to be discussed in the next sections, clearly are much harder to handle, and thus, to our knowledge, only the present "toy problem" has received such a complete treatment, where selfdiffusion as well as Onsager coefficients as well as the interdiffusion constant have been obtained for the same model. It remains a challenge for the future to perform a similar study for an off–lattice model of a real system.

3 Estimation of Selfdiffusion Coefficients in Various Models of Fluids

While in the case of the noninteracting mixture on a lattice considered in the previous section the use of (6) to estimate selfdiffusion coefficients is straightforward, this task is much more demanding for chemically realistic models of fluids: firstly, it is often difficult to reach times that are sufficiently large that the Einstein relation actually holds, while a different behavior is detected for short and intermediate time scales.

We first exemplify these considerations with the case of molten silica (SiO_2) and its mixtures with sodium oxide (Na_2O). These systems are of great technical interest as prototypes of network glassformers, and also relevant in the context of earth sciences and mineralogy. Furthermore, a very well tested pair potential well suited for Molecular Dynamics simulations has been developed from quantum–chemical calculations [63,64]. This so–called BKS potential [63] and its generalization [64] has the form (e is the charge of a proton)

$$\phi_{\alpha\beta}(r) = \frac{q_\alpha q_\beta e^2}{r} + A_{\alpha\beta} \exp(-B_{\alpha\beta} r) - \frac{C_{\alpha\beta}}{r^6} \quad , \qquad (7)$$

α, $\beta \in$ [Si, Na, O]. Here r is the distance between an ion of type α and an ion of type β. The values of parameters $A_{\alpha\beta}$, $B_{\alpha\beta}$ and $C_{\alpha\beta}$ can be found in Refs. [63,64]. Note that here the q_α are partial charges with values 2.4, -1.2 and 0.6 for Si, O, and Na. Furthermore the potential given by (7) is slightly modified at short r if α or β are Na. More details on this can be found in Ref. [40]. It turns out that the potential gives a good agreement between the simulated static structure factor [40] and corresponding experiments for molten $Na_2Si_2O_5$ (NS2) [65].

The simulations have been done at constant volume, fixing the density at $\varrho = 2.37$ g/cm^3 and using 8064 atoms for NS2 and 8016 atoms for $Na_2Si_3O_7$ (NS3) [40]. Since the long range Coulomb interactions, which require the use of Ewald summation techniques, need very large computing times for these relatively large system sizes, it would be convenient if one could work with smaller sizes. However, it turns out that finite size effects for smaller sizes affect systematically the dynamical properties [66] (although no such size effects were detected in static properties). For the integration of the equations of motion, the velocity form of the Verlet algorithm [4–8] was used, with a time step of 1.6 fs. The temperature of the system was controlled by coupling it to a stochastic heat bath, i.e. by substituting periodically the velocities of the particles with the ones from a Maxwell–Boltzmann distribution with the correct temperature. After the system was equilibrated at the target temperature, the run was continued in the microcanonical ensemble (i.e. the heat bath is switched off).

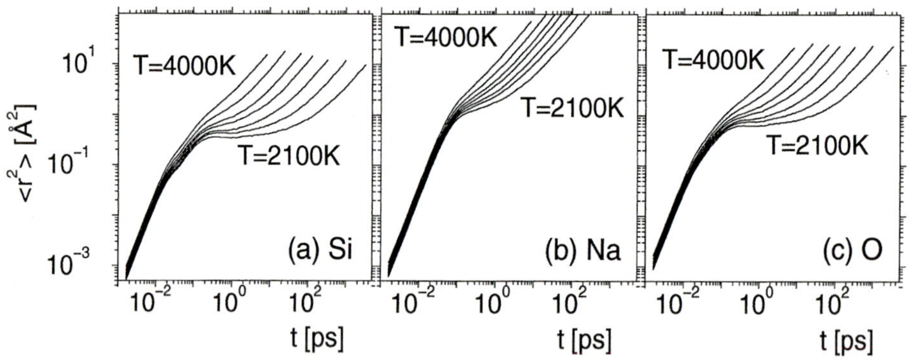

Fig. 6. Mean square displacements $\langle r^2(t) \rangle$ of (a) Si ions, (b) Na ions, and (c) O ions for sodium trisilicate (NS3) on log–log plots versus time (in ps) at the temperatures $T = 4000$ K, 3400 K, 3000 K, 2750 K, 2500 K, 2300 K, and 2100 K (from top to bottom). Note that $\mathbf{r}(t = 0)$ is taken as the coordinate origin for each tagged particle. From Horbach [67].

It should be noted that many oxide glass formers are already very viscous at relatively high temperatures (e.g. for SiO_2 the experimental glass transition temperature is around 1450 K). Therefore the lowest temperature at which the SiO_2 system could be equilibrated (in a run over a physical time of 20 ns [37,38]) was $T = 2750$ K, while for NS3 $T = 2100$ K and for NS2 $T = 1900$ K could be reached [40].

Fig. 6 shows typical data for the mean square displacements of the various ions obtainable in such simulations [67]. One recognizes three different regimes: for times up to about 10 fs, there is a ballistic regime, $\langle r^2(t) \rangle = 3k_BTt^2/m_\alpha$, where m_α is the mass of the ion. For the corresponding very small displacements, the variation of the potential due to the neighboring particles is not yet felt, and thus a free flight results. In the regime from about 10 to about 100 fs, the curves bend over gradually to a slower motion, or even a flat region (in particular for the lowest temperatures displayed in the figure). These "plateaus" in the $\langle r^2(t) \rangle$ vs. t curves can be interpreted in terms of the "cage effect" [68]: the ions are confined in "cages" formed by their nearest neighbors, and for low temperatures it takes a long time until they can "escape from the cage". This "cage effect" leads to a slowing down of all motions, and can (at least to some extent) be understood in terms of the so–called mode coupling theory [68] for the glass transition. Only on the time scale of the structural relaxation time τ this cage breaks up, and the slow crossover to diffusive motion, in accord with (6) sets in. At $T = 2100$ K, a time of 1 ns in NS3 is hardly enough to reach this regime. Of course, only when $\sqrt{\langle r^2(t) \rangle}$ is of the order of many interatomic distances, one can be sure that the asymptotic regime where (6) holds has been reached. Note that there is an obvious problem of statistical accuracy, since the subsequent system configurations generated in the simulation are correlated over a time scale τ needed to reach the diffusive regime which implies that one has to simulate over a time which is significantly larger than τ.

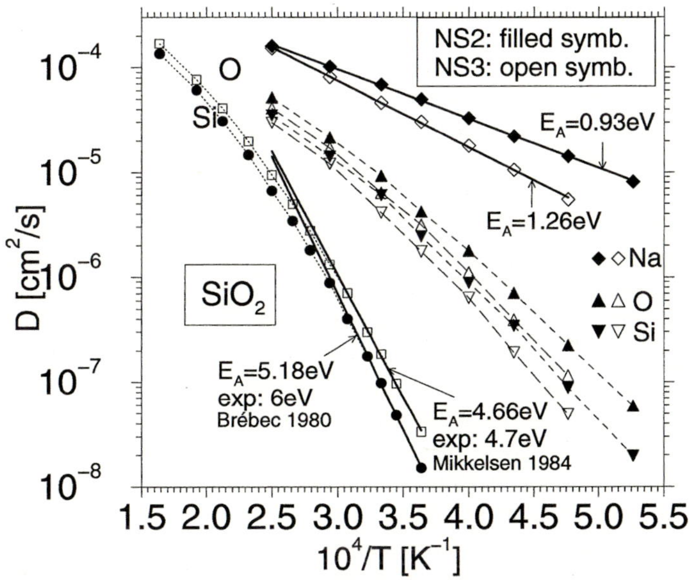

Fig. 7. Selfdiffusion constants for pure SiO_2, sodium disilicate (NS2) and sodium trisilicate (NS3) plotted vs. inverse temperature. The bold straight lines are fits with Arrhenius laws, indicating the respective activation energies E_A. The dashed curves are guides to the eye. Experimental results for $E_A^{(\mathrm{Si})}$ are taken from Brébec et al. [69] and for $E_A^{(\mathrm{O})}$ from Mikkelsen [70]. From Horbach et al. [40].

A further problem are systems in which the diffusion constants for the various species are very different, such as in the discussed sodium–silicate systems. Fig. 7 shows that at low T the diffusion constant of Na is much larger than the one of Si or O. Hence one will find a mean square displacement which increases quickly with time, see Fig. 6b, and one might be tempted to read off a diffusion constant from runs that are relatively short. This might, however, give rise to erroneous results, since the time scale might be insufficient to allow the Si–O matrix to relax and to reconstruct. Hence it is necessary to make runs that extend at least over the α–relaxation time τ of the matrix before the diffusion constant can be measured reliably. Note that with decreasing temperature this problem becomes more and more pronounced since the activation energy of Na is significantly smaller than the one of Si or O (see Fig. 7).

Fig. 7 shows also that for pure SiO_2 the results for the activation energies agree reasonable well with experiment [69,70]. Note, however, that the experimental values are extracted from a regime of much smaller temperatures ($5.5 < 10^4/T < 7.5$) where no simulations are possible for pure SiO_2. In the case of SiO_2–Na_2O mixtures, the Na ions cause a partial breakup of the network of SiO_2 tetrahedra (each O sits at a corner of such a tetrahedron and is shared between two tetrahedra). Therefore, the diffusion of Si and O in the mixtures is enhanced, in comparison with pure SiO_2. Unfortunately, we are not aware of any experimental data on diffusion constants for either NS2 or NS3; but there exist data for $Na_2Si_4O_9$ [72], e.g. for $T = 2800$ K and $p = 10$ GPa $D_{\mathrm{Si}} = 1.22 \times 10^5$ cm^2/s and $D_{\mathrm{O}} = 1.53 \times 10^5$ cm^2/s. Simulating this system at a density $\rho = 2.9$ g/cm^3 at $T = 2800$ K (then $p = 10.13$ GPa in the simulation, using 7680 particles), one finds [40] $D_{\mathrm{Si}} = 0.78 \times 10^5$ cm^2/s, $D_{\mathrm{O}} = 1.17 \times 10^5$ cm^2/s, and $D_{\mathrm{Na}} = 2.35 \times 10^5$ cm^2/s. Thus, the simulation results underestimate the corresponding experimental data by less than 40%, which is within the error bars of experiment. This agreement shows that the model used in [40] gives a quite reasonable description of the diffusion in sodium silicate melts.

In order to gain insight into the microscopic mechanism of the diffusion of the different atomic species, it is useful to analyze the probability $P_{\alpha\beta}(t)$ that a bond between an atom of type α and an atom of type β is present at time t, if it was present at time $t = 0$. Two atoms are defined as bonded if their distance is less than the location of the first minimum r_{min} in the corresponding pair correlation function $g_{\alpha\beta}(r)$. One finds [40] $r_{\mathrm{min}} = 3.6$ Å, 5.0 Å, 2.35 Å, 5.0 Å, 3.1 Å, and 3.15 Å, for the Si–Si, Si–Na, Si–O, Na–Na, Na–O, and O–O correlations, respectively. If one plots $P_{\alpha\beta}(t)$ versus the logarithm of time one finds a plateau at intermediate time scales, which corresponds to the plateau seen in Fig. 6, while at late times there is a decay according to a stretched exponential function, $P_{\alpha\beta}(t) \propto \exp[-(t/\tau)^{\beta_{\mathrm{KWW}}}]$, with $\beta_{\mathrm{KWW}} \approx 0.54$ [40] for Na, while for the other ions $P_{\alpha\beta}(t)$ decays almost like a simple exponential function (which would have $\beta_{\mathrm{KWW}} = 1$, while the best fit actually yields $\beta_{\mathrm{KWW}} \approx 0.9$ [38]). Therefore, it is convenient to define a lifetime $\tau_{\alpha\beta}$ of a bond from the condition $P_{\alpha\beta}(t = \tau_{\alpha\beta}) = e^{-1}$. As expected, the selfdiffusion of oxygen is simply controlled by the lifetime of Si–O bonds (Fig. 8): only if such a bond breaks, can the associated O–atom move. The

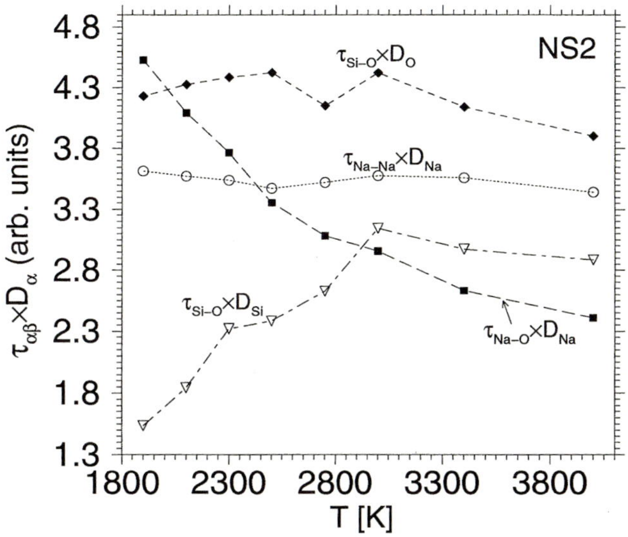

Fig. 8. Products $\tau_{\alpha\beta} D_\alpha$ for sodium disilicate plotted vs. temperature. From Horbach et al. [40].

O–diffusion can in fact be interpreted in terms of a kind of defect mechanism: while normally each Si atom has 4 oxygen neigbors, defects can occur where one Si "catches" a 5th O–atom, leaving a Si with only 3 O–atoms bound to it in its neighborhood. These coordination defects diffuse through the random networks, much like vacancy–interstitial defects diffuse in a crystal lattice.

However, Fig. 8 also shows that there is no simple relation between D_{Si} and the relaxation time $\tau_{\text{Si-O}}$, and surprisingly D_{Na} seems to be linked to $\tau_{\text{Na-Na}}$, and not to $\tau_{\text{Na-O}}$, as one might naively expect. This result gives evidence that the sodium diffusion is relatively independent of the matrix and that the Na atoms hop independently through it, a picture that has been confirmed by the investigation of the van Hove correlation function (see [3] for a definition) and other time dependent correlation functions [40,71].

The above remark that the regime where the Einstein relation, (6), is reached when atoms have moved over a few interatomic distances holds only for simple liquids, of course, but is not true for complex fluids such as polymer melts. For a flexible macromolecule, one must distinguish between the motion of a single monomeric unit and the motion of the chain as a whole, which is formed from N such units (Fig. 9a) [42]. While both the single (effective) monomer and the center of mass of the chain show a regime of ballistic motion at very short times in Molecular Dynamics simulations, as found for SiO_2 (Fig. 2), there occurs already at high temperatures (far above any glass transition) a regime of anomalous diffusion, before (6) holds. The coarse–grained

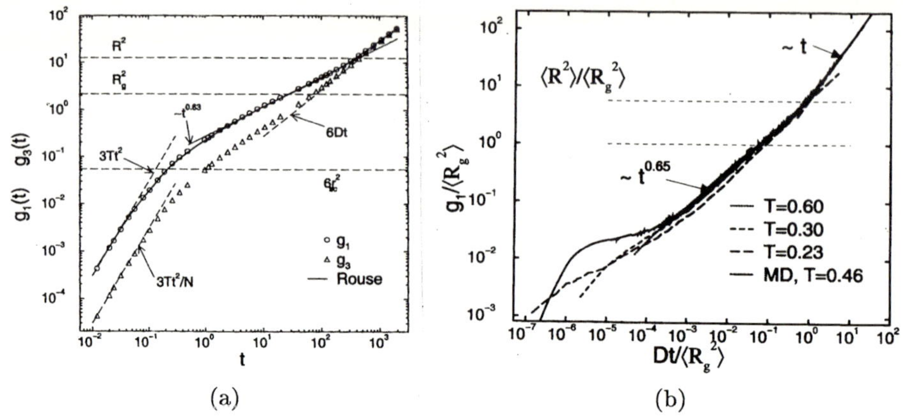

Fig. 9. a) Log–log plot of the mean square displacement of an inner monomer, $g_1(t)$, and of the chains' center of mass, $g_3(t)$, versus time for a temperature $T = 1$, pressure $p = 1$, for a bead–spring model of a polymer chain with chain length $N = 10$. Beads interact with a Lennard–Jones potential given by (8). The thick solid line is based on the Rouse model, using the amplitudes of the Rouse modes and the associated Rouse times observed in the same simulation. Characteristic lengths indicated are r_{sc}, the size of the nearest neighbor cage, R_{g}, the gyration radii of the chains, and R_{e}, the end–to–end distance. Both the initial ballistic regime ($3Tt^2$ and $3Tt^2/N$, respectively), the regime of anomalous diffusion of the monomers ($g_1(t) \propto t^{0.63}$) and the diffusion regime ($g_1(t) = g_3(t) = 6Dt$) are indicated by straight lines. From Bennemann et al. [73]. b) Comparison of the mean square displacement $g_1(t)$, rescaled with the mean square radius of gyration, plotted vs. the scaled time $Dt/\langle R_{\mathrm{g}}^2\rangle$, for the off–lattice model (curve marked MD) with the corresponding Monte Carlo results for the bond fluctuation model at three temperatures. Straight lines indicate power laws of anomalous diffusion ($t^{0.65}$) and ordinary diffusion (t^1). From Binder et al. [74].

model used in Fig. 9a is a bead–spring model of a flexible polymer, where all beads interact with a truncated and shifted Lennard Jones potential,

$$U_{\mathrm{LJ}}(r) = 4\varepsilon[(\sigma/r)^{12} - (\sigma/r)^6)] + U_0, \quad r \le r_{\mathrm{c}} = 2 \times 2^{1/6}\sigma, \tag{8}$$

while $U_{\mathrm{LJ}}(r \ge r_{\mathrm{c}}) = 0$, and U_0 is chosen such that $U_{\mathrm{LJ}}(r)$ is continuous at $r = r_{\mathrm{c}}$. The parameters $\sigma = 1$ and $\varepsilon = 1$ define units of length scale and temperature scale, respectively, and also $k_B \equiv 1$ and $m = 1$ (time is measured in units of $\tau_0 = \sigma(m/\epsilon)^{1/2}$). Bonded neighbors along a chain interact through a FENE potential

$$U_{\mathrm{FENE}}(\ell) = -15R_0^2 \, \ln[1 - (\ell/R_0)^2] \tag{9}$$

with $R_0 = 1.5$. Eqs. (8), (9) lead to an equilibrium bond length of $\ell_0 = 0.96$, different from the distance $r_0 = 2^{1/6}\sigma \approx 1.12$ where $U_{\rm LJ}$ has its minimum. This competition between the two lengths preferred by bonded and non-bonded interactions has the effect that this model of a polymer melt also tends to glass–like freezing, and crystallization is avoided. For pressure $p = 1$, however, the critical temperature of mode coupling theory is $T_{\rm c} = 0.45$ [74], and thus at the temperature $T = 1$ shown in Fig. 9a one is far away from the temperatures where slowing down due to the glass transition occurs. In fact, the anomalous diffusion seen in Fig. 9a can be well understood as an effect of chain connectivity, and can be accounted for by the Rouse model [75], at least to some extent. In the regime of displacements exceeding the size of the cages confining the monomers but smaller than the end–to–end distance, $r_{\rm sc}^2 < g_1(t) < R_{\rm e}^2$ where $g_1(t) \equiv \langle[\mathbf{r}_i(t) - \mathbf{r}_i(0)]^2\rangle$, one observes $g_1(t) \propto t^{0.63}$ (for an ideal chain in a heat bath, the Rouse model would yield [75] $g_1(t) \propto t^{1/2}$). Similarly, the center of mass diffusion $g_3(t) = \langle[\mathbf{r}_{\rm CM}(t) - \mathbf{r}_{\rm CM}(0)]^2\rangle$ behaves as $g_3(t) \propto t^{0.8}$ as long as $g_3(t) < R_{\rm g}^2 \approx \frac{R_{\rm e}^2}{6}$, and (6) holds only for even larger displacements.

When this model is studied at lower T closer to the glass transition, then a plateau near $g_1(t) \approx r_{\rm sc}^2$ develops, similar to the mean square displacement shown in Fig. 6, see also Fig. 9b [74] and the data in [17,73,74]. The above Rouse–like region then follows at those times where the “cages” confining the monomeric units have decayed.

Fig. 9b also illustrates the fact that the mesoscopic scale motions of polymers can also be modelled by even cruder lattice models of polymers, such as the bond fluctuation model [14,15]. In this latter model the elementary dynamical steps consist of attempted hopping moves of monomeric units by one lattice spacing in a randomly chosen direction. While such a model clearly cannot capture the small–scale ballistic regime, and thus also the initial part of the dynamics in the cage differs from the corresponding off–lattice results, the Rouse–type regime already coincides with the latter, if appropriately rescaled quantities are compared. Thus, not all lattice models behave as simple as the model (Fig. 5) discussed in Sec. 2. At this point, we also mention that by fine–tuning of effective interactions of a lattice model one even can approximately fit the behavior of atomistic off–lattice models for real polymeric materials, such as polyethylene melts [16]. The accuracy of this “mapping” has been tested by a comparison with Molecular Dynamics simulations for a melt of $C_{100}H_{202}$ at $T = 509$ K [16].

Finally, we emphasize that another case, where the approach to the regime where (6) holds is very slow, is the diffusion of small particles in a frozen environment, e.g. gas molecules diffusing in a polymeric matrix [76–80]. It is found that (6) holds only if $\langle r^2(t)\rangle$ is sufficiently large, such that in the explored region the locally random system is already homogeneous. This fact is clear when one considers the related problem of diffusion on percolation clusters (“ant in the labyrinth”) near the percolation threshold [81]: The crossover

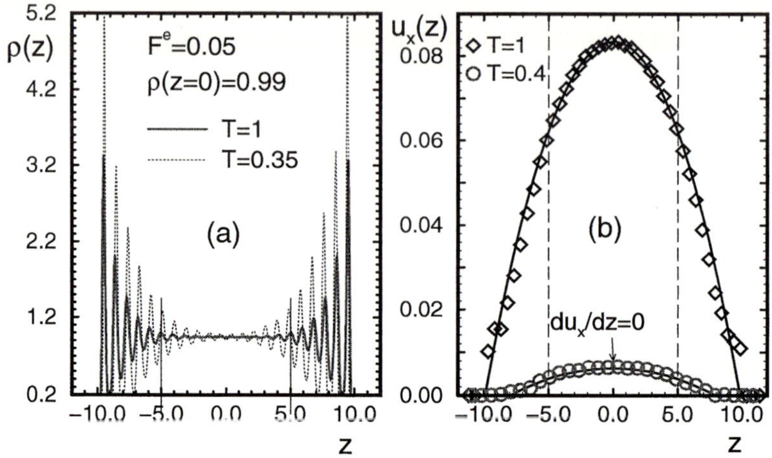

Fig. 10. Density profile of a polymer melt confined between two walls (at $z_{\mathrm{wall}} = \pm 10$), for a force $F^{\mathrm{e}} = 0.05$, density $\rho_0 = 0.99$, and two temperatures as indicated (left part). All quantities are measured in units of the Lennard–Jones potential, (8) that acts between the monomers. Only the region $-5 \le z \le 5$, where the density oscillations ("layering") are weak, is used for the fit of (10) to the observed velocity profile (right part). Solid lines are results of a fit to $u_x(z) = a - bz^2$ and η_{eff} results then from b as $\eta_{\mathrm{eff}} = \rho_0 F^{\mathrm{e}}/(2b)$. From Varnik and Binder [59].

from anomalous to ordinary diffusion is observed only if $\sqrt{\langle r^2(t)\rangle}$ exceeds the correlation length ξ_p which diverges at the percolation threshold [81].

4 Estimation of the Shear Viscosity

Both for molten oxides and for polymer melts the shear viscosity $\eta(T)$ is of great interest, and a large variety of techniques to extract $\eta(T)$ from simulations has been proposed. The first NEMD method that we shall discuss considers the flow properties of the fluid between parallel solid walls. The flow is created by a constant external force F^{e} that points in the x–direction parallel to the walls on all fluid particles, and the resulting velocity profile $u_x(z)$ in the z–direction perpendicular to the walls is recorded.

While for a study of a fluid at rest confined between walls it makes perfect sense to consider the case of perfectly smooth, repulsive walls [82,83], for a fluid flowing between walls a different treatment of the boundaries is needed, in order to obtain a nontrivial profile $u_x(z)$. In [59] structured walls were assumed that are atomistically rough, choosing a triangular lattice (with equilibrium sites $\mathbf{r}_{i,\mathrm{eq}}$) for each wall, to which the wall atoms are harmonically bound, $U_{\mathrm{harm}}(\mathbf{r}) = \frac{1}{2}K_{\mathrm{h}}(\mathbf{r}_i - \mathbf{r}_{i,\mathrm{eq}})^2$, with $K_{\mathrm{h}} = 100$. The beads of the polymers interact with the wall atoms also with a Lennard–Jones potential, and in addition a repulsive wall potential ($\propto z^{-9}$) is used to represent the effect of the layers other than the topmost one of the solid forming the

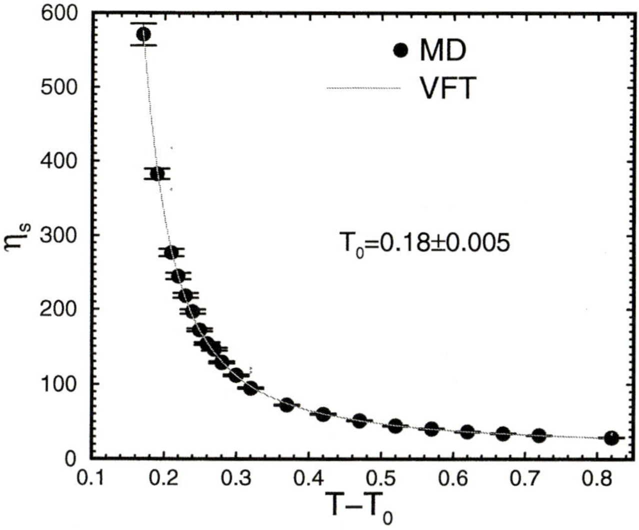

Fig. 11. Viscosity of a polymer melt plotted vs. the temperature distance from the Vogel–Fulcher temperature $T_0 = 0.194 \pm 0.012$, obtained from a least–square fit (full curve) to the Molecular Dynamics results (circles with error bars). From Varnik and Binder [59].

walls [84]. The parameters of the wall atom–monomer interaction are chosen $\sigma_{\rm wm} = 1/\sqrt{2}$, $\epsilon_{\rm wm} = 2$ [59]. In general, partial slip at the position $z = z_{\rm wall}$ of the wall is observed, $\partial u_x(z)/\partial z|_{z=z_{\rm wall}} = u_x(z_{\rm wall})/\delta$, δ being a slip length. Choosing the position $z = 0$ in the center of the film, hydrodynamics then implies the well–known Poisseuille type flow [52–60,84,85]

$$u_x(z) = -\rho_0 F^{\rm e}(z^2 - z_{\rm wall}^2 - 2z_{\rm wall}\delta)/(2\eta_{\rm eff}) \qquad (10)$$

where $\rho_0 = \rho(z = 0)$ is the fluid density in the center of the fluid film, and $\eta_{\rm eff}$ is an effective viscosity, which reduces to the shear viscosity of the bulk fluid for $F^{\rm e} \to 0$ and if $|z_{\rm wall}|$ is large enough. In practice, this holds already if $|z_{\rm wall}|$ is a few atomic diameters, see Fig. 10. (Note that since the system is driven one has to apply a thermostat in order to avoid that it heats up.) Similar to the diffusion constant of the melt [17,44], the viscosity $\eta(T)$ can be described by a Vogel–Fulcher relation $\eta(T) \propto \exp(\text{const}/(T - T_0))$ (Fig. 11). Thus neither $\eta(T)$ nor D can be described by simple Arrhenius laws [17,42] which could be expected, since polymers are "fragile" glass formers. On the other hand, the simple Stokes–Einstein relation $k_BT/(\eta D) = \text{const}$ which holds for the simple Rouse model [43,75] is not verified here at low temperatures [59].

An alternative method to estimate the viscosity does not use any NEMD nor considers any confinement, but rather studies appropriate dynamical correlation functions of a system in equilibrium, where periodic boundary conditions are applied as usual. The shear viscosity η can be obtained as a time

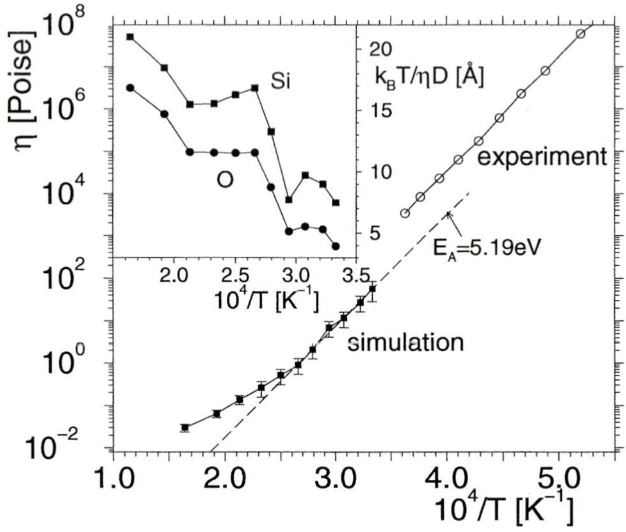

Fig. 12. Molecular Dynamics results for the viscosity of the BKS model (7) for SiO_2 plotted vs. inverse temperature. The dashed straight line indicates an Arrhenius fit with an activation energy $E_A = 5.19$ eV. Experimental data [86] are compatible with this value but would suggest a slightly different pre–exponential factor. Note that for SiO_2 the analysis of other correlation functions at very high temperature suggests a critical temperature of mode coupling theory $T_c \approx 3330$ K, and $\eta(T_c) \approx 8$ Poise. The insert shows the failure of Stokes–Einstein relations. From Horbach and Kob [38].

integral over the autocorrelation function of the off–diagonal elements of the pressure tensor [3] ("Kubo formula")

$$\eta = (k_B T V)^{-1} \int_0^\infty dt \, \langle \Pi_{\alpha\beta}(t) \Pi_{\alpha\beta}(0) \rangle \; , \quad \alpha \neq \beta \tag{11}$$

where V is the volume of the simulation box, and

$$\Pi_{\alpha\beta}(t) = \sum_{i=1}^{N} m_i u_i^\alpha(t) u_i^\beta(t) + \sum_{i=1}^{N} \sum_{j>i} F_{ij}^\alpha(t) r_{ij}^\beta(t) \; . \tag{12}$$

In (12) the sum runs over all N particles i (which may have different masses m_i) in the system, u_i^α is the αth cartesian component of the velocity $\mathbf{u}_i, \mathbf{r}_{ij}$ is the distance between particles i, j and $\mathbf{F}_{ij}$ the force acting between them. While (11) and (12) have not been used for the model of the polymer melt described by (8) and (9), they were used for the model for SiO_2 described by the BKS potential, (7). Fig. 12 shows a plot of $\eta(T)$ [on a logarithmic scale] vs.

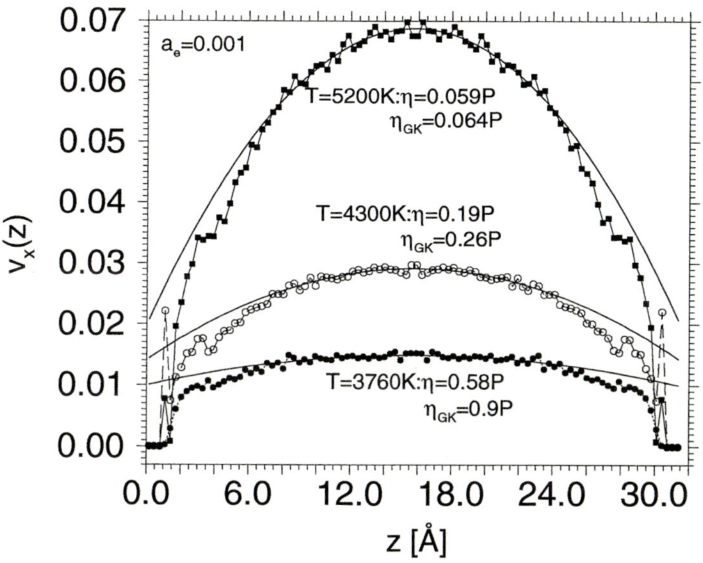

Fig. 13. Velocity profiles of SiO_2 melts obtained from NEMD simulations between parallel walls. Three temperatures are included, as shown in the figure; solid curves show fits to (10), and the corresponding viscosity estimates are quoted. Results from the Green–Kubo formula, (11), are included for comparison. From Horbach et al. [60].

$1/T$ [38], to demonstrate the extent to which an Arrhenius law holds. (Note that it is very remarkable that at sufficiently high T strong deviations from the Arrhenius dependence are observed. This is discussed in more detail in Ref. [38].) Also experimental data [86] are included, demonstrating reasonable agreement. The inset shows that again no Stokes–Einstein relation holds.

If one compares $\eta(T)$ and corresponding data for the self diffusion constants for the same model (Figs. 7 and 12) it is evident that the statistical accuracy of $\eta(T)$ is much less. This must be expected, however, since for $D_{\mathrm{Si}}(D_{\mathrm{O}})$ every Si(O)–atom contributes an independent "measurement" in (6), while η as computed from (11) is a collective quantity of the whole system. Thus the question arises whether $\eta(T)$ can be estimated more accurately from the NEMD technique, described above. This question was followed up in [60], and as Fig. 13 shows, the data obtained from both methods are compatible with each other, but there is no indication that the accuracy of the NEMD approach is really better than that of the Kubo formula, (11). (Note that a comparable amount of computer time had to be used to obtain η via the two methods.)

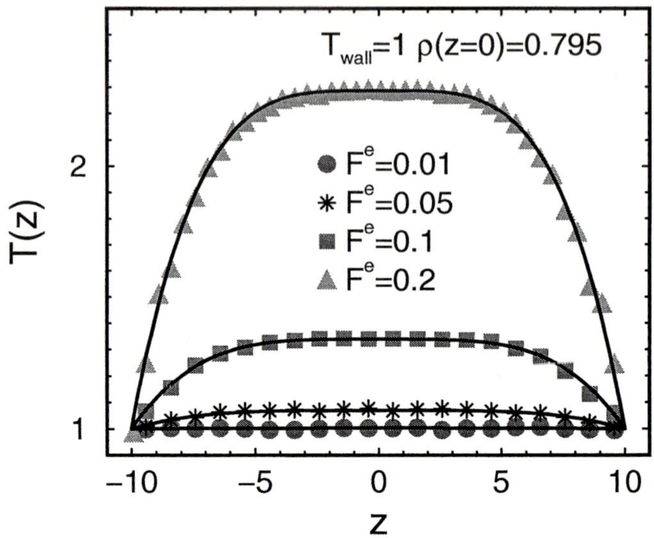

Fig. 14. Temperature profile across a polymer film (chain length $N = 10$, density $\rho(z = 0) = 0.795$, $z_{\rm wall} = \pm 10$) at a wall temperature $T_{\rm wall} = 1$, for four different choices of the external force $F^{\rm e}$ driving the flow. Curves are the result of a fit with (13), with λ being the single fitting parameter for each curve. From Varnik and Binder [59].

5 Thermal Conductivity

Another transport phenomenon of interest in fluids is transport of heat. Again one can estimate the corresponding transport coefficient either by NEMD methods or analyze suitable time correlation functions in thermal equilibrium.

In fact, the above simulation of a polymer melt flowing in x–direction between walls can be used to estimate the thermal conductivity λ, if one lets the thermostat act only on the wall atoms, while the fluid flow in between the walls is simulated in a strictly microcanonical ensemble. Due to the flow, heat is generated in the fluid film, which is transported to the walls. This gives rise to a temperature profile across the wall [87]

$$T(z) = T_{\rm wall} + \frac{(\rho F^{\rm e} D^2)^2}{192\lambda\eta_{\rm eff}}[1 - (z/z_{\rm wall})^4] \,, \tag{13}$$

$T_{\rm wall}$ being the temperature maintained (via the Nosé–Hoover thermostat, for instance [4–8,59]) at the wall. Fig. 14 shows results of a corresponding simulation, together with a fit to (13) using data for the viscosity $\eta_{\rm eff}$ such as shown in Fig. 10. For the range of temperatures reached in the film in the case shown in Fig. 14, the thermal conductivity lies in the range $3 \le \lambda \le 4.4$.

An alternative method to estimate λ, which considers the autocorrelation function of temperature fluctuations in the microcanonical ensemble, was used in [61], again for the case of SiO_2 melts. One needs to measure the autocorrelation function $\phi_{TT}(q,t)$ of temperature fluctuations $\delta T_q(t)$ at finite wave vectors. The relaxation time $\tau(q)$ extracted from this correlator can be shown to behave as [61]

$$\tau(q) = \frac{\rho C_p}{\lambda} \frac{1}{q^2} + \text{const} , \tag{14}$$

where C_p is the heat capacity at constant pressure in thermal equilibrium. For 3000 K $\leq T \leq$ 4700 K, a nearly temperature independent value $\lambda \approx$ 2.4 W/(Km) resulted [61], in good agreement with experimental estimates ($2 \leq \lambda \leq 3$ [W/(Km)] if $T \geq 1000$ K [88]).

6 Concluding Remarks

For simple lattice models it has been possible for a long time to estimate reliably the transport coefficients and thus to describe interdiffusion in mixed crystals (Sec. 2), or surface diffusion in adsorbed registered monolayers at surfaces [89]. This is in contrast to MD methods for off–lattice models of practically relevant fluids (fluid SiO_2, polymer melts, etc.) which only recently have reached a level of accuracy that allows a meaningful comparison with experimental data, and a significant test of theoretical concepts.

By examples drawn from the research group of the authors, it has been shown that only the estimation of self–diffusion coefficients is rather straightforward, while the estimation of truly collective quantities (shear viscosity, thermal conductivity, etc.) still is at a stage where "feasability–type" exploratory simulations are made. However, the results are fairly encouraging, and it is hoped that in the near future much broader applications will appear, as well as more refined techniques to extract all transport coefficients of interest from suitably designed simulations. At the same time, there is often a surprising scarcity of experimental data to compare with. Thus it is hoped that the present article will stimulate both corresponding theoretical and experimental investigations.

Acknowledgments: One of us (K. B.) is greatly indebted to the late K. W. Kehr for stimulating his interest in the Monte Carlo study of diffusion phenomena, and for a longstanding collaboration [13,31], including the development of NEMC methods [31]. We are grateful to J. Baschnagel, C. Bennemann and W. Paul for their fruitful collaboration on the studies of diffusion in glassforming polymer melts [15,17,42,44,73] and to A. Latz and P. Scheidler for their fruitful collaboration on the methods to estimate the thermal conductivity of molten silica [61]. We acknowledge partial financial support from the BMBF and generous grants of computer time from the NIC Jülich.

References

1. Alder, B. J., Wainwright, T. E.: J. Chem. Phys. **31**, 459 (1959); Phys. Rev. A **1**, 18 (1970); Rahman, A.: Phys. Rev. A **136**, 405 (1964).
2. Erpenbeck, J. J., Wood, W. W.: In *Statistical mechanics*; Part B, Vol. 6 of Modern theoretical chemistry ed. by Berne, B. J., (Plenum Press, New York, 1977) 1–40.
3. Boon, J. P., Yip, S.: *Molecular hydrodynamics.* (Mc Graw Hill, New York, 1980).
4. Ciccotti, G., Hoover, W. G., (Eds.): *Molecular dynamics simulations of statistical mechanical systems.* (North–Holland, Amsterdam, 1986).
5. Allen, M. P., Tildesley, D. J.: *Computer simulation of liquids.* (Clarendon Press, Oxford, 1987).
6. Rapaport, D. C.: *The art of molecular dynamics simulation* (Cambridge Univ. Press, Cambridge, 1995).
7. Binder, K., Ciccotti, G. (Eds.): *Monte Carlo and Molecular Dynamics of Condensed Matter Systems* (Societa Italiana di Fisica, Bologna, 1996).
8. Frenkel, D., Smit, B.: *Understanding molecular simulation. From algorithms to applications* (Academic Press, San Diego, 1996).
9. Binder, K. (Ed.): *Monte Carlo methods in statistical physics* (Springer, Berlin, 1979).
10. Binder, K. (Ed.): *Applications of the Monte Carlo method in statistical physics* (Springer, Berlin, 1984).
11. Landau, D. P., Binder, K.: *A guide to Monte Carlo simulations in statistical physics* (Cambridge University Press, Cambridge, 2000).
12. Murch, G. E.: *Atomic diffusion theory in highly defective solids* (Trans. Tech. House, Adermannsdorf, 1980).
13. Kehr, K, W., Binder, K.: In Ref. [10], 181–221.
14. Binder, K. (Ed.): *Monte Carlo and Molecular dynamics simulations in polymer science* (Oxford University Press, New York, 1995).
15. Binder, K., Paul, W.: J. Polymer Sci., Part B **35**, 1 (1997).
16. Tries, V., Paul, W., Baschnagel, J., Binder, K.: J. Chem. Phys. **106**, 738 (1997).
17. Binder, K., Baschnagel, J., Paul, W.: Progr. Polymer Sci. (in press).
18. Gunton, J. D., San Miguel, M., Sahni, P. S.: In *Phase transitions and Critical phenomena, Vol. 8*, ed. by C. Domb and J. L. Lebowitz (Academic Press, London, 1983), p. 267.
19. Binder, K., Fratzl, P.: In *Phase transformations in materials*, ed. by G. Kostorz (Wiley–VCH, Weinheim, 2001) 409–480.
20. Binder, K., Kalos, M. H.: In Ref. [9], 225–260.
21. Reister, E., Müller, M., Binder, K.: Phys. Rev. E **64**, 041804 (2001).
22. Binder, K., Stauffer, D.: Adv. Phys. **25**, 343 (1976).
23. Novotny, M. A.: Phys. Rev. Lett. **74**, 1 (1995); Kolesik, M., Novotny, M. A., Rikvold, P. A.: Phys. Rev. Lett. **80**, 3384 (1998).
24. Müller–Krumbhaar, H.: In Ref. [9], 261–299.
25. Müller–Krumbhaar, H., Kurz, W., Brener, E.: In *Phase transformations in materials*, ed. by Kostorz, G. (Wiley–VCH, Weinheim, 2001) 81–170.
26. Mon, K. K., Binder, K., Landau, D. P.: Phys. Rev. B **35**, 3683 (1987).
27. Binder, K.: In *Kinetics of ordering and Growth at surfaces*, ed. by Lagally, M. (Plenum Press, New York, 1990) 31–44.
28. Patrykiejew, A., Binder, K.: Surface Sci. **273**, 413 (1992).

29. Barabasi, A. L., Stanley, H. E.: *Fractal concepts in surface growth* (Cambridge Univ. Press, Cambridge, 1995).
30. Pal, S., Landau, D. P.: Physica A **267**, 406 (1999).
31. Kehr, K. W., Binder, K., Reulein, S.: Phys. Rev. B **39**, 4891 (1989).
32. Murch, G. E.: In *Phase transformations in materials*, ed. by Kostorz, G. (Wiley–VCH, Weinheim, 2001) 171–238.
33. Brochard, F., Jouffray, J., Levinson, P.: Macromolecules **16**, 2638 (1983).
34. Kramer, E. J., Green, P., Palmstrom, C. J.: Polymer **25**, 437 (1984).
35. Binder, K.: J. Chem. Phys. **79**, 6387 (1983); Colloid & Polymer Sci. **265**, 273 (1987).
36. For a review, see e.g. Binder, K., Sillescu, H.: In *Encyclopedia of polymer sci. & eng. Suppl. Vol.*, ed. by Mark, H. (Wiley, New York, 1989) 297–315.
37. Horbach, J., Kob, W., Binder, K.: Phil. Mag. B **77**, 297 (1998).
38. Horbach, J., Kob, W.: Phys. Rev. B **60**, 3169 (1999).
39. Horbach, J., Kob, W.: Phil. Mag. B **79**, 1981 (1999).
40. Horbach, J., Kob, W., Binder, K.: Chemical Geology **174**, 87 (2001).
41. Winkler, A.: *Dissertation* (Johannes Gutenberg Universität Mainz, 2002).
42. Paul, W., Baschnagel, J.: In *Monte Carlo and Molecular dynamics simulations in polymer science*, ed. by Binder, K. (Oxford Univ. Press, New York, 1995) 307–355.
43. Tries, V., Paul, W., Baschnagel, J., Binder, K.: J. Chem. Phys. **106**, 738 (1997).
44. Bennemann, C., Paul, W., Binder, K., Dünweg, B.: Phys. Rev. E **57**, 843 (1998).
45. Binder, K., Baschnagel, J., Paul, W.: Progr. Polymer Sci. (2002, in press).
46. Ashurst, T. W., Hoover, W. G.: Phys. Rev. Lett. **31**, 206 (1972).
47. Lees, A. W., Edwards, S. F.: J. Phys. C **5**, 1921 (1972).
48. Hoover, W. G.: Ann. Rev. Phys. Chem. **34**, 103 (1983).
49. Evans, D. J.: J. Chem. Phys. **78**, 3297 (1983).
50. Evans, D. J., Morris, G. P.: Comp. Phys. Rep. **1**, 287 (1984).
51. Loose, W., Hess, S.: Phys. Rev. A **37**, 2099 (1988); Physica A **162**, 138 (1989).
52. Thompson, P. A., Robbins, M. D.: Phys. Rev. A **41**, 6830 (1990).
53. Todd, B. D., D. Evans, D., Daivis, P.: Phys. Rev. E **52**, 1627 (1995).
54. Travis, K. P., Todd, B. D., Evans, D. J.: Phys. Rev. E **55**, 4288 (1997).
55. Koplik, J., Banavar, J. R.: Phys. Rev. Lett. **80**, 5115 (1998).
56. Barrat, J. -L., Bocquet, L.: Phys. Rev. Lett. **82**, 4671 (1999).
57. Travis, K. P., Gubbins, K. E.: J. Chem. Phys. **112**, 1984 (2000).
58. Sokhan, V. P., Nicholson, D., Quirke, N.: J. Chem. Phys. **115**, 3878 (2001).
59. Varnik, F., Binder, K.: preprint.
60. Horbach, J., Kob, W., Binder, K.: in preparation.
61. Scheidler, P., Kob, W., Latz, A., Horbach, J., Binder, K.: Phys. Rev. B **67**, 104204 (2001).
62. De Groot, S. R., Mazur, P.: *Non–Equilibrium thermodynamics* (Amsterdam, North–Holland, 1962).
63. van Beest, B. W. H., Kramer, G. J., van Santen, R. A.: Phys. Rev. Lett. **64**, 1955 (1990).
64. Kramer, G. J., de Man, A. J. M., van Santen, R. A.: J. Amer. Chem. Soc. **64**, 6435 (1991).
65. Misawa, M., Price, D. L., Suzuki, K.: J. Non–Cryst. Sol. **37**, 85 (1980).
66. Horbach, J., Kob, W., Binder, K., Angell, C. A.: Phys. Rev. E **54**, R5897 (1996).
67. Horbach, J.: *Dissertation* (Johannes Gutenberg Universität, Mainz, 1998).

68. Götze, W.: In *Liquids, Freezing, and the Glass transition*, Ed. by Hansen, J.-P., Levesque, D., and Zinn-Justin, J. (North–Holland, Amsterdam, 1991) p. 287.
69. Brébec, G., Seguin, R., Sella, C., Bevenot, J., Martin, J. C.: Acta Metall. **28**, 327 (1980).
70. J. C. Mikkelsen: Appl. Phys. Lett. **45**, 1187 (1984).
71. Jund, P., Kob, W., Jullien, R.: Phys. Rev. B **64**, 134303 (2001).
72. Poe, B. T., McMillan, P. F., Rubie, D. C., Chakrabarty, S., Yarger, J., Diefenbacher, J.: Science **267**, 1245 (1997).
73. Bennemann, C., Baschnagel, J., Paul, W., Binder, K.: Comp. Theor. Polymer Sci. **9**, 217 (1999).
74. Binder, K., Baschnagel, J., Bennemann, C., Paul, W.: J. Phys.: Condensed Matter **11**, A47 (1999).
75. Doi, M., Edwards, S. F.: *The theory of polymer dynamics* (Oxford, Clarendon Press, 1986).
76. Müller–Plathe, F., Rogers, S. C., van Gunsteren, W. F. : Macromolecules **25**, 6722 (1992); J. Chem. Phys. **98**, 8995 (1993).
77. Gusev, A. A., Suter, U. W.: J. Chem. Phys. **99**, 2221 (1993); J. Chem. Phys. **99**, 2228 (1993).
78. Gusev, A. A., Müller–Plathe, F., van Gunsteren, W. F., Suter, U. W.: Adv. Polymer Sci. **116**, 207 (1994).
79. Weber, H., Paul, W.: Phys. Rev. E **54**, 3999 (1996).
80. Paul, W., Weber, H., Binder, K.: Ann. Phys. Leipzig **7**, 554 (1998).
81. Stauffer, D., Aharony, A: *Introduction to percolation theory* (Taylor and Francis, London, 1992).
82. Varnik, F., Baschnagel, J., Binder, K.: J. Chem. Phys. **113**, 444 (2000).
83. Varnik, F., Baschnagel, J., Binder, K.: Phys. Rev. E **65**, 021507 (2002).
84. Jabbarzadeh, A., Atkinson, J., Tanner, R.: J. Chem. Phys. **110**, 2612 (1999).
85. Bocquet, L., Barrat, J.-L.: Phys. Rev. E **49**, 3079 (1993).
86. Urbain, G. : Geochim. Cosmochim. Acta **46**, 1061 (1982).
87. Todd, B., Evans, D.: Phys. Rev. E **55**, 2800 (1997).
88. Mazurin, O. V., Streltsina, M. V., Shvaikovskaya, T. P.: *Handbook of glass data, Part A: Silica, Glass and Binary silicate glasses* (Elsevier, Amsterdam, 1983).
89. Sadiq, A., Binder, K.: Surface Sci. **128**, 350 (1983).

Unusual Viscosity Feature in Spinodal Decomposition Under Shear Flow

Jian *Wang*[†,‡], Wolfram *Gronski*[‡], Christian *Friedrich*[‡], Peter *Galenko*[†] and Dieter *Herlach*[†]

†: Institute for Space Simulation, German Space Research Center, D-551147 Koeln, Germany
‡: Freiburger Materialforschungszentrum, D-79104 Freiburg i. Br. Germany

1 Introduction

The process of phase separation of quenched binary systems, for example quenched binary alloys or polymer blends, is generally known as spinodal decomposition [1]. It is well known that domain growth in spinodal decomposition follows scaling power-law. Hydrodynamic effect plays an important role in spinodal decomposition. For example, without hydrodynamic effect, domain grows following a scaling power-law with exponent 1/3, with hydrodynamics the exponent increases to 1 [2].

Under steady shear flow, the domain is usually deformed along the shear direction. The anisotropoic domain morphology leads to noval rheological properties of the system. Both theoretical analysis and computer simulation in 2d [3] showed that the extra viscosity due to the deformed domains undergoes a maximum and then decreases. The maximum was supposed to be the point when the stretched domains start to break up into small pieces. Experimental results confirmed the signle peaked shear viscosity [4].

However, there were also experimental results that the viscosity developed into two peaks [5]. Since early computer simulation by Ohta, Nozaki and Doi [3] did not consider the hydrodynamic effect, in view of the strong effect of the hydrodynamics on the speed of domain growth, there was great expectation that the hydrodynamic effect might also be responsible to the two peaks in shear viscosity. Zhang, Zhang and Yang [2] recently made simulation combining the phase seperation equation with the Navier-Stokes equation. They observed two peaks in their simulations. However, it seems that their simulation stoped too soon to identify the second peak, and one might as well expect additional peak after the second. In theory[6], it has been shown that even without the hydrodynamic equation, the phase seperation alone can result in multiple peaks in shear viscosity. The reason is that stretching and and breakup can occur cyclically. In experiemnt, wether a peak shows up also depends on the resolution.

2 Theory

Spinodal decomposition is well described by the Cahn-Hilliard equation, or the time-dependent Ginzburg-Landau equation by choosing a phenomenological free energy, for example [6,7]

$$H[\psi] = \int dr[-\frac{1}{2}\psi^2 + \frac{1}{4}\psi^4 + \frac{1}{2}(\nabla\psi)^2] \tag{1}$$

The order parameter ψ is defined as the volume fraction difference between the two components of the binary system. The function $f = -\frac{1}{2}\psi^2 + \frac{1}{4}\psi^4$ is a typical double-well potential, which implies that the two seperated phases are more stable than the mixed one. The term $\frac{1}{2}(\nabla\psi)^2$ is the contribution to the free energy due to the domain interface. The Langevin equation for the evolvlution of order parameter is

$$\frac{\partial\psi(\mathbf{r},t)}{\partial t} = M\nabla^2\mu(\mathbf{r},t) \tag{2}$$

where $\mu(\mathbf{r},t) = -\psi + \psi^3 - \nabla^2\psi$ is the chemical potential, M is the mobility.

Considering convection, the above phase seperation equation may be modified as

$$\frac{\partial\psi(\mathbf{r},t)}{\partial t} + \nabla(\psi(\mathbf{r},t)\mathbf{v}) = M\nabla^2\mu(\mathbf{r},t) \tag{3}$$

Efficient algorithm, such as cell-dynamical system [8], has been developed to solve the phase seperation equation.

For liquid, the velocity field may be derived from Navier-Stokes equation,

$$\rho\frac{\partial\mathbf{v}}{\partial t} = \eta\nabla^2\mathbf{v} + \psi\nabla\mu - \nabla p \tag{4}$$

where ρ is the mass density, η is the liquid viscosity, and p is the pressure. In the study of phase seperation of polymer blends, the Navier-Stokes equation is usually approximated by neglecting the left hand side density term [2,7],

$$\eta\nabla^2\mathbf{v} = -\psi\nabla\mu + \nabla p \tag{5}$$

Applying the incompressible condition,

$$\nabla\cdot\mathbf{v} = 0 \tag{6}$$

one obtains a Poisson equation for the pressure

$$\nabla^2 p = \nabla\cdot(\psi\nabla\mu) \tag{7}$$

Once the pressure is solved, the velocity field can be obtained from the approximate Navier-Stokes equation above, which is also in the form of Poisson equation for each component of the velocity.

3 Simulation Detail

In case of steady shear experiment, there is an additional external velocity field, $v_x = sy$, where s is the shear rate and shear is applied in the x-direction. Shear causes strain as time goes $\gamma = st$, where t is the time. To model the strain effect in the simulation, Ohta, Nozaki and Doi [3] have proposed a shear boundary condition,

$$(x, y, z) \rightarrow (x + \gamma L, y + L, z) \tag{8}$$

The shear is applied in two (x,z) planes seperated by a distance L and in the x direction. The strain will eventually increase to such a high level as time passes, that the domians will be stretched into the shear direction. Periodic boundary condition is applied in the x and z directions in simulation.

For periodic boundary system, one can solve the Poisson equation either with Fourier transform method or real-space numerical method. For system under shear boundary condition, the real-space numerical method is more natural. In addition, quite efficient real-space algorithms, such as multigrid methods [9], are available to solve the Poisson equation. In earlier simulations [2], the phase seperation equation was solved under shear boundary condition, while the Navier-Stokes equation was solved with Fourier transform method which still implying using of periodic boundary condition. This might be inconsistent. For periodic system of spinodal decomposition, we found the same scaling growth power-law of exponent 1 when the hydrodynamics equation was solved with the real-space method, as Zhang, Zhang and Yang [2] observed when the hydrodynamics equation was solved with Fourier transform method.

To check how important the Navier-Stokes equation in explaining the shear data, we made simulations with different shear rates and initial equilibrium time steps. But we were still unable to convince ourselves an immediate and apparent correlation between hydrodynamics and the two peaks in shear viscosity. Often the second peak was too small to compare with that in experiment.

The shear viscosity in computer simulation is defined as [3],

$$\Delta\eta = -\frac{1}{s} < \frac{\partial\psi}{\partial x}, \frac{\partial\psi}{\partial y} > \tag{9}$$

where s is the shear rate. $\Delta\eta$ is different from the liquid viscosity η introduced in the Navier-Stokes equation above. For uniform field, $\Delta\eta$ will be zero, but η may not be zero. $\Delta\eta$ is a quantity being able to characterize the anisotropic morphology.

After switching on the shear, the domains start to strech in the shear direction, which cause the morphology to change from uniform to anisotropy, see Fig. 1. The black domain is $\psi > 0$, and the white domain is $\psi < 0$. (or vice versa). As the shear viscosity reaches the maximum, the fully stretched domains break up, which causes the shear viscosity to decrease. At this stage, stretching and breakup coexist. It is likely that a second peak will occur, but its magnitude can be quite different depending on the portions of stretching and breakup domains. In experiment [5], the second peak appeared if the shear was applied at a delayed time after quench. That inital equilibrium allowed the domains to grow to such a narrow size distribution that it favours most domains to stretch and break up synchronously. This synchronizing will enhance the second peak to rise.However, in our computer simulations, simple adjustment of the initial equilibrium time steps did not readily lead to an apparent enhanced second peak.

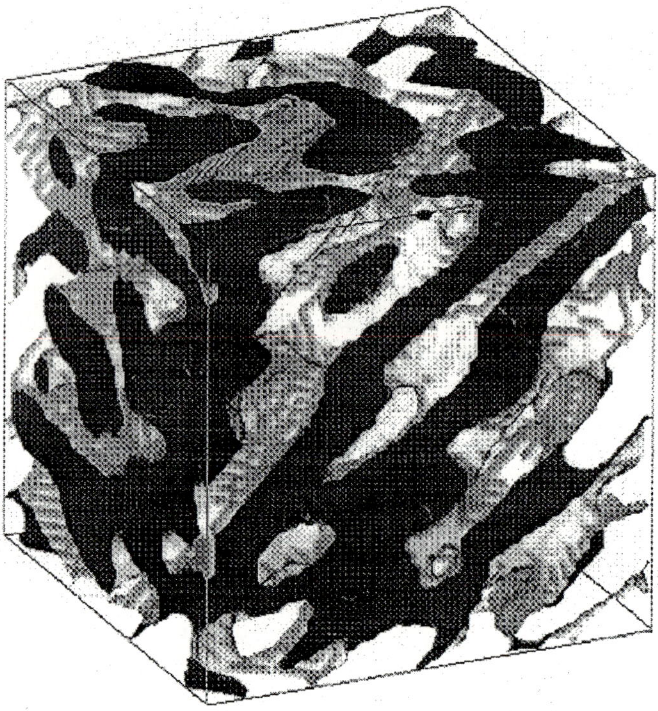

Fig. 1. Typical anisotropical growth of the domain in spinodal decomposition under shear.

In shear experiment, the distance between the shear planes (in y direction) is usually much smaller than the dimension in other direction (x and z directions). Observing this geometric setup of experiment, we designed simulations with a number of selected thicknesses. Fig. 2 displayes result with

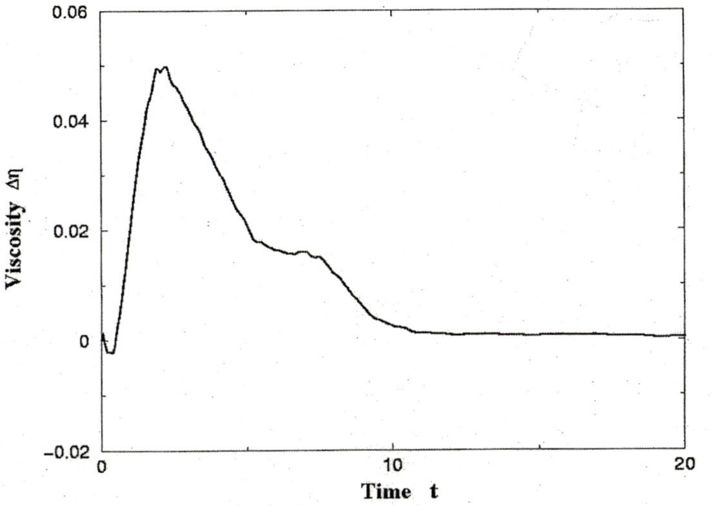

Fig. 2. Shear viscosity for a system of 100×100×100.

a grid of 100×100×100. The shear rate was 0.001 and initial equilibrium time steps was 300. The results was obtained by solving the phase seperation equation only without including the Navier-Stokes equation.

Fig. 3 displays simulation result on a grid 100×50×100. The difference of Fig. 3 from Fig. 2 is the domain size. It is obvious that the boundary has a significant effect on the peaks in shear viscosity. Reducing distance between the shear plane L has also resulted in reduction of the periodicity in the shear direction ($x = x + \gamma L$). That accelerated the second peak to appear in experiment. The 3d simulation result is in agreement with other theoretical prediction [6] that the phase seperation alone, without the hydrodynamic equation, can result in multiple peaks in shear viscosity. Experiments only observed two peaks because the resulotion was not sensitive enough at the beginning and the tail.

4 Conclusion

We simulate spinodal decomposition under the shear flow and explain the unusual multiple peaks observed in viscosity measurement. Previously the extra peaks have been attributed to the hydrodynamics effect of the liquid. According to the scaling exponent of 1, hydrodynamics tends to enhance the growth of domain during spinodal decomposition. But this effect is more or less isotropic, so one may expect little contribution to the peaks in shear vis-

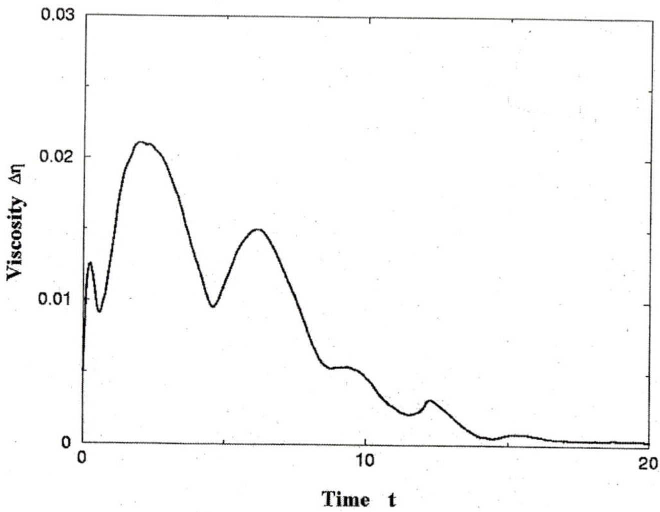

Fig. 3. Shear viscosity for a system of 100×50×100.

cosity. We found, on the other hand, the distance between the shearing planes plays a significant role in the appearence of peaks observed in experiment.

This work was made under support from DFG-German Research Foundation, research project He 1601/13.

References

1. K. Binder, in: Phase Transition in Materials, edited by R. W. Cahn, P. Haasen and E. J. Kramer, Materials Science and Technology Vol. 5 (VCH Weinheim, New York, 1990).
2. Z. Zhang, H. Zhang and Y. Yang, J. Chem. Phys. 113, 8348 (2000).
3. T. Ohta, H. Nozaki and M. Doi, , J. Chem. Phys. 93, 2664 (1990); Phys. Lett. A145, 304 (1990).
4. J. Läuger, Chr. Laubner and W. Gronski, Phys. Rev. Lett. 75, 3576 (1995)
5. W. Gronski, J. Läuger, and Chr. Laubner, J. Mol. Struct. 383, 23 (1996) Chr. Laubner, Diplomarbeit, Albert-Ludwigs University, Freiburg, 1996
6. F. Corberi, G. Gonnella and A. Lamura, Phys. Rev. Lett. 81, 3852 (1998)
7. H.Chen, and A. Chakrabarti, J. Chem. Phys. 108, 6006 (1998).
8. A. Shinozaki and Y. Oono, Phys. Rev. A45, R2161 (1992); Phys. Rev. E48, 2622(1993).
9. A. Brandt, Mathematics in Computation **31**, 333 (1977)

Micro-macro Approach to Cluster Formation in Granular Media

S. Luding

Institute for Computer Applications 1, University of Stuttgart,
Pfaffenwaldring 27, 70569 Stuttgart, Germany
Particle Technology, DelftChemTech, TU Delft,
Julianalaan 136, 2628 BL Delft, The Netherlands
e-mail: lui@ica1.uni-stuttgart.de or s.luding@tnw.tudelft.nl
homepage: www.ica1.uni-stuttgart.de/~lui

Abstract

Dissipation in granular media leads to interesting phenomena as there are cluster formation and crystallization in non-equilibrium dynamical states. The freely cooling system is examined concerning the energy decay and the cluster evolution with time. Finally, an example is given for a possible micro-macro transition with respect to the flow variables density, velocity, and fluctuation kinetic energy.

Keywords:
inhomogeneous free cooling, clustering, cooperative phenomena, event-driven molecular dynamics, micro-macro transition

1 Introduction

Interesting phenomena can be observed when granular media are studied [1–4]. The subject of this paper is the pattern formation via clustering in a dissipative, freely cooling system [5–8]. The special behavior of granular media is connected to its ability to form a hybrid state between a fluid and a solid: Energy input leads to a reduction of the density due to more collisions, so that the material can flow, i.e. it becomes 'fluid'. On the other hand, in the absence of energy input, granular materials 'solidify' due to dissipation. This makes granular media an interesting multi-particle system with a rich phenomenology, however, theoretical approaches are non-classical and appear often extreemly difficult, so that there is still active research directed towards the understanding of granular media.

The basic ingredients of a granular model system are discussed briefly in section 2. The inhomogeneous cooling and clustering are described in detail in section 3. The basic idea is that in a freely cooling granular gas, fluctuations in density and temperature cause a position dependent energy loss. Due to strong local dissipation, pressure and energy drop rapidly and material moves from 'hot' to 'cold' regions, leading to even stronger dissipation and thus causing the density instability with ever growing clusters.

In order to describe and understand such inhomogeneous, non-equilibrium systems, a particle simulation approach may be insufficient due to the enormous number of particles involved. However, a hydrodynamic theory of such dissipative granular flows is not available yet. Therefore, a first step towards a micro-macro transition that it neccessary to understand the hydrodynamic theory of granular media is presented. The density-, velocity-, and fluctuation kinetic energy-fields will be computed and discussed for the case of strong clustering.

2 Models for Multi-particle Simulations

The constituents of granular media are mesoscopic particles. When those objects interact (collide) the attractive potentials of the individual grains can be neglected. Two models for the repulsive particle-particle interactions are discussed in the following. They account for the excluded volume of the particles via a repulsive potential, either 'hard' or 'soft' and also account for dissipation in collisions via some coefficient of restitution. The third ingredient of a model granular material is friction which couples the rotational degrees of freedom to the linear motion, but it is not discussed in this paper.

The difference between the two most frequently used discrete element methods is the repulsive interaction potential. For the molecular dynamics (MD) method, soft particles with a power-law interaction potential are assumed, whereas for the event-driven (ED) method perfectly rigid particles are used. The consequence is that the duration of the contact of two particles, t_c, is finite for MD, but vanishes for ED.

2.1 The Event-driven, Rigid Particle Method

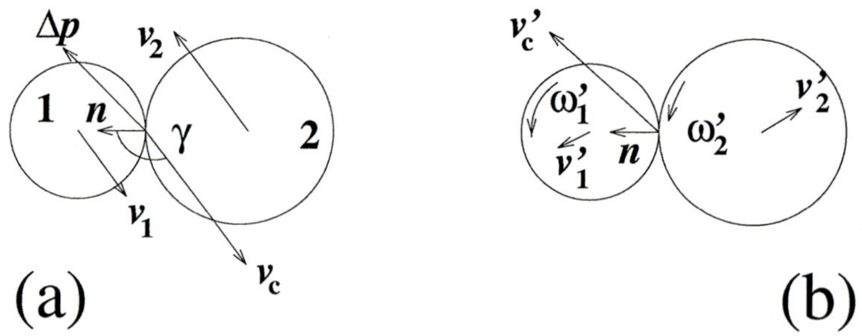

Fig. 1. Typical velocities of two particles immediately before (a) and just after (b) collision.

Consider two particles with diameter d_1 and d_2 and masses m_1 and m_2, The normal unit vector for their contact is $\boldsymbol{n}$, and $\boldsymbol{r}_i$ is the vector to the position of the center of particle i ($i = 1, 2$). The relative velocity of the contact points is $\boldsymbol{v}_c = \boldsymbol{v}_1 - \boldsymbol{v}_2$, with the velocity $\boldsymbol{v}_i$ of particle i. From momentum conservation it follows

$$\boldsymbol{v}_1' = \boldsymbol{v}_1 + \Delta\boldsymbol{P}/m_1 \quad \text{and} \quad \boldsymbol{v}_2' = \boldsymbol{v}_2 - \Delta\boldsymbol{P}/m_2 \ , \tag{1}$$

where $\boldsymbol{v}_i'$ is the unknown velocity of particle i after collision. The change of linear momentum of particle 1 is a function of the coefficient of restitution r:

$$\Delta\boldsymbol{P} = -m_{12}(1+r)\boldsymbol{v}_c^{(n)} \ , \tag{2}$$

with the reduced mass $m_{12} = m_1 m_2/(m_1 + m_2)$.

For the simulation of rigid particles, an event-driven method is used [9,10,7]. The particles undergo an undisturbed motion until an event occurs. An event can be either the collision of two particles or the collision of a particle with a wall. From the velocities just before contact, the particle velocities after a contact are computed following (1). Lubachevsky [9] introduced an efficient scalar ED algorithm which updates only those particles involved in the previous collision. The original algorithm is implemented and generalized to take into account dissipation.

2.2 The Time Driven, Soft Particle Technique

Even without using the soft particle method [11–13] in this study, it is convenient to discuss briefly the standard approach. Replacing $\Delta\boldsymbol{P}$ in (1) by $\boldsymbol{f}(t)\Delta t_{\mathrm{MD}}$, with the molecular dynamics time step Δt_{MD}, allows the integration of the corresponding, discretized equations of motion with standard numerical methods [12].

Since the modeling of realistic deformations of the particles would be much too complicated, let us assume that the overlap of two particles is the only quantity important for the interaction potential. The interaction of two particles can be split into (at least) three independent forces, and is typically short range, i.e. the particles interact only when they are in contact. The first force, an elastic repulsive force proportional to the overlap, accounts for the excluded volume which each particle occupies. In the simplest case, a linear spring can be used. The second force, a viscous damping force, models the dissipation in the normal direction and is proportional to the relative velocity. The simplest possible dashpot is again linear. This linear spring-dashpot model can be solved analytically and leads to a constant contact duration t_c and a constant restitution coefficient r [14]. The third force, accounting for friction, acts in the tangential direction, but will not be discussed here; for more information see Refs. [1].

2.3 The Connection Between Hard- and Soft-sphere Models

In the ED method, the time during which two particles are in contact is implicitly zero. The consequence is that exclusively pair contacts occur and the instantaneous momentum change $\Delta \boldsymbol{P}$ in (1) suffices to describe the collision. However, ED algorithms with constant r run into difficulties when the time between events, t_n, becomes too small – typically in systems with strong dissipation – and the so-called "inelastic collapse" occurs [6,15], i.e. the collision rate diverges for a few particles in the system. Since this is an artefact of the hard sphere model, it is unphysical and has to be avoided. Because a diverging number of collisions is only possible if the contact duration vanishes, the physical contact duration t_c has to be reintroduced in order to allow for realistic ED simulations. In MD simulations, on the other hand, one has $t_c > 0$, since every contact takes some finite time. Therefore, only a limited amount of kinetic energy ($\Delta E \propto 1 - r^2$) can be dissipated per collision. A finite contact duration implies a finite energy dissipation rate. In contrast, the consequence of a diverging collision rate would be a diverging energy dissipation rate.

In a dense system of real particles, energy dissipation becomes ineffective, i.e. the 'detachment effect' occurs [16,17]. This effect is not obtained with hard particles and a constant coefficient of restitution r. Therefore, in the framework of the so-called TC model, the restitution becomes elastic in nature, $r_n^{(i)} = 1$, if collisions occur too frequently, i.e. $t_n^{(i)} \leq t_c$, for the collision n of particle i. The time since the last collision is $t_n^{(i)}$ and the cut-off parameter t_c for elastic contacts can be identified with the contact duration. Thus, an additional material parameter is defined for the hard sphere model, that leads to qualitative agreement between ED and MD simulations and, in addition, avoids the inelastic collapse artefact. Recently, it has been shown that the TC model does not affect physical observables of the system, like the energy, as long as it is reasonably small [15].

3 Freely Cooling Granular Media

The simulations presented in the following involve $N = 99856 = 316^2$ dissipative particles with the restitution coefficients $r = 0.9$, 0.8, and 0.6, in a periodic, quadratic system with volume fraction $\nu = 0.25$. The system size is $l = Ld$ with dimensionless size $L = 560$ and particle diameter $d = 1\,\mathrm{mm}$. In order to reach an equilibrated initial condition, the system is first allowed to equilibrate with $r = 1$ for several hundreds of collisions per particle, so that a Maxwellian velocity distribution and a homogeneous density can be found. Then, at $t = 0\,\mathrm{s}$, dissipation is activated and the quantities of interest are examined. Snapshots of the simulation with $r = 0.8$ are presented in Fig. 2 at different, rescaled times τ (see below).

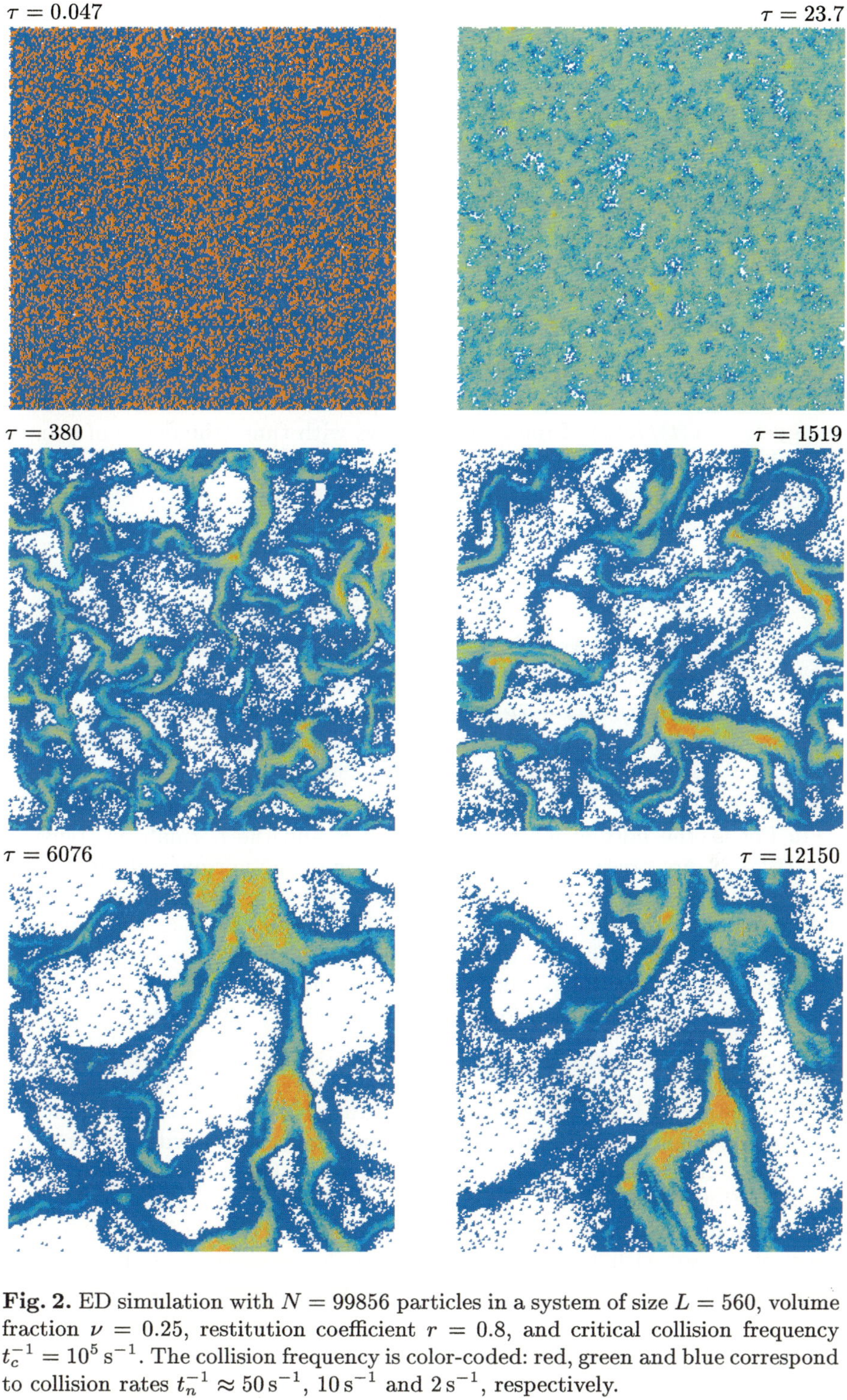

Fig. 2. ED simulation with $N = 99856$ particles in a system of size $L = 560$, volume fraction $\nu = 0.25$, restitution coefficient $r = 0.8$, and critical collision frequency $t_c^{-1} = 10^5\,\mathrm{s}^{-1}$. The collision frequency is color-coded: red, green and blue correspond to collision rates $t_n^{-1} \approx 50\,\mathrm{s}^{-1}$, $10\,\mathrm{s}^{-1}$ and $2\,\mathrm{s}^{-1}$, respectively.

The first picture in Fig. 2 is taken in the initally homogeneous cooling regime, whereas the next pictures show the different stages of the cluster growth regime. The final snapshots are taken in the limiting state, where the cluster has reached the system size. The particles are colored spots, where the green/red areas in the cluster centers correspond to particles with collision rate $t_n^{-1} \geq 50\,\mathrm{s}^{-1}$. This is much smaller than the critical collision rate $t_c^{-1} = 10^5\,\mathrm{s}^{-1}$, so that only a very small number of particles will be affected by the TC model.

3.1 Homogeneous and Inhomogeneous Cooling

In the homogeneous cooling state [18,6,19], one can expect that the kinetic energy $E = K(t)/K(0)$ of the system decays with time (the decay of energy with time is also evidenced by the change of color from red over orange to green and blue in Fig. 2) and follows the master-curve

$$E(\tau) = \frac{1}{(1+\tau)^2} , \tag{3}$$

with the dimensionless time $\tau = (1-r^2)t/(4t_E)$, the collision rate $t_E^{-1} = 8\nu g(\nu)\bar{v}/(\sqrt{\pi}d)$, the mean velocity $\bar{v} = \sqrt{K(t)/Nm}$, and the increased contact probability $g(\nu) = (1-7\nu/16)/(1-\nu)^2$ due to excluded volume effects at finite volume fractions ν. Inserting the parameters from the simulation, $1-r^2 = 0.36$, $g(\nu) \approx 1.5833$, and $\bar{v} = 0.02883\,\mathrm{m/s}$, one obtains an initial collision rate $t_E^{-1}(0) = 51.5\,\mathrm{s}^{-1}$ (corresponding to the red color in Fig. 2).

In Fig. 3, the normalized kinetic energy E and the normalized collision rate t_n^{-1}/t_E^{-1}, are presented, both as a function of the normalized time τ. At the beginning of the simulation we observe a perfect agreement between the theory for homogeneous cooling and the simulations. At $\tau \approx 10-40$ substantial deviations from the homogeneous cooling behavior become evident, i.e. the decay of energy is slowed down earlier for stronger dissipation. The deviation from the analytical form increases until $\tau \approx 10^3$ where the clusters reach system size and the behavior of E changes again to a slightly more rapid decrease.

This change in behavior is evident from the collision rate t_n^{-1}. At first, in the homogeneous cooling regime, the collision rate decays with $t_n^{-1} \propto \sqrt{E}$. Then the collision rate is almost constant (for $r = 0.9$ and $r = 0.8$) or even increases (for $r = 0.6$), until at $\tau \approx 10^3$ it becomes very noisy, indicating another change in the collective behavior. The long-time power law for the decay of energy with time is -2 in the homogeneous cooling case. In the cluster growth regime, however, we obtain slopes slightly smaller than -1 (the best fit leads to -0.920, -0.927, and -0.941 for $r = 0.9$, 0.8, and 0.6, respectively, with errors ± 0.002).

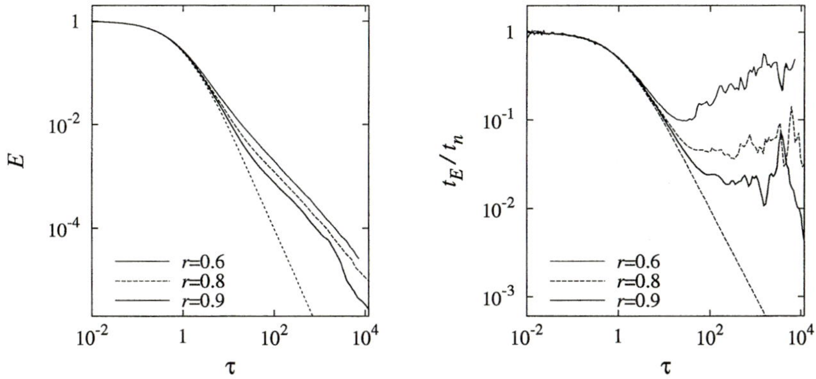

Fig. 3. (Left) Kinetic energy E plotted against τ for different values of $r = 0.9$, 0.8, and 0.6. The dotted line represents (3). (Right) Normalized collision rate t_E/t_n plotted against τ. The dashed line represents the collision rate $\sqrt{E}$ in the homogenously cooling case.

3.2 Cluster Structure

In Fig. 4, the pair-correlation function during the clustering is plotted in order to visualize the short-range structure in the clusters. In Fig. 5, zooms into the bottom-right area of the system in Fig. 2 are presented in order to give a more direct picture. In the initial state, the system is rather disordered and homogeneous. In the cluster growth regime, some particles approach closer and form loose clusters, however, the structure is still disordered, fluid-like. Only in the very late stage of the simulation, where the clusters are very large, one obtains crystalline, triangular lattice structures with a peculiar distribution of collision rates as color-coded.

This more qualitative picture can also be verified by computing the particle correlation function $g(r/d)$, see Fig. 4. The data are displayed for different τ-values, showing that the structure in the system becomes more and more pronounced with increasing time. Like the zoom-in in Fig. 5, also the correlation function shows that the crystalline structure occurs only in the late regime where the clusters are very large, with 'frozen' cores. Pronounced peaks at 1, $\sqrt{3}$, 2, ... indicate a triangular order of the particles.

3.3 Cluster Growth

The cluster growth can be studied quantitatively in the spirit of Luding and Herrmann [7]. All particle pairs with a distance smaller than some cut-off distance $\delta < (1 + S)d$, with an arbitrary cut-off parameter $S = 0.1$, are assumed to belong to the same cluster. After all particle pairs are examined, one obtains a cluster-size distribution and its moments. The first moment, the

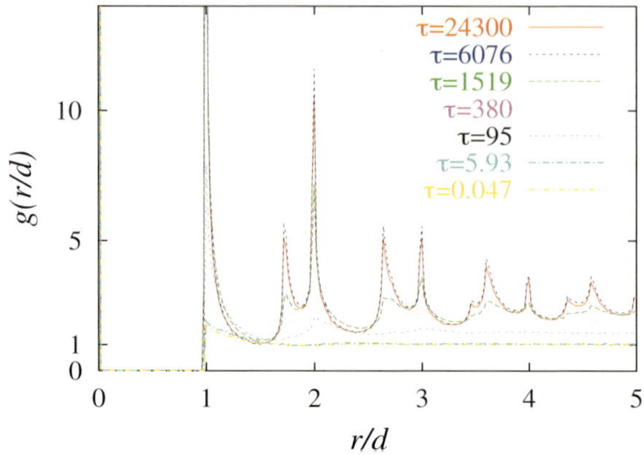

Fig. 4. Correlation function $g(r/d)$ as obtained from the simulation in Fig. 2 with τ as given in the inset.

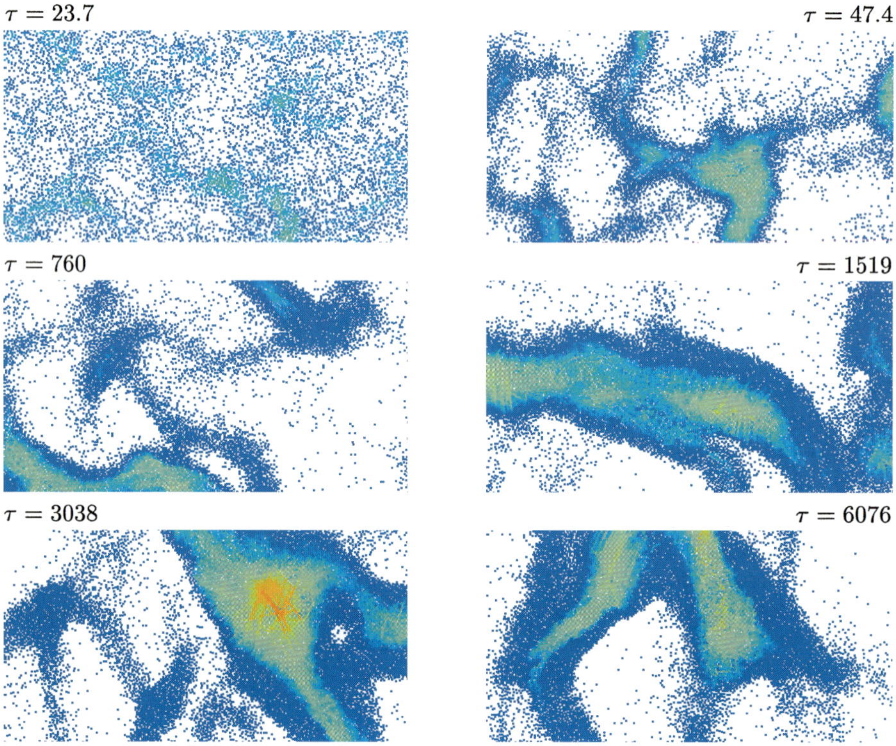

Fig. 5. Zooms into the lower right part of the ED simulation from Fig. 2.

mean cluster size $\langle M \rangle$, and the size of the largest cluster M_{max} are plotted in Fig. 6 against the time τ.

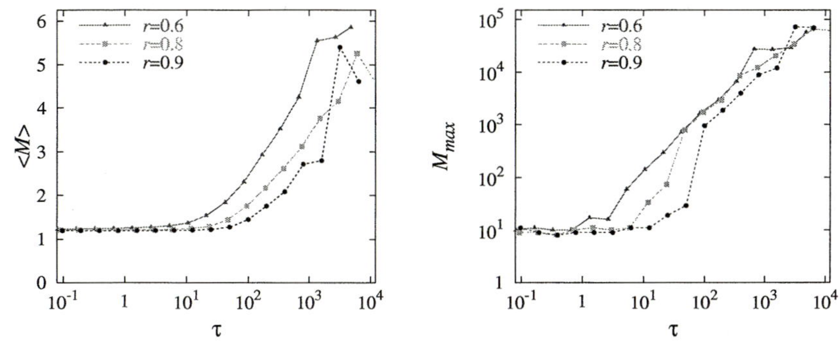

Fig. 6. (Left) Mean cluster size and (Right) Maximum cluster size as funtions of time τ for different coefficients of restitution r.

Both values are almost constant in the initial, homogeneous cooling regime. In the cluster growth regime a rapid increase of both $\langle M \rangle$ and M_{max} is evidenced until, at larger τ, the values reach a maximum and seemingly saturate or even decrease in the final regime where the clusters have reached system size. The cluster growth startes earlier for stronger dissipation, but the largest cluster seems to grow more rapidly for weaker dissipation, however, at a later time.

3.4 Micro-macro Transition

In this subsection, the macroscopic field quantities density, velocity, and fluctuation kinetic energy are discussed. In order to obtain the fields from the simulation data, the system is arranged on a 50×50 grid and the field quantities are computed for every cell. The density $\nu(x,y) = n_c \pi a^2 / V_c$ and the fluctuation kinetic energy $e(x,y) = (m/2) \sum_{i \in c} [\mathbf{v}_i - \mathbf{v}(x,y)]^2 / n_c$, are plotted in Fig. 7. The symbols are the number of particles per cell n_c, the cell-volume V_c, the particle radius a, mass m, and the mean velocity $\mathbf{v}(x,y) = \sum_{i \in c} \mathbf{v}_i / n_c$ in the cell.
The iso-lines of the density are chosen such that the dilute area ($\nu < 0.1$ – pink) is visible as well as the coexistence area between fluid and solid regions ($0.6 < \nu < 0.8$ – between blue and green). The green islands are solid clusters with triangular arrangement, whereas the pink islands are almost empty regions. Between pink and blue, the fluid phase exists at almost all densities.

The iso-lines for the fluctuation kinetic energy are chosen in arbitrary units. However, it is evident, that the temperature and the density fields are not correlated in a simple fashion. Dense regions are not neccessarily cold neither are dilute regions neccessarily hot. The structure of the e-field appears more detailed than the structure of the density field.

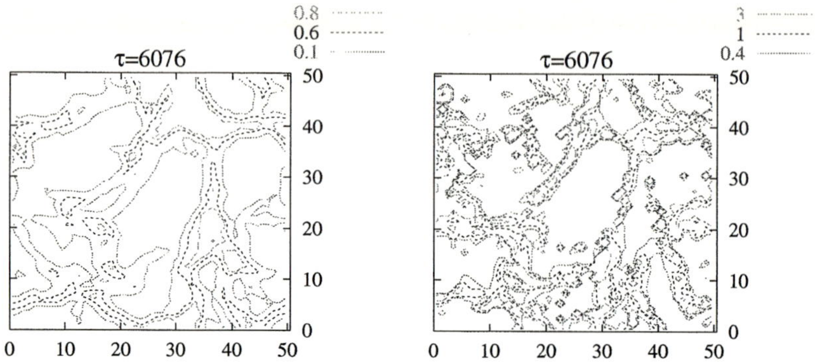

Fig. 7. Density and fluctuation kinetic energy iso-lines for the snapshot at time $\tau = 6076$ in Fig. 2. The density is given as volume fraction, whereas the energy is given in arbitrary units.

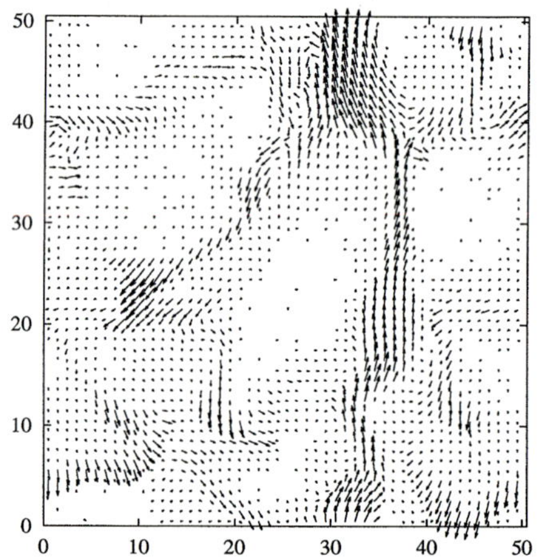

Fig. 8. Momentum flux field $\nu\mathbf{v}$ for the snapshot with $\tau = 6076$ in Fig. 2.

Finally, the velocity vector field is plotted as momentum flux field in Fig. 8 and shows the mass-flux in the system. A closer examination leads to the conclusion that the clusters are not stable, but have remaining internal motion due to shear- or compressive modes. Note that also the velocity field (energy due to flux motion) is not directly correlated to the fluctuation kinetic energy.

4 Summary and Conclusion

With a rather simple description of a granular material as an ensemble of inelastic hard disks (spheres), we have investigated the interesting effect of clustering in freely cooling systems. For short times the system is disordered and gas-like, whereas the structures at larger times are dense, crystalline clusters. The clusters grow until they reach the system size. Simulations at very long times were possible with the TC model, which reduces dissipation when contacts become too frequent.

As a first step towards a macroscopic, hydrodynamic description of clustering in freely cooling granular media, the density, the velocity and the fluctuation kinetic energy were computed from the simulation snapshots. The density field is directly correlated with the snapshot, whereas neither velocity nor fluctuation kinetic energy are directly correlated with the density. The latter fact implies that all three quantities are independent variables, neccessary for a hydrodynamic description of the system.

Further investigations concerning the transport parameters pressure, viscosity and heat conductivity in similar systems are in progress in two dimensional model systems and are to be applied to a successful hydrodynamic theory of granular flow.

Achnowledgements

We acknowledge the financial support of the Deutsche Forschungsgemeinschaft (DFG) and of the Max-Planck PKS in Dresden.

References

1. Herrmann, H. J., Hovi, J. -P. and Luding, S.: editors. *Physics of dry granular media - NATO ASI Series E 350*, Dordrecht, 1998. Kluwer Academic Publishers.
2. Pöschel, T. and Luding, S.: editors. *Granular Gases* (Berlin, 2001. Springer) Lecture Notes in Physics 564.
3. Vermeer, P. A., Diebels, S., Ehlers, W., Herrmann, H. J., Luding, S. and Ramm, E.: editors. *Continuous and Discontinuous Modelling of Cohesive Frictional Materials* (Berlin, 2001. Springer) Lecture Notes in Physics 568.
4. Kishino, Y.: editor. *Powders & Grains 2001*, Rotterdam, 2001. Balkema.
5. Goldhirsch, I. and Zanetti, G.: Clustering instability in dissipative gases. *Phys. Rev. Lett.*, 70(11):1619–1622, 1993.
6. McNamara, S. and Young, W. R.: Dynamics of a freely evolving, two-dimensional granular medium. *Phys. Rev. E*, 53(5):5089–5100, 1996.
7. Luding, S. and Herrmann, H. J.: Cluster growth in freely cooling granular media. *Chaos*, 9(3):673–681, 1999.
8. Cafiero, R., Luding, S. and Herrmann, H. J.: Two-dimensional granular gas of inelastic spheres with multiplicative driving. *Phys. Rev. Lett.*, 84:6014–6017, 2000.

9. B. D. Lubachevsky, B. D.: How to simulate billards and similar systems. *J. of Comp. Phys.*, 94(2):255, 1991.
10. Luding, S.: Granular materials under vibration: Simulations of rotating spheres. *Phys. Rev. E*, 52(4):4442, 1995.
11. Cundall, P. A. and Strack, O. D. L.: A discrete numerical model for granular assemblies. *Géotechnique*, 29(1):47–65, 1979.
12. Allen, M. P. and Tildesley, D. J.: *Computer Simulation of Liquids.* Oxford University Press, Oxford, 1987.
13. Lätzel, M., Luding, S. and Herrmann, H. J.: Macroscopic material properties from quasi-static, microscopic simulations of a two-dimensional shear-cell. *Granular Matter*, 2(3):123–135, 2000. cond-mat/0003180.
14. Luding, S.: Collisions & contacts between two particles. In H. J. Herrmann, J.-P. Hovi, and S. Luding, editors, *Physics of dry granular media - NATO ASI Series E350*, page 285, Dordrecht, 1998. Kluwer Academic Publishers.
15. Luding, S. and McNamara, S.: How to handle the inelastic collapse of a dissipative hard-sphere gas with the TC model. *Granular Matter*, 1(3):113–128, 1998. cond-mat/9810009.
16. Luding, S., Clément, E., Blumen, A., Rajchenbach, J. and Duran, J.: Anomalous energy dissipation in molecular dynamics simulations of grains: The "detachment effect". *Phys. Rev. E*, 50:4113, 1994.
17. Luding, S., Clément, E., Blumen, A., Rajchenbach, J. and Duran, J.: Interaction laws and the detachment effect in granular media. In *Fractal Aspects of Materials*, volume 367, pages 495–500, Pittsburgh, Pennsylvania, 1995. Materials Research Society, Symposium Proceedings.
18. Haff, P. K.: Grain flow as a fluid-mechanical phenomenon. *J. Fluid Mech.*, 134:401–430, 1983.
19. Luding, S., Huthmann, M., McNamara, S., and Zippelius, A.: Homogeneous cooling of rough dissipative particles: Theory and simulations. *Phys. Rev. E*, 58:3416–3425, 1998.

Part III

TRANSPORT

Urban Transport Phenomena in the Street Canyon

Maciej M. Duras

Cracow University of Technology, Institute of Physics, ulica Podchorażych 1, PL-30084 Cracow, Poland

Abstract. A field deterministic model of the vehicular dynamics in a generic urban street canyon with two neighboring canyons is considered. The assumed hydrodynamical model of vehicular movement is coupled to the gasdynamical model of the air and emitted pollutants. The vehicles are assumed to move on distinct left lanes and right ones. At the upstream and downstream ends there are coordinated traffic lights which introduce control parameters to the field model. The model of optimal control of street canyon dynamics is based on the two optimal multi-criteria control problems. The problems consist of minimization of the dimensionless functionals of the cumulative total travel time, global emissions of pollutants, and global concentrations of pollutants, both in the studied street canyon, as well as together with the two nearest neighbor substitute canyons, respectively.

1 The Field Models of Vehicles and Pollutants

In the present article we consider two coupled deterministic field models of the urban street canyon. The vehicular fields are one-dimensional in spatial variable x and they depend on time variable t. The considered fields are the vehicular number density $k^{s}_{l,vt}(x,t)$, vehicular velocity $\mathbf{w}^{s}_{l,vt}(x,t)$, emissivity of exhaust gases $e^{s}_{l,ct,vt}(x,t)$, and vehicular heat emissivity $\sigma^{s}_{l,vt}(x,t)$, where vt is emission type's number, ct is number of the constituent of emitted exhaust gases. It is assumed that there are $n_1 = n_L$ left lanes $s = 1$ and l is the left lane's number, $l = 1, ..., n_L$, whereas for $n_2 = n_R$ right lanes $s = 2$ and l is the right lane's number, $l = 1, ..., n_R$, $vt = 1, ..., VT$, $ct = 1, ..., CT$. VT is total number of types of vehicular emissions, and CT is total number of emitted exhaust gases. The above mentioned model consists of the equations of balances of vehicular numbers [5] and vehicular equations of state (Greenshields' equilibrium speed-density u-k model [6]) as well as exhaust gas emissivities and heat emissivities. The vehicular flow in the canyon is multilane bidirectional one-level rectilinear, and it is considered with two systems of coordinated signalized junctions at the upstream and downstream ends of lanes [1],[2], [3], [4]. The considered vehicles belong to distinct vehicular classes: passenger cars, and trucks. Emissions from the vehicles are based on technical measurements of the following types of pollutants: carbon monoxide CO, hydrocarbons HC, nitrogen oxides NO_x. The field model of pollutants is gasdynamical. The gases are Newtonian viscous perfect and noninteracting. The considered fields are three-dimensional in spatial variables (x, y, z) and one-dimensional in temporal variable t and they are mixture's density $\rho(x, y, z, t)$, velocity $\mathbf{v}(x, y, z, t)$, temperature $T(x, y, z, t)$, pressure $p(x, y, z, t)$, as well as constituents' mass concentrations $c_i(x, y, z, t)$, and partial pressures $p_i(x, y, z, t), i = 1, ..., N$, where N is total number of mixture's elements. $(N = N_E - 1 + N_A)$. The first $N_E - 1 = 3$ gases are the exhaust gases emitted by vehicle engines during combustion $\mathrm{CO}, \mathrm{CH}, \mathrm{NO_x}$. The remaining $N_A = 9$ gases are the constituents of air: $\mathrm{O_2}, \mathrm{N_2}, \mathrm{Ar}, \mathrm{CO_2}, \mathrm{Ne}, \mathrm{He}, \mathrm{Kr}, \mathrm{Xe}, \mathrm{H_2}$. The field equations are the balances of mixture's mass, linear momentum, energy, and equation of state (Clapeyron's law) as well as the balances of masses of constituents and constituents' equations of states (Dalton's law). The studied sources are mixture mass source $S(x, y, z, t)$, mixture energy source $\sigma(x, y, z, t)$, and constituents' mass sources $S_i^E(x, y, z, t)$ (exhaust gases and consumed oxygen). The boundary-initial problems are of mixed types (Dirichlet's and von Neumann's types). The equations of dynamics are solved by the finite difference scheme. The two separate multi-criteria optimization problems are posed by defining the dimensionless functionals of cumulative total travel time, global emissions of pollutants, and global concentrations of pollutants, either in the studied street canyon or in the canyon and its two nearest neighbor substitute canyons. The vector of control is a five-tuple composed of two cycle times, two green times, and one offset time between the traffic lights. The optimal control problems consists of minimization of the two functionals over the admissible control domain manifold.

2 The Governing Equations

Under the abovementioned specifications, the set of governing equations is formulated as follows [7], [8]:

$$\frac{\partial k_{l,vt}^s}{\partial t} + \mathrm{div}(k_{l,vt}^s \mathbf{w}_{l,vt}^s) = 0$$

$$\mathbf{w}_{l,vt}^s(x,t) = (w_{l,vt,f}^s \cdot (1 - \frac{k_{l,vt}^s(x,t)}{k_{l,vt,\mathrm{jam}}^s}), 0, 0)$$

$$\frac{\partial \rho}{\partial t} + \mathrm{div}(\rho \mathbf{v}) = S$$

$$\rho(\frac{\partial \mathbf{v}}{\partial t} + (\mathbf{v} \circ \nabla)\mathbf{v}) + S\mathbf{v} = -\nabla p + \eta \Delta \mathbf{v} + (\xi + \frac{\eta}{3})\nabla(\mathrm{div}\mathbf{v}) + \mathbf{F}$$

$$\rho(\frac{\partial \epsilon}{\partial t} + \mathbf{v} \circ \nabla \epsilon) = -(-\frac{1}{2}\mathbf{v}^2 + \epsilon)S + \mathbf{T} : \nabla \mathbf{v} + \mathrm{div}(-\mathbf{q}) + \sigma$$

$$\frac{p}{\rho} = \frac{R}{m_{\mathrm{air}}} \cdot T$$

$$\rho(\frac{\partial c_i}{\partial t} + \mathbf{v} \circ \nabla c_i) = S_i^E - c_i S +$$
$$+ \sum_{m=1}^{N-1} \{(D_{im} - D_{iN}) \cdot \mathrm{div}[\rho \nabla (c_m + \frac{k_{T,m}}{T} \nabla T)]\}$$
$$p_i = c_i \cdot \frac{m_{\mathrm{air}}}{m_i} \cdot p,$$

where $w_{l,vt,f}^s$ is maximum free flow speed, $k_{l,vt,\mathrm{jam}}^s$ is jam vehicular density, η is the first viscosity coefficient for the air, ξ is the second viscosity coefficient, $\mathbf{F} = \rho \mathbf{g}$ is the gravitational body force density, $\mathbf{g}$ is the gravitational acceleration, $D_{im} = D_{mi}$ is the mutual diffusivity coefficient from the i-th constituent to m-th one, and D_{ii} is the autodiffusivity coefficient of the i-th constituent, and $k_{T,m}$ is the thermodiffusion ratio of the m-th constituent, ϵ is the mass density of intrinsic (internal) energy of the air mixture, $\mathbf{T}$ is the stress tensor, symbol : denotes the contraction operation, $\mathbf{q}$ is the vector of flux of heat. We assume that [2]:

$$\epsilon = \sum_{i=1}^{N} \epsilon_i,$$
$$\epsilon_i = \frac{1}{m_i} \left\{ c_i k_B T \exp(-\frac{m_i |\mathbf{g}| z}{k_B T}) \cdot \left[(-\frac{z}{c}) \cdot (1 - \exp(-\frac{m_i |\mathbf{g}| c}{k_B T})) - \exp(-\frac{m_i |\mathbf{g}| c}{k_B T}) \right] \right\} + \tilde{\mu}_i c_i,$$
$$\tilde{\mu}_i = \frac{\mu_i}{m_i},$$
$$\mu_i = k_B T \cdot \left\{ \ln[(c_i p)(k_B T)^{-\frac{c_{p,i}}{k_B}} (\frac{m_{\mathrm{air}}}{m_i})(\frac{2\pi h^2}{m_i})^{\frac{3}{2}}] \right\} + m_i |\mathbf{g}| z,$$
$$\mathrm{T}_{mk} = -p\delta_{mk} + \eta \cdot \left[(\frac{\partial v_m}{\partial x_k} + \frac{\partial v_k}{\partial x_m} - \frac{2}{3}\delta_{mk} \mathrm{div}(\mathbf{v})) \frac{\partial v_k}{\partial x_m} \right] + \xi \cdot [(\delta_{mk} \mathrm{div}(\mathbf{v}))^2], m, k = 1, ..., 3,$$
$$\mathbf{T} : \nabla \mathbf{v} = \sum_{m=1}^{3} \sum_{k=1}^{3} \mathrm{T}_{mk} \frac{\partial v_m}{\partial x_k},$$
$$\mathbf{q} = \sum_{i=1}^{N} \left\{ \left[(\frac{\beta_i T}{\alpha_{ii}} + \tilde{\mu}_i) \mathbf{j}_i \right] + [(-\kappa) \nabla T] \right\},$$
$$\mathbf{j}_i = -\rho D_{ii} (c_i + \frac{k_{T,i}}{T} \nabla T),$$
$$\alpha_{ii} = \frac{[\rho D_{ii}]}{[(\frac{\partial \tilde{\mu}_i}{\partial c_i})_{(c_n)_{n=1,\dots,N, i \neq n}, T, p}]},$$
$$\beta_i = [\rho D_{ii}] \cdot \left\{ \frac{k_{T,i}}{T} - \frac{[(\frac{\partial \tilde{\mu}_i}{\partial T})_{(c_n)_{n=1,\dots,N}, p}]}{[(\frac{\partial \tilde{\mu}_i}{\partial c_i})_{(c_n)_{n=1,\dots,N, i \neq n}, T, p}]} \right\},$$

where ϵ_i is the mass density of intrinsic (internal) energy of the ith constituent of the air mixture, m_i the molecular mass of the ith constituent, k_B is Boltzmann's constant, μ_i is the complete partial chemical potential of the ith constituent of the air mixture (it is complete since it is composed of chemical potential without external force field and of external potential), $m_{\text{air}} = 28.966$ [u] is the molecular mass of air, δ_{mk} is Kronecker's delta, $c_{p,i}$ is the specific heat at constant pressure of the ith constituent of air mixture, h is Planck's constant, $\mathbf{j}_i$ is the vector of flux of mass of the ith constituent of the air mixture, and κ is the coefficient of thermal conductivity of air. These magnitudes were derived from Grand Canonical ensemble with external gravitational Newtonian field.

3 Optimal Control Problems

Let us consider the measurs of the total travel time (TTT): J_{TTT} [5], emissions (E): J_{E} , and concentrations (C): J_{C} of exhaust gases in the street canyon as well as in the canyon and two its nearest neighbor substitute canyons, $J_{\text{TTT,ext}}$ $J_{\text{E,ext}}$ $J_{\text{C,ext}}$, respectively. Therefore the appropriate optimization problems may be formulated as follows [2]. The measures depend on traffic lights at the upstream and downstream ends of the lanes. The vector of boundary control $\mathbf{u}$ reads:

$$\mathbf{u} = (g_1, C_1, g_2, C_2, F) \in U^{\text{adm}},$$

g_m are green times, C_m are cycle times, F is offset time, and U^{adm} is a set of admissible control variables. The TTT, E, C functionals are given by:

$$J_{\text{TTT}}(\mathbf{u}) = \sum_{s=1}^{2} \sum_{l=1}^{n_s} \sum_{vt=1}^{VT} \int_0^a \int_0^{T_S} k_{l,vt}^s(x,t) dx\, dt.$$

$$J_{\text{E}}(\mathbf{u}) = \sum_{s=1}^{2} \sum_{l=1}^{n_s} \sum_{ct=1}^{CT} \sum_{vt=1}^{VT} \int_0^a \int_0^{T_S} e_{l,ct,vt}^s(x,t) dx\, dt.$$

$$J_{\text{C}}(\mathbf{u}) = \rho_{\text{STP}} \cdot \sum_{i=1}^{N_E - 1} \int_0^a \int_0^b \int_0^c \int_0^{T_S} c_i(x,y,z,t) dx\, dy\, dz\, dt.$$

$$J_{\text{TTT,ext}}(\mathbf{u}) = J_{\text{TTT}}(\mathbf{u}) + a \cdot \sum_{s=1}^{2} \sum_{l=1}^{n_s} \sum_{vt=1}^{VT} k_{l,vt,\text{jam}}^s \cdot (C_s - g_s).$$

$$J_{\text{E,ext}}(\mathbf{u}) = J_{\text{E}}(\mathbf{u}) + a \cdot \sum_{s=1}^{2} \sum_{l=1}^{n_s} \sum_{ct=1}^{CT} \sum_{vt=1}^{VT} e_{l,ct,vt,\text{jam}}^s \cdot (C_s - g_s).$$

$$J_{\text{C,ext}}(\mathbf{u}) = J_{\text{C}}(\mathbf{u}) + \rho_{\text{STP}} \cdot a \cdot b \cdot c \cdot \sum_{i=1}^{N_E - 1} c_{i,\text{STP}} \cdot \sum_{s=1}^{2} (C_s - g_s).$$

The integrands $k_{l,vt}^s, e_{l,ct,vt}^s, c_i$ in the functionals depend on the control vector $\mathbf{u}$ through the boundary conditions, through the equations of dynamics, as well as, through the sources. The value of the vector of control u directly affects the boundary conditions for vehicular densities, velocities, and emissivities. It also affects the sources. Next, it propagates to the equations of dynamics and then it influences the values of functionals. ρ_{STP} is the density of air at standard temperature and pressure STP, $c_{i,\mathrm{STP}}$ is concentration of the ith constituent of air at standard temperature and pressure, a, b, c, are dimensions of the street canyon, T_S is time of simulation.

We define two additional functionals. The first one is the composite total travel time, emissions and concentrations of the pollutants in the single canyon, whereas the second one in the canyon and two its neighbours:

$$\begin{aligned} H(\mathbf{a},\mathbf{u}) &= \alpha_{\mathrm{TTT}} F_{\mathrm{TTT}}(\mathbf{u}) + \alpha_{\mathrm{E}} F_{\mathrm{E}}(\mathbf{u}) + \alpha_{\mathrm{C}} F_{\mathrm{C}}(\mathbf{u}). \\ H_{\mathrm{ext}}(\mathbf{a},\mathbf{u}) &= H(\mathbf{a},\mathbf{u}) + \\ &\quad + \alpha_{\mathrm{TTT,ext}} J_{\mathrm{TTT,ext}}(\mathbf{u}) + \alpha_{\mathrm{E,ext}} J_{\mathrm{E,ext}}(\mathbf{u}) + \alpha_{\mathrm{C,ext}} J_{\mathrm{C,ext}}(\mathbf{u}). \end{aligned}$$

The scaling parameters $\alpha_{\mathrm{TTT}}, \alpha_{\mathrm{E}}, \alpha_{\mathrm{C}}, \alpha_{\mathrm{TTT,ext}}, \alpha_{\mathrm{E,ext}}, \alpha_{\mathrm{C,ext}}$ make the two functionals dimensionless, and

$$\begin{aligned} \mathbf{a} &= (\alpha_{\mathrm{TTT}}, \alpha_{\mathrm{E}}, \alpha_{\mathrm{C}}) \in A^{\mathrm{adm}}, \\ \mathbf{a}_{\mathrm{ext}} &= (\alpha_{\mathrm{TTT}}, \alpha_{\mathrm{E}}, \alpha_{\mathrm{C}}, \alpha_{\mathrm{TTT,ext}}, \alpha_{\mathrm{E,ext}}, \alpha_{\mathrm{C,ext}}) \in A^{\mathrm{adm}}_{\mathrm{ext}}, \end{aligned}$$

where A^{adm}, $A^{\mathrm{adm}}_{\mathrm{ext}}$ are the sets of admissible scaling parameters. We formulate two separate multi-criteria optimization problems consisting in minimization of the functionals H, H_{ext} with respect to control vector $\mathbf{u}$ over admissible domain and with respect to scaling parameters, while the equations of dynamics are fulfilled:

$$\begin{aligned} H^\star &= H(\mathbf{a}^\star, \mathbf{u}^\star) = \min\{\mathbf{a} \in A^{\mathrm{adm}}, \mathbf{u} \in U^{\mathrm{adm}} : H(\mathbf{a},\mathbf{u})\} \\ H^\star_{\mathrm{ext}} &= H_{\mathrm{ext}}(\mathbf{a}^\star_{\mathrm{ext}}, \mathbf{u}^\star_{\mathrm{ext}}) = \min\{\mathbf{a}_{\mathrm{ext}} \in A^{\mathrm{adm}}_{\mathrm{ext}}, \mathbf{u} \in U^{\mathrm{adm}} : H_{\mathrm{ext}}(\mathbf{a},\mathbf{u})\}, \end{aligned}$$

where $H^\star, H^\star_{\mathrm{ext}}$, are the minimal values of the two functionals, $(\mathbf{a}^\star, \mathbf{u}^\star)$, and $(\mathbf{a}^\star_{\mathrm{ext}}, \mathbf{u}^\star_{\mathrm{ext}})$, are scaling and control vectors at which the functionals reach the minima, respectively.

4 Conclusions

The proecological urban traffic control idea and the model of the street canyon have been developed. It is assumed that the proposed model represents the main features of complex air pollution phenomena.

5 Acknowledgements

The author would like to gratefully thank the organizers of the Comphy02 Computational Physics of Transport and Interface Dynamics conference:

Heike Emmerich, Britta Nestler and Michael Schreckenberg, held in the Max-Planck-Institut für Physik komplexer Systeme, Dresden, Germany, for the invitation and creation of scientific atmosphere. The present work was partly covered by the Polish State Commitee for Sientific Research KBN grant number F-1/63/BW/02.

References

1. M. M.Duras. Continuum Field Model of Street Canyon: Numerical Examples. Engng. Trans. **48** (2000) 433-448.
2. M. M. Duras. Road Traffic Control in Street Canyons, Ph. D. Thesis, University of Mining and Metallurgy, Cracow. (1998) 1-363.
3. A. Adamski, M. M. Duras. Environmental Traffic Control Issues in Street Canyons. Polish J. Environ. Stud. **6** (1997) 13-20.
4. A. Adamski, M. M. Duras. Air Pollution Optimal Traffic Control Issues in Integrated Street. Polish J. Environ. Stud. **8** (1999) 7-17.
5. P. G. Michalopoulos, D. E. Beskos, J.-K. Lin. Analysis of Interrupted Traffic Flow by Finite Difference Methods. Transp. Res.: Part B: **18B** (1984) 409-421.
6. B. D. Greenshields. A study of traffic capacity. Proceedings of Highway Research Board **14** (1934) 448-477.
7. A. Cemal Eringen, G. A. Maugin. Electrodynamics of Continua. Springer-Verlag, New York (1990) Vol. I, Chapters 3, 5; Vol. II, Chapters 9, 10.
8. L. D. Landau, E. M. Lifshitz. Theoretical Physics: Hydrodynamics. Nauka, Moscow (1986)Chapters 2, 6.
9. M. M. Duras. Continuum Field Model of Street Canyon: Theoretical Description. Engng. Trans. **48** (2000) 415-432.
10. M. Brzeżański. Evaluation of Transport Emission of the Amount on a Chosen Section of the Road Network in Cracow. KONMOT'98 (1998) Table 2.

Walker Behaviour Modelling by Differential Games*

Serge P. Hoogendoorn[1]

Delft University of Technology, Faculty of Civil Engineering and Geosciences

Abstract. Gaining insights into pedestrian flow operations and assessment tools for pedestrian walking speeds and comfort is important for e.g. planning and geometric design of infrastructural facilities, as well as for management of pedestrian flows under regular and safety-critical circumstances. Pedestrian flow operations are complex, and vehicular flow simulation modeling approaches are generally not applicable to pedestrian flow modeling.

This paper focusses on pedestrian walking behavior. It is assumed that pedestrians are autonomous predictive controllers that minimize the subjective predicted cost of walking. Pedestrians predict the behavior of other pedestrians based on their observations of the current state as well as predictions of the future state, given the assumed walking strategy of other pedestrians in their direct neighborhood. As such, walking can be represented by a (non-cooperative or cooperative) differential game, where pedestrians may or may not be aware of the walking strategy of the other pedestrians.

1 Introduction

In optimizing the design of airports, public transit stations, stadiums, shopping malls, etc., theory and models predicting spatiotemporal pedestrian flow patterns and individually experienced walking conditions, in terms of walking times, comfort levels, etc., are important. This holds equally for timetable design, where walking times experienced by public transit riders when accessing or transferring play an important role. Analysis tools can support infrastructure designers as well as public transport planners to optimize their design, making it more efficient, more comfortable, and safer. In this respect, we emphasize that pedestrian traffic and car-based traffic are very different. This is why flow theories and models of different modes generally are not interchangeable, justifying the development of a dedicated comprehensive dynamic theory for pedestrians with related models.

In the theory underlying the model development described in this contribution, key elements in predicting pedestrian behavior are *activity scheduling* (choice to execute a particular activity, the order in which activities are performed, the location where to perform an activity), *route choice* between activities in the two-dimensional continuous space, and *multi-directional walk-*

* This research is funded by the Social Science Research Council (MaGW) of the Netherlands Organization for Scientific Research (NWO).

ing behavior. The main theoretical assumption is that of the "pedestrian economicus". I.e. pedestrians are subjective expected cost minimizers.

The focus of this contribution will be on pedestrian walking behavior, assuming that pedestrian routes, described by continuous trajectories, are given. The theory is operationalized in terms of behavioral models, determined by application of differential game theory. The walker model is continuous in time and space, and describes walking behavior considering the pedestrian state (position and velocity) and control (acceleration). Under the assumption of expected (predicted) cost minimization, pedestrians choose the acceleration optimizing the predicted subjective walking cost, thereby considering the future actions (and reactions) of the other pedestrians (so-called opponents). The subjective walking cost depends on the costs of drifting from the planned route, of interacting with other pedestrians and obstacles, and the cost of applying the control. Assuming that pedestrians do not take into account the anticipated reactions of the other pedestrians, it turns out that the resulting feedback models are similar to the social-forces model of [7]. The modelling framework thus provides a different interpretation of the latter model, and its implied behavioral assumptions.

This contribution is outlined as follows. Section 2 presents the overall theory, distinguishing different levels of pedestrian behavior. The third section focusses on the behavioral assumptions underlying walker model derivation. Section 4 discusses the conceptual walker model. The fifth section presents the mathematical model derivation using differential game theory. Furthermore, the analogy with the social forces model is discussed. Section 6 presents the approach to model calibration, while in section 7 some application results are discussed briefly. The final section presents the general conclusions as well as future research directions.

2 Pedestrian Behaviour Framework

The main behavioral assumption is that *all actions*, let it be performing an activity, walking along a certain route, or bypassing a slower pedestrian performed within the infrastructure facility *will provide utility, which the pedestrian aims to optimize*. Several empirical studies have shown the applicability of utility-based approaches to modelling pedestrian behavior [3], [9], [16], even though most decisions are made subconsciously. Therefore, we argue that its use is justified.

The theory distinguishes the following mutually dependent behavioral levels:

1. Departure time choice, and activity set choice (strategic level);
2. Activity scheduling and activity area choice; route-choice to activity areas (tactical level);
3. Walking behavior (operational level).

Figure 1 depicts the chosen overall framework of pedestrian behavior. Given the departure time t_0 and the prior activity set Σ a pedestrian aims to perform, the theory asserts that at the tactical level, pedestrians make a simultaneous route-choice / activity schedule decision minimizing expected subjective dis-utility. These choices at this level are influenced by both external factors (e.g. presence of obstacles, stimulation of the environment), and internal factors (e.g. time-pressure, attitudes of the pedestrian). Together with the expected traffic conditions, the result of the tactical level (i.e. optimal schedule S^* and route $r^*(\cdot)$) serves as input for the pedestrian walking process. In turn, pedestrian walking behavior will be a result of predicted cost minimization at the operational level, where pedestrians incur costs due to straying from their intended route, walking close to other pedestrians and obstacles, and frequent or severe accelerations and decelerations.

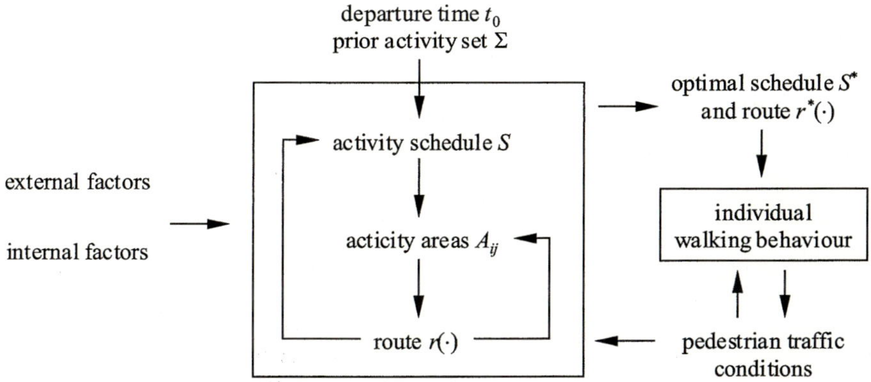

Fig. 1. Pedestrian behaviour framework

While staying in the facility, pedestrians either perform activities, or walk from one activity area to the next. Each pedestrian entering the facility at t_0 has a prior activity set Σ. This set consists of the activities i the pedestrian may perform. Some of these activities are discretionary, others are mandatory; some can be performed in random order; other activities can only be performed once others are completed. In theory, optimal activity choice, activity area choice, and route choice are mutually dependent. To describe this dependency, we hypothesize that the total cost C can be defined by a function of the route $r(\cdot)$, the schedule $S = \{i\}$, and the times $\{T_i\}$ at which activities $i \in S$ are performed. We hypothesize that pedestrians simultaneously choose the optimal activity schedule S^*, the activity times T_i^* (and thus the activity areas A_{ij}^*) for each $i \in S^*$, and the route $r^*(\cdot)$ optimizing the total cost C. The latter route costs will among other things depend on travel time, proximity of obstacles, number of sharp turns and rapid directional changes, and stimulation of environment, and route attractiveness [4], [3], [7]. The importance of these factors in route-choice will vary substantially between different pedestrian groups.

Contrary to network-based models where travellers can choose between a finite number of alternatives, the theory allows pedestrians to choose from an infinite set of routes (or paths) through the walking facility. This is achieved by considering the so-called value function $W_i(t,r)$, describing the minimal cost to perform activity i as a function of time t and location r. The value function depends on the prevailing flow conditions, and as such defines a mapping from these predicted conditions to the pedestrian route choice behavior. Hoogendoorn and Bovy [11] present an approach to determine optimal continuous routes and activity scheduling in continuous space under uncertainty. In the remainder of this contribution, we assume that the planned path - described by the desired velocity $v^*(t,r)$ - has been determined.

3 Theory of Walking

This section concerns walking behavior. The theory and models aim to describe the essential microscopic processes in walking (e.g. speed choice and acceleration behavior, mutual interaction between pedestrians, reaction to obstacles, etc.). The model must be able to predict and explain the phenomena in pedestrian flow that are important from both a theoretical and practical point of view, given the envisaged model applications. Examples are the relation between density and speed, and the formation of lanes and clusters, for a specific infrastructure design.

In the considered approach, pedestrians choose their way in time and space according to subjective utility maximization by continuously observing their surroundings and taking appropriate actions. This essentially means that pedestrians are modeled as *feedback-oriented dynamic optimal control systems*. Amongst the theoretical hypotheses underlying the walker model are that pedestrians:

1. continuously reconsider their walking choices by using current observations and the resulting predictions into the subjective utility optimization (*rolling horizon*).
2. are anisotropic particles that react mainly to stimuli in front of them;
3. anticipate on the behavior of other pedestrians by predicting their walking behavior according to *non-cooperative* or *cooperative* strategies;
4. have limited predicting possibilities, reflected by *discounting cost* of actions over time and space, implying that they mainly consider pedestrians in their direct environment;
5. will be more evasive when encountering a group of pedestrians than when encountering a single pedestrian (described by assuming additivity of the proximity costs);

6. a) minimize predicted discounted costs resulting from b) drifting from the planned route, the proximity of other pedestrians and obstacles, and applying the control (acceleration);

Let us briefly reflect upon these hypotheses by considering some empirical and experimental research findings from literature. Goffman [6] describes how the environment of the pedestrian (infrastructure and pedestrians) is continuously observed through a mostly *subconscious process called scanning* in order to side-step small obstructions on the flooring and handling encounters with other pedestrians (hypothesis 1). The scanning area is an ellipse, narrow to either side of the individual and longest in front of him (hypothesis 2). Moreover, the area of the ellipse changes constantly according to the prevailing traffic density. Goffman [6] also describes how pedestrians react to one another by bilateral, subconscious communication. This is how pedestrians are generally aware of the future behavior of other pedestrians in their direct environment (hypothesis 3). Other researchers also stressed the importance of cooperation between pedestrians. In illustration, Sobel and Lillith [17] report that pedestrians are reluctant to unilaterally withdraw from an encounter until the last moment, possibly even resulting in physical contact between the pedestrians. This brushing sends a signal to the 'offender' to cooperate. Naturally, the predictions are limited with respect to the time-horizon as well in a spatial sense (hypothesis 4). The latter is also due to the limited observation range. Not all encounters will result in a cooperative decision, as the amount of space granted by a pedestrian depends on the cultural, social, and demographic characteristics of the interacting pedestrians [2], [17]. Willis *et al.* [19] found similar results for one-to-many interactions: individuals tend to move for groups (hypothesis 5).

Hill [9] convincingly argues that a pedestrian (or in fact, his brain) can be seen as a "*pilot of a very special vehicle*". In the remainder, we argue that the task of guiding this vehicle (the body) is similar to the task of guiding a car, and can hence be described in an analogous manner, e.g. see [15]. Experience and knowledge have skilled the pedestrian in optimally performing the guidance subtask (hypothesis 6a). It is obvious that pedestrians aim to walk along the route that best meets their walking goals. However, pedestrians generally dislike walking too close to other pedestrians, yielding tension and irritation [18]. This holds equally for frequent or severe accelerations, since these will yield additional cost in terms of energy consumption. It is plausible that the resulting walking behavior is a trade-off between these factors (hypothesis 6b).

Important behavioral parameters that will largely determine pedestrian behavior are:

- the acceleration time τ (reflecting the acceleration and direction changing delay)

- temporal discount factor η (rate at which costs are discounted over time), and the spatial discount factor R_0 (proximity cost reduction rate as a function of the distance between two pedestrians);
- anisotropy factors c_0^+ and c_0^- reflecting the difference in behavior when reacting to stimuli in front or behind, relative to stimuli from the side;
- relative weighting factors c_k for the respective walking cost components;
- physical dimensions of the pedestrians (radius ρ_p), and;
- control restrictions (maximum speed $v_0(t, r)$, given specific infrastructure and prevailing traffic conditions).

These parameters are to be estimated from either microscopic data (section 6) or macroscopic data (reproducing emerging flow characteristics, i.e. speed density curves, spatiotemporal patterns, etc.). The parameters are discussed in detail in the ensuing of this contribution.

4 Conceptual Pedestrian Walking Task Model

In the remainder of the contribution we describe how, given the planned activity schedule and route, pedestrian behavior at the operational level can be described. This section discusses a conceptual model to describe walking behavior in terms of a closed feedback control system. Similar approaches have been proposed to describe execution of other human tasks, such as car-driving. Several authors have proposed using utility optimization for modelling driving-task execution. We will show that utility optimization is also a fruitful approach to describe walking task execution.

4.1 Walking Subtask Hierarchy

Pedestrians are operators performing walking tasks. Using the terminology of [15], the walking task comprises both the *control* and the *guidance* of the 'pedestrian vehicle' [9]. The pedestrian vehicle control subtask includes all activities pertaining to the split-second exchange of information between the brain (the 'pilot'), the senses (eyes, ears), and the actuators (legs, arms). These actions are nearly always skill-based and are performed automatically with no conscious effort. The pedestrian vehicle guidance subtask describes the collection of decisions required to guide the pedestrian safely and comfortable over the available walking infrastructure and its elements, as well as the proper behavior when encountering other pedestrians.

The pedestrian is assumed to use an *internal model for determining an appropriate guidance control decision*, thereby aiming to optimize his task performance. Experience and knowledge have skilled the pedestrian in performing this walking task, and hence the assumption of optimal behavior is justified. The optimization includes the limitations of the pedestrians in terms of observation, information processing and internal state estimation, as well as processing times and reaction times. Moreover, the optimization includes the pedestrian walking strategy. The final walking task level pertains to navigation (route-choice). This level is beyond the scope of this contribution (see [11], [13] for details).

4.2 Walking as a Feedback-oriented Control System

A pedestrian is constantly monitoring his position and velocity relative to the other pedestrians in the flow, as well as relative to obstacles and walls (hypothesis 1). This monitoring allows the pedestrian to perform corrective actions, implying that walking is *feedback-oriented.* A continuous feedback control system consists of a comparison between the input (positions, velocities) and the controlled output (acceleration, deceleration and direction changes).

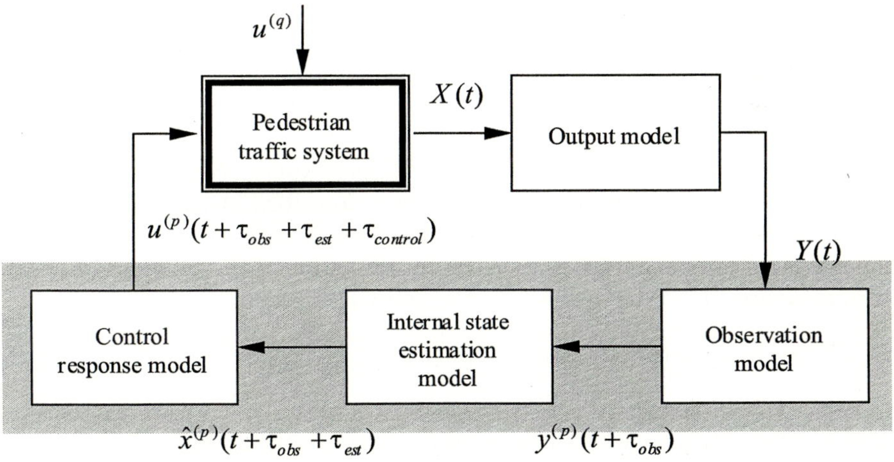

Fig. 2. Continuous control loop of walking task execution (for single pedestrian p)

Figure 2 depicts the continuous control loop for pedestrian $p \in Q$, where $Q = \{1, ..., n\}$ denotes the set of all pedestrians in the walking facility. The pedestrian traffic system describes the actual pedestrian traffic operations. The state $X(t)$ of the system encompasses all information needed to summarize the system's history, typically including the positions $r_q(t) \in \Omega$ and the velocities $v_q(t) \in V_a$ of all pedestrians $q \in Q$.

$$X(t) := (r, v) \text{ where } r := (r_1, ..., r_n) \text{ and } v := (v_1, ..., v_n) \tag{1}$$

The dynamics of the pedestrian traffic system are described by kinematic equals described the changes in the pedestrian system state over time. These are reflected by a coupled system of ordinary differential equations

$$\dot{X}(t) = F(t, X(t), u(t)) \tag{2}$$

where $u = (u_1, ..., u_n)$ denotes the control vector, reflecting how the pedestrians influence the state X (e.g. by accelerating and changing directions). We emphasize that the pedestrian traffic system state dynamics (2) describe the true changes in the system.

The output model maps the state $X(t)$ onto the observable system output vector $Y(t)$. Since both the location and the velocity can be observed externally, in the remainder of the contribution, we will assume that state can be observed directly, i.e. $Y(t) = X(t)$.

The observation model of pedestrian p describes which of the elements of the pedestrian traffic system's output can be observed by p, as a function of the state $X(t)$, the time τ_{obs} required to process the observations, and the observation errors ε_{obs}. In mathematical terms:

$$y^{(p)}(t + \tau_{obs}) = G^{(p)}(X(t))\, Y(t) + \varepsilon_{obs} \tag{3}$$

where $G^{(p)}$ is a matrix describing the observable elements of the system's output $Y(t)$. The output $y^{(p)}(t)$ typically describes the locations and velocities of pedestrians that p can observe.

Pedestrian p will use his observation $y^{(p)}$ and experiences to estimate the state of the pedestrian traffic system, which is described by the internal state estimation model of figure 2. Without going into details, we assume that the internal state estimate can be described by a function of the previous internal state estimate, the observation, and the estimation delay τ_{est}

$$\hat{x}^{(p)}(t + \tau_{est}) = h^{(p)}\left(\hat{x}^{(p)}(t), y^{(p)}(t)\right) \tag{4}$$

In the model described here, describes estimates of the locations and velocities of pedestrians , where denotes the subset of Q describing which pedestrians q are included in the internal state $x^{(p)}$ and internal state estimate of p. Note that these estimates generally contains more information than the observations $y^{(p)}(t)$. E.g., p may be aware of the presence of a pedestrian q, who is blocked from p's view (either by an obstacle or another pedestrian) by previous observations and prediction q's walking behavior.

The control response model in figure 2 reflects the process of determining the control actions of pedestrian p. The control actions describe the acceleration vector $a_p(t)$ applied by p. We hypothesize that the control decision depends on the internal state estimate, the walking strategy, reflecting the walking objectives c of the pedestrians, and the control delay $\tau_{control}$

$$u^{(p)}(t + \tau_{control}) = v^{(p)}\left(\hat{x}^{(p)}(t), c\right) \tag{5}$$

where the control delay describes both the time needed to determine the control decision, and the time needed to start the control movement. In determining the walking actions, p will predict and optimize the walking cost

$J^{(p)}$ incurred during some time interval $[t, t+T)$. The prediction of the internal state $x^{(p)}$ will be described by the internal state dynamics

$$\dot{x}^{(p)} = f^{(p)}\left(s, x^{(p)}, u^{(p)}\right) \text{ subject to } x^{(p)}(t) = \hat{x}^{(p)}(t) \tag{6}$$

This state dynamics include the reactions of the pedestrians q as well. For any control path applied during $[t, t+T)$, p predicts the state dynamics using eqn. (6), as well as the walking cost

$$J^{(p)}(t) = J^{(p)}(u^{(p)}_{[t,t+T)}|t, \hat{x}^{(p)}(t)) \tag{7}$$

and chooses the optimal control path satisfying

$$u^{*}_{[t,t+T)} = \arg\min J^{(p)}(u^{(p)}_{[t,t+T)}|t, \hat{x}^{(p)}(t)) \tag{8}$$

The cost optimization process, its parameters, and its resulting control decisions differ between the pedestrians, due to different objectives, preferences, and physical abilities.

To derive mathematical models, the control framework discussed here will be used. To this end, the following section specifies the pedestrian kinematics as well as the walking cost $J^{(p)}$ as functions of the internal state $x^{(p)}$ and control. In doing so, both observation errors and delays will be neglected. Furthermore, the observation and estimation models are specified such that pedestrian p is assumed to be able to estimate the locations and velocities of all pedestrians $q \in Q$, implying that p has perfect information regarding the current state $x(t)$ of the pedestrian traffic system. The model is specified such that p will only react to pedestrians in his direct environment and directly in front of him.

4.3 Pedestrian Kinematics (Internal Model)

In order to apply the Pontryagin's Maximum Principle to establish the walker model, the internal state dynamics model (6) as well as the factors determining the walking disutility in terms of the cost function (7). To predict the evolution of the dynamics of the state, p will use an internal dynamic model reflecting the kinematics of the pedestrians $q \in Q$ in terms of ordinary differential equations that describe the dynamics of the location r_q and velocity v_q of pedestrians $q \in Q$

$$\dot{r}_q = v_q \text{ and } \dot{v}_q = a_q \tag{9}$$

subject to the initial conditions stemming from the internal estimation model (4)

$$r_q(t) = \hat{r}_q(t) \text{ and } v_q(t) = \hat{v}_q(t) \tag{10}$$

where a_q denotes the acceleration vector. In this formulation, a_p denotes the controllable acceleration vector of p, while a_q for $q \neq p$ denotes the predicted acceleration of pedestrians q.

The internal state of pedestrian p consists of the locations r_q and the velocities v_q of all pedestrians $q \in Q$ (including p), i.e.

$$x := (r, v) \text{ where } r := (r_1, ..., r_n) \text{ and } v := (v_1, ..., v_n) \tag{11}$$

Pedestrian p can influence the internal state x directly via the control vector u defined by the acceleration a_p, i.e.

$$u(t) := a_p(t) \tag{12}$$

We can easily determine the internal state dynamics (6) using the state definition (11), the kinematics (9), and the control definition (12)

$$\dot{x} = (\dot{r}, \dot{v}) = (v, a) = f(x, u) \tag{13}$$

We hypothesize that pedestrian p has knowledge pertaining to the walking strategies of the other pedestrians (referred to as 'opponents'). Assuming that the opponents q are also modelled by continuous feedback controllers, p presumes that the acceleration a_q of opponents q are a function of p's internal state x, i.e.

$$a_q(t) = a_q(x(t)) \tag{14}$$

4.4 Walking Discomfort (Resistance)

Pedestrians incur different types of discomfort (disutility or cost) while walking. These costs can be expressed as a function $J^{(p)}$ of the control applied by p during the planning period $[t, t+T)$, where t denotes the current time, and T denotes the prediction horizon. The cost function is determined by *integrating the running cost* L over the planning period $[t, t+T)$. We assume that the pedestrian is aware of his limited predictive abilities. This is why cost incurred in the near future will be of higher value than the costs that may be incurred in the long run. To model this, the concept of discounted costs is included in the modelling. Let η be the (temporal) discount factor. For $\eta > 0$, and T sufficiently large, the following specification of the cost function is applicable

$$J^{(p)}\left(u_{[t,\infty)}; t, \hat{x}^{(p)}(t)\right) = \int_t^\infty e^{-\eta z} L(z, x(z), u(z))\, dz \tag{15}$$

where z denotes the integration variable. In the remainder of the contribution, we assume that p will apply the control function minimizing eqn. (15)

4.5 Cost Components

The running cost L reflects a variety of factors L_k. For the sake of simplicity, we assume that the running cost is *linear-in-parameters* (although this restriction is not required upon application of the approach)

$$L = \sum_k c_k L_k \tag{16}$$

where c_k denotes the relative weight of cost component L_k. In this section, we will consider three walking cost components L_1, L_2, and L_3, respectively reflecting discomfort due to straying from the optimal velocity (speed and direction), walking too close to another pedestrian (proximity costs), and high acceleration or deceleration rates in the longitudinal as well as the lateral direction. Dealing with pedestrian's reaction to obstacles and walls is similar, and is not discussed explicitly in this contribution.

4.6 Cost L_1 due to Deviation from Planned Route

We have hypothesized that pedestrians at the tactical planning level determine an optimal velocity function $v^* = v^*(t, r_p)$ minimizing the expected cost of walking and performing activities (hypothesis 6b). We assume that the disutility incurred due to straying from the velocity v^* (e.g. due to interactions) is described by a quadratic running cost function

$$L_1 = \frac{1}{2} \left\| v^*(t, r_p) - v_p \right\|^2 \tag{17}$$

4.7 Proximity Cost L_2

The cost caused by the presence of other pedestrians in the walking area will be referred to as the proximity discomfort or proximity cost (hypothesis 6b). The discomfort incurred due to walking too close to pedestrian q is given by a function of the (relative) location and (relative) velocity of p with respect to q. will also depend on parameters specific for pedestrians p and q, such as their size (required to determine the net distance between pedestrian p and q). Moreover, the fact that pedestrians are anisotropic (react mostly to stimuli in front of them) is included in the proximity cost as well. Since the approach is applicable to any cost function L_2 that is not time dependent, we will not further specify the proximity cost function here. However, we will assume that the total proximity cost L_2 is determined by the sum of proximity costs caused by pedestrians q, i.e.

$$L_2 = \sum_{q \neq p} L_2^{(q)} \tag{18}$$

Adding the cost incurred due to different pedestrians q causes p to be more evasive when a group is encountered, than when in case of a single pedestrian (hypothesis 5).

4.8 Acceleration and Deceleration Cost L_3

Finally, we assume that p incurs cost when accelerating (hypothesis 6b). We distinguish between cost of *longitudinal acceleration* (increasing / decreasing speed) and of *lateral acceleration* (side-stepping perpendicular to the current direction). Let us define vectors α and β

$$\alpha = (\alpha_1, \alpha_2) := \frac{v}{||v||} \text{ and } \beta := (-\alpha_2, \alpha_1) \tag{19}$$

respectively denoting the unit vectors in the direction and perpendicular to the current direction of p. The total cost of acceleration is found by the following weighted sum

$$L_3 = \frac{\theta}{2} \left(a_p' \alpha\right)^2 + \frac{1-\theta}{2} \left(a_p' \beta\right)^2 \tag{20}$$

where $0 \leq \theta \leq 1$ is a weighting factor reflecting the relative costs of longitudinal acceleration compared to side-stepping.

5 Derivation of Walker Model

In this section, we derive a pedestrian walker model, by application of *Pontryagin's Maximum Principle* to find the optimal acceleration $u^* = a_p^*$, satisfying eqn. (8). To this end, let us define the so-called Hamiltonian $\mathcal{H}$ by

$$\mathcal{H}(t, x, u, \Lambda) = e^{-\eta t} L(t, x, u) + \Lambda' f(t, x, u) \tag{21}$$

$$= e^{-\eta t} \frac{c_1}{2} \left\| v_p^* - v_p \right\|^2 + e^{-\eta t} c_2 L_2(x) \tag{22}$$

$$+ e^{-\eta t} \frac{c_3}{2} \left[\frac{\theta}{2} \left(a_p' \alpha\right)^2 + \frac{1-\theta}{2} \left(a_p' \beta\right)^2 \right] + \lambda' v + \gamma' a \tag{23}$$

$\mathcal{H}$ is an auxiliary function to determine the equations for the state and the so-called co-state Λ, and required optimality conditions for the control u. $\Lambda := (\lambda, \gamma)$ denotes the marginal cost (shadow cost, co-state vector, or adjoint vector) of the internal state vector x: the vector $\lambda = (\lambda_1, ..., \lambda_n)$ describes the marginal change in the total cost (15) subject to eqn. (8) resulting from a small change in the position vector $r = (r_1, ..., r_n)$; $\gamma = (\gamma_1, ..., \gamma_n)$ describes the marginal change in the total cost (15) due to small changes in the velocity vector $v = (v_1, ..., v_n)$.

5.1 Application of the Maximum Principle

The *required conditions for optimality* (for unconstrained control), state that the partial derivative of $\mathcal{H}$ to u equals zero

$$D_u \mathcal{H} = 0 \tag{24}$$

It can then be shown that the optimal control (acceleration) satisfies [10]

$$u^*(t,x) = -\frac{1}{c_3} e^{\eta t} M^{-1} \gamma_p(t,x) \tag{25}$$

where the matrix M is defined by

$$M^{-1} = \frac{1}{\theta} \begin{bmatrix} \alpha_1^2 & \alpha_1\alpha_2 \\ \alpha_1\alpha_2 & \alpha_2^2 \end{bmatrix} + \frac{1}{1-\theta} \begin{bmatrix} \alpha_2^2 & -\alpha_1\alpha_2 \\ -\alpha_1\alpha_2 & \alpha_1^2 \end{bmatrix} \tag{26}$$

Clearly, the optimal acceleration u^* is a linear function of the marginal costs γ_p of the velocity v_p.

Dynamic equations for the co-states Λ can be determined by considering the partial derivatives of $\mathcal{H}$ to the elements of the state x. Given that the state-dynamics are time-invariant, as is the running cost L, we can show easily that the co-states Λ satisfy

$$\eta \Lambda = D_x \mathcal{H} \tag{27}$$

Using eqn. (27), we can derive expressions for the marginal cost λ_p and γ_p of the position r_p and velocity v_p of pedestrian p, as well as the marginal costs λ_q and γ_q of his opponents q. In turn, these expressions can be used to express the marginal cost γ_p in eqn. (25) in terms of the internal state x. The resulting expression for the marginal cost γ_p can then be substituted in the optimal acceleration law (25), yielding the following expression

$$\begin{aligned} u_p^* = {} & M^{-1}\left[I - \frac{1}{\eta}\left(D_{r_p} v_p^*\right)'\right]\left(\frac{v_p^* - v_p}{\tau}\right) - A_0 M^{-1}\left[D_{r_p} L_2 + \eta D_{v_p} L_2\right] \\ & - A_0 M^{-1}\left(D_{r_p} a_{Q\backslash p} + \eta D_{v_p} a_{Q\backslash p}\right)' S_p^{-1}\left[D_{r_{Q\backslash p}} L_2 + \eta D_{v_{Q\backslash p}} L_2\right] \end{aligned} \tag{28}$$

where the acceleration time τ and the interaction constant A_0 are defined by

$$\tau = \eta \frac{c_3}{c_1} \text{ and } A_0 = \frac{c_2}{\eta^2 c_3} \tag{29}$$

We have employed the subscript notation Q/p to indicate the $2(n-1)$-vectors pertaining to all opponents $q \neq p$, e.g.

$$r_{Q/p} = (r_1, \dots, r_{p-1}, r_{p+1}, \dots, r_n) \tag{30}$$

and defined the $2(n-1) \times 2(n-1)$ matrix S_p as follows

$$S_p = \left(\eta^2 I_{2(n-1)} - \eta D_{v_{Q\backslash p}} a'_{Q/p} - D_{r_{Q\backslash p}} a'_{Q\backslash p}\right) \tag{31}$$

The optimal acceleration u_p^* of pedestrian p is governed by three terms. The first term

$$\left(u_p^*\right)_{accel} = M^{-1}\left[I - \frac{1}{\eta}\left(D_{r_p} v_p^*\right)'\right]\left(\frac{v_p^* - v_p}{\tau}\right) \tag{32}$$

can be considered an *acceleration term*, and describes how p aims to keep walking along the planned path described by the optimal velocity v_p^*. From eqn. (32) we can see that this is in part determined by the weights c_1 and c_3 determining the acceleration time τ_p, reflecting respectively the cost of drifting from the planned velocity v_p^* and of applying control u_p. The ratio c_1/c_3 reflects that when the cost of acceleration is relatively low (i.e. c_3 relatively small) or when the cost of drifting from the optimal velocity is relatively high (i.e. c_1 large), the acceleration time τ will be small while the acceleration term will be large. In this case, the pedestrian will be more inclined to keep walking at the optimal speed and in the optimal direction, even though this may imply large accelerations from time to time. The matrix M^{-1} describes the preference for longitudinal acceleration and braking over side-stepping. Neglecting the role of M^{-1}, the term

$$\left[I - \frac{1}{\eta}\left(D_{r_p} v_p^*\right)'\right]\left(\frac{v_p^* - v_p}{\tau}\right) \tag{33}$$

shows that the pedestrian will generally aim to walk in the direction $\left(v_p^* - v_p\right)$ i.e. into the direction where the drifting cost L_1 will decrease most rapidly. The matrix $I - \frac{1}{\eta}\left(D_{r_p} v_p^*\right)'$ corrects this direction in line with changes in the optimal velocity that are caused by changes in the location r_p of the pedestrian. In other words, it corrects for the fact that when the pedestrian's location r_p is changed, in general so will his optimal velocity $v_p^* = v_p^*(r_p)$.

The second term

$$\left(u_p^*\right)_{int,1} = -A_0 M^{-1}\left[D_{r_p} L_2 + \eta D_{v_p} L_2\right] \tag{34}$$

$$= -A_0 M^{-1} \sum_{q \neq p} \left[D_{r_p} L_2^{(q)} + \eta D_{v_p} L_2^{(q)}\right] \tag{35}$$

of eqn. (28) reflects what we will call *first-order interactions.* showing how pedestrian p reacts to the other pedestrians $q \neq p$. Eqn. (34) shows that the effects of multiple interactions is additive.

Considering the interaction of pedestrian p and q, expression (34) shows that p will make the following trade-off in determining his walking direction and speed. On the one hand, the term $D_{r_p} L_2^{(q)}$ reflects the direction in which the proximity cost $L_2^{(q)}$ changes most rapidly. On the other hand, the term $\eta D_{v_p} L_2^{(q)}$ reflect changes in the proximity cost $L_2^{(q)}$ due to changes in the velocity v_p of p. For instance, this can express the fact that the proximity cost will be relatively large when two pedestrians approach each other very quickly, even though the distances between two pedestrians is still considerable.

Finally, the third term

$$\left(u_p^*\right)_{int,2} = -A_0 M^{-1}\left(D_{r_p} a_{Q\backslash p} + \eta D_{v_p} a_{Q\backslash p}\right)' S_p^{-1}\left[D_{r_{Q\backslash p}} L_2 + \eta D_{v_{Q\backslash p}} L_2\right] \tag{36}$$

describes what we will can *second-order interactions.* More precisely, it reflects the effects of changes in the locations r_q and velocities v_q of pedestrians $q \neq p$ that are expected by p, given the anticipated acceleration strategy $a_q = a_q(x)$ due to changes in the velocity v_p of p caused by acceleration. At this point, note that when $a_q = a_q^0$ (p presumes constant acceleration on the part of his opponents q), the second-order interactions are equal to zero.

5.2 Cooperative Walker Model

Empirical studies have shown the importance of cooperation between pedestrians in walking [6], [9]. So far, we have assumed that pedestrians only optimize their own walking objectives, thereby taking into account the presumed decisions of the opponents. As such, control decisions can be described in terms of non cooperative differential games.

To described cooperation, we assume that pedestrians will not only consider their own predicted walking cost, but also the cost incurred by their opponent. We hypothesize that the interaction pedestrians will make a decision that optimizes the *joint predicted cost.* Let us consider the case of two pedestrians p_1 and p_2 interacting. Neglecting the influence of other, non-cooperative pedestrians q, we assume that the p_1 and p_2 will make a simultaneous decision optimizing the following objective function

$$J = J^{(p_1)} + J^{(p_2)} \tag{37}$$

where $J^{(p_1)}$ and $J^{(p_2)}$ are defined by eqn. (15). The objective of the pedestrians is to find the joint control $(u_1, u_2) = (a_{p_1}, a_{p_1})$ optimizing the objective function eqn. (37). Let us assume that both the weighting factors c_k and the discount factors η are equal for both pedestrians p and q. The necessary conditions for optimality $D_u\mathcal{H} = 0$ now yield

$$(u_1^*(t,x), u_2^*(t,x)) = -\frac{1}{c_3} e^{\eta t} \left(M^{-1}\gamma_{p_1}(t,x), M^{-1}\gamma_{p_2}(t,x) \right) \tag{38}$$

which is equal to the expression for the non-cooperative interaction (25). Expressions for the marginal costs γ_{p_i} of the velocities v_{p_i} for $i = 1, 2$ can be determined using eqn. (27). If can be verified by the reader that the marginal cost γ_{p_i} depends on the derivative of the joint running cost $L = L^{(p_1)} + L^{(p_2)}$ to v_{p_i}, and on the marginal cost λ_{p_i} of the location r_{p_i}. In turn, the latter depends on the derivative of the joint running cost L to the location r_{p_i}. In sum, the following optimal acceleration law for cooperative encounters between p_1 and p_2 results

$$u_i^* = M^{-1} \left[I - \frac{1}{\eta} \left(D_{r_{p_i}} v_{p_i}^* \right)' \right] \left(\frac{v_{p_i}^* - v_{p_i}}{\tau} \right) - A_0 M^{-1} \left[D_{r_{p_i}} L_2 + \eta D_{v_{p_i}} L_2 \right] \tag{39}$$

for $i = 1, 2$, where $L_2 = L_2^{(p_1)} + L_2^{(p_2)}$ is the joint proximity cost. The optimal acceleration law eqn. (39) for cooperative encounters is very similar to the optimal acceleration law for non-cooperative encounters eqn. (28). The main difference is that instead of considering the proximity cost of the individual pedestrian, in the cooperative case the joint proximity cost is considered. When the individual proximity cost are 'symmetric', i.e. when they satisfy $D_{r_{p_i}} L_2^{(p_1)} = D_{r_{p_i}} L_2^{(p_2)}$ and $D_{v_{p_i}} L_2^{(p_1)} = D_{v_{p_i}} L_2^{(p_2)}$, the optimal acceleration laws for the cooperative encounter and the non cooperative encounter are equal. When anisotropy is considered in the proximity cost specification, the assumption of symmetry of the proximity cost does not hold, and the non-cooperative model and cooperative model are not equal.

5.3 Relation to the Social-forces Model

Let us assume that the proximity cost is described by the following simplified expression

$$L_2^{(q)} = \exp\left(- \left\| r_q - r_p \right\| / R_0\right) \tag{40}$$

Furthermore, let us assume that the planned velocity v_p^* is independent of the position r_p. Finally, we assume $\theta = \frac{1}{2}$, yielding $M = I$. Furthermore, let us assume that p presumes no reaction on the side of his opponents q, i.e. $a_q = 0$. The optimal acceleration law (32) then simplifies to

$$u_p^* = \left(\frac{v_p^* - v_p}{\tau}\right) - \frac{A_0}{R_0} \sum_{q \neq p} \frac{r_q - r_p}{\left\| r_q - r_p \right\|} e^{-\frac{\left\| r_q - r_p \right\|}{R_0}} \tag{41}$$

which equals the basic social-forces model of [7] (i.e. not considering anisotropy, bodily dimensions, and physical contact). This analogy provides a behavior interpretation of the social-forces model, namely that pedestrians are dynamic feedback controllers who *presume* that the other pedestrians in their vicinities will not react to their control decisions. Since the specification of the running cost function eqn. (40) is symmetric, the resulting optimal control law is equal for both cooperative and non cooperative encounters. When anisotropy is considered, the social-forces model implicitly presumes non cooperative encounters. Furthermore, the proximity costs are additive.

Furthermore, since $M = I$, it appears that the social-forces model does not differentiate between longitudinal and lateral acceleration. Furthermore, the social-forces model implicitly assumes that the planned direction of the pedestrians does not change smoothly ($D_{r_p} v_p^{*\prime} = 0$).

5.4 Model Refinements

The preceding sections have discussed the basic model, which does not fulfill several important pedestrian model requirements. Among the most important requirements are the finite dimensions of the pedestrians and the effect of physical contact, and the anisotropy requirement. The former describes the fact that given the scale of the applications, pedestrian dimensions need to be accounted for explicitly - contrarily to particles in fluidic or gas flows. To this end, pedestrian p is described by a circle with radius ρ_p. The inclusion of the pedestrian dimensions in the model involves two steps. For one, the effect of collisions between pedestrians is to be reflected in the pedestrian traffic system dynamics eqn. (2). These effects may involve repellent forces as well as frictional forces [8]. Secondly, pedestrians' behaviors will be influenced by the finite space requirements as well as the effect of collisions between pedestrians. This can be described by including a collision disutility into the cost function eqn. (15). For details, we refer to [12].

Anisotropy can be included easily by proposing the correct proximity cost term L_2, yielding increased cost for pedestrians walking in front of p. At the same time, the proximity cost function can be adapted to include pedestrian grouping behavior [12].

6 Approach to Model Calibration

Model calibration and validation (determining the model parameters and comparing model results with real-life observations) is an important issue for any model application, and various approaches have been put forward in the past. For microscopic pedestrian simulation models, model calibration is complicated due to the absence of good microscopic pedestrian data. This is why in most cases, model calibration is based on cross-comparing macroscopic model predictions with real-life observations. In illustration, the simulation model predicts the relation between density and speed as a function of the model's parameters. This relation can then be compared to empirical results and the model parameters can be adapted accordingly. In illustration, Hoogendoorn and Bovy [12] show how the speed density curve depends on the anisotropy parameters and propose parameters ensuing realistic macroscopic behavior. Similar analysis can be performed by comparing spatiotemporal patterns in the traffic flow, e.g. how congestion propagates upstream of a bottleneck, etc.

Ideally however, microscopic data is used for model calibration. This is why Delft University of Technology is performing experimental research with the aim of collecting individual pedestrian data in a controlled environment. During a pilot study, the behavior of twenty subjects was monitored using video, from which pedestrian trajectories were derived automatically using dedicated software tools. The subjects conducted a number of experiments, mostly aimed at assessing the technical aspects of the study set-up, as well as the experimental design. Among the experiments were walking in unidirectional and bidirectional flow, walking through a bottleneck, and walking in random directions.

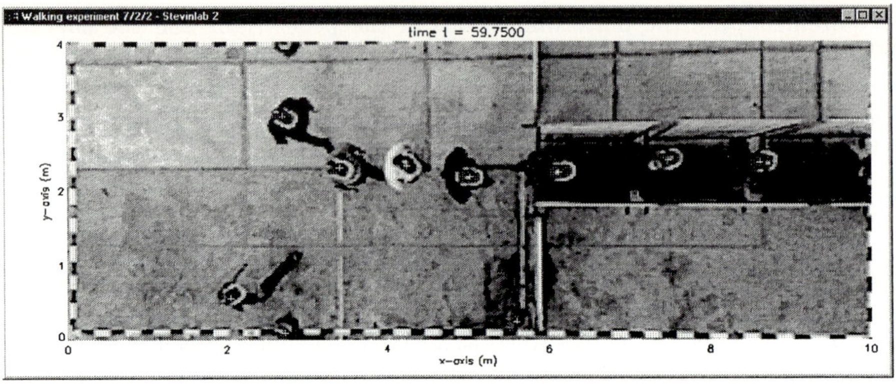

Fig. 3. Example of pedestrians moving through a 80*cm* bottleneck . The heads of the pedestrians are recognised and tracked. The observation area is $10m \times 4m$.

To derive the pedestrian trajectories from the collected video data, a tracking tool was developed. This tracking algorithm performs the following steps:

1. Map the picture coordinates to the real-life coordinates.
2. Recognize the caps of the pedestrians, based on the colors of the caps.
3. Cluster the pixels identified as the cap and determine its centre, describing the candidate pedestrian locations.
4. Use the candidate pedestrian locations identified for the consecutive video images to establish the most likely trajectory, based on Kalman filtering techniques.

It turned out that the pedestrians' trajectories could be determined from the data accurately. Moreover, it was concluded that the experiments should be as simple as possible to ensure the desired pedestrian behavior and reduce the number of faults to a minimum.

In the end, the microscopic data will be used to determine the model parameters directly (e.g. the acceleration time τ, the spatial discount factor R_0, the anisotropy parameters c_0^+ and c_0^-). This can be done by comparing the microscopic model predictions of the dynamics of a single pedestrian to the measurements (i.e. the filtered accelerations) and determining the parameters that minimize the difference.

7 Application Results

This contribution is aimed at showing the derivation of a walker model from behavioral hypothesis pertaining to the way pedestrians perform their walking task. It turns out that for correct model parameters, the model can correctly replicate important macroscopic pedestrian flow characteristics, such as the relation between speed and density [10], [12].

Having shown the equivalence with the social-forces model, it is deemed obvious that the model behaves in a similar way and hence will be able to simulate pedestrian flow features such as dynamic lane formation, the formation of so-called bubbles, and oscillatory behavior in bidirectional flows [7], [8]. Besides dynamic lanes, the model also shows the formation of stripes in crossing pedestrian flows, in line with Japanese empirical studies [14]. It turns out that the trade-off between longitudinal and lateral acceleration cost (not featured in the social-forces model) substantially influence these self-formed characteristics [10], [12]. The variability of the formed patterns increases with descreasing cost of sidestepping due to the fact that pedestrians are more inclided to look for free spaces in the pedestrian flow, rather than to decelerate and wait their turn. In other words, the inclination to make an evasive manoeuvre (side-stepping) rather than to decelerate upon interaction has a negative influence of the lane formation, as well as on the formation of strips in crossing pedestrian flows.

This holds equally for the heterogeneity of the pedestrian population (in terms of the free speed distribution).

8 Conclusions and Future Research Directions

In this contribution, we have put forward a new theory and models for pedestrian walking behavior. The theory is based on the assumption that pedestrians are cost minimizers. Important factors are the predicted cost due to applying control (accelerating and braking, side-stepping), walking too close to other pedestrians and obstacles, drifting from the planned path, and the anticipated behavior of the other pedestrians.

In the approach, pedestrians are described by optimal predictive controllers that minimize the predicted walking cost, given the presumed walking strategies of the other pedestrians. As such, different presumed strategies have been considered, varying from 'no reaction of the other pedestrians' to 'cooperative optimization of the joint walking cost'. To operationalize the theory, mathematical optimal control theory has been applied. The resulting model equations bear strong resemblance to the social-forces model.

References

1. W. Daamen, P.H.L. Bovy, and S.P. Hoogendoorn. Modelling Pedestrians in Transfer Stations. *Proceedings of the Pedestrian and Evacuation Dynamics Conference.* Duisburg (2001).
2. J. M. Dabbs and N.A. Stokes. Beauty is Power: the Use of Space on the Sidewalk. *Sociometry.* Vol. 38, no. 4 (1975) 551-557.
3. C. Y. Cheung and W.H.K. Lam. Pedestrian Route Choices between Escalators and Stairways in Metro Stations. *Journal of Transportation Engineering* 124(3), 277-285.

4. M. T. Chiolek. Spatial Behaviour in Pedestrian Areas. *Ekistics* 268 (1978) 120-121.
5. W. H. Fleming and H.M. Soner. *Controlled Markov Processes and Viscosity Solutions.* Applications of Mathematics 25. Springer-Verlag 1993.
6. E. Goffman. *Relations in Public: Microstudies in the Public Order.* New York. Basic Books 1971.
7. D. Helbing. *Verkehrsdynamik.* Springer-Verlag 1997.
8. D. Helbing, I. Farkas, and T. Vicsek. Simulating Dynamical Features of Escape Panic. *Nature 407* (2000) 487-490.
9. M. R. Hill. *Spatial Structure and Decision-Making of Pedestrian Route Selection Through an Urban Environment.* Dissertation Thesis. University Microfilms International 1982.
10. S. P. Hoogendoorn. *Normative Pedestrian Flow Behaviour.* Report VK2001.002, Transportation and Traffic Engineering Section, Delft University Of Technology 2001.
11. S. P. Hoogendoorn and P.H.L. BovyṖedestrian Route-Choice and Activity Scheduling Theory and Models. *Submitted to Transportation Research B* 2002.
12. S. P. Hoogendoorn and P.H.L. Bovy. Normative Pedestrian Behaviour Theory and Modelling. *15th International Symposium on Transportation and Traffic Theory*, Sydney 2002.
13. S. P. Hoogendoorn. Pedestrian travel behavior in walking areas by subjective utility optimization. *Transportation Research Annual Meeting.* Paper nr. (2002) 02-2667.
14. R. Hughes. The Flow of Large Crowds of Pedestrians. Mathematics and Computers in Simulation 53 (2001).
15. J. Reason. *Human Error.* New-York: Cambridge University Press 1990.
16. P. N. Senevarante and J.F. Morall. Analysis Of Factors Affecting the Choice of Route of Pedestrians. *Transportation Planning And Technology,* 10 (1986) 147-159.
17. Sobel, R.S., and N. Lillith. Determinant of Nonstationary Personal Space Invasion. *Journal of Social Psychology.* Vol. 97 (1975) 39-45.
18. I.B. Stilitz. Pedestrian Congestions. In: D. Canter (ed.). Architectural Phsychology. London Royal Institute of British Architects (1970) 61-72.
19. Willis et al. (1979). Stepping aside: correlates of Displacements in Pedestrians. *Journal of Communication.* Vol. 29, no 4., 34-39.

Investigations of Vibrations in the Complex Dynamical Systems of Transmission Pipelines

Elena Mul[1] and Vladimir Kravchenko[2]

[1] Institute of Computer Information Technologies, Ternopil Academy of National Economy, 3 Peremoga Sq., 46004 Ternopil, Ukraine
[2] Institute of Cybernetics of National Academy of Sciences of the Ukraine, 40 Glushkov avenue, 03650 Kyiv, Ukraine

Abstract. Mathematical models for complex dynamical systems of transmission pipelines with distributed and discrete parameters are formulated in this paper, which consist of a partial differential equation and nonlinear boundary conditions. The approximate analytical method of small parameter is developed for investigation of vibrations in the considered systems. The conditions of vibrations self-excitation were obtained in analytical form as well as a formula for amplitudes of possible vibrations. It is shown that it is possible to change purposefully frequencies and amplitudes of vibrations and avoid an excitation of undesirable vibration processes, including self-excited vibrations.

1 Introduction

Let us consider dynamics of transmission pipelines intended for lifting of minerals from great depth. A pipe of significant length with big mass platform on the end is a basic constructive element of such deep-water plants. Intensive dynamical processes of different physical nature may be excited under the influence of forced wave disturbances and nonlinear hydrodynamical forces in the system. Water flow, which washes the pipe, can continuously transport the energy into the system, exciting significant and dangerous vibrations, as a result breaking of the plant is possible.

Hydrodynamical load is nonconservative, as it does not have potential (force function) in contrast to gravitation. Changing of energy of system, which interacts with fluid flow, is explained by the presence of energy source in the flow and not by of dissipation of energy, which is always present in real dynamical systems at the expense of internal friction. If load on the pipe is nonconservative, i.e. applied forces can not be written as gradient of some potential function, energy statical stability criterion has not applied and the possibility of appearance of dynamical bifurcations and different self-oscillating modes appears.

Whereas the platform of the mass M has high-drag form in contrast to the pipe, the force of resistance appears under interaction of mass M with environment and this force is nonlinear as a rule. It is necessary to solve the considered and similar problems of instability during design and creation of different constructions for work in the ocean.

2 Model

The mathematical model of vibrations excitation in the case, when the pipe is washed by water flow, is a dissipative wave equation of longitudinal vibrations. Accounting energy dissipation in the material we will write the next partial differential equation [1]:

$$\frac{\partial^2 u}{\partial t^2} - a^2 \frac{\partial^2}{\partial x^2}(u + \beta \frac{\partial u}{\partial t}) = 0, \tag{1}$$

where $u(x,t)$ is longitudinal displacement of pipe points; $a^2 = \frac{E}{\rho}$; E is modulus of elasticity of the material; ρ is mass of material volume unit; β is a coefficient characterizing internal friction in the material.

For dynamical system of transmission pipeline with elastically attached pipe, boundary conditions for equation (1) are the next [2]:

$$x = 0, \ ES\frac{\partial}{\partial x}(u(0,t) + \beta \frac{\partial u(0,t)}{\partial t}) = ku(0,t), \tag{2}$$

$$x = L, \ M\frac{\partial^2 u(L,t)}{\partial t^2} + ES\frac{\partial}{\partial x}(u(L,t) + \beta \frac{\partial u(L,t)}{\partial t}) = \alpha_1 \frac{\partial u(L,t)}{\partial t} - \alpha_3 (\frac{\partial u(L,t)}{\partial t})^3, \tag{3}$$

where M is a platform mass; S is pipe cross-section area; k is longitudinal rigidity of the elastic hanger at the point of pipe attaching; L is pipe length; $\alpha_1 \frac{\partial u(L,t)}{\partial t} - \alpha_3 (\frac{\partial u(L,t)}{\partial t})^3$ is force of non-linear resistance, which appears as a result of interaction between the body and environment; α_1, α_3 are constants characterizing resistant environment.

3 Method

In this paper the approximate analytical method of small parameter is developed for problem investigation.

First of all for estimation of order of smallness of components from the boundary problem (1)-(3), as well as some simplification of bulky calculations, the boundary problem (1)-(3) was written in the dimensionless form [2,3]:

$$\frac{\partial^2 \overline{u}}{\partial \tau^2} - \frac{\partial^2 \overline{u}}{\partial \overline{x}^2} = \varepsilon_1 \frac{\partial^3 \overline{u}}{\partial \overline{x}^2 \partial \tau}, \tag{4}$$

$$\overline{x} = 0, \ -r\overline{u} + \frac{\partial \overline{u}}{\partial \overline{x}} = -\varepsilon_1 \frac{\partial^2 \overline{u}}{\partial \overline{x} \partial \tau}, \tag{5}$$

$$\overline{x} = 1, \ \frac{\partial^2 \overline{u}}{\partial \tau^2} + \mu \frac{\partial \overline{u}}{\partial \overline{x}} = -\varepsilon_1 \mu \frac{\partial^2 \overline{u}}{\partial \overline{x} \partial \tau} + \varepsilon_2 \frac{\partial \overline{u}}{\partial \tau} - \varepsilon_3 (\frac{\partial \overline{u}}{\partial \tau})^3, \tag{6}$$

where dimensionless variables and dimensionless parameters are determined by the expressions:

$$\overline{x} = \frac{x}{L}, \quad \overline{u} = \frac{u}{L}, \quad \tau = t\frac{a}{L}, \tag{7}$$

$$\varepsilon_1 = \frac{a\beta}{L}, \quad \varepsilon_2 = \frac{\alpha_1 L}{Ma}, \quad \varepsilon_3 = \frac{\alpha_3 a}{ML}, \quad \mu = \frac{\rho SL}{M}, \quad r = \frac{kL}{ES}. \tag{8}$$

In most cases the dimensionless parameters $\varepsilon_1, \varepsilon_2, \varepsilon_3$ in dynamical systems of transmission pipelines are significantly smaller than unit. This allows to assume that these parameters are directly proportional to some common small parameter ε ($\varepsilon \ll 1$):

$$\varepsilon_1 = \varepsilon\overline{\varepsilon_1}, \quad \varepsilon_2 = \varepsilon\overline{\varepsilon_2}, \quad \varepsilon_3 = \varepsilon\overline{\varepsilon_3}. \tag{9}$$

And now let us begin from the problem of self-excitation of vibrations. It is known [2–4], that in that case it is sufficient to account only the linear part in the expression for the resistance force, i.e. to assume that $\varepsilon_3 = 0$. Then we will search for the solution of equation (4) in the form

$$\overline{u}(\overline{x}, \tau) = (A\cos\lambda\overline{x} + B\sin\lambda\overline{x})\exp(-\varepsilon\overline{\varepsilon_1}\frac{\lambda^2}{2} \pm i\lambda)\tau, \tag{10}$$

where λ is an eigenvalue of the nonconservative boundary problem; A, B are arbitrary constants; e is the basis of natural logarithm.

It is easy to show, that solution (10) satisfies partial differential equation (4) accurate to ε^2. Satisfying boundary conditions (5)-(6) as well accurate to ε^2, we will have the next transcendental equation:

$$\cot\lambda = \frac{\lambda(\mu + r)}{-\lambda^2 + \mu r} \pm i\varepsilon\frac{\overline{\varepsilon_1} r\lambda^2(\lambda^2 + \mu^2) + \overline{\varepsilon_2}\mu(\lambda^2 + r^2)}{(-\lambda^2 + \mu r)^2}. \tag{11}$$

The solution of this equation at $\varepsilon = 0$ is known and it determines eigenvalues ω of the corresponding conservative boundary problem. Accounting smallness of damping forces, we will approximately determine eigenvalues of the nonconservative boundary problem by expression

$$\lambda \approx \omega + \varepsilon\lambda_1, \tag{12}$$

where $\varepsilon\lambda_1$ is correction factor of system nonconservatism.

After substitution of expression (12) into transcendental equation (11), we will obtain the next approximate formula for complex eigenvalues of the nonconservative system:

$$\lambda \approx \omega \mp i\varepsilon\frac{\overline{\varepsilon_1} r\omega^2(\omega^2 + \mu^2) + \overline{\varepsilon_2}\mu(\omega^2 + r^2)}{\omega^4 + \omega^2 r(r+1) + \omega^2\mu(\mu+1) + \mu r(\mu r + \mu + r)}. \tag{13}$$

Then such that system increment should be greater than its decrement, we will obtain from the exponential function of solution (10) the condition of excitation of self-oscillations in the considered dynamical system in the form:

$$\varepsilon_1 r\omega^2(\omega^2 + \mu^2) + \varepsilon_2\mu(\omega^2 + r^2) \geq \frac{\varepsilon_1\omega^2}{2}(\omega^4 + \omega^2 r(r+1) + \omega^2\mu(\mu+1) + \mu r(\mu r + \mu + r)). \tag{14}$$

And now let us analyse not only the conditions of self-excitation of vibrations, but their amplitudes as well, that is much more complicated problem.

In the conservative system (at $\varepsilon = 0$) periodic motions appear with any amplitude dependent on initial conditions. But in the nonconservative system (at $\varepsilon \neq 0$) they have only fixed amplitudes, which conform to equality of the energy present at the expense of negative resistance and energy dissipation which is always present in existent dynamical systems [1]. So the solution of the nonconservative boundary problem (4)-(6) we will search in the form

$$\overline{u}(\overline{x},\tau) = \overline{u_0}(\overline{x},\tau) + \varepsilon\overline{u_1}(\overline{x},\tau) + \dots , \tag{15}$$

where $\overline{u_0}(\overline{x},\tau)$ is one from the solutions of the corresponding conservative problem (at $\varepsilon = 0$).

It is easy to obtain the next expression for such solution:

$$\overline{u_0}(\overline{x},\tau) = A(\frac{\omega}{r}\cos\omega\overline{x} + \sin\omega\overline{x})\cos\omega\tau, \tag{16}$$

where A is sought-for amplitude, and it is necessary to select it close to amplitude of boundary cycle of self-sustained oscillation system.

Taking into consideration relationship

$$(\mu r - \omega^2)\cos\omega = (r + \mu)\omega\sin\omega, \tag{17}$$

which determines eigenvalues ω of conservative boundary problem (4)-(6) at $\varepsilon = 0$, we will have the next problem:

$$\frac{\partial^2\overline{u_1}}{\partial\tau^2} - \frac{\partial^2\overline{u_1}}{\partial\overline{x}^2} = A\overline{\varepsilon_1}\omega^3(\sin\omega\overline{x} + \frac{\omega}{r}\cos\omega\overline{x})\sin\omega\tau, \tag{18}$$

$$\overline{x} = 0, \ -r\overline{u_1} + \frac{\partial\overline{u_1}}{\partial\overline{x}} = A\overline{\varepsilon_1}\omega^2\sin\omega\tau, \tag{19}$$

$$\begin{aligned}\overline{x} = 1, \ &\frac{\partial^2\overline{u_1}}{\partial\tau^2} + \mu\frac{\partial\overline{u_1}}{\partial\overline{x}} = A\omega(\sin\omega + \frac{\omega}{r}\cos\omega) \times \\ &(\overline{\varepsilon_1}\omega^2 - \overline{\varepsilon_2})\sin\omega\tau + A^3\overline{\varepsilon_3}\omega^3(\sin\omega + \frac{\omega}{r}\cos\omega)^3\sin^3\omega\tau.\end{aligned} \tag{20}$$

Now let us transform the last addend in condition (20) and neglect the component of the force with harmonic 3ω (as in the case of equation (17) frequency spectrum is not equidistant, i.e. $3\omega \neq \omega_3$, therefore we assume, that the component of the force with harmonic 3ω does not cause considerable amplitudes during vibrations). So we will write condition (20) in the form:

$$\begin{aligned}\overline{x} = 1, \ &\frac{\partial^2\overline{u_1}}{\partial\tau^2} + \mu\frac{\partial\overline{u_1}}{\partial\overline{x}} = A\omega(\sin\omega + \frac{\omega}{r}\cos\omega) \times \\ &(\overline{\varepsilon_1}\omega^2 - \overline{\varepsilon_2} + 0.75A^2\overline{\varepsilon_3}\omega^2(\sin\omega + \frac{\omega}{r}\cos\omega)^2)\sin\omega\tau.\end{aligned} \tag{21}$$

The boundary problem (18), (19), (21) may be considered as a mathematical model of the conservative system with natural frequency ω, on which the external force of resonance frequency effects. It is known [1] that the stationary solution exists only on condition of orthogonality of external force and normal mode of the system. This condition determines the amplitude, which does not increase with time. Therefore for our problem let us choose the next form of solution:

$$\overline{u_1}(\overline{x},\tau) = U(\overline{x}) \sin \omega\tau, \tag{22}$$

and write the problem as follows:

$$\frac{d^2U}{d\overline{x}^2} + \omega^2 U = -A\overline{\varepsilon_1}\omega^3(\sin \omega\overline{x} + \frac{\omega}{r}\cos \omega\overline{x}), \tag{23}$$

$$\overline{x} = 0, \qquad -rU + \frac{dU}{d\overline{x}} = A\omega^2\overline{\varepsilon_1}, \tag{24}$$

$$\begin{aligned} \overline{x} = 1, \qquad & \mu\frac{dU}{d\overline{x}} - \omega^2 U = A\omega(\sin\omega + \frac{\omega}{r}\cos\omega) \times \\ & (\overline{\varepsilon_1}\omega^2 - \overline{\varepsilon_2} + 0.75\overline{\varepsilon_3}A^2\omega^2(\sin\omega + \frac{\omega}{r}\cos\omega)^2. \end{aligned} \tag{25}$$

Searching for general solution of equation (23) in the form

$$U(\overline{x}) = (B_1 + B_2\overline{x})\cos\omega\overline{x} + (C_1 + C_2\overline{x})\sin\omega\overline{x}, \tag{26}$$

where B_1, B_2, C_1, C_2 are arbitrary constants, and satisfying boundary conditions (24)-(25), we can write:

$$C_1\omega - rB_1 = 0.5A\omega^2\overline{\varepsilon_1}, \tag{27}$$

$$\begin{aligned} &-0.5A\overline{\varepsilon_1}r\omega^2(\mu^2 + \mu + \omega^2) - \tfrac{A\mu\omega^2}{2r}(\overline{\varepsilon_1}\mu\omega^2 + \overline{\varepsilon_1}\omega^2 - 2\overline{\varepsilon_2}) - (\mu^2 + \omega^2)rB_1 \\ &+A\overline{\varepsilon_2}\mu r + (\mu^2 + \omega^2)\omega C_1 - \tfrac{A\overline{\varepsilon_1}\omega^6}{2r} = \tfrac{3\overline{\varepsilon_3}\mu^3 A^3\omega^2(r^2+\omega^2)^3\sin^2\omega}{4r^3(\mu r-\omega^2)^2}. \end{aligned} \tag{28}$$

This allows to find quadrate of amplitude A of stationary vibrations as follows:

$$\begin{aligned} A^2 = & \frac{4r^2(\mu r - \omega^2)^2}{3\varepsilon_3\mu^3\omega^2(r^2+\omega^2)^3\sin^2\omega}(\varepsilon_2\mu(r^2+\omega^2) - \\ & 0.5\varepsilon_1\omega^2((\omega^2 + r^2)(\mu^2 + \mu + \omega^2) - r(\omega^2 + \mu^2))). \end{aligned} \tag{29}$$

It is easy to see from the expression (29) that condition of excitation of vibrations will be obtained in the form (14) just as for linear problem. The first approximation of stationary solution of the problem will be the next:

$$\begin{aligned} \overline{u} = & A(\frac{\omega}{r}\cos\omega\overline{x} + \sin\omega\overline{x})\cos\omega\tau + \\ & \varepsilon\frac{\omega}{r}(C_1 - 0.5A\omega\overline{\varepsilon_1} + 0.5A\omega\overline{\varepsilon_1}r\overline{x})\cos\omega\overline{x}\sin\omega\tau + \\ & \varepsilon(C_1 - \frac{A\overline{\varepsilon_1}\omega^3\overline{x}}{2r})\sin\omega\overline{x}\sin\omega\tau. \end{aligned} \tag{30}$$

4 Discussion of Results and Conclusions

Analysis of condition (14) of vibrations excitation allows to draw the next conclusion. Such boundary value ω_b of frequency always exists in the dynamical systems of transmission pipelines with discrete and continuous parameters, that on the frequencies, which are greater than this boundary value ω_b, excitation of vibrations is impossible. The value of such boundary frequency ω_b is easy calculated from inequality (14).

So it becomes possible to obtain easy qualitative estimations for determination of the effect of different pipe parameters on frequency domain, inside which excitation of vibrations is possible.

The numerous calculations showed that at small dimensionless parameters of the system only the first eigenvalue ω_1 of the conservative boundary problem was included in domain of vibrations excitation. Increase of the parameter of external friction ε_2 causes increasing of the corresponding value of boundary frequency ω_b and, as a result, vibrations can be excited now on the first two frequencies in most cases of practical values of parameter μ.

Thus, for each definite system of transmission pipeline it is possible to define a character of possible vibration processes. It allows to avoid an excitation of undesirable vibration processes, including self-excited vibrations.

The results of complex dynamical systems investigations, which were obtained with the help of the approximate analytical method of small parameter, were verified with the help of numerical method of normal fundamental systems of solutions [5].

The developed approximate analytical method can be applied also to solution of different boundary value problems, which describe various distributed systems, for example, dynamical systems of controlled machine units.

References

1. Philipov, A.P.: Oscillations of Mechanical Systems [in Russian]. Kiev, Naukova Dumka (1965) 706.
2. Kuhta, K.Y., Kravchenko, V.P., V.A.Krasnoshapka, V.A.: Qualitative Theory of Controlled Dynamical Systems with Distributed and Discrete Parameters [in Russian]. Kiev, Naukova Dumka (1986) 224.
3. Mul, E.V.: On Conditions of Excitation of Self-Oscillations in a Nonconservative Dynamic System with Distributed Parameters. Cybernetics and Computing Technology, Complex Control Systems, New York, Allerton Press, Inc., **111** (1998) 70–72.
4. Svetlitskiy, V.A.: Pipe-Lines and Hoses Mechanics [in Russian]. Moscow, Mashinostroenie (1982) 279.
5. Kravchenko, V.P., Mul, E.V., Shut M.I.: Possibilities to Control Oscillations in Systems of Machine Units with Distributed and Discrete Parameters [in Russian and in English]. Moscow, Mekhanika Kompozitsionnykh Materialov i Konstruktsii Vol.5 **4** (1999) 77–86.

Information in Intelligent Transportation System

J. Wahle[1] and M. Schreckenberg[2]

[1] TraffGo GmbH, Grabenstr. 132, 47057 Duisburg
[2] Physics of Transport and Traffic, Univ. of Duisburg, 47048 Duisburg, Germany

Abstract. Since Advanced Traveler Information Systems (ATIS) have been introduced, their potential benefits as well as their drawbacks have been discussed controversially. This will continue as long as the drivers' reactions upon current or even predictive information about the traffic situation are not known. Thus, traffic models that also consider this feedback-loop are necessary. In this paper, we address a basic two-route scenario with different types of information and study the impact of it using simulations. Different ways of generating current information are tested. It is pointed out that the nature of the information very much influences the potential benefits of the ATIS.

1 Introduction

In the recent decades, Intelligent Transportation Systems were developed aiming at improving the travel efficiency and mobility, enhancing safety and thus providing economic and environmental benefits. An integral part of these are Advanced Traveler Information Systems (ATIS) which provide current or even predictive information about the traffic flow to the road users [1–3].

Although such systems have reached a high technical standard and have a high potential to alleviate traffic congestion, the impact of the systems on traffic patterns is not known to a sufficient extend [5–12]. Furthermore, some phenomena or behavior patterns are observed that may disprove the benefits of the given information. In [5] the following three phenomena are listed:

Oversaturation ITS can offer a huge amount of information to the user. But if people are confronted with too much information, they are oversaturated, tend to ignore the information and develop simple heuristics to solve the problem. Therefore, especially ATIS should offer personalized information [1].

Overreaction A more fundamental problem is the overreaction which occurs if too many drivers respond to the information. Suddenly, a congestion may be transferred from the original area to the alternative routes. People start to anticipate the behavior of the other drivers'. This anticipation gets even worse if predictive information is given. Thus, the behavior of the drivers has to be incorporated in the forecast [9,11]. Otherwise, this behavior might cause oscillations of traffic flow among the alternatives, since fast routes attract traffic and, in the process, become slower, whereas the previously slower ones turn faster [5,6].

Concentration A set of drivers who go from one origin to one destination tend to use different routes, since they have different preferences or perceive the situation in different ways. This leads to a natural distribution among the routes. ATIS can reduce these variations if all the drivers use the objective information broadcasted. As a result a greater number of them may select the best alternative and consequently drivers with similar preferences will choose the same route, leading to congestion on this [13,14].

Although the effects of overreaction and concentration are similar, the nature of both phenomena is different [5]. Concentration is intrinsic to any system with information, whereas overreaction is a consequence of the fact that drivers cannot forecast the behavior of the others. However, as intelligent devices will provide even more information about link travel times, densities, or route guidance to the road user, this ability will be crucial for the utility of the ATIS. Therefore, understanding travelers' route choice behavior is an important consideration for the development and effectiveness of such systems (for an overview see [5,6,12]).

Many classical microscopic traffic flow models lack the description of this kind of behavior. In this work, we try to model the impact of real-time information on the traffic patterns. Special attention is paid to study different kinds of information and their specific impact on the traffic patterns.

The remainder of this paper is structured as follows: In order to examine different types of information and their impacts a two route scenario is proposed in the next section. The following sections discuss different kinds of information and their individual impact on the traffic patterns.

2 Two-route scenario

For the simulation studies, a scenario with two routes is used. The drivers are modeled according to cellular automaton model by Nagel and Schreckenberg [15]. The setup is the following: Suppose that there are two routes **A** and **B** with the same length, i.e., the number of cells L is identical (Fig. 1). With a constant rate α road users enter the system. They have to choose between the two alternative routes. Their decision is based on some kind of information. In order to study the impact of real-time information on traffic patterns, dynamic data has to be generated. Here, the data stem from the simulation itself. The process of generating the information will be discussed later. In the literature, such a scenario is a standard one used to study the impact of ATIS, e.g., [6,7,16]. In general, two types of road users are identified: *static* and *dynamic* ones:

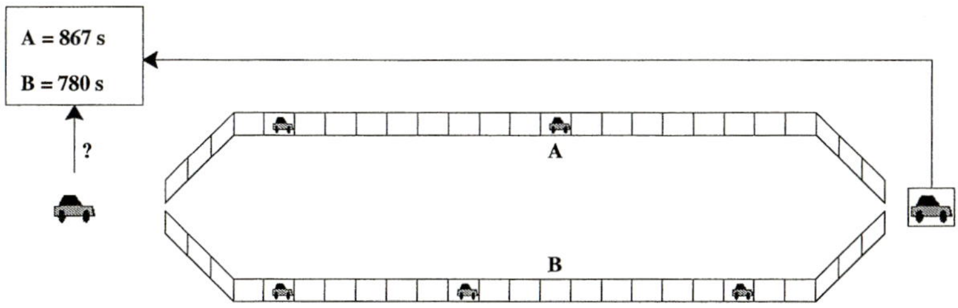

Fig. 1. Traffic scenario with dynamic information. Agents enter the system and have to select a route. Each route comprises L cells. Static drivers choose their route with a probability $p_{\mathbf{AB}}$, whereas dynamic drivers decide with regard to the information displayed on the board, e.g., travel time. The information is transmitted to the board by Floating Cars when they leave the routes.

Static Agents Static agents can be either non-equipped with in-vehicle devices or just ignore the dynamic information, e.g., they do not read information boards or do not listen to the radio. Therefore, they decide on the basis of static information, e.g., the length of the routes or road signs. In the simulation experiment this is represented by the variable $p_{\mathbf{AB}}$, i.e., the probability that a static driver chooses the route **A**.

Dynamic Agents Dynamic drivers are equipped with ATIS or other devices. Thus, they receive and process traffic messages. In this scenario, the messages are displayed on a board (Fig. 1). The board can represent a message sign at the beginning of the routes or an in-vehicle device. Using this information a dynamic driver always tries to optimize his trip, i.e., select the route with the minimal travel time. In that sense they behave "rational". Note that this is a very particular kind of rationality: the agents trust the information on the board and decide based on it. They do not *over-react* , i.e., there is no reasoning about the behavior of the others. If the information for both routes is the same, a route is chosen randomly. The amount of dynamic drivers is represented by the variable s_{dyn}. After the route choice, both kinds of vehicles move according to the CA-rules [15].

Generation of Dynamic Data The decision of the dynamic agents is based on real-time information, which is for instance provided by *Floating-Cars* (FCs). FCs are employed to investigate traffic conditions on the routes, e.g., to measure speed or link travel times. In this paperwork, we study different kinds of information. In the first scenario described, the FCs transmit their travel times to a board when leaving the system (Fig. 1). Thus, they generate dynamic traffic information, which is the basis of the decision of new agents entering the system. The amount of FCs is represented by the variable s_{FC}. The remaining vehicles do not transmit information. Additionally, the impact of different kinds of information, like average speed or density on the route, will also be studied (see Sect. 4).

2.1 Simulation technique

Simulation run The simulations are performed in the following way: at the beginning both routes are empty and the information on the board is set to travel time on free routes $t_{\mathbf{A}} = t_{\mathbf{B}} = L/(v_{\max} - p)$. Note that the mean speed in the free-flow region is $v_{\max} - p$. Dynamic drivers choose their route at random and static drivers with regard to $p_{\mathbf{AB}}$. With the probability s_{FC}, a static or dynamic driver is a FC. After a certain time FCs finish their trip and transmit their travel time to the board. Now, the dynamic drivers decide on the basis of the information available. In general, every simulation lasts for 50,000 time-steps. For the analyses the first 5,000 steps are ignored as only then the system has reached a steady state. The results of a typical experiment are depicted in Fig. 2.

Open boundaries Because of the model employed, special care has to be taken when injecting new vehicles on the routes. In most investigations of the dynamic properties of traffic flow models, periodic boundary conditions are employed to study equilibrium dynamics (for an overview [17]).

Recently, systems with open boundary conditions have been discussed in literature [18,19]. Open systems are characterized by an injection rate α and an extinction rate β, i.e., a vehicle enters the system with the probability α or leaves it at the end with the rate β. Various investigations of systems with open boundary conditions yield that there are three different phases: the free-flow, the congested traffic and the maximum current phase [19]. Note that these phases are only stable if the rates are steady, i.e., $\frac{d}{dt}\alpha(t) = \frac{d}{dt}\beta(t) = 0$. For convenience, we set $\beta_{\mathbf{A}}(t) = \beta_{\mathbf{B}}(t) = 1$ and $\alpha(t) = const.$ For the injection rates of route **A** and **B**, the following relation holds $\alpha_{\mathbf{A}}(t) + \alpha_{\mathbf{B}}(t) = \alpha$, with $\frac{d}{dt}\alpha_{\mathbf{A}}(t) \neq 0$ and $\frac{d}{dt}\alpha_{\mathbf{B}}(t) \neq 0$. The time dependence represents the dynamics of the decisions.

The process of injection is carried out in the following way [18]: in front of the system, an injection area with the size $v_{\max} + 1$ is prepared. To add a new vehicle, the position of the first car on the road is determined. Then the additional vehicle is put into the injection area with a minimum gap of $v_{\max}$ to the detected car, and a speed of $v_{\max}$. Then the update is carried out according to the rules presented in [15]. If a vehicle remains in the injection area after the update procedure it is deleted. This is necessary to avoid jams close to that area, since otherwise the injection rate is replaced by the flow out $\alpha = j_{jam}$ and high-flow rates cannot be established.

There are two sets of parameters which stem from either the model used or the scenario itself. The parameters are the following: $v_{\max}$ maximum speed of vehicles, p_{dec} deceleration probability, α input rate, L length of the routes, $p_{\mathbf{AB}}$ decision of static drivers, s_{FC} percentage of FCs, and s_{dyn} share of dynamic/equipped drivers.

Obviously, s_{dyn} and s_{FC} are interpreted as probabilities for creating a dynamic road user or a FC, respectively. With the probability $p_{\mathbf{AB}}$ a static driver is put on route **A**. The system is symmetric and thus only values of $p_{\mathbf{AB}} \leq 0.5$ are studied. However, in most simulations the default value of $p_{\mathbf{AB}} = 0.5$ is used, since the route length L is the only information static drivers have. Note that for $p_{\mathbf{AB}} = 0.5$ a natural equilibrium is reached, i.e., an optimal solution of the problem. In all simulations the maximum speed is set to $v_{\max} = 3$ cells/time-step. In this work, we are only interested in the parameter s_{dyn} share of dynamic/equipped drivers and $p_{\mathbf{AB}}$ decision of static drivers. The other parameters are studied in [20].

In order to reduce the stochastic noise in the quantities, a correlation analysis is used. Such an analysis is especially useful to study oscillations, which will occur for some quantities in the next section, it is valuable to employ correlation functions. They detect and quantify coupling effects between variables. The auto-correlation for a variable ψ is defined in the following way:

$$ac_{\psi}(\tau) = \frac{1}{\sigma(\psi)} \left[\langle \psi(t)\psi(t+\tau) \rangle - \langle \psi(t) \rangle \langle \psi(t+\tau) \rangle \right] . \tag{1}$$

It provides information about the temporal evolution of ψ ($\sigma(\psi)$ is the variance of ψ).

3 Floating-Car Data

In this section, FCs are used to provide dynamic data. They are inserted into the system with the rate s_{FC}. They flow with the traffic and transmit their travel time to the information board, when they leave the system. The dynamic drivers which enter the system receive the moving average (20 time-steps) of the travel times collected.

A typical experiment is depicted in Fig. 2. The parameters are set to $p_{\text{dec}} = 0.25$, $\alpha = 1$, $L = 2{,}000$, $p_{\mathbf{AB}} = 0.5$, $s_{\text{dyn}} = 0.5$, and $s_{\text{FC}} = 1$, i.e., every second driver processes dynamic information and every vehicle serves as a FC. The number of vehicles, travel times, and the moving average of the speeds oscillate on both routes. Note that unlike other quantities the fluctuations of the number of vehicles are not very strong since every time-step only one vehicle can enter and two vehicles can leave the system (one on route **A** and **B**).

The oscillations are due to the dynamic drivers who receive the information on the board. Because they behave rational and trust the information, they aim at minimizing their own travel time. Therefore, they use the route with the shorter travel time. Since every dynamic driver behaves in the same way the route with the smaller number of vehicles is haunted more (Fig. 2a,b). At the same time, the speed on this route decreases and thus the travel time increases (Fig. 2c). Nevertheless, it takes some time for the FCs to transmit this to the information board since they have to reach the end of the route. During this time even more road users enter this route.

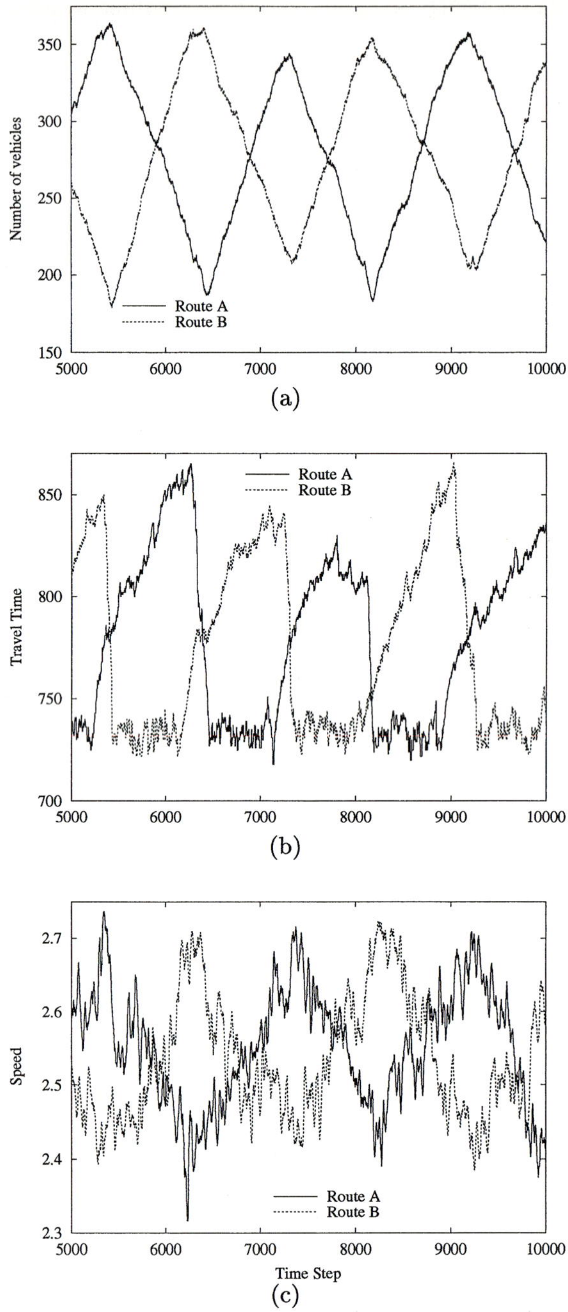

Fig. 2. (a) Number of vehicles on each route; (b) travel time on the board; (c) moving average (20 time-steps) of the mean speed vs. time. The parameters are $p_{\text{dec}} = 0.25$, $\alpha = 1$, $L = 2,000$, $p_{\mathbf{AB}} = 0.5$, $s_{\text{dyn}} = 0.5$, and $s_{\text{FC}} = 1$. Every quantity exhibits characteristic oscillations.

Simultaneously, the situation on the other route gets better since it is not used anymore by dynamic agents. The travel time on this route slowly decreases and the situation turns around – this route gets more crowded again. This oscillatory behavior is well known in the literature, e.g., Bonsall sees it as some kind of interaction:

fast links attract traffic but are, in the process, made slower [6].

. In [5] this effect is named concentration:

Information tends to reduce the variations among drivers because it increases uniformity of perceptions of network conditions around the true value.

However, this effect seems to be intrinsic in most information systems. Nevertheless, the information influences the dynamics of the system in an undesirable way: it would be much more efficient to distribute the vehicles on both routes randomly. This is not only due to the effect of concentration but also due to the fact that the information is too old, the drivers tend to react too late to the new situation. To quantify this behavior in more detail, the influence of the parameter s_{dyn} will be investigated in the following.

3.1 Influence of the dynamic drivers

Since the oscillations are due to the behavior of the dynamic drivers, it is interesting to study the influence of the parameter s_{dyn}, the rate of dynamic drivers. Obviously, the amplitude of the oscillations depends on s_{dyn}. A dramatic situation occurs for $s_{\mathrm{dyn}} = 1$, i.e., every road user follows the information displayed on the board. In this scenario, it may happen that one route is empty and there are no FCs on this route. Therefore, the travel time remains constant, although the route is empty. In order to avoid such negative effects, it is advisable to remove or slowly change information, which is too old. This can be done by, e.g., setting default values after a certain time-out.

In Fig. 3 the auto-correlation for a number of cars on route **A**, $ac_{N_{\mathbf{A}}}(\tau)$, is depicted. For small values of s_{dyn}, the correlation is not strong, whereas systems with a higher number of dynamic drivers s_{dyn} are strongly correlated. Similar results are found for the travel time and the speed.

Travel time distribution The effect of mixing the two types of agents can be understood better by studying the travel time distribution of both species. In order to provide the distribution, a histogram is used, i.e., the number of drivers in a certain interval of travel time Δt is aggregated. We have set $\Delta t = 5$ time-steps. A distribution for different s_{dyn} is depicted in Fig. 4.

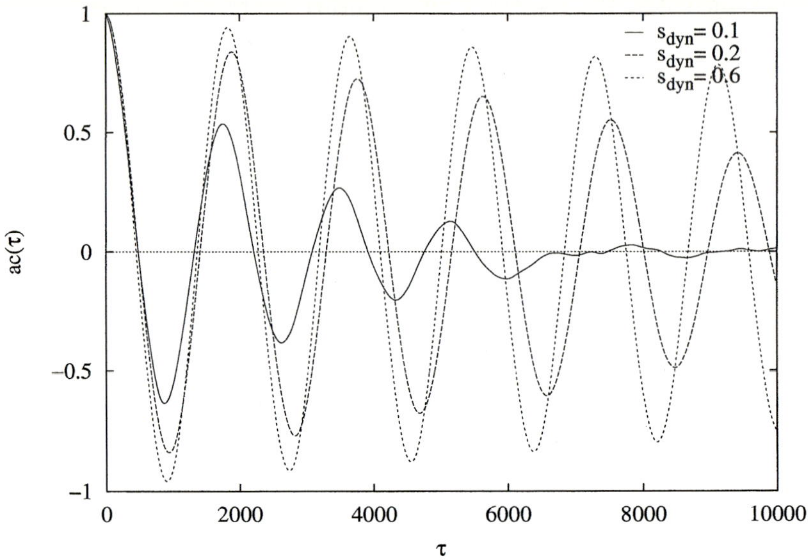

Fig. 3. Auto-correlation $ac_{N_{\mathbf{A}}}$ (1) vs. τ, for different values of s_{dyn}. The parameters are $L = 2,000$, $p_{\mathrm{dec}} = 0.25$, $\alpha = 1$, $p_{\mathbf{AB}} = 0.5$, $s_{\mathrm{dyn}} = 0.5$, and $s_{\mathrm{FC}} = 1$. For small values of s_{dyn} the strong correlations vanish.

In a society of static agents, $s_{\mathrm{dyn}} = 0$, the agents visit both routes with the same frequency and the travel time distribution is more or less Gaussian, with a small standard deviation (Fig. 4a). Note that using $p_{\mathbf{AB}} = 0.5$ an optimal solution is reached. For a mixture of the two species it is changing. The dynamic drivers introduce fluctuations into the system. Their average travel time increases, whereas some static drivers benefit from emptier routes. A second peak at higher travel times is established due to the dynamic drivers (Fig. 4b). For higher values of s_{dyn}, this peak broadens and both peaks separate (Fig. 4c). The majority of the static drivers profit from the dynamic ones. For $s_{\mathrm{dyn}} = 1$, every driver decides with regard to the current information on the board. A broad peak is found (Fig. 4d).

The behaviour of the dynamic drivers influences existing traffic patterns in an undesirable way and leads to a sub-optimal situation, i.e., the overall flow of the system is reduced due to concentration of the equipped drivers (see Fig. 9). The travel time distribution implies that the static drivers profit from this behaviour (see Fig. 4). Therefore, the question can be raised:

Is it good to be ignorant to information?

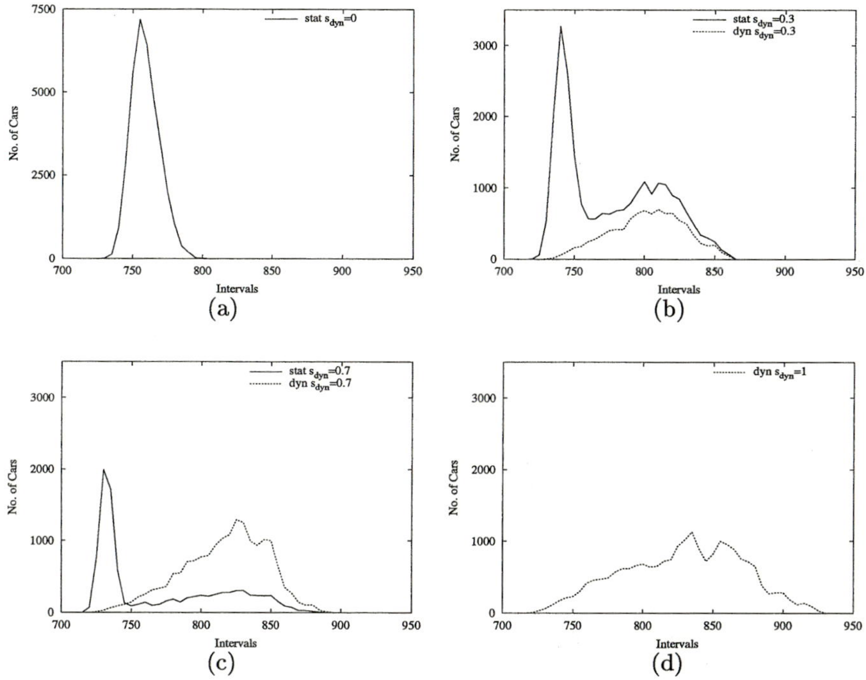

Fig. 4. Travel time distribution for static and dynamic drivers for different values of s_{dyn}. The parameters are $L = 2,000$, $p_{\text{dec}} = 0.25$, $\alpha = 1$, $p_{\mathbf{AB}} = 0.5$, and $s_{\text{FC}} = 1$. (a) A homogeneous society of static drivers arrange in an optimal way for $p_{\mathbf{AB}} = 0.5$. (b) A mixture between dynamic and static drivers is unfavourable for the dynamic ones. They introduce a second, broader peak at higher travel times. (c) This effect is amplified for higher s_{dyn}. The peak of the dynamic ones is broadened, whereas the peak for the static ones gets sharper, i.e., the minority of static drivers profits even more. (d) For a homogeneous society of dynamic drivers $s_{\text{dyn}} = 1$ the distribution is broadened further.

3.2 Influence of the static drivers

So far, the static agents have chosen their route randomly, i.e., $p_{\mathbf{AB}} = 0.5$. This means that they behave in an optimal way. In Fig. 5 the travel time distribution for $p_{\mathbf{AB}} < 0.5$ is shown. For $p_{\mathbf{AB}} = 0.2$ the dynamic agents are not really able to profit from the information (cf. Fig. 4a), since the static agents on route **A** have lower travel times. Only for $p_{\mathbf{AB}} \ll 1$, i.e., all static drivers use the same route, the dynamic agents can profit from the information and there are still agents, who use route **B** (see Fig. 5b). Besides, such a behaviour is quite unrealistic.

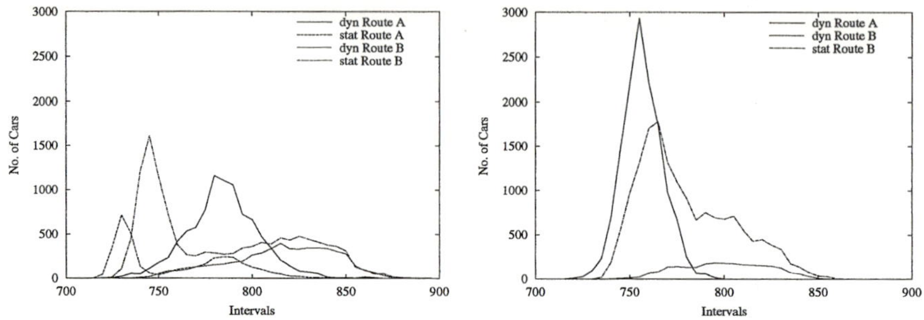

(a) (b)

Fig. 5. Travel time distribution for different values of $p_{\mathbf{AB}}$. The parameters are set to $L = 2,000$, $p_{\text{dec}} = 0.25$, $s_{\text{dyn}} = 0.5$, and $s_{\text{FC}} = 1$. (a) $p_{\mathbf{AB}} = 0.2$, i.e., 80 % of the static drivers use route **B**. (b) $p_{\mathbf{AB}} = 0$, i.e., all static drivers use route **B**. Although, the static agents are far away from the optimal coordination, the dynamic agents do not profit very much and the advantage of using the information is small.

4 Different criteria

It becomes clear that the effect of concentration is intrinsic to information systems, which provide the same information to all users, since faster links attract more traffic and in the process are made slower [6]. Additionally, FC travel time is not a good criterion to control and distribute traffic in a network, since it reflects the history of the network. In this section, different kinds of information are tested: the gradient of the travel time, the global density and the global speed.

4.1 Gradient of travel time

In control theory, it is well known that in order to control some systems it is better to use the gradient of a quantity than the absolute values themselves [21]. Here, the moving average of the difference between two consecutive travel times of FCs $\Delta\tau_i = (t_i - t_{i-1})$ is used, with t_i being the travel time of the ith FC on a route. Hence, the information displayed on the board is $\Delta\tau$, the moving average of the last twenty $\Delta\tau_i$, i.e., the dynamic drivers do not follow the absolute travel time but the trend of the measured values. Thus, changes of the traffic states on both routes are reported much faster.

Sub-Optimal Static Agents Nevertheless, in some situations the trend can be misleading. In Fig. 6 the travel time distribution for a sub-optimal behaviour of the static drivers is depicted. The parameters are set to $L = 2,000$, $p_{\text{dec}} = 0.25$, $\alpha = 1$, $s_{\text{dyn}} = 0.5$, $s_{\text{FC}} = 1$, and $p_{\mathbf{AB}} = 0$, i.e., all static drivers use route **B**. In this situation, the absolute travel time is the better information. For $p_{\mathbf{AB}} = 0$, the bulk of the dynamic agents are on route **A**. Thus, they improve the situation for the static drivers (Fig. 6a). If the gradient of the travel time is used as source of information a lot of dynamic agents use route **B** and increase the travel time further (Fig. 6b).

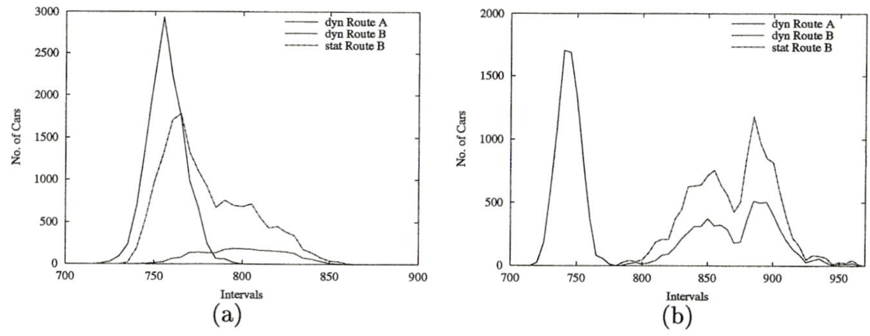

Fig. 6. Travel Time Distribution for (a) absolute travel times and (b) the gradient of the travel time. The parameters are $L = 2,000$, $p_{\text{dec}} = 0.25$, $\alpha = 1$, $s_{\text{dyn}} = 0.5$, $s_{\text{FC}} = 1$, and $p_{\mathbf{AB}} = 0$, i.e., all static drivers use route **B**. In this situation providing the gradient of the travel time is misleading to the dynamic drivers. They introduce a distinct peak at higher travel times.

This negative effect is also reflected by the average travel time for the dynamic and static drivers. In the first scenario, $t_{\text{dyn}} = 760.15$ time-steps and $t_{\text{stat}} = 776.74$ time-steps, whereas for the second scenario $t_{\text{dyn}} = 798.05$ time-steps and $t_{\text{stat}} = 866.79$ time-steps are necessary. The problem using the gradient is that the absolute value of the travel time is neglected. Of course a criterion which combines the absolute value and its gradient, would provide better results.

4.2 Global density and speed

Up to now, the information provided by FCs has been discussed. However, one main problem is that this information is delayed. Better quantities to characterise the current traffic state of a route are necessary, like the global density and the mean speed. They are defined as follows: the density $\rho_{\mathbf{A,B}} = N_{\mathbf{A,B}}/L$, and the mean speed $v_{\mathbf{A,B}} = \sum_{i=0}^{N_{\mathbf{A,B}}} v_{i;\mathbf{A,B}}$. Note that in real systems such precise quantities cannot be provided, since most measurements are local.

Simulation experiments with global density and mean speed as information displayed on the board are carried out. In the case of the density criterion, the dynamic drivers choose the route with the lowest mean density. If the mean speed is given, the dynamic agents select the route with the highest value.

However, in both cases the correlations vanish. In Fig. 7 the travel time distribution is shown for $s_{\mathrm{dyn}} = 0.3$ and $p_{\mathbf{AB}} = 0.5$. Using the global average speed it is possible to coordinate the dynamic as well as static agents in a more efficient way. Thus, giving a quantity which represents the current state of a route as information, e.g., mean speed or density, helps to coordinate the behaviour of the two different types of road users, so that all drivers are able to profit.

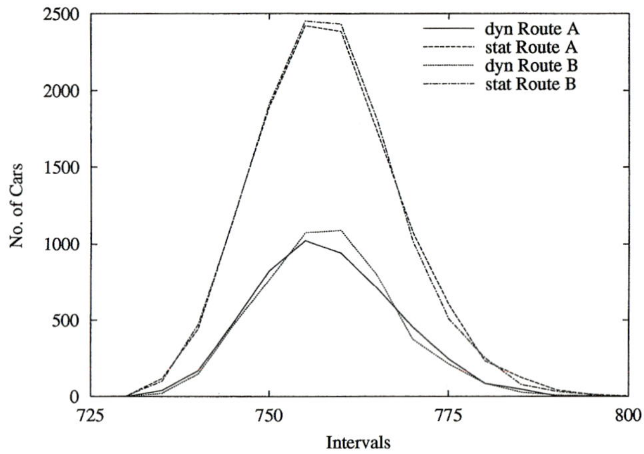

Fig. 7. Travel time distribution for using the global speed as traffic information. The parameters are $L = 2,000$, $p_{\mathrm{dec}} = 0.25$, $\alpha = 1$, $s_{\mathrm{dyn}} = 0.3$, $s_{\mathrm{FC}} = 1$, and $p_{\mathbf{AB}} = 0.5$. Using this information the traffic demand is distributed on both routes in the same way.

Obviously, the difference between the two quantities is that if density is the information, the speed fluctuates, and vice-versa. In the case density is provided, a higher number of vehicles is injected into the system. If speed is given, the mean speed on both routes is quite high.

So far, it has been shown that the dynamic drivers can hardly profit from the information provided. Only in very extreme cases, e.g., $p_{\mathbf{AB}} = 0$, the situation for some dynamic agents is better than for most static ones. However, using a global quantity as information the dynamic road users outperform the static ones. Such a situation is shown in Fig. 8, where the travel time distribution for a sub-optimal behaviour of the static drivers ($p_{\mathbf{AB}} = 0$) is given. In contrast to the previous experiments, all dynamic drivers profit from the provided information since no one uses route **A**. This shows that information in principle can help to improve traffic patterns by coordinating the behaviour among the agents. Note that in this experiment $s_{\text{dyn}} = 0.25$, i.e., there is a market penetration of 25 %. With a higher market penetration this effect is shortened.

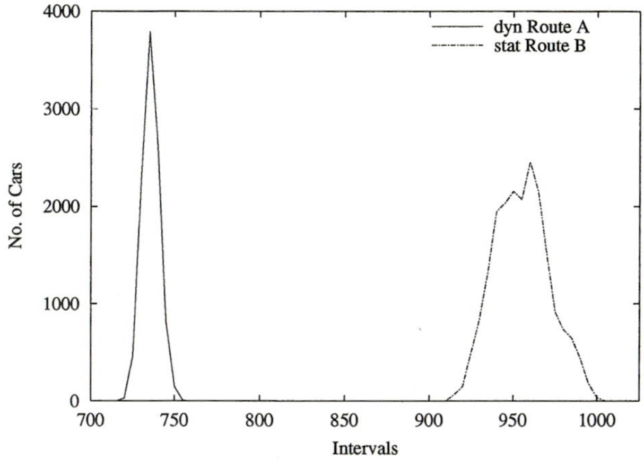

Fig. 8. Travel time distribution for using the global density as traffic information. The parameters are $L = 2,000$, $p_{\text{dec}} = 0.25$, $\alpha = 1$, $s_{\text{dyn}} = 0.25$, $s_{\text{FC}} = 1$, and $p_{\mathbf{AB}} = 0$, i.e., all static drivers use route **B**. All dynamic drivers select route **A**.

4.3 Comparison

In the previous sections, four different kinds of dynamic information have been discussed: FC travel time and its trend, the mean speed and density. All types of information are evaluated from the road usert́s point of view. From the traffic control point of view, an important quantity to evaluate is the capacity or the flow: $j = N_{\mathbf{A}} v_{\mathbf{A}} + N_{\mathbf{B}} v_{\mathbf{B}}$. The objective of the traffic controller is to maximise the capacity of the network by providing information. In Fig. 9 the four different types of information are compared with each other for different market penetrations. The parameters are $L = 2,000$, $p_{\text{dec}} = 0.25$, $\alpha = 1$, $s_{\text{FC}} = 1$, and $p_{\mathbf{AB}} = 0.5$, i.e., the static drivers behave optimal.

It is clear that with a higher share of dynamic drivers s_{dyn} and the FC travel time provided, the flow is reduced. This is due to the oscillations, which destroy the traffic patterns (see Sect. 3). Providing the gradient of the travel time is slightly better but also leads to a decrease in the flow with increasing s_{dyn}. This is also true for the average speed as information, since it leads to a high average speed but reduces the density. Thus, the flow is slightly reduced.

The only type of information, which enhances the efficiency of the system is the average density. Therefore, this is the best type of information for traffic control purposes, since it helps to coordinate the agents in a sufficient way so that even the natural equilibrium of the static drivers ($p_{\mathbf{AB}} = 0.5$) can be optimised by reducing the fluctuations. This discussion makes clear that the efficiency of an information system depends heavily on the nature of the information provided.

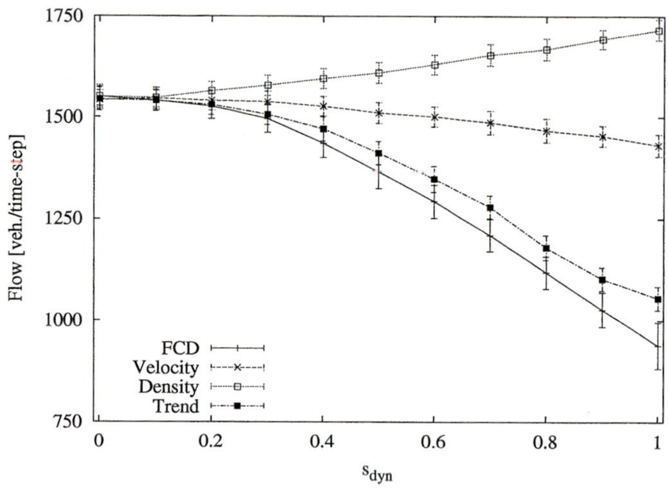

Fig. 9. Flow vs. s_{dyn} for different kinds of information. The parameters are $L = 2,000$, $p_{\mathrm{dec}} = 0.25$, $\alpha = 1$, $s_{\mathrm{FC}} = 1$, and $p_{\mathbf{AB}} = 0.5$. For the FC data and the trend the flow goes down rapidly. This even holds for the speed. From the traffic control point of view, the density is the best criterion.

5 Summary and conclusion

In this paper, the impact of en-route information is studied by simulating a simple two route scenario. For the analysis static and dynamic agents are introduced – representing conservative or flexible route selection. Dynamic travel times for both routes are generated by traffic flow simulations. The data are transmitted by floating cars (FCs) to an information board, which is the basis of the route choice of dynamic road users.

The en-route information leads to a concentration of drivers on the recommended routes, which is intrinsic to many information systems and leads to a negative impact on the traffic pattern, like oscillations. These oscillations are studied using a correlation analysis and travel time distributions. Different parameters are investigated systematically. It is found that the information transmitted by the FCs is often too old and reflects the previous states of the network. Thus, the dynamic agents do not profit from the information received.

Therefore, different types of message are tested: the gradient of the FC travel time, the global density and speed. The analysis yields that beside the amount and the validity of the information an important question is the nature of it. By its nature, travel time is not a good quantity since it introduces not only concentration but also oscillations into the system. For traffic control purposes, the best criterion is the global density since it optimises the flow. Nevertheless, the concentration of drivers on the recommended routes is intrinsic to many information systems, especially for a high market penetration of information systems the benefits of providing information are questionable.

The analysis yields that one should be very careful to provide dynamic information, since concentration is intrinsic to all information systems and the overall performance of the system can get worse [22,7,16,10,14,23]. One key question is: what kind of and how much information do road users need? Another interesting issue is to study the effect of providing various types of messages to different drivers. This may introduce a natural diversity which avoids the effect of concentration.

References

1. J.L. Adler and V.J. Blue. Toward the design of intelligent traveler information systems. *Transp. Res. C*, 6 (1998) 157–172.
2. ITS International, editor. *6th World Congress on Intelligent Transport Systems*, CD-ROM, Toronto, 1999. ITS World Congress.
3. D. Helbing, H.J. Herrmann, M. Schreckenberg, and D.E. Wolf, editors. *Traffic and Granular Flow '99*, Heidelberg, 2000. Springer.
4. A. Ceder, editor. *Proceedings of the 14th International Symposium on Transportation and Traffic Theory*, Amsterdam, 1999. Pergamon.
5. M. Ben-Akiva, A. de Palma, and I. Kaysi. Dynamic network models and driver information systems. *Transp. Res. A*, 25 (1991) 251–266.
6. P. Bonsall. The influence of route guidance advice on route choice in urban networks. *Transportation*, 19 (1992) 1–23.
7. R.W. Hall. Route choice and advanced traveler information systems on a capacitated and dynamic network. *Transp. Res. C*, 4 (1996) 289–306.
8. H. Dia and H. Purchase. Modelling the impacts of advanced traveller information systems using intelligent agents. *Road and Transport Research*, 8 (1999) 68–73.
9. J. Bottom, M. Ben-Akiva, M. Bierlaire, I. Chabini, H. Koutsopoulos, and Q. Yang. Investigation of route guidance generation issues by simulation with DynaMIT. In Ceder [4], 577–600.
10. K. Lee, P.M. Hui, B.H. Wang, and N.F. Johnson. Effects of announcing global information in a two-route traffic flow model. *J. Phys. Soc. Japan*, 70 (2001) 3507–3510.
11. J. Wahle, A.L. Bazzan, F. Klügl, and M. Schreckenberg. Anticipatory traffic forecast using multi-agent techniques. In Helbing et al. [3], 87–92.

12. H.S. Mahmassani and R.-C. Jou. Transferring insights into commuter behavior dynamics from laboratory experiments to field surveys. *Transp. Res. A*, 34 (2000) 243–260.
13. R. Arnott, A. de Palma, and R. Lindsey. Does providing information to drivers reduce traffic congestion? *Transp. Res. A*, 25 (1991) 309.
14. J. Wahle, A.L. Bazzan, F. Klügl, and M. Schreckenberg. Decision dynamics in a traffic scenario. *Physica A*, 287 (2000) 669–681.
15. K. Nagel and M. Schreckenberg. A cellular automaton model for freeway traffic. *J. Physique I*, 2 (1992) 2221–2229.
16. K. Nagel and S. Rasmussen. Traffic at the edge of chaos. In R.A. Brooks and P. Maes, editors, *Artificial Life IV: Proc. of the 4th Int. Workshop on the Synthesis and Sim. of Living Systems*, pages 222–235, Cambridge, 1994. MIT Press.
17. D. Chowdhury, L. Santen, and A. Schadschneider. Statistical physics of vehicular traffic and some related systems. *Physics Reports*, 329 (2000) 199–329.
18. R. Barlovic, T. Huisinga, A. Schadschneider, and M. Schreckenberg. Effects of open boundaries in a cellular automaton model. *Physica A*, 2001.
19. V. Popkov, L. Santen, A. Schadschneider, and G.M. Schütz. Boundary-induced phase transitions in traffic flow. *J. Phys. A*, 34 (2000) L45–L52.
20. J. Wahle. *Information in Intelligent Transportation Systems.* PhD thesis, Gerhard-Mercator-University, 2002.
21. S. Sastry. *Nonlinear Systems: Analysis, Stability, and Control.* Springer, New York, 1999.
22. W. Barfield and T.A. Dingus. *Human Factors in Intelligent Transportation Systems.* Lawrence Erlbaum Associates Inc., Mahwah, New Jersey, 1998.
23. J. Wahle, A.L. Bazzan, F. Klügl, and M. Schreckenberg. The impact of real-time information in a two-route scenario using agent-based simulation. *Transp. Res. C*, 10, 2001.

Experiments on Route Choice Behaviour

Reinhard Selten[1], Michael Schreckenberg[2], Thomas Pitz[1], Thorsten Chmura[1], and Joachim Wahle[2]

[1] Laboratory of Experimental Economics, Adenauerallee 24-42, 53113 Bonn, Germany
[2] Physics of Transport and Traffic, Gerhard Mercator University Duisburg, Lotharstr. 1, 47048 Duisburg, Germany

Abstract. The paper reports laboratory experiments on a day-to-day route choice game with two routes. Subjects had to choose between a main road M and a side road S. The capacity was greater for the main road. 18 subjects participated in each session. In equilibrium the number of subjects is 12 on M and 6 on S.

Two treatments with 6 sessions each were run at the computerized Laboratory of Experimental Economics at Bonn University. Feedback was given in treatment I only about own travel time and in treatment II on travel time for M and S. Money payoffs increase with decreasing time. The main results are as follows: Though mean numbers on M and S are very near to the equilibrium, the fluctuations persist until the end of the sessions in both treatments. Fluctuations and the total number of changes are significantly smaller under treatment II. Subjects' road changes and payoffs are negatively correlated in all sessions.

1 Introduction

Understanding individual travel behaviour is essential for the design of Advanced Traveler Information Systems (ATIS), which provide real-time travel information, like link travel times [2]. However, the response of road users to information is still an open question [3], [4]. Moreover it is not clear whether more information is beneficial [3]. Drivers confronted with too much information may become oversaturated in the sense that information processing becomes to difficult and users develop simple heuristics to solve the problem [5].

Drivers may also overreact to information and thereby cause additional fluctuations. Thus, the behaviour of the drivers has to be incorporated in the forecast [3], [4], [8] ATIS can reduce fluctuations only if behavioural effects are correctly taken into account.

The Literature reports a number of experiments on route choice behaviour [4]. Here we focus on the route choice in a generic two route scenario, which already has been investigated in the literature [7]. However our aim is to present experiments with a large number of periods and with sufficiently many independent observations for meaningful applications of non-parametric significance tests.

If one wants to investigate results of day to day route choice which can be transferred to more realistic environments, it is necessary to explore individual behaviour in an interactive experimental set-up. Does behaviour converge to equilibrium? Does more feedback reduce fluctuations? What is the structure of individual responses to recent experiences? Our experimental study tries to throw light on these questions.

2 Experimental Setup

Subjects are told that in each of 200 periods they have to make a choice between a main road M and a side road S for travelling from A to B. They were told that M is faster if M and S are chosen by the same number of people. No further information was given to the subjects. The number of subjects in each session was 18. The time and depends on the numbers n_S and n_M of participants choosing M and S, respectively:

$$t_M = 6 + 2n_M$$
$$t_S = 12 + 3n_S$$

The period payoff was $40 - t$ with $t = t_M$ if M was chosen and $t = t_S$ if S was chosen. The total payoff of a subject was the sum of all 200 period payoffs converted to money payoffs in DM with a fixed exchange rate of .015 DM for each experimental money unit. One session took roughly one and a half hours. All pure equilibria of the game are characterized by $n_M = 12$ and $n_S = 6$.

Two treatments have been investigated. In treatment I the subjects received feedback only about the time on the road chosen after each period. In treatment II feedback was provided about both times. Six sessions were run with treatment I and six with treatment II.

3 Equilibrium Predictions and Observed Behaviour

For each time series of participants on the side road it can be seen that there is no convergence to the theoretical equilibrium. There are substantial fluctuations until the end of the session. The same is true for all sessions of both treatments. The overall average of numbers of participants on S is very near to the equilibrium prediction. The standard deviation of the session average from 6 is never greater than .17 and it is only .07 on the average. The fluctuations can be measured by the standard deviation of the number of participants choosing S per period. This standard deviation is between 1.53 and 1.94. In view of these numbers one can speak of substantial fluctuations in each of the 12 sessions.

The fluctuations are a little larger under treatment I than under treatment II. The effect is significant. The null-hypothesis is rejected by a Wilcoxon-Mann-Whitney-Test on the significance level of 5% (one sided).

The game underlying the experiment has many pure strategy equilibrium points. In all of them the number of participants on the side road is 6, but the set of players who choose S can be any set of 6 players. The multiplicity of pure strategy equilibria poses a coordination problem which may be one of the reasons for non-convergence and the persistence of fluctuations. Feedback on both travel times vs. feedback on only own travel time has a beneficial effect by the reduction of fluctuations, but this effect is relatively small.

The fluctuations are connected to the total number of road changes within one session. The Spearman-rank-correlation between the total number of road changes and the standard deviation of the number of participants per period on S is .795. This is significant on the level of 1% (one sided). The total number of road changes under treatment I is greater than under treatment II. A Wilcoxon-Mann-Whitney-Test rejects the null-hypothesis only on a very weak significance level of 8.98% (one sided).

Under treatment I subjects who mainly choose only one of the roads feel the need to travel on the other road from time to time in order to get information on both roads. Under treatment II there is no necessity for such information gathering. This seems to be the reason for the greater number of changes and maybe also for the stronger fluctuations under treatment I.

4 Response Mode

A participant who had a bad payoff on the road chosen may change his road in order to travel where it is less crowded. We call this the direct response mode. A road change is the more probable the worse the payoff is.

The direct response mode is the prevailing one but there is also a contrarian response mode. Under the contrarian response mode a road change is more likely the better the payoff is. The contrarian participant expects that a high payoff will attract many others and that therefore the road chosen will be crowded in the next period.

The equilibrium payoff is 10. Payoffs perceived as bad tend to be below 10 and payoffs perceived as good tend to be above 10. Accordingly we classified the response of a subject as *direct* if the road is changed after a payoff smaller than 10 or not changed after a payoff greater than 10. An opposite response is classified as *contrarian*. Table 1 shows the numbers of times in which a subject changes roads (c_- for a payoff below 10 and c_+ for a payoff above 10), or stays at the same road (s_- for payoff below 10 and s_+ for a payoff above 10).

For each subject such a 2x2 table has been determined and a Yule coefficient Q has been computed as follows.

$$Q = \frac{c_- s_+ - c_+ s_-}{c_- s_+ - c_+ s_-} \tag{1}$$

Table 1. 2x2 table for the computation of Yule coefficients.

	change	stay
payoff < 10	c_-	s_-
payoff > 10	c_+	s_+

The Yule coefficient has a range from -1 to $+1$. In our case a high Yule coefficient reflects a tendency towards direct responses and a low one a tendency towards contrarian responses.

In each of four sessions, one of them in Treatment I and three in treatment II, there was one player for whom no Yule coefficient could be determined since these four subjects never change roads. These subjects are not considered in the evaluation of Yulc cocfficicnts.

Evidence for the importance of both response modes can be found in the distributions of Yule coefficients within a session. If the two response modes were not present in behaviour one would expect distributions of Yule coefficients concentrated around 0. However the number of subjects with extreme Yule coefficients below $-.5$ or above $+.5$ tends to be greater than the number of subjects with Yule coefficients in the middle range between $-.5$ and $+.5$. A Wilcoxon one sample test supports this alternative hypothesis by rejecting the null-hypothesis that none of both numbers tends to be greater than the other, on the significance level of 0.01 (two sided).

If one classifies subjects with Yule coefficients above $+.5$ as direct responders and subjects with Yule coefficients below $-.5$ as contrarian responders, then one receives 44% direct responders, 14% contrarian responders and 42% unclassified subjects.

5 Payoffs and Road Changes

In all sessions the number of road changes of a subject is negatively correlated with the subject's payoff. Even if subjects change roads in order to get higher payoffs, they do not succeed in doing this on the average. This suggests that it is difficult to use the information provided by the feedback to one's advantage.

6 Conclusion

The study has shown that the mean numbers on both roads tend to be very near to the equilibrium. Nevertheless, fluctuations persist until the end of the sessions in both treatments. This is of particular interest in view of the fact that the experiments run over 200 periods which is unusually long and should be enough to show a tendency of convergence to equilibrium, if there is one.

Feedback on both road times significantly reduces fluctuations in treatment II compared to treatment I. However the effect is small. There is a significant rank correlation between the total number of road changes and the size of fluctuations. In treatment I road changes may serve the purpose of information gathering. This motivation has no basis in treatment II. However road changes may also be attempts to improve payoffs. The finding of a negative correlation between a subject's payoff and number of road changes suggests that on the average such attempts are not successful.

Two response modes can be found in the data, a direct one in which road changes follow bad payoffs and a contrarian one in which road changes follow good payoffs. One can understand these response modes as due to different views of the causal structure of the situation. If one expects that the road which is crowded today is likely to be crowded tomorrow one will be in the direct response mode but if one thinks that many people will change to the other road because it was crowded today one has reason to be in the contrarian response mode. We have presented statistical evidence for the importance of the two response modes.

References

1. Abbink, K., Sadrieh A., RatImage - Research Assistance Toolbox for Computer-Aided Human Behavior Experiments. SFB Discussion Paper B-325, University of Bonn, (1995)
2. Adler, J., Blue, V., Toward the design of intelligent traveler information systems, Transpn. Res. C 6, (1998), 157
3. Ben-Akiva, M., de Palma, A., Kaysi, I., Dynamic network models and driver information systems, Transpn. Res. A 25, (1991), 251
4. Bonsall, P., The influence of route guidance advice on route choice in urban networks, Transportation 19, (1992), 1-23
5. Gigerenzer, G., Todd, P.M., and ABC Research Group (eds.), Simple heuristics that make us smart, Oxford University Press, (1999)
6. Hall, R., Route choice and advanced traveller information systems on a capacitated and dynamic network, Transpn. Res. C 4, (1996), 289-306
7. Iida Y., Akiyama T., Uchida T., Experimental analysis of dynamic route choice behaviour, Trans. Res. B 26, (1992), 17-32
8. Wahle, J., Bazzan, A., Klügl, F., Schreckenberg, M., Decision dynamics in a traffic scenario, Physica A 287, (2000), 669-681

Transport Out of a Gravitationally Stable Layer with the Help of a Faster Diffusing Substance: PDE Simulations and Scaling Laws

Karsten Koetter, Malte Schmick, and Mario Markus

Max-Planck-Institut fuer molekulare Physiologie, Postfach 500247, 44202 Dortmund, Germany

Abstract. Hydrodynamic equations of an incompressible fluid with two dissolved substances are integrated to simulate previously reported observations of double diffusive convection. These observations had been performed in a setup that is extremely easy to implement and to observe: a layer consisting of a mixture of a surfactant, glycerine and water is placed below water; fingers arising at the interface transport the surfactant upwards. The calculated mean distance between fingers, as well as their emergence time, both as functions of the initial concentration of glycerine, are described by power laws. These results are in agreement with analytical approximations and with the experimental observations.

1 Introduction

In the present contribution we examine double-diffusive convection [1–4] at the horizontal interface between water (upper layer) and a mixture (lower layer) of a surfactant (S), glycerine (T) and water. (In a previous work [5], we have reported on experiments with this system). S is the alcohol ethoxylate CH_3-$(CH_2)_3$-O-$[CH_2$-CH_2-$O]_{12}$-H. The densities are $\rho_S = 0.95$ g/cm^3 and $\rho_T = 1.26$g/cm^3. A lower layer containing only S would rise, while it would stay at the bottom if it contained T alone. We choose the mixtures in such a way that they are gravitationally stable in the lower layer. However, an infinitesimal, finger-like protrusion emerging upwards from the lower layer is rapidly depleted of T and retains S owing to the difference in diffusion coefficients ($D_S \approx 3 \cdot 10^{-12}$ m^2/s, $D_T \approx 10^{-10}$ m^2/s). Therefore, the protrusion grows and leads to an upwards transport of S.

An analogous process is considered to be the most important mechanism for transport of salt in oceans (see [1]). In fact, in low- and midlatitude regions, upper layers of water are heated by the sun and become richer in salt due to evaporation. Since the diffusion of salt is slower than that of heat, salt fingers grow downwards in a configuration that is upside down if compared to the one described in the preceeding paragraph [6,7].

An example of ecological importance is that of sewage that is injected by a pipe on the bottom of the sea and is assumed to be safely deposited because it is heavier than water. However, being colder and (eventually) less salty than the layers above, salt fingers grow downwards and push dangerous pollutants towards the surface of the sea [2,8]. Other examples of phenomena involving double-diffusive convection are: unwanted finger-like distributions of components in alloy manufacture out of the melt [3,9], layering of magma [2,10,11], "helium-fingers" and "magnetic-moment-fingers" in stars [3,12–14], and polymer transport in biological tissues [15,16].

We have shown experimentally [5] that the setup involving the surfactant-glycerine-water mixture mentioned above has several advantages: i) if there is sufficient space for the lower layer to expand due to gravity, a bulk flow induced in the water straightens and confines the fingers quasi-two-dimensionally, so as to allow comfortable visualization; ii) the extremely low diffusion coefficient D_S causes long-lasting optical definition of the borders of the observed fingers; iii) the micelles formed by the surfactant are so large (length $\sim$ 4000 Å) that they strongly scatter or absorb visible light and are thus clearly detectable by the bare eye or by a video camera; thus, no dyes (see e.g. [2,17,18]) or special optical eqipment, e.g. to measure the refractive index [9,19–21], are needed; and iv) the medium is not distorted by invasive techniques, such as withdrawing samples or inserting conductivity probes or thermocouples [9,17,18,20,21]. On the other hand, a disadvantage of our setup is that the layer of the mixture is so thin that only features involving the horizontal direction are observable (such as the distance between fingers) but not those in the vertical direction (such as transport velocity).

In this contribution we will report on PDE simulations and analytical estimates of the finger emergence time T_{em} and the mean distance between fingers (λ) versus initial concentrations. The results will be compared with measurements.

2 Analytical Estimates

A perturbation ξ grows as

$$\xi \sim \exp\left(\int_0^t \Lambda(t')\, dt'\right), \tag{1}$$

where

$$\Lambda \sim \bar{T}_z^{\frac{1}{2}} f(R_\rho), \tag{2}$$

see [4]. $R_\rho = [(\rho_T - \rho_0)\bar{T}_z]/[(\rho_S - \rho_0)\bar{S}_z]$ (ρ_0: density of water). $\bar{T}_z$ and $\bar{S}_z$ are the vertically averaged gradients $\partial T/\partial z$ and $\partial S/\partial z$, respectively. If we assume times small enough to approximate $T \approx T_0$, $S \approx S_0$ at $z = 0$ (T_0, S_0 : initial concentrations in the lower layer), then $\bar{T}_z = T_0/h(t)$ and $\bar{S}_z = S_0/h(t)$ ($h(t) \sim t^{1/2}\, T_0$ is the height of the layer, growing vertically by diffusion). Thus, $\Lambda \sim T_0\, t^{-1/2}\, f(R_\rho)$ and $R_\rho \sim R'_\rho = T_0/S_0$. Each set of measurements was performed holding $R'_\rho =$ const. Therefore

$$\int_0^t \Lambda(t')\, dt' \sim \int_0^t \left(T_0 t'^{-\frac{1}{2}}\right)^{\frac{1}{2}} dt' \sim T_0^{\frac{1}{2}} t^{\frac{3}{4}}. \tag{3}$$

The measurements were performed at a time $t = T_{em}$, such that ξ has an approximately fixed value, namely that at which the minima of the absorbed light are within a given percentage p of the intensity measured with a void dish. Therefore, Eqs. (1) and (2) yield $T_0^{\frac{1}{2}} t^{\frac{3}{4}} = \text{const}$, or

$$T_{em} \sim T_0^{-\gamma_T} \tag{4}$$

with $\gamma_T = 2/3$.

For the wavelength λ it has been shown [22] that $\lambda \sim \bar{T}_z^{-\frac{1}{4}}$. Thus $\lambda \sim (T_0\, t^{-\frac{1}{2}})^{-\frac{1}{4}}$. Setting $t = T_{em}$ and considering (4) we obtain

$$\lambda \sim T_0^{-\gamma_\lambda} \tag{5}$$

with $\gamma_\lambda = 1/3$.

3 Experiments

Measurements were performed as described in Ref. 5. We now varied the initial concentration T_0 of glycerin in the mixture, while holding the ratio $R'_\rho = T_0/S_0$ constant, so as to fulfill the conditions required by the estimates in Sec 2. The mixture was stirred for 1-2 minutes and then 50 μl were injected with a pipette over a time of 20 s at a constant rate at the center of the bottom of a cylindrical dish (diameter: 2 cm) containing water ($\approx$3 mm high). For this injection we used a Precidor HT-pump (Infors, Basel). The mixture in the dish was illuminated with white light from above, while the formation of fingers was monitored (by determining the light absorbtion of S) from below by a light microscope connected to an image acquisition equipment. λ was determined by two-dimensional Fourier transforms of the images, averaging (at fixed T_0 and R'_ρ) over 50 independent measurements. We determined λ at the time $t = T_{em}$ ("emergence time") when p=5%.

Fig. 1 shows T_{em} and λ vs. T_0 for different values of R'_ρ. For R'_ρ=1, we obtain $\gamma_\lambda = 0.34 \pm 0.02$ and $\gamma_T = 0.42 \pm 0.03$. The fact that the plots in Fig. 1 are approximately parallel reveals that these values of γ_λ and γ_T are nearly independent of R'_ρ. Substituting the surfactant-glycerine-water mixture by a non-dairy coffee-whitener, which also contains a surfactant, as well as a component that takes up the role of glycerine (see [5]), we obtain Fig. 2. This plot yields $\gamma_\lambda = 0.38 \pm 0.02$ and $\gamma_T = 0.51 \pm 0.03$. Unfortunately, the whitener has the disadvantage that we do not know quantitatively its chemical composition.

4 PDE Simulations

We solved the equations for an incompressible fluid with two dissolved substances S and T:

$$\partial_t \mathbf{v} + (\mathbf{v}\,\nabla)\mathbf{v} = \frac{\rho}{\rho_0}\mathbf{g} - \frac{1}{\rho_0}\nabla p + \nu \Delta \mathbf{v} \tag{6}$$

$$\nabla \mathbf{v} = 0 \tag{7}$$

$$\partial_t S + (\mathbf{v}\,\nabla)S = D_S \Delta S \tag{8}$$

$$\partial_t T + (\mathbf{v}\,\nabla)T = D_T \Delta T \tag{9}$$

$$\rho = \rho_0 + (\rho_T - \rho_0)T + (\rho_S - \rho_0)S \tag{10}$$

For initialization of the instability, we set the velocity $\mathbf{v}$ at t=0 equal to noise vectors with randomly and homogeneously distributed directions and magnitudes (the latter between 0 and 10^{-2} mm/s). We set the viscosity $\nu = 1.1 \cdot 10^{-6}$ m^2/s as a constant throughout this work, following our capillary measurements. We assumed cyclic boundary conditions in the horizontal directions; at the bottom and the top, we set no-slip boundary conditions for $\mathbf{v}$ and no-flux boundary conditions for S and T. We used a finite volume method on a colocated grid with an implicit three time level scheme, central difference scheme and the SIMPLE algorithm for pressure correction [23]. We set the time step to Δt=0.1 s. When considering the injection process, we set the grid size to $\Delta x = 4 \cdot 10^{-2}$ mm and the domain size to $10 \times 10 \times 3$ mm^3 ; else, we set $\Delta x = 10^{-2}$ and the domain to $1 \times 1 \times 1$ mm^3.

We assumed injection to occur out of a cylinder (vertical axis; diameter: 0.8 mm ; base centered and 0.5 mm above the bottom). For $t < 20$ s, we set all variables equal to zero in a thin region around the cylinder (cylinder wall; thickness: 0.16 mm). Inside the cylinder for $t < 20$ s, we set S_0 =const, T_0 =const and $\mathbf{v}$=(0,0,-1.6 mm/s). At t=20 s, we restarted calculations, setting $S = T$ =0 in the cylinder.

Fig. 3 shows results, not by considering injection, but by starting with a vertically constant distribution of T and S in the lower layer. For R'_ρ=1, we obtain $\gamma_\lambda = 0.3 \pm 0.01$ and $\gamma_T = 0.67 \pm 0.02$. As in Fig. 1, the fact that the plots in Fig. 3 are approximately parallel reveals that these values of γ_λ and γ_T are nearly independent of R'_ρ. Simulations considering injection yield $\gamma_\lambda = 0.33 \pm 0.02$ and $\gamma_T = 0.43 \pm 0.02$, also independently of R'_ρ.

We also integrated the PDEs starting with T ans S homogeneously distributed within with a cone (height: 1 mm, diameter: 4.8 mm) that expands due to gravity in a larger domain. In this way, we simulated the experiments [5] in which a drop is injected at the center of a dish. An example of such a simulation is shown in Fig. 4a, as compared with a corresponding measurement (Fig. 4b). The gravitational expansion of the mixture induces a torodial bulk flow of the surrounding water. This flow confines the mixture near the bottom of the medium, as explained in [5]. Thus, views from above (as in Fig. 4) permit to clearly visualize single fingers.

Note that in Fig. 4 only subsets (containing the centers of the total domains) of the simulated and the observed areas are shown. This is sufficient because - disregarding small fluctuations - the structures are rotationally symmetric (see [5]). The distance λ between fingers is visible here around the centers of these pictures.

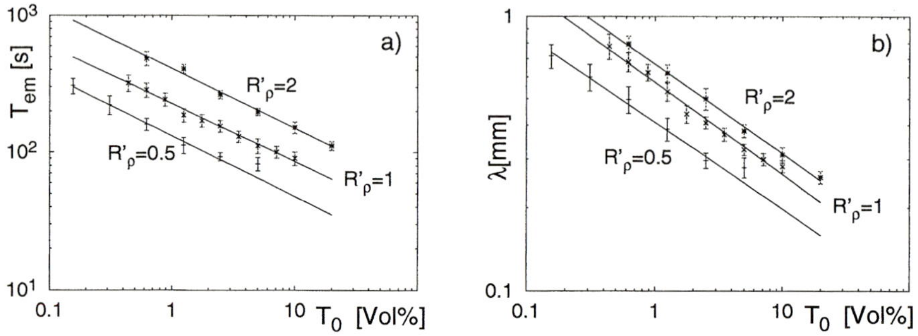

Fig. 1. Finger emergence time T_{em} and mean distance between fingers λ vs. the initial glycerine concentration. $R'_\rho = T_0/S_0$.

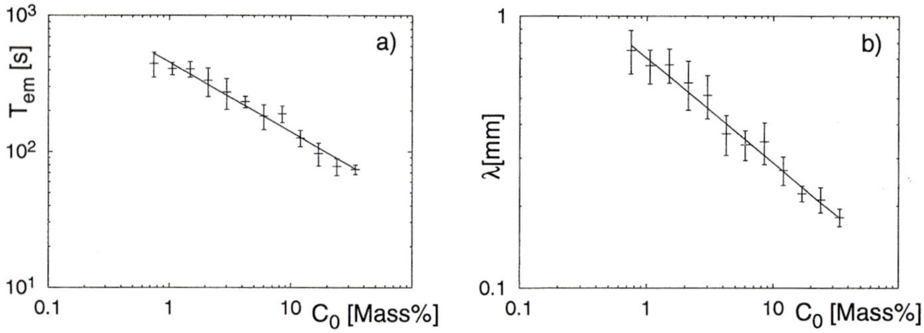

Fig. 2. T_{em} and λ vs. the initial concentration of the whitener.

5 Conclusion

Varying T_0 within factors of 20-30 in experiments (Fig. 1 and Fig. 2) and simulations (Fig. 3) yields power laws comparable to analytical estimations (Eqs. (4) and (5)). Moreover, the exponents are independent of R'_ρ, as predicted analytically and as indicated by the parallel displacements of the plots

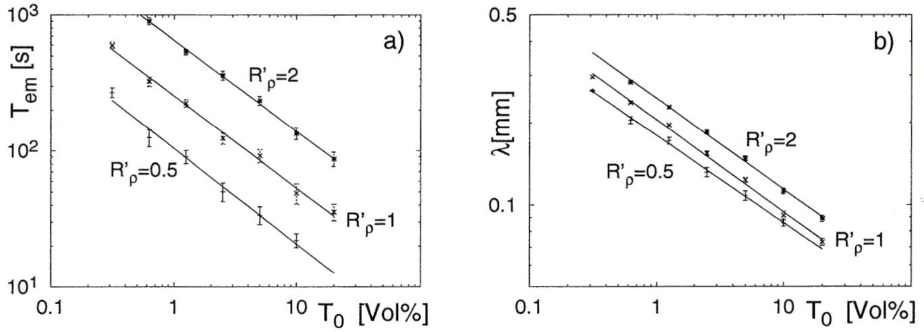

Fig. 3. As Fig. 1, but from PDE simulations. (Initialization was done by a step profile of the concentrations).

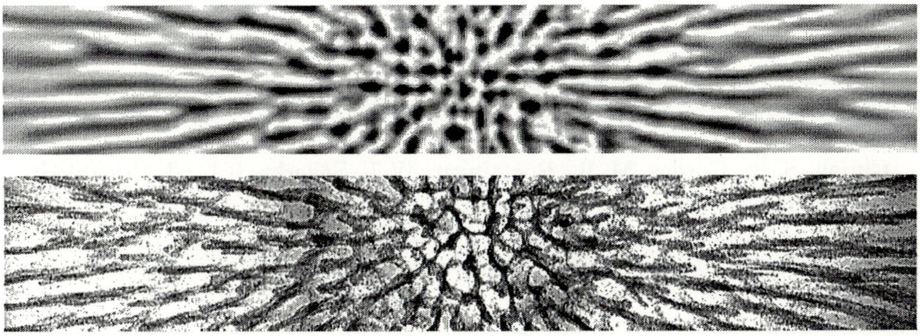

a)

b)

Fig. 4. a) PDE simulations of surfactant fingers straightend and confined quasi-two-dimensionally by the flow of water that is induced by the radial expansion of the lower layer (view from above). For comparison with the experimental observations, the darkest grey level corresponds to the maximum of S. b) Experimental observation corresponding to the simulation shown above. a,b) Simulation and experiments were performed in larger domains; here only slices of these domains passing through the center (12×2 mm^2 for simulations 6×1 mm^2 for experiments) are shown.

in Figs. 1 and 3. For the exponent γ_λ (describing the mean distance between fingers), the agreement between experiments, PDE simulations and the analytical estimate of 1/3 is excellent. For the exponent γ_T (corresponding to the time of emergence of fingers) the analytical estimate of 2/3 agrees well with the result obtained from PDE integration initialized with a step profile. A lower value is obtained in experiments initialized by injection ($\gamma_T \approx 0.42$), and this value agrees with that resulting from PDE simulations with injection (≈ 0.43).

All in all, we obtain consistency between experiments, PDE simulations and analytical estimates. The fact that rough analytical approximations agree with PDE simulations, with well controlled experiments (Fig. 1) and with experiments using quantitavely unknown mixtures (Fig. 2) points to a general validity of the power laws reported here.

Acknowledgement

We thank the Deutsche Forschungsgemeinschaft (Grant Ma 629/5) for financial support.

References

1. Chen, C. F., Johnson, D. H.: J. Fluid Mech. **138** (1984) 405–416
2. Huppert, H. E., Turner, J. S.: J. Fluid Mech. **106** (1981) 299–329
3. Turner, J. S.: Ann. Rev. Fluid Mech. **6** (1974) 37–56
4. Schmitt, R. W.: Phys. Fluids **26** (1983) 2373–2377
5. Koetter, K., Markus, M.: Europhys. Lett. **55** (2001) 807-813
6. Williams, A. J.: Science **185** (1974) 941–943
7. Tait, R. I., Howe, M. R.: Nature **231** (1971) 178–179
8. Fisher, H.: Water Res. **5** (1971) 909–915
9. Coriell, S. R., Cordes, M. R., Boettinger, W.J., Sekerka, R.F.: J. Crystal Growth **49** (1980) 13–28
10. McBirney, A. R., Noyes, R. M.: J. Pet. **20** (1979) 487–554
11. McBirney, A. R.: J. Volcanol. Geotherm. Res. **7** (1980) 357–371
12. Schmitt, J., Rossner, R.: Astrophys. J. **265** (1983) 901–924
13. Goldreich, P., Schubert, G.: Astrophys. J. **150** (1967) 571–587
14. Spiegel, E. A.: Ann. Rev. Astron. Astrophys. **10** (1972) 261–304
15. Laurent, T. C., Preston, B. N., Sundelof, L.O.: Nature **279** (1979) 60–62
16. Harper, G. S., Comper, W. D., Preston, B. N.: J. Biol. Chem. **259** (1984) 582-589
17. Linden, P. F.: Deep-Sea Res. **20** (1973) 325-340
18. Turner, J. S.: Deep-Sea Res. **14** (1973) 599-611
19. Huppert, H. E., Manins, P. C.: Deep-Sea Res. **20** (1973) 315-323
20. Shirtcliffe, T. G. L.: J. Fluid. Mech. **57** (1973) 27-43
21. Taylor, J. R., Veronis, G.: Fluid Mech. **321** (1996) 315-333
22. Stern, M. E.: Tellus **12** (1960) 172–175
23. Ferziger, J. H., Peric, M.: *Computational methods for fluid dynamics* (Springer, New York, 1997)

Microscopic Parameters and Macroscopic Features of Traffic Flow

Peter Berg[1] and Eddie Wilson[2]

Department of Mathematics, Simon Fraser University, 8888 University Drive, Burnaby V5A 1S6, Canada, www.peterberg.net

Abstract. A major task of traffic modelling is to investigate how traffic parameters on the microscopic scale, such as reaction time and inertia, influence the macroscopic flow phenomena. In this article, we present a transformation that relates microscopic car-following models to their macroscopic continuum counterpart. For a specific type, the optimal-velocity model, it turns out that the related model is analogous in the sense that it fulfills the same linear stability criterion. Moreover, it predicts similar flow patterns for similar traffic situations as they occur at on-ramps or between platoons of vehicles of different fluxes. However, the analysis suggests that reaction time and multi-species flows may have a profound effect on the flow pattern. To date, this has rarely been taken into account in continuum models. This paper points out some of the problems and leaves some open questions regarding macroscopic traffic modelling.

1 Microscopic Parameters of Traffic Flow

From personal experience we know that car drivers react mainly to traffic events ahead. On fairly congested motorways, drivers are forced into a certain reaction pattern rather than making their own decisions. Yet on a macroscopic scale, the coupling of many individual vehicles can lead to certain traffic states (*phases*) like *stop-and-go* or *synchronised* flow [1]. It is, therefore, a very interesting task to investigate how microscopic reaction algorithms and parameters influence the overall macroscopic flow situation.

Figure 1 shows a general sketch of a one-lane highway. In what follows, vehicles will carry integer numbers n, which indicate their position in the convoy. Generally speaking, the system is characterised by six parameters, some of which vary from vehicle to vehicle:

1. **Reaction time**: T_n
2. **Inertia** of cars: $I_n = 1/a_n$
3. Distance to the cars in front also referred to as **headway**: h_n, (h_n+h_{n-1}), $(h_n + h_{n-1} + h_{n-2})$, ...
4. Own **current speed**, v_n, and those of others, v_{n-1}, v_{n-2}, ...
5. **Top speed** as $h_n \to \infty$: $V_n^{\max}$
6. **Minimum distance** to the car in front in a traffic jam: $h_n^{\min}$

It is at the heart of traffic flow modelling that one finds reasonable models based on these parameters to reproduce real features of the system. One can approach this problem from both the macroscopic and microscopic side. We will start with the latter and try to link it to the former.

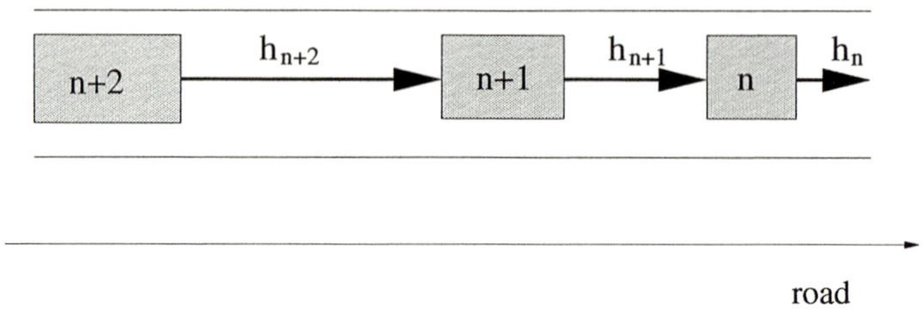

Fig. 1. An idealised sketch of road traffic. Vehicles are separated by *headway* h_n and carry speed v_n.

2 Stochastic and Deterministic Microscopic Description

At first it seems very questionable whether a complex system containing so many variables and different parameters, like car traffic, can be described by simple flow models. Often the simplifications lie in the following assumption: identical vehicles (*single-species traffic*), one lane only and the neglect of reaction time, varying road conditions as well as ramps.

Another fundamental question that arises from the modelling aspect is whether traffic can be treated as a deterministic system, despite variations in driver behaviour temporally and amongst other drivers. We will come back to this issue in more detail in Chap. 6. However, deterministic models match some of the traffic features quite well [2] and that is why we will focus on them in this publication. For reasons of completeness though, a popular representative of the class of stochastic models shall be briefly presented before we proceed.

2.1 A Cellular Automata Model

A radical approach representing the road and the vehicles, would be to divide it into cells whose lengths equal roughly the stretch that a vehicle occupies in a traffic jam. The assignment of either of these two numbers to the cells indicates whether it is occupied ('1') by a car or empty ('0'), as shown in Fig. 2. Moreover, each occupied cell holds an integer number between zero and $v^{\max}$ that stands for the current speed of that car.

Nagel and Schreckenberg [3] proposed such a cellular automata (CA) model based on a small number of update rules, but allows for stochasticity via some sort of "dawdling". The updates transform the state of the system into a new state at a time interval, ΔT, later via cell hopping of the cars.

This ansatz is very interesting for its discretisation of both space and time, but will not be pursued here any further. It might also be mentioned that some refer to CA models as car-following models, since the update rules also include the distance between the vehicles.

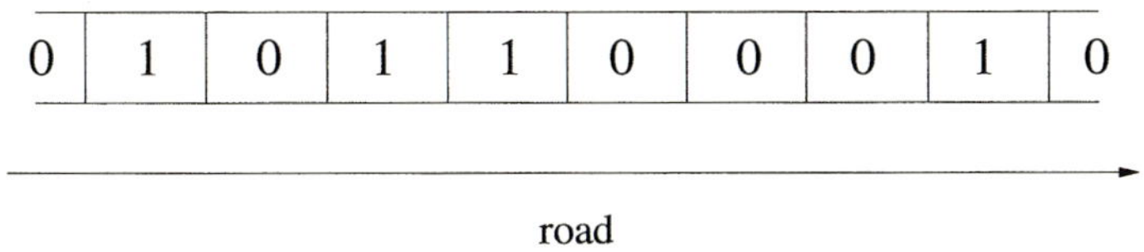

Fig. 2. In a cellular automata model, the road is cut into blocks which may ('1') or may not ('0') be occupied by the vehicles.

2.2 Car-following Models

Car-following models are generally based on ordinary differential delay equations (ODDE) of the form

$$\dot{v}_n(t) = F(v_n(t - T_1), h_n(t - T_2), h_{n-1}(t - T_3)) \ . \tag{1}$$

It means that the reaction of a driver depends somehow on his speed, and his distance to the cars in front at some point in the past due to the reaction time. This discrete ansatz used to be considered as a different description than the macroscopic continuum models.

So far, many car-following models have been proposed. Some include reaction time and some neglect this effect which leads to a set of coupled ordinary differential equations (ODE). Some are based on macroscopic traffic data and some on speed-headway measurements [4].

However, linear stability analyses and numerical simulations reveal how close each suggestion comes to reality. Currently, this leaves the scientific community with the optimal-velocity model, also referred to as the Bando model, which we will discuss next.

3 Optimal-velocity Car-following Models

In this paper, we concentrate on the most promising car-following model to date with respect to the right instability criterion, and stop-and-go simulations, the optimal-velocity (OV) model [5]

$$\dot{v} = a\,[V_{\mathrm{B}}(h_n) - v_n] \ . \tag{2}$$

It clearly neglects delay which might be a major criticism. Cars tend to adjust to an optimal speed

$$V_{\mathrm{B}}(h_n) = \tanh(h_n - 2) + \tanh(2), \tag{3}$$

depending on the current headway h_n as shown in Fig. 3. Here, we choose the dimensionless form based on driver data [5].

The relaxation term resembles those of earlier proposed continuum models like Payne's [6], Kühne's [7] and Kerner-Konhäuser's [1]. This makes us wonder whether they are somehow related, especially since the numerical simulations show similar features to the latter model [2].

Figure 4 represents the *fundamental diagram*, the plot of the flow q versus the density ρ, where $\rho = 1/h$ to first order. Two regions can be clearly distinguished: the linear stable regime where the flow is given by $q = \rho V(\rho)$, and the linear unstable regime in which small perturbations of a stationary, homogeneous and uniform flow lead eventually to congested stop-and-go traffic.

Bando et al. found that the system becomes linear unstable when

$$\frac{2V_{\mathrm{B}}'(h_0)}{a} > 1 \tag{4}$$

where h_0 is the initial headway of the platoon [5]. This defines the boundary of the congested and the homogeneous flow in Fig. 4. Depending on the OV function chosen, the model is generally stable in the free flow regime and unstable in the denser region. Moreover, it shows meta-stabilities [4] and other traffic features in almost full accordance with the Kerner-Konhäuser model [2]. This further pushes the question of whether there is a link between these models, or whether this has another explanation.

4 Derivation of Macroscopic Counterparts

In this section, we give a brief summary of how to relate car-following models to continuum analogues. For a more detailed discussion, see Berg et al. [9], Lee et al. [10] and Helbing [11]. It is striking that the headway h is the variable representing distances in the car-following model, whereas it is the density ρ in the continuum models. Therefore, we seek for a relation between these two variables. One way to do so is to change formally to continuum variables

$$v_n(t) \to v(x,t), \tag{5}$$
$$h_n(t) \to h(x,t), \tag{6}$$

and require

$$\int_0^{h(x,t)} \rho(x+y,t)\,dy \equiv 1 \ . \tag{7}$$

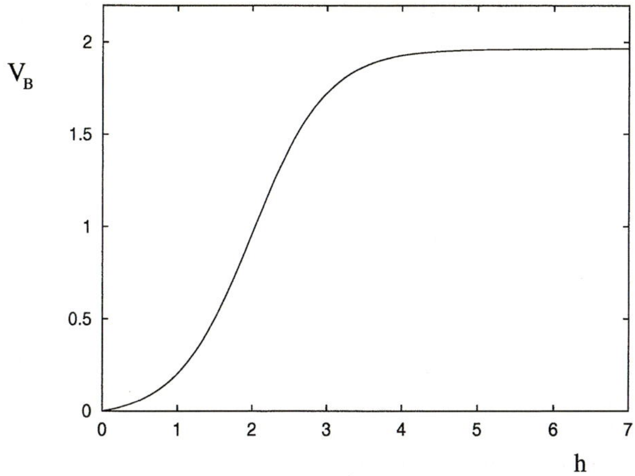

Fig. 3. The optimal-velocity function of the Bando model. Assumption of point-like vehicles leads to a minimum headway $h^{\min} = 0$.

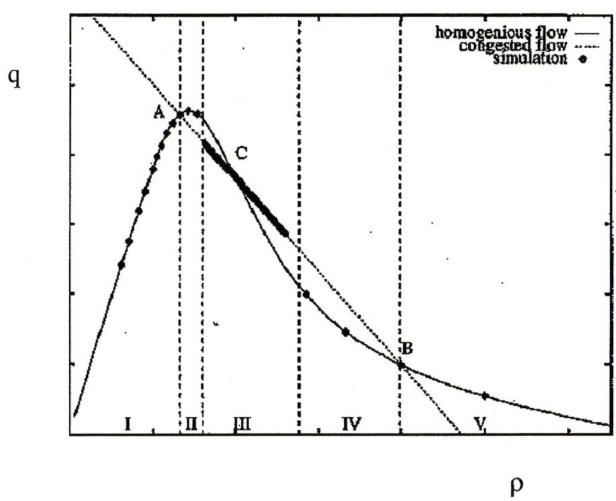

Fig. 4. A qualitative sketch of the fundamental diagram of the OV model (2) contains regions of linear stable and linear unstable flow [8]

Here, ρ defines h uniquely, but not vice versa. However, this is what we are looking for.

A Taylor expansion of the integral yields

$$h\rho + \frac{1}{2!}h^2\rho_x + \frac{1}{3!}h^3\rho_{xx} + ... = 1 \ . \tag{8}$$

Solving up to second order gives

$$h \simeq \frac{1}{\rho} - \frac{\rho_x}{2\rho^3} - \frac{\rho_{xx}}{6\rho^4} + \frac{\rho_x^2}{2\rho^5} \ . \tag{9}$$

Here, the series is assumed to converge sufficiently fast and hence, that higher order derivatives play a minor role.

We assume that the expansion (9) can now be used to transform any car-following model into a continuum version. However, we will focus on the OV model for a start, and comment on general transformations later on. The substitution of (9) into (2) and a Taylor expansion of $V_\mathrm{B}(h(x,t))$ delivers

$$v_t + vv_x \simeq a\left[\bar{V}(\rho) - v\right] + a\bar{V}'(\rho)\left[\frac{\rho_x}{2\rho} + \frac{\rho_{xx}}{6\rho^2} - \frac{\rho_x^2}{2\rho^3}\right] \ . \tag{10}$$

Here we set

$$\bar{V}(\rho) = V_\mathrm{B}(1/\rho) \ . \tag{11}$$

Moreover, the system is closed by the conservation of cars

$$\rho_t + (v\rho)_x = 0, \tag{12}$$

which holds via

$$\frac{d}{dt}\int_0^{h(x,t)} \rho(x+y,t)\,dy = 0 \ . \tag{13}$$

In order to speak of an analogous model, we have to establish that it shows similar dynamical behaviour in its numerical simulations (next section), and that it fulfills the same stability criterion. In fact, the latter can be proven explicitly [9], and (4) reads now ($\bar{V}' < 0$)

$$-\frac{2\rho_0^2\bar{V}'(\rho_0)}{a} > 1 \ . \tag{14}$$

Strictly speaking, the stability criterion is only conserved for long wavelengths, but the accuracy increases with the number of terms in the approximation of h in (9).

5 Similarity of Flow Patterns

One suitable comparison of the dynamical behaviour of discrete car-following and continuum models, are travelling waves. In the car-following model, this is achieved by running a simulation for a sufficiently long time until the structure changes only minisculely. In contrast, the real advantage of continuum models is shown by their reduction to a single ODE [9].

The boundary conditions of both simulations are linked by $h(x \to -\infty) = h_- = 1/\rho_- = 1/\rho(x \to -\infty)$, and similarly $h(x \to \infty) = h_+ = 1/\rho_+ = 1/\rho(x \to \infty)$, where the speed equals the optimal velocity.

Both numerical results are then drawn in the same diagram. One example is Fig. 5, which reveals how the wave structure of the car-following model approaches the travelling wave solution of the continuum model. Note that the solutions of the former have been shifted along the x-axis for reasons of comparison. After $t = 1000$, the points lie almost perfectly on the curve predicted by the continuum model.

When we increase the change in headway across the wave structure we find, eventually, oscillatory behaviour [12] as illustrated in Fig. 6. We now observe slight deviations between the two solutions. This comes at no surprise since larger gradients are now involved and would require higher order terms in the asymptotic series (9) in order to be matched more accurately. Another example arises from the simulation of an on-ramp in the discrete OV model (2) [13]. Here, an initial homogeneous and stationary flow of uniform headway h_0 is perturbed by a constant ramp flux that sets in at $t = 0$.

Figure 7 shows a ramp flux, $q_{\mathrm{ramp}} = 0.01$ vehicles per unit time, that is placed onto a lane of maximum flux at $h_0 = 2.78$. The structure that develops reminds us of *cnoidal* water waves, a solitary wave train.

An analysis in the corresponding continuum picture reveals that this profile can be explained to first order by a sech^2-solution of a Kortweg-de Vries (KdV) equation via the neglect of dissipation [13]. In Fig. 8, the comparison between this approximation, a sech^2-like fit and the numerical data reveals two things: firstly, the analytical solution of the KdV equation does not quite reach the right height of the density hump and secondly, the numerical data of the car-following model is not exactly symmetric unlike the sech^2-shape. Both effects can be explained by the steep gradients and the dissipation involved, which can only be matched by further derivatives of ρ in the analysis. Strictly speaking, the comparison of the two model types (2) and (10) should be extended to non-stationary solutions. Then the dependence of the linear stability criterion in the continuum model on the wavelength might have a significant impact and might show the limits of this transformation.

6 More Realistic Models

We now turn to the main part of this publication. We have seen that car-following models can be related to a continuum counterpart via the neglect of certain aspects: reaction time, multi-lane flow, different driver and vehicle types, as well as the influence of vehicles further ahead.

In this section, we address these issues to reveal where the problems of a more realistic description of traffic flow lie.

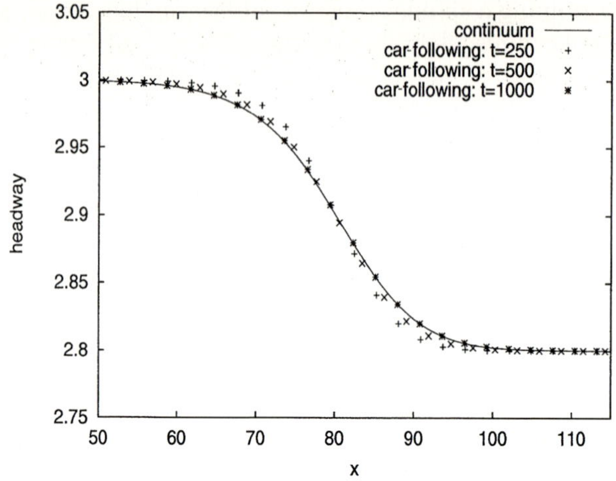

Fig. 5. A comparison of the asymptotic behaviour of the car-following OV model (2) and the travelling wave solution of the continuum counterpart (10). The car-following model evolves from a Heaviside-like jump in headway at $t = 0$ with boundary conditions as given in the text [4].

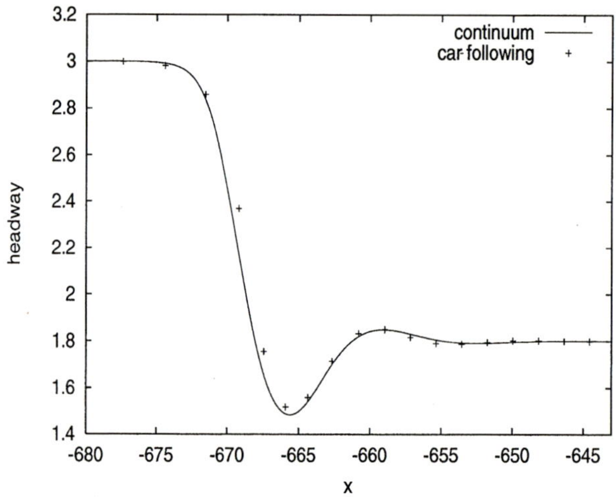

Fig. 6. For certain changes in headway, the wave profile assumes oscillatory behaviour. This is still matched by the continuum model (10) although with less accuracy.

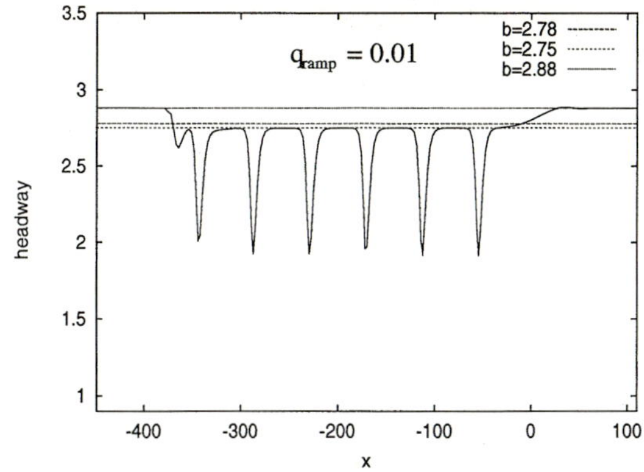

Fig. 7. An on-ramp simulation of the OV model (2) shows solitary wave structures in certain parameter regimes.

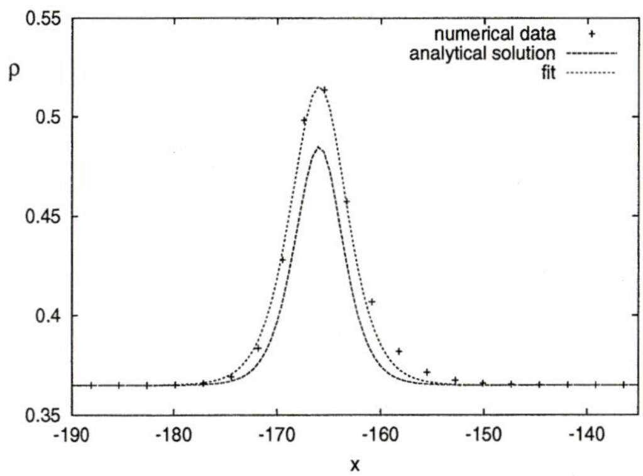

Fig. 8. The wave solution of the discrete car-following model can be described approximately by a KdV equation derived from the continuum counterpart.

6.1 Multi-species Traffic; Temporal and Spatial Variations of Driver Behaviour

A first objection to simplified car-following models like the OV model (2) might be that it does not incorporate the diversity of driver types on the road. A straightforward way to take this into account is to allow the parameter a and the OV function V_B to depend on the vehicle number

$$a \to a(n), \quad (15)$$

$$V_B \to V_B(n) \ . \quad (16)$$

A macroscopic (continuum) counterpart for **multi-species flow** would now read [4]

$$v_t + vv_x = a\left[\bar{V}(\rho) - v\right] + a\bar{V}'(\rho)\left[\frac{\rho_x}{2\rho} + \frac{\rho_{xx}}{6\rho^2} - \frac{\rho_x^2}{2\rho^3}\right], \tag{17}$$

$$\rho_t + (\rho v)_x = 0, \tag{18}$$

$$\frac{d}{dt}a(x,t) = a_t + va_x = 0, \tag{19}$$

$$\frac{d}{dt}\bar{V}(x,t) = \bar{V}_t + v\bar{V}_x = 0\ . \tag{20}$$

Equations (19) and (20) mean that the vehicle's and driver's characteristics do not change with time and flow with the vehicle. It is worth pointing out that the system (17)–(20) reminds one of Helbing's model [14], whose modelling approach is based on the introduction of higher order statistical variables, like the speed variance. In accordance, we also find a set of three coupled partial differential equations, if we assume that either $\bar{V}$ or a does not depend on time or space.

The initial conditions might be

$$\rho(x,0) = \rho_0(x)\ ,\ v(x,0) = v_0(x),$$
$$a(x,0) = a_0(x)\ ,\ \bar{V}(x,0) = \bar{V}_0(x)\ .$$

In the car-following model, we observed phase transitions caused by a small amount of vehicles whose characteristics differ from the main platoon [4]. Hence, it remains an interesting question as to whether the continuum counterpart above would show similar results. However, in real traffic flow we do not only find different driver and vehicle types. Generally, those **characteristics also change with time and space** due to road layout, speed limits and weather conditions, for example. It can be expected that these variations might cause phase transitions of the flow, also.

The modelling aspect reads similarly to multi-species flow. In addition, we have to add two functions, $A(x,t)$ and $B(x,t)$ say, to the right hand sides of (19) and (20). The crucial problem is to determine their form, which remains an unresolved issue when the vehicle characteristics change with time. It is a challenging task to link these functions to the car-following counterparts, where this effect is taken into account straightforwardly via the dependencies $a_n = a(n,t)$ and $V_{\mathrm{B},n} = V_{\mathrm{B}}(n,t)$.

In fact, it is much easier to prescribe spatial variations both for the car-following model

$$a \to a(x), \tag{21}$$

$$V_{\mathrm{B}} \to V_{\mathrm{B}}(x), \tag{22}$$

and equivalently for the macroscopic counterpart

$$a \to a(x), \tag{23}$$

$$\bar{V} \to \bar{V}(x)\ . \tag{24}$$

In this case, the kinematic equations (19) and (20) are superfluous, and the system consists again of only two partial differential equations (PDEs).

Nevertheless, it remains a challenging task to investigate what traffic simulations predict for temporal and spatial variations.

6.2 Reaction Time

An advantage of car-following models over continuum models clearly shows in the simulation of reaction time, which can easily be incorporated into car-following models via differential delay equations (DDE).

As an example, the OV model (2) could be modified to give an ordinary differential delay equation (ODDE) of the form

$$\dot{v}_n(t+T) = a\,[V_{\mathrm{B}}(h_n(t)) - v_n(t)] \quad . \tag{25}$$

Hooper [15] analysed the linear stability of this model, and observed oscillations and even backward moving vehicles in his numerical simulations, depending on the parameters chosen.

Generally, the reaction depends on the history of a past time interval, which leads to an integro-differential equation. However, we focus on a discrete delay here for reasons of simplicity.

The macroscopic counterpart now turns into a partial differential delay equation (PDDE)

$$\frac{d}{dt}v(x+\Delta x, t+T) = a\left[\bar{V}(\rho) - v\right] + a\bar{V}'(\rho)\left[\frac{\rho_x}{2\rho} + \frac{\rho_{xx}}{6\rho^2} - \frac{\rho_x^2}{2\rho^3}\right], \tag{26}$$

where $x_i + \Delta x_i$ at $t_i + T$ is obtained by integrating

$$\dot{x}(t) = v(x,t),\; x(t_i) = x_i \tag{27}$$

from $t_i \to t_i + T$. Suffice to say that the integration of this PDDE closed by the conservation of cars (12), is not only a hard task but also very time consuming in terms of the numerical requirements.

However, time delay might have a relevant impact on the flow and should be taken into account. Unfortunately, this is only practical on the microscopic level.

6.3 Multiple Look-ahead Models

A major problem of the dimensional OV model

$$\dot{v}_n(t) = a\,[V_{\mathrm{data}}(h_n(t)) - v_n(t)] \tag{28}$$

are the extreme values of acceleration and deceleration. Here, V_{data} is a function derived from real traffic data [5]. Hooper [15] suggested various ways to fix this problem: limiting the maximum acceleration and deceleration by cut-offs, introducing time delay and multiple look-ahead models, which are based on reactions towards numerous drivers ahead. Only the last approach appeared as being successful. Multiple look-ahead in the dimensionless OV model (2) can be most easily simulated in two different ways:

$$\text{Model A:} \quad \dot{v}_n = a \left[\sum_{j=0}^{k} \beta_j V_{\mathrm{B}}(b_{n-j}) - v_n \right], \tag{29}$$

$$\text{Model B:} \quad \dot{v}_n = a \left[\sum_{j=0}^{k} \beta_j V_{\mathrm{B}} \left(\frac{\sum_{l=0}^{j} b_{n-l}}{j+1} \right) - v_n \right], \tag{30}$$

with $\beta_j > 0$ and $\sum_{j=0}^{k} \beta_j = 1$. Both approaches are reduced to the standard OV model (2) in case of reactions to the vehicle ahead only. The unresolved issue here is how to obtain a macroscopic counterpart. Could downstream information be incorporated by higher order spatial derivatives of ρ and v or is a non-local model like Nagatani's [17]

$$(\rho v)_t = a \left[\rho_0 V_{\mathrm{N}}(\rho(x+1)) - \rho v \right], \tag{31}$$

with $V_{\mathrm{N}}(\rho) = V_{\mathrm{B}}(1/\rho)$, necessary? Here, non-locality refers to the spatial coordinate rather than time. Similarly to reaction time, we may deal with an important effect on the flow that cannot be simulated on the macroscopic level.

6.4 Multi-lane Flow

Like time delay and multiple look-ahead, the simulation of multi-lane flow poses another unresolved problem to macroscopic flow simulations. This is because the discrete nature of traffic flow becomes most apparent upon lane-changing, for two reasons. Firstly, headways and hence densities do not change continuously. Secondly, lane-changing is a process that depends on many conditions, e.g. speed and headway differences to cars in front and behind, on the same lane and on the other lane.

How could these discrete processes be incorporated into a macroscopic counterpart? So far, models have only taken density criteria into account, but not speed criteria.

An example is the asymmetric lane-coupling [18] described by the following coupling term on the right hand side of the equation for the conservation of cars

$$s(x,t) = \alpha \left[\rho_1 \frac{1-\rho_2}{c_2 + W\rho_2^4} - \rho_2 \frac{1-\rho_1}{c_1 + W\rho_1^4} \right] . \tag{32}$$

Macroscopic flow models will never be able to match the random and discrete process of lane-changing, regardless of the inclusion of speed in the coupling term [4]. Unfortunately, the interaction of lanes might also be of such impact on the flow that it should enter the modelling process, which can again only be done realisticly in microscopic models.

7 Synchronised Flow

In the unstable regime of real motorway traffic ($\rho > 50$ veh./km), there is no definite allocation of the density to the flow anymore. This leads to scattered data points in the fundamental diagram obtained from flow measurements. This area contains two phases: stop-and-go traffic and the *synchronised flow* regime [19]. In the latter case, the lanes are strongly coupled, which leads to a homogeneous movement of vehicles on all lanes, with an overall low velocity but high throughput. Overtaking is impossible, and lane-changing rare compared to free flow or traffic jams. It typically occurs at on-ramps where a localised perturbation causes a linearly stable free flow of high speed and flux to break down and move more slowly. Motorway data shows that synchronised flow consists of a great number of steady state flows along the road. Each state is characterised by its density and its flux. It has been claimed of various models that they explain synchronised flow. On the other hand, it is very challenging to distinguish between real synchronised flow and a similar numerical phenomena.

One fundamental question is, therefore, whether we need more complex models in which, for example, lane-coupling and ramps play a role. Perhaps we have to go even further and describe synchronised flow as a phenomena of a non-symmetrical[1] delay system consisting of many particles whose microscopic features are different from each other and vary randomly with time and space (e.g. road conditions and layout)? This is indeed how traffic flow could be characterised most generally.

Considering this highly complicated system, it is surprising how successful some simple flow models are.

8 Outlook

To date, it remains a great problem to work out whether traffic flow can be described by simple models, neglecting stochasticity, reaction time, and different vehicles and drivers on the road. This article shows and addresses how microscopic features could enter the macroscopic description. Here, reaction time and the simulation of multi-lane flow pose serious hurdles since the corresponding continuum models are neither practical nor exist as such. In particular, the much discussed and unresolved issue of synchronised flow might lead to a more comprehensive ansatz of modelling traffic flow [14].

Acknowledgement Peter Berg would like to thank Professor Woods, University of Cambridge, for his support throughout the author's PhD. Moreover, we thank Steven Hooper who delivered vital results in his MSc thesis.

[1] The term "non-symmetrical" refers to the violation of *actio=reactio* between the vehicles due to the reaction towards downstream events only.

References

1. Kerner, B.S., Konhäuser, P.: Structure and parameters of clusters in traffic flow. Phys. Rev. E **50** (1994) 54–83
2. Herrmann, M., Kerner, B.S.: Local cluster effect in different traffic flow models. Physica A **255** (1998) 163–188
3. Nagel, K., Schreckenberg, M.: A cellular automaton model for freeway traffic. J. Phys. I. France **2** (1992) 2221–2229
4. Berg, P.: Optimal-velocity models of motorway traffic. PhD Thesis, University of Bristol (2001)
5. Bando, M., Hasebe, K., Nakayama, A., Shibata, A., Sugiyama, Y.: Dynamical model of traffic congestion and numerical simulation. Phys. Rev. E **51** (1995) 1035–1042
6. Payne, H.J.: FREFLO: a macroscopic simulation model of freeway traffic. Transp. Res. Record **722** (1979) 68–77
7. Kühne, R.: Macroscopic freeway model for dense traffic - stop-start waves and incident detection. 9th Int. Symp. on Transp. and Traffic Theory, VNU Science Press (1984) 21–42
8. Sugiyama, Y.: Dynamical model for congestion of freeway traffic and its structural stability. In: Bachem, A., Schreckenberg, M., Wolf, D.E., Traffic and granular flow, World Scientific (1996) 137–150
9. Berg, P., Mason, A., Woods, A.W.: Continuum approach to car-following models. Phys. Rev. E **61** (2000) 1056–1066
10. Lee, H.K., Lee, H.W., Kim, D.: Macroscopic traffic models from microscopic car-following models. Phys. Rev. E **64** (2001) 056126
11. Helbing, D.: From microscopic to macroscopic traffic models. In: Parisi, J., Müller, S.C., Zimmermann, W., A perspective look at nonlinear media, Springer (1998) 122–139
12. Berg, P., Woods, A.W.: Travelling waves of an optimal-velocity model of freeway traffic. Phys. Rev. E **63** (2001) 036107
13. Berg, P., Woods, A.W.: On-ramp simulations and solitary waves of a car-following model. Phys. Rev. E **64** (2001) 035602
14. Helbing, D.: Traffic modelling by means of physical concepts. In: Bachem, A., Schreckenberg, M., Wolf, D.E., Traffic and granular flow, World Scientific (1996) 87–104
15. Hooper, S.: Modification of Bando's car-following model of highway traffic. MSc Thesis, University of Bristol (2000)
16. Bando, M., Hasebe, K., Nakanishi, K., Nakayama, A.: Analysis of optimal velocity model with explicit delay. Phys. Rev. E **58** (1998) 5429–5435
17. Nagatani, T.: Density waves in traffic flow. Phys. Rev. E **61** (2000) 3564–3570
18. Lee, H.Y., Kim, D., Choi, M.Y.: Continuum model for two-lane traffic flow. In: Schreckenberg, M., Wolf, D.E., Traffic and granular flow '97, Springer (1999) 433–438
19. Kerner, B.S., Rehborn, H.: Experimental properties of phase transitions in traffic flow. Phys. Rev. Lett. **79** (1997) 4030–4033

An Adaptive Smoothing Method for Traffic State Identification from Incomplete Information

Martin Treiber and Dirk Helbing

Institute for Economics and Traffic, Faculty of Traffic Sciences "Friedrich List", Dresden University of Technology, D-01062 Dresden, Germany

Abstract. We present a new method to obtain spatio-temporal information from aggregated data of stationary traffic detectors, the *"adaptive smoothing method"*. In essential, a nonlinear spatio-temporal lowpass filter is applied to the input detector data. This filter exploits the fact that, in congested traffic, perturbations travel upstream at a constant speed, while in free traffic, information propagates downstream. As a result, one obtains velocity, flow, or other traffic variables as smooth functions of space and time. Applications include traffic-state visualization, reconstruction of traffic situations from incomplete information, fast identification of traffic breakdowns (e.g., in incident detection), and experimental verification of traffic models.

We apply the adaptive smoothing method to observed congestion patterns on several German freeways. It manages to make sense out of data where conventional visualization techniques fail. By ignoring up to 65% of the detectors and applying the method to the reduced data set, we show that the results are robust. The method works well if the distances between neighbouring detector cross sections do not exceed 3 km.

1 Introduction

During the last decades, traffic dynamics and pedestrian flows have been intensively studied. Regarding the observed phenomena and simulation approaches we refer the reader to some recent reviews [1–4]. Presently, scientists are more and more getting interested in detailed empirical studies. Reasons for this are the availability of better data and the need to verify and calibrate models. Both, theoretical [5–7] and empirical [8–17] studies indicate that the phenomenology of congested traffic is more complex than originally expected. There seems to be a rich spectrum of traffic states [5,6,12,13], hysteretic or continuous temporal or spatial transitions among them [7,11–13,18,19], and fluctuations or a erratically appearing dynamics play a significant role [13,16,20–22]. In order to make sense out of this, it is increasingly important to have suitable ways of data processing, in order to extract the information relevant for scientific investigations or specific applications. Here, we will propose a three-dimensional data-evaluation method allowing to visualize the spatio-temporal dynamics of traffic patterns along freeways.

The developed *"adaptive smoothing method"* filters out small-scale fluctuations and adaptively takes into account the main propagation direction of the information flow (i.e. the dominating characteristic line), which have been determined by means of a spatio-temporal correlation analysis in other studies [23]. The temporal length scale of the smoothing procedure can be as small as the sampling interval, while the spatial length scale is related to the distance between successive detectors, which can be up to 3 kilometers long.

By "filter" we just mean a transformation of the data with specific properties. Here, we use a spatio-temporal lowpass filter, i.e., only (Fourier) components of low frequency can pass the filter, while high-frequency contributions are considered as fluctuations and smoothed out. One particular feature of our filter is that it is nonlinear and adaptive to the traffic situation in distinguishing free and congested traffic, as the propagation direction of perturbations differs. The results are three-dimensional visualizations of traffic patterns, which are quite robust with respect to variations of the filter parameters and very helpful in obtaining a clear picture of the systematic spatio-temporal dynamics.

Therefore, this method is suitable for the reconstruction of traffic situations from incomplete information, fast identification of traffic breakdowns (incident detection), and experimental verification of traffic models. First results support the phase diagram of traffic states occuring at bottlenecks [17,24], which, apart from free traffic, predicts pinned or moving localized clusters, spatially extended patterns such as triggered stop-and-go waves, and oscillating or homogeneous congested traffic, or a spatial coexistence of some of these states [5,7,1]. The respectively occuring spatio-temporal pattern depends on the specific freeway flow and bottleneck strength, but also on the level of fluctuations, as these can trigger transitions from, for example, free traffic to localized cluster states [5,17,1]. We do not see sufficient support for one "generalized pattern" [25] that would always be observed when traffic flow breaks down.

2 Description of the Method

The adaptive smoothing method is a data processing method for obtaining traffic variables as smooth functions of space and time out of stationary traffic data. It has following heuristically motivated properties:

1. In case of free traffic, perturbations (of, e.g., velocity or flow) move essentially into the direction of traffic flow [26]. More specifically, they propagate with a characteristic velocity c_{free} at about 80% of the *desired* velocity V_0 on empty roads [23]. Therefore, at locations with free traffic, perturbations with propagation velocities near c_{free} should pass the filter.
2. In case of congested traffic, perturbations propagate against the direction of traffic flow with a characteristic and remarkably constant velocity $c_{\text{cong}} \approx -15$ km [12]. With modern data analysis techniques, it has been

shown that such propagation patterns persist even in "synchronized" congested traffic flow, where they are hardly visible in the time series due to a wide scattering of the data in this state [26]. So, for high traffic densities or low velocities, the filter should transmit spatio-temporal perturbations propagating with velocities near c_{cong} more or less unchanged.

3. The filter should smooth out all high-frequency fluctuations in t on a time scale smaller than τ and spatial fluctuations in x on a length scale smaller than σ. The parameters τ and σ of the smoothing method can be freely chosen in a wide range (cf. Tab. 1).

Let us assume that aggregated detector data $\mathbf{z}_{ij}^{\text{in}}$ are available from n cross sections i at positions x_i where $j \in \{j_{\min}, \cdots, j_{\max}\}$ denotes the index of the aggregation intervals.

Usually, the aggregation interval

$$\Delta t = t_j - t_{j-1} \tag{1}$$

is fixed (between 20 s and 5 min, depending on the measurement device). A typical value for German highways is $\Delta t = 1$ min.

The components of $\mathbf{z}_{ij}^{\text{in}}$ represent the desired aggregated quantities as obtained from cross section i during the time interval j. Typical examples include the average velocity V_{ij}, the vehicle flow Q_{ij}, the occupancy O_{ij}, or some derived quantity such as the traffic density ρ_{ij}. The input data $\mathbf{z}_{ij}^{\text{in}}$ can represent either averages over all lanes or quantities on a given lane.

The adaptive smoothing method provides estimates $\mathbf{z}(x,t)$ for *all* locations $x \in [x_1, x_n]$ between the positions of the first and the last detector, and for all times $t \in [t_{\min}, t_{\max}]$. Extrapolations (in the sense of a short-term traffic forecast) will also be discussed. Without loss of generality, we assume $x_1 < x_2 < \cdots < x_n$, and a traffic flow in positive x-direction.

The core of our *"adaptive smoothing method"* is a nonlinear filter transforming the discrete input detector data $\mathbf{z}_{ij}^{\text{in}}$ into the smooth spatio-temporal functions $\mathbf{z}(x,t)$. To satisfy the first two requirements mentioned above, we write the filter as

$$\mathbf{z}(x,t) = w(\mathbf{z}_{\text{cong}}, \mathbf{z}_{\text{free}})\mathbf{z}_{\text{cong}}(x,t) + \left[1 - w(\mathbf{z}_{\text{cong}}, \mathbf{z}_{\text{free}})\right]\mathbf{z}_{\text{free}}(x,t). \tag{2}$$

This is a superposition of two linear anisotropic lowpass filters $\mathbf{z}_{\text{cong}}(x,t)$ and $\mathbf{z}_{\text{free}}(x,t)$ with an adaptive weight factor $0 \le w \le 1$ which itself depends nonlinearly on the output of the linear filters as discussed later on.

The filter $\mathbf{z}_{\text{cong}}(x,t)$ for congested traffic is given by

$$\mathbf{z}_{\text{cong}}(x,t) = \frac{1}{N_{\text{cong}}(x,t)} \sum_{i=1}^{n} \sum_{j=j_{\min}}^{j_{\max}} \phi_{\text{cong}}(x_i - x, t_j - t)\mathbf{z}_{ij}^{\text{in}}, \tag{3}$$

with

$$t_j = t_{\min} + j\Delta t, \tag{4}$$

and the normalization factor

$$N_{\text{cong}}(x,t) = \sum_{i=1}^{n} \sum_{j=j_{\min}}^{j_{\max}} \phi_{\text{cong}}(x_i - x, t_j - t). \tag{5}$$

This normalization guarantees that a constant input vector $\mathbf{z}_{ij}^{\text{in}} = \mathbf{z}_0$ for all i, j is transformed into a constant output function with the same components: $\mathbf{z}(x,t) = \mathbf{z}_0$. Notice that the normalization depends on both, location x and time t.

The analogous expression for the lowpass filter $\mathbf{z}_{\text{free}}(x,t)$ for free traffic is

$$\mathbf{z}_{\text{free}}(x,t) = \frac{1}{N_{\text{free}}(x,t)} \sum_{i=1}^{n} \sum_{j=j_{\min}}^{j_{\max}} \phi_{\text{free}}(x_i - x, t_j - t)\mathbf{z}_{ij}^{\text{in}} \tag{6}$$

with

$$N_{\text{free}}(x,t) = \sum_{i=1}^{n} \sum_{j=j_{\min}}^{j_{\max}} \phi_{\text{free}}(x_i - x, t_j - t). \tag{7}$$

The kernels $\phi_{\text{cong}}(x,t)$ and $\phi_{\text{free}}(x,t)$ of the linear homogeneous filters do the required smoothing and are particularly transmissible for perturbations propagating with the typical velocities c_{cong} and c_{free} observed in congested and free traffic, respectively (cf. Fig. 1).

If the propagation velocity of the perturbations was zero, the filter should only perform the smoothing. We chose the (non-normalized) exponential function

$$\phi_0(x,t) = \exp\left(-\frac{|x|}{\sigma} - \frac{|t|}{\tau}\right). \tag{8}$$

Instead, one could apply other localized functions such as a two-dimensional Gaussian. However, it turned out that the exponential had more favourable properties in our application.

The linear filter for nonzero propagation velocities can be mapped to $\phi_0(x,t)$ by the coordinate transformations (cf. Fig. 1)

$$x = x', \quad t = t'_{\text{cong}} + \frac{x}{c_{\text{cong}}} = t'_{\text{free}} + \frac{x}{c_{\text{free}}}. \tag{9}$$

Thus, we obtain for the kernels of the linear anisotropic filters

$$\phi_{\text{cong}}(x,t) = \phi_0(x', t'_{\text{cong}}) = \phi_0\left(x, t - \frac{x}{c_{\text{cong}}}\right), \tag{10}$$

$$\phi_{\text{free}}(x,t) = \phi_0(x', t'_{\text{free}}) = \phi_0\left(x, t - \frac{x}{c_{\text{free}}}\right). \tag{11}$$

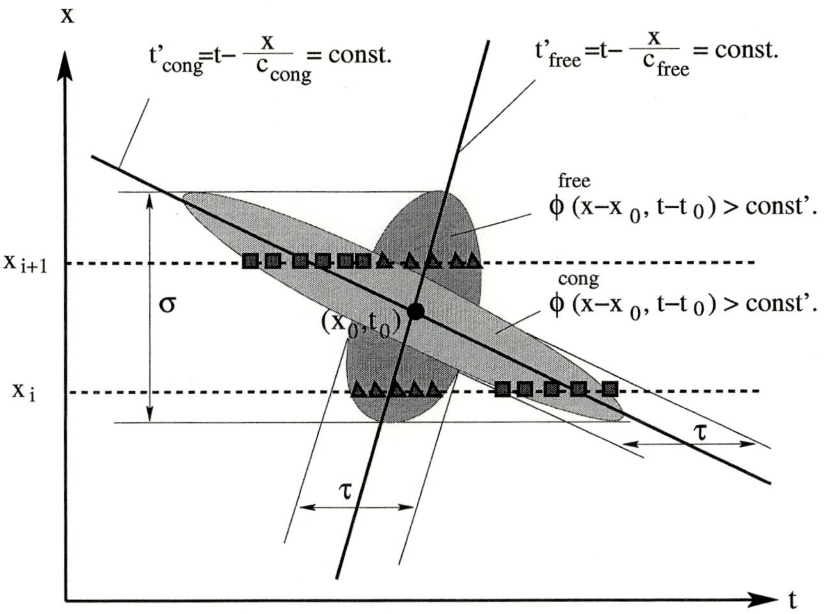

Fig. 1. Visualization of the effects of linear homogeneous filters with the kernels $\phi_{\text{free}}(x,t)$ and $\phi_{\text{cong}}(x,t)$, respectively. The shaded areas denote the regions considered in the calculation of a data point at (x,t). Triangles denote the mainly contributing input data sampled in free traffic, squares the ones sampled in congested traffic.

Figure 2 shows the action of the filters $\mathbf{z}_{\text{cong}}(x,t)$ and $\mathbf{z}_{\text{free}}(x,t)$ for the velocity fields $V_{\text{cong}}(x,t)$ and $V_{\text{free}}(x,t)$.

Finally, we define the nonlinear adaptive weight function $w(\mathbf{z}_{\text{cong}}, \mathbf{z}_{\text{free}}) \in [0,1]$. Obviously, we must have $w \approx 1$ for congested traffic, and $w \approx 0$ for free traffic, so we need some *a priori* estimate of the traffic situation at the point (x,t).

Congested traffic is characterized by a high traffic density and low average velocity. Since, in contrast to the density, the velocity can be directly measured with stationary detectors, we chose the velocity to determine the *a priori* estimate. Different possibilities to estimate the velocity at point (x,t) are:

- The measured velocity V_{ij} of the detector cross section whose position x_i is nearest to x in the time interval j containing the actual time t,
- the velocity $V_{\text{cong}}(x,t)$ as calculated with the "congested-traffic" filter $\mathbf{z}_{\text{cong}}$ according to Eq. (3) with kernel (10) (assuming *a priori* congested traffic),

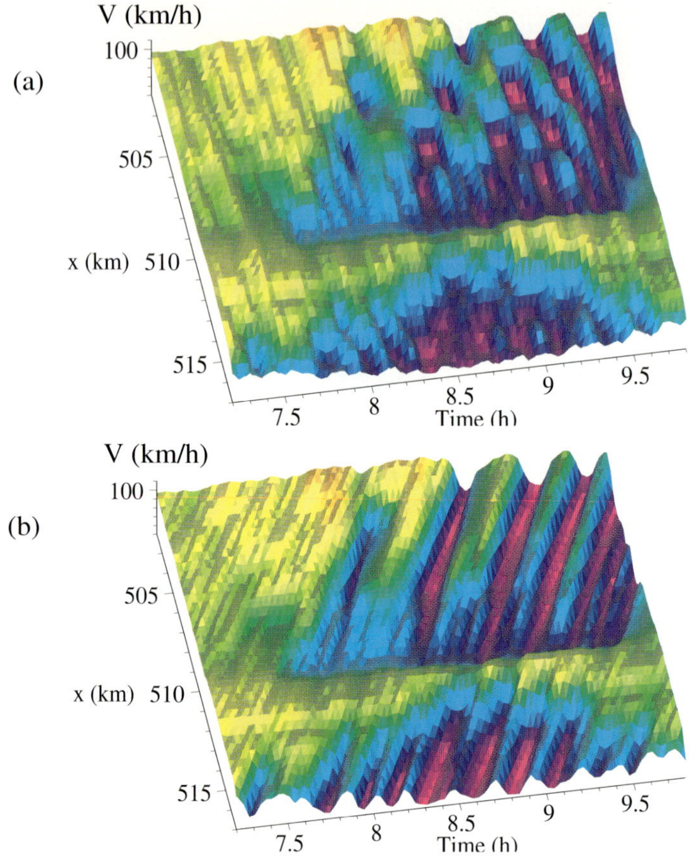

Fig. 2. Typical velocity fields $V_{\text{free}}(x,t)$ (top) and $V_{\text{cong}}(x,t)$ (bottom) obtained by application of the filters $\mathbf{z}_{\text{free}}(x,t)$ and $\mathbf{z}_{\text{cong}}(x,t)$ to traffic data of a section of the German freeway A9 South.

- the velocity $V_{\text{free}}(x,t)$ as calculated with the "free-traffic" filter $\mathbf{z}_{\text{free}}$ according to Eq. (6) with the kernel (11),
- or some combination of the above estimates.

The first way to estimate the velocity is subject to errors, if the typical length scale λ of occuring stop-and-go structures is not larger than the distance Δx_i between two neighbouring detectors. At this point, it is crucial that propagating structures in congested traffic, especially stop-and-go waves, are very persistent. It has been shown [27] that they can propagate through freeway intersections or other inhomogeneities nearly unchanged, passing all perturbations of free traffic on their way (see, e.g., Fig. 9). Therefore, whenever at least one of the estimates indicates congested traffic, the weight function should favour the filter for congested traffic. Specifically, we assume

$$w(\mathbf{z}_{\text{cong}}, \mathbf{z}_{\text{free}}) = w(V_{\text{cong}}(x,t), V_{\text{free}}(x,t)) = \frac{1}{2}\left[1 + \tanh\left(\frac{V_c - V^*(x,t)}{\Delta V}\right)\right], \tag{12}$$

where

$$V^*(x,t) = \min\left(V_{\text{cong}}(x,t), V_{\text{free}}(x,t)\right). \tag{13}$$

V_c and ΔV are parameters that can be varied in a wide range.

If not explicitely stated otherwise, for all simulations in the following section we will use the parameters specified in Tab. 1. We will also show that the four parameters c_{cong}, c_{free}, V_c and Δv can be varied in a wide range without great differences in the output. In this way, we show that the proposed adaptive smoothing method does not need to be calibrated to the respective freeway. One can take the values from Tab. 1 as a global setting.

The smoothing parameters σ and τ have the same meaning and the same effect as in standard smoothing methods, e.g., Eqs. (22) and (23) in Ref. [17].

Table 1. Parameters of the *"adaptive smoothing method"* defined by Eqs. (2)–(13), their interpretation, and their typical values.

Parameter	Typical Value	Meaning
σ	0.6 km	Range of spatial smoothing in x
τ	1.1 min	Range of temporal smoothing in t
c_{free}	80 km/h	Propagation velocity of perturbations in free traffic
c_{cong}	−15 km/h	Propagation velocity of perturbations in congested traffic
V_c	60 km/h	Crossover from free to congested traffic
ΔV	20 km/h	Width of the transition region

In summary, the proposed *"adaptive smoothing method"* is given by Eq. (2) with the nonlinear weight function (12), the filter (3) with normalization (5) and kernel (10) for congested traffic, the filter (6) with normalization (7) and kernel (11) for free traffic, and the smoothing filter (8). Table 1 gives an overview of the six parameters involved and typical values for them. The adaptive smoothing method includes the following special cases:

- Isotropic smoothing resulting in the limits $c_{\text{free}} \to \infty$ and $c_{\text{cong}} \to \infty$ (for practical purposes, one may chose $c_{\text{free}} = c_{\text{cong}} = 10^6$ km/h);
- only filtering for structures of congested traffic in the limit $V_c \to \infty$ (for practical purposes, one may set $V_c \gg V_0$ with V_0 being the desired velocity);
- only filtering for structures of free traffic for $V_c = \Delta V = 0$;
- consideration of the data of the nearest detector only, if $\sigma = 0$;
- application of the actual sampling interval of the detectors, if $\tau = 0$, $c_{\text{free}} \to \infty$ and $c_{\text{cong}} \to \infty$. (To avoid divisions by zero, zero values of σ, τ, and ΔV are replaced by very small positive values in the software).

3 Application to German Freeways

We will now discuss data from the German freeways A8-East and A9-South near Munich, and of the freeway A5-North near Frankfurt. In all cases, the traffic data were obtained from several sets of double-induction-loop detectors recording, separately for each lane, the passage times and velocities of all vehicles. Only aggregate information was stored with an aggregation interval of $\Delta t = 1$ min. We will use the following input data $\mathbf{z}_{ij}^{\text{in}}$:

- The lane-averaged vehicle flow

$$Q_{ij} = \sum_{l=1}^{L} \frac{n_{ij}^l}{L \Delta t}, \tag{14}$$

  where n_{ij}^l is the vehicle count at cross section i during time interval j on lane l. The considered sections of the freeways have $L = 3$ lanes in most cases.
- The lane-averaged mean velocity

$$V_{ij} = \sum_{l=1}^{L} \frac{Q_{ij}^l V_{ij}^l}{Q_{ij}}, \tag{15}$$

  where V_{ij}^l is the average velocity at cross section i during time interval j on lane l.
- The traffic density determined via the formula

$$\rho_{ij} = \frac{Q_{ij}}{V_{ij}}, \tag{16}$$

  as occupancies were not available for all freeways.

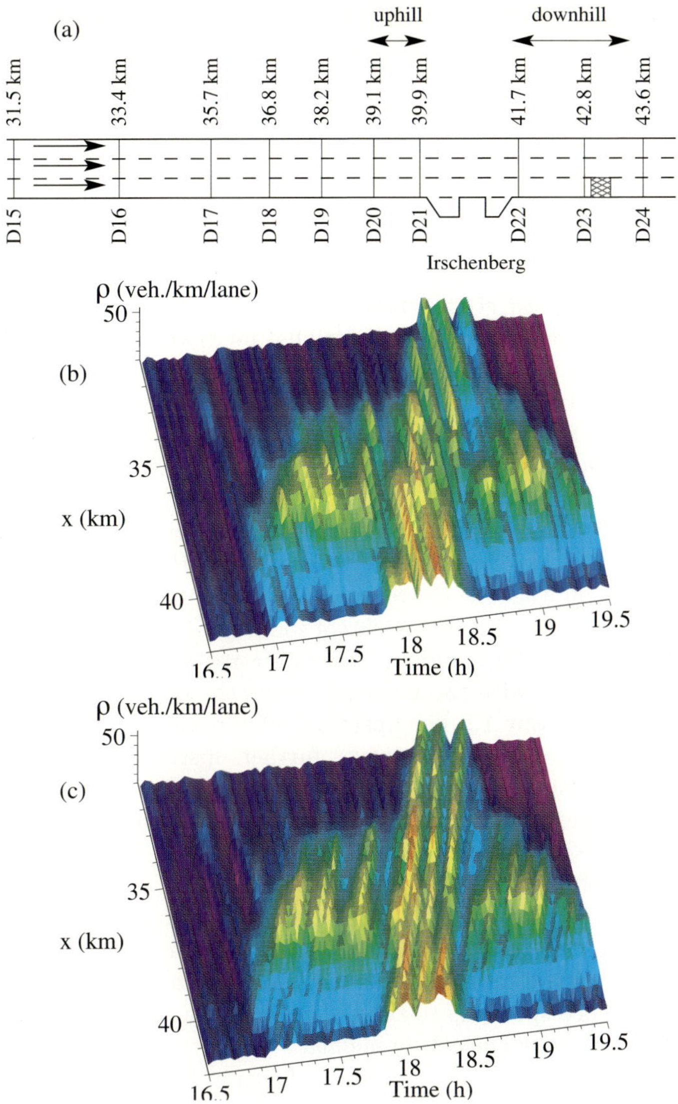

Fig. 3. Complex congested traffic pattern on the German freeway A8-East from Munich to Salzburg during the evening rush hour on November 2, 1998. (a) Sketch of the freeway. (b) Plot of the spatio-temporal density $\rho(x,t)$ using conventional smoothing (resulting for the setting $c_{\text{free}} = c_{\text{cong}} = 10^6$ km/h). (c) Plot of the same data as calculated with the adaptive smoothing method.

Figure 3 shows an example of a complex traffic breakdown that occurred on the freeway A8 East from Munich to Salzburg during the evening rush hour on November 2, 1998. Two different kinds of bottlenecks were relevant, (i) a relatively steep uphill gradient from $x = 38$ km to $x = 40$ km (the "Irschenberg"), and (ii) an incident leading to the closing of one of the three lanes between the cross sections D23 and D24 between $t = $17:40 h and $t = $18:10 h. For further details, see Ref. [17]. Figure 3(c) shows the traffic density using the proposed data processing method with the parameters specified in Tab. 1. For comparison, (b) shows the result of conventional isotropic smoothing ignoring the speed of information propagation, cf. Ref. [17]. While the structures in the free-traffic regions are nearly the same in both plots, the new smoothing method gives a better reconstruction of the congested traffic patterns. In contrast to the conventional method, our proposed method can resolve oscillations with wavelengths λ comparable to or smaller than the distance Δx_i between two neighbouring cross sections as illustrated for the three stop-and-go waves propagating through the whole displayed section ($\lambda \approx 2$ km). Notice that the conventional method generates artefacts for this example.

The congestion patterns caused by the uphill gradient (the "wings" in Fig. 3(b), (c)) have larger characteristic wavelengths and are resolved, at least partially, also with the conventional method.

For comparison, we plot time series of some cross sections in Fig. 4. Let us consider more closely the jammed traffic patterns caused by the temporary bottleneck associated with the incident. While the fluctuations are irregular at detector D22 (about 1.2 km upstream of the aforementioned incident), they grow to three stop-and-go waves further upstream, cf. the arrows in Fig. 5. However, due to measurement errors and other errors in determining the density [1], the time series of the density is very noisy, especially that of cross section D20. The adaptive smoothing method suppresses this noise without suppressing the structure of the pattern.

To check, whether there are significant patterns in the data of detector D20 at all, we plot the velocity data at D20 as well (Fig. 5). In contrast to the density, the velocity is measured directly resulting in less noise. One clearly sees three dips corresponding to three stop-and-go waves which confirms the results of the *"adaptive smoothing method"*.

As a further check that the structures shown in Fig. 3(c) are not artefacts of the data processing, we tested the method for a reduced data set. Figure 6 shows the result using only the even-numbered detectors as input, whose time series are plotted in Fig. 4. The main structures remain nearly unchanged, particularly, the three stop-and-go waves propagating through the whole section. Notice that the whole plot is based on data of only four cross sections, and that the data are extrapolated in the upstream direction over a distance of nearly 2 km.

We now apply the processing method to data from a section of the freeway A9-South near Munich. There are two major intersections I1 and I2 with other freeways. Virtually every weekday, traffic broke down to oscillatory congested traffic upstream of each of these intersections. (For details, see Ref. [17]).

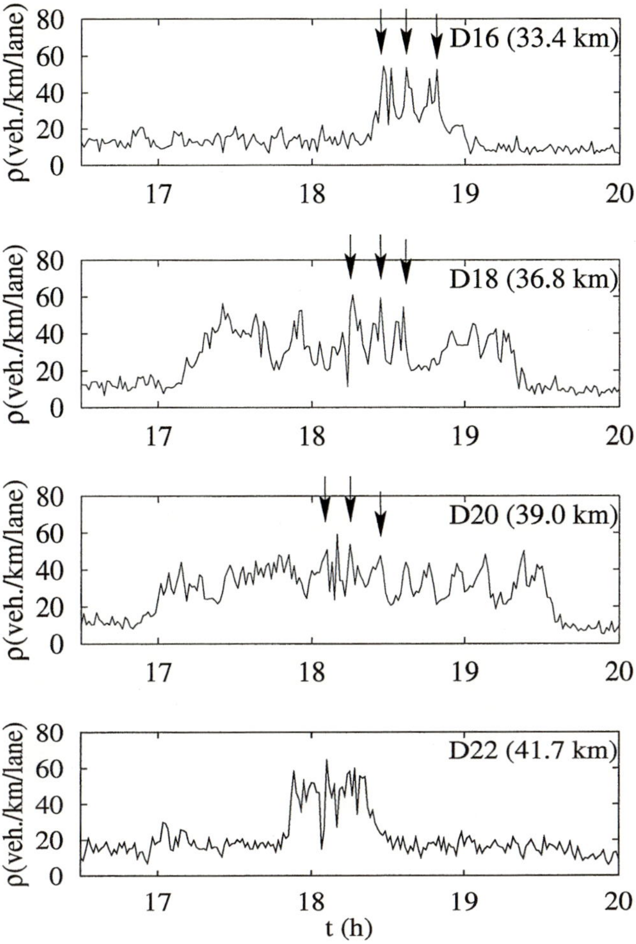

Fig. 4. Time series of the empirical density, approximated by means of Eq. (16), for several cross sections.

Figure 7 (b) shows the inverse of the velocity, $1/V(x,t)$ for a typical congested situation. When plotting $1/V(x,t)$, the structures of congested traffic comes out more clearly than for the density. The method resolves small density clusters in the region 508 km $\leq x \leq$ 510 km between about 8:30 am and 9:30 am which disappear around $x = 508$ km. These structures cannot be identified with the conventional smoothing method (see Fig. 7(c)), which just shows a hilly pattern.

Figure 7(b) is an example for the spatial coexistence of homogeneous congested traffic (a short stretch around $x = 510$ km), oscillating congested

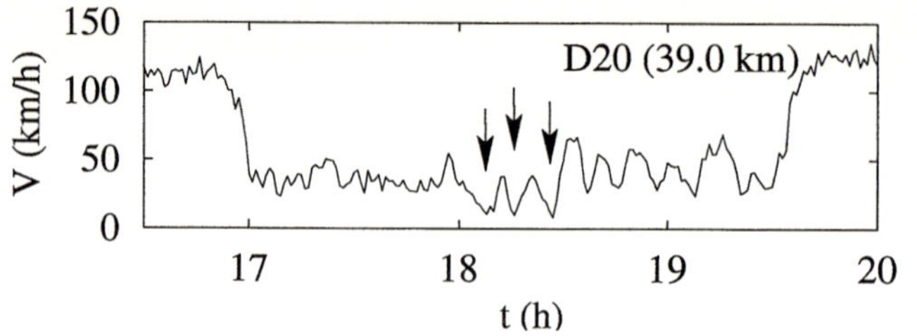

Fig. 5. Time series of the empirical velocity data determined via Eq. (15) for cross section D20.

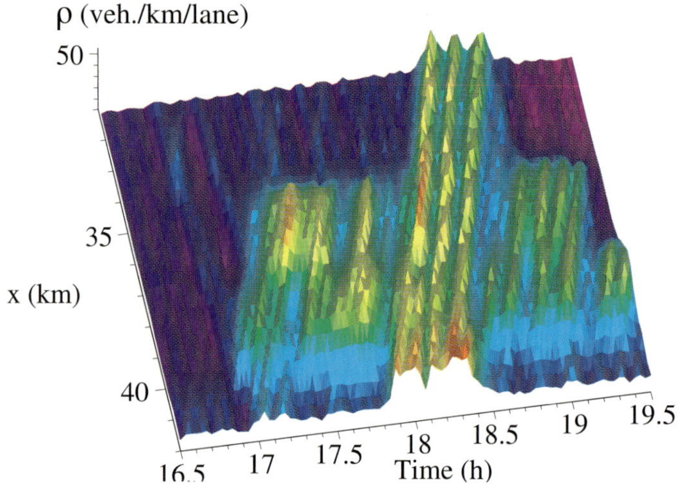

Fig. 6. Spatio-temporal density $\rho(x, t)$ as in Fig. 3, but using only the data of cross sections D16, D18, D20, and D22 as input (cf. Fig. 4).

traffic (around $x = 509$ km), and stop-and-go waves (around $x = 507$ km). This may be a three-dimensional illustration of the so-called "pinch effect" [11,14], but we do not observe a merging of narrow clusters to form wide moving jams. The narrow structures of short wavelengths rather disappear. This may be an artefact of our smoothing method, so that video data are required to get a more detailed picture. Simulations of this spatial coexistence, however, indicate that narrow clusters rather disappear than merge, namely when they do not exceed the critical amplitude in an area of metastable traffic [7].

Notice that the bottleneck causing this congestion is located at about $x = 510$ km implying that it is caused by weaving traffic and slowing down

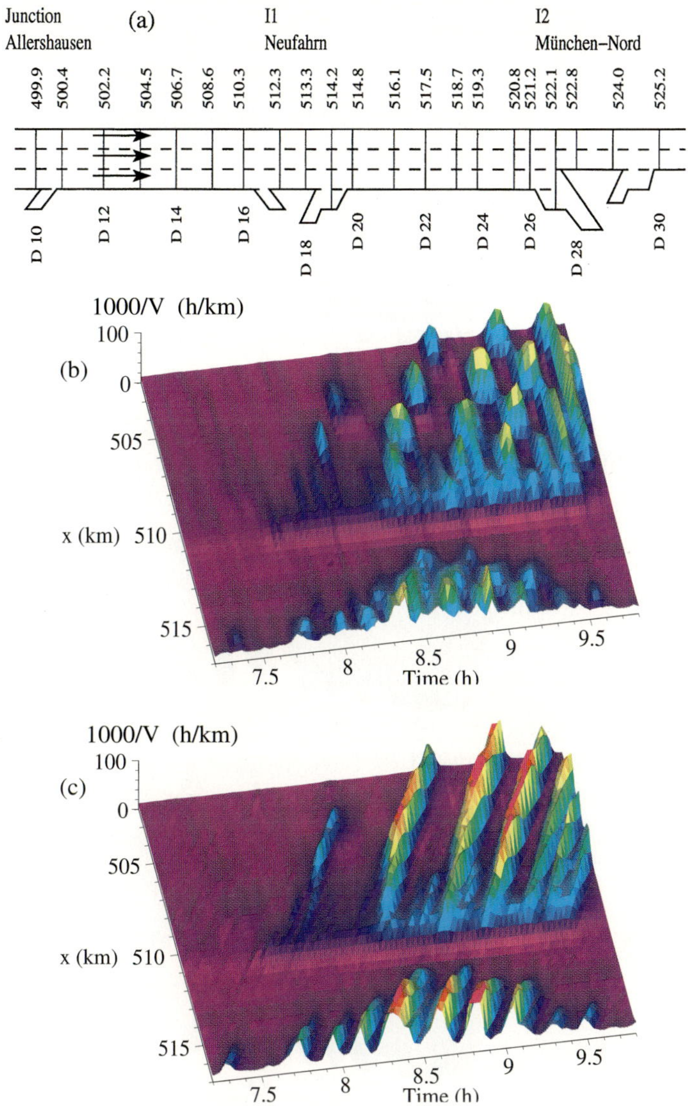

Fig. 7. Two regions of stop-and-go waves on the German freeway A9-South near Munich occurring upstream of the freeway intersections I1 and I2. (a) Sketch of the freeway. (b) Plot of the inverse $1/V(x,t)$ of the spatio-temporal velocity $V(x,t)$, using conventional smoothing. (c) Plot of the same data using the adaptive smoothing method.

at an off-ramp, not by an on-ramp as is most often the case on German highways.

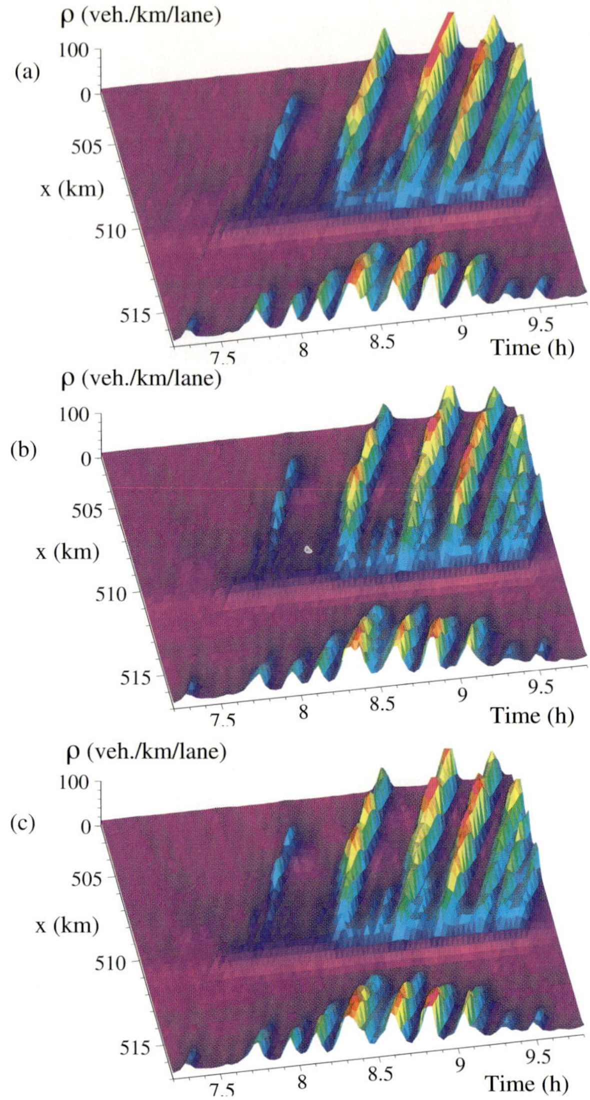

Fig. 8. Same data and same processing method as in Fig. 7(c), but with (a) a propagation velocity of $c_{\text{cong}} = -12$ km/h instead of -15 km/h (b) $c_{\text{cong}} = -20$ km/h, (c) with the crossover parameter $V_c = 40$ km/h instead of 60 km/h.

Now we demonstrate that the method is robust with respect to reasonable parameter changes. Figure 8(a) shows the result for an assumed propagation velocity of $c_{\text{cong}} = -12$ km/h instead of -15 km/h. In plot (b), we assumed $c_{\text{cong}} = -20$ km/h instead. In both cases, the results are systematically better than with the isotropic procedure. Since the propagation velocity of perturbations in congested traffic is always about $c_{\text{cong}} = -15$ km [12,23,28,29], one can take this value as a global setting.

Due to the comparatively high magnitude of the propagation velocity c_{free}, the related time shifts in the transformation (9) for free traffic are small. Therefore, the method is insensitive to changes with respect to this parameter. It turned out that taking the isotropic limit $c_{\text{free}} \to \infty$ nearly gives the same results.

The crossover parameters V_c and ΔV can be varied within reasonable limits as well. As an example, Fig. 8(c) shows the result after changing the crossover velocity from $V_c = 60$ km/h to $V_c = 40$ km/h.

Finally, we applied the method to isolated moving localized clusters and pinned localized clusters observed on the A5-North near Frankfurt. This freeway is particularly well equipped with detectors, so that the typical length scales of the observed structures are always larger than the distance between neighbouring detectors. However, we obtained nearly the same results when using only the even-numbered detectors or only the odd-numbered detectors as input. Figure 9 shows that even a further reduction of the information to only 35% of the detectors yields good results.

4 Summary and Outlook

We have proposed a new *"adaptive smoothing method"* for the three-dimensional visualization of spatio-temporal traffic patterns, which takes into account the characteristic propagation velocities observed in free and congested traffic. The method is robust with respect to variations of its parameters, so that it can be applied to new freeway sections without calibration. In principle, it would be possible to determine the parameters (such as the propagation velocity of perturbations) locally (e.g., by means of a correlation analysis). However, the results would look less smooth and regular, as the small number of data to determine the local parameters would be associated with considerable statistical errors. Consequently, a large part of the variations in the local parameters would *not* reflect *systematic* variations of the parameters. So, both the local and global parameter calibration may produce artefacts, but it is advantageous to use global parameter settings. Moreover, there is empirical support for surprisingly constant propagation velocities c_{free} and c_{cong} of perturbations in free and congested traffic. The parameters V_c and ΔV are related to the transition from free to congested traffic and, therefore, can also be well determined. The spatial and temporal smoothing parameters σ and τ can be specified according to the respective requirements. Suitable parameters allow a good representation of traffic patterns even when the distances between successive detectors are about 3 kilometers.

We point out that the suggested *"adaptive smoothing method"* itself can be applied to calibrate the characteristic propagation velocities c_{free} and c_{cong}.

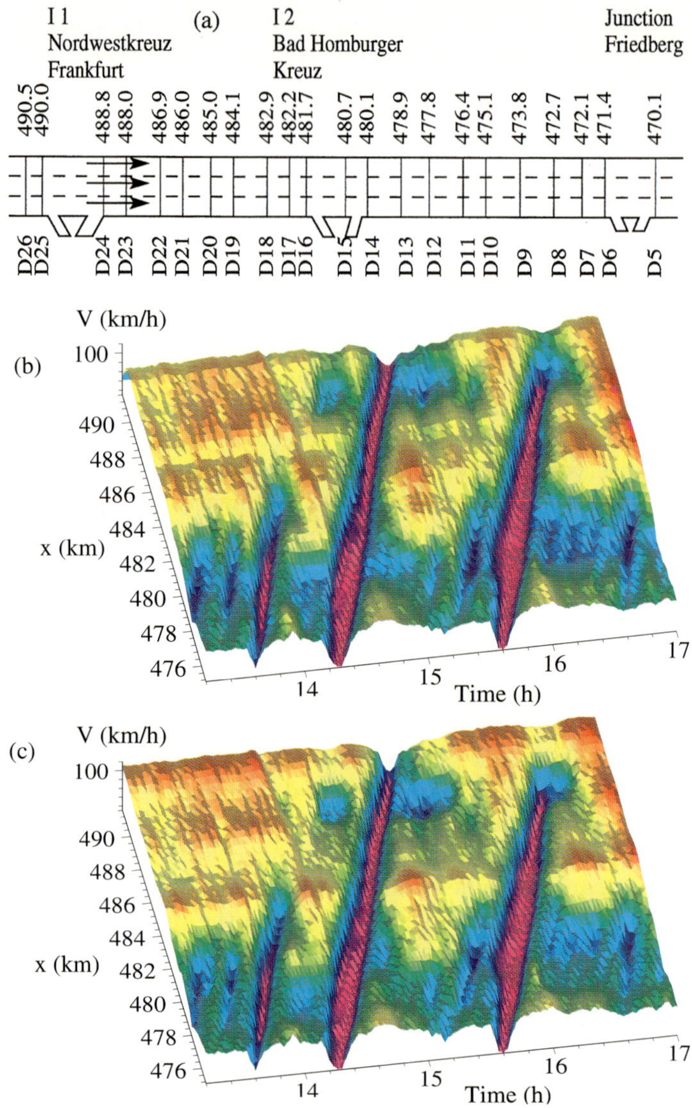

Fig. 9. Freeway A5-North from Frankfurt to Kassel during the evening rush hour on August 7, 1998. (a) Sketch of the freeway. (b) Data reconstruction of the velocity using all detectors D10–D26. (c) Reconstruction using a reduced data set of only 6 of all 17 detectors used in (b).

For this purpose, the ranges σ and τ of spatial and temporal smoothing are chosen small. The optimal propagation velocities minimize the offsets in the propagation patterns.

The aim of the *"adaptive smoothing method"* is to reconstruct the spatio-temporal traffic data from incomplete information as good as possible to allow a better understanding of the complex traffic dynamics. Potential applications are, for example, traffic state visualization, incident detection, or the experimental verification of traffic models. Our method could be further improved by taking into account information about the traffic dynamics such as the continuity equation or a suitable equation for the average vehicle velocity as a function of space and time. It could, then, be used to determine short-term traffic forecasts.

Acknowledgments

The authors would like to thank the German Research Foundation (DFG) for financial support through the grant He 2789/2-1. They are also grateful to the *Autobahndirektion Südbayern*, and the *Hessisches Landesamt für Straßen und Verkehrswesen* for providing the traffic data.

References

1. Helbing, D.: *Traffic and related self-driven many-particle systems*, Reviews of Modern Physics **73**, 1067–1141 (2001).
2. Chowdhury, D., Santen, L. and Schadschneider, A.: *Statistical physics of vehicular traffic and some related systems*, Physics Reports **329**, 199–329 (2000).
3. Helbing, D., Farkas, I. J., Molnár, P. and Vicsek, T.: *Simulation of pedestrian crowds in normal and evacuation situations*, in *Pedestrian and Evacuation Dynamics*, edited by M. Schreckenberg and S. D. Sharma (Springer, Berlin, 2002), pp. 21–58.
4. Helbing, D., Molnár, P., Farkas, I. and Bolay, K.: *Self-organizing pedestrian movement*, Environment and Planning B **28**, 361–383 (2001).
5. Helbing, D., Hennecke, A. and Treiber, M.: *Phase diagram of traffic states in the presence of inhomogeneities*, Phys. Rev. Lett. **82**, 4360–4363 (1999).
6. Lee, H. Y., Lee, H. -W. and Kim, D.: *Dynamic states of a continuum traffic equation with on-ramp*, Phys. Rev. E **59**, 5101–5111 (1999).
7. Treiber, M. and Helbing, D.: *Explanation of observed features of self-organization in traffic flow*, e-print cond-mat/9901239 (1999).
8. Cassidy, M. J. and Bertini, R. L.: *Some traffic features at freeway bottlenecks*, Transpn. Res. B **33**, 25–42 (1999).
9. Daganzo, C. F., Cassidy, M. J. and Bertini, R. L.: *Possible explanations of phase transitions in highway traffic*, Transpn. Res. A **33**, 365–379 (1999).
10. Hall, F. L. and Agyemang-Duah, K.: *Freeway capacity drop and the definition of capacity*, Transpn. Res. Rec. **1320**, 91–108 (1991).
11. Kerner, B. S.: *Experimental features of self-organization in traffic flow*, Phys. Rev. Lett. **81**, 3797–3800 (1998).
12. Kerner, B. S. and Rehborn, H.: *Experimental features and characteristics of traffic jams*, Phys. Rev. E **53**, R1297–R1300 (1996).
13. Kerner, B. S. and Rehborn, H.: *Experimental properties of complexity in traffic flow*, Phys. Rev. E **53**, R4275–R4278 (1996).

14. Koshi, M., Iwasaki, M. and Ohkura, I.: *Some findings and an overview on vehicular flow characteristics*, in *Proceedings of the 8th International Symposium on Transportation and Traffic Flow Theory*, pp. 403–426, V. F. Hurdle, E. Hauer, and G. N. Stewart (Eds.) (University of Toronto, Toronto, Ontario, 1983).
15. Lee, H. Y., Lee, H. -W. and Kim, D.: *Phase diagram of congested traffic flow: An empirical study*, Phys. Rev. E **62**, 4737–4741 (2000).
16. Neubert, L., Santen, L., Schadschneider, A. and Schreckenberg, M.: *Single-vehicle data of highway traffic: A statistical analysis*, Phys. Rev. E **60**, 6480–6490 (1999).
17. Treiber, M., Hennecke, A. and Helbing, D.: *Congested traffic states in empirical observations and microscopic simulations*, Phys. Rev. E **62**, 1805–1824 (2000).
18. Helbing, D. and Treiber, M.: *Gas-kinetic-based traffic model explaining observed hysteretic phase transition*, Phys. Rev. Lett. **81**, 3042–3045 (1998).
19. Lee, H. Y., Lee, H. -W. and Kim, D.: *Origin of synchronized traffic flow on highways and its dynamic phase transitions*, Phys. Rev. Lett. **81**, 1130–1133 (1998).
20. Banks, J. H.: *An investigation of some characteristics of congested flow*, Transpn. Res. Rec. **1678**, 128–134 (1999).
21. Leutzbach, W.: *Introduction to the Theory of Traffic Flow* (Springer, Berlin, 1988).
22. Treiber, M. and Helbing, D.: *Macroscopic simulation of widely scattered synchronized traffic states*, J. Phys. A: Math. Gen. **32**, L17–L23 (1999).
23. Sollacher, R., Kerner, B. S., Konhäuser, P., Rehborn, H., Kühne, R., Schreckenberg, M. and Helbing, D.: *SANDY - Nichtlineare Dynamik im Straßenverkehr*, in *Technische Anwendungen von Erkenntnissen der Nichtlinearen Dynamik* (VDI - Technologiezentrum Physikalische Technologien, Düsseldorf, ISBN 3-931384-25-X, 1999).
24. Helbing, D. and Treiber, M.: *Critical discussion of "sychronized flow"*, submitted to Cooper@tive Tr@nsport@tion Dyn@mics (2002).
25. Kerner, B. S. and Klenov, S. L.: *A micropscopic model for phase transitions in traffic flow*, J. Phys. A: Math. Gen. **35**, L31–L43 (2002).
26. Daganzo, C. F.: *A behavioral theory of multi-lane traffic flow, Part I: Long homogeneous freeway sections*, (ITS Working Paper, UCB-ITS-RR-99-5, revised June 20, 2000).
27. Kerner, B. S.: *Theory of breakdown phenomenon at highway bottlenecks*, Transpn. Res. Rec. **1710**, 136–144 (2000).
28. Cassidy, M. J. and Mauch, M.: *An observed traffic pattern in long freeway queues*, Transpn. Res. A **35**, 143–156 (2001).
29. Mika, H. S., Kreer, J. B. and Yuan, L. S.: *Dual mode behavior of freeway traffic*, Highw. Res. Rec. **279**, 1–13 (1969).

Probabilistic Description of Nucleation in Vapours and on Roads

Reinhard Mahnke

Universität Rostock, Fachbereich Physik, D–18051 Rostock, Germany

Abstract. The aggregation of particles out of an initially homogeneous situation is well known in physics. Depending on the system under consideration and its control parameters the cluster formation in a supersaturated (metastable or unstable) situation has been observed in nucleation physics as well as in other branches. We investigate the well–known example of condensation (formation of liquid droplets) in an undercooled vapour to conclude that the formation of bound states as a phase transition is related to transportation science. We present a comparison of nucleation in an isothermal–isochoric container with traffic congestion on a circular one–lane freeway. The analysis is based, in both cases, on the probabilistic description by stochastic master equations. The construction of physically motivated transition probabilities plays the central role in our analysis and comparison.

1 Introduction

The term *cluster* as a *group of similar things* is used in science in many different fields. It is common on all scales from subatomic physics up to astronomy. The existance of bound states of atoms or molecules as clusters with different sizes in gases, liquids and solid matter is an intermediate step between the elementary isolated particle and the macroscopic condensed state. The formation of bound states as an aggregation process is related to self–organization phenomena [5,28]. In general the properties of clusters, especially the binding energy, depend on the cluster size and shape [4,9]. The nucleation and growth of atomic and molecular clusters is one of the basic topics in nucleation theory [28,29]. An example of experimental verification is given in [27].

The formation of clusters takes place on different scales in physical (nucleation of droplets, condensation of clouds) and nonphysical systems (insect societies, socioconfigurations, transport communities). Nonphysical systems are generally a reflection of the interaction of people or animals. We, however, do not discuss here the social aspects of human interaction. We are particularly interested in traffic flow theory where the car drivers (or, formally, the vehicles) are considered as interacting particles. Physical systems like spin glasses are one of the most investigated complex systems to understand the clustering of states. In chemically reacting systems the domain formation is simulated by Monte Carlo experiments and by lattice–gas automata. Monte Carlo methods are well established in statistical physics [6,22].

The clustering behaviour can be considered as a phase transition between different states of matter. In dependence on the boundary conditions a quenched system as for instance an undercooled vapour or an overcrowded road undergoes a relaxation process to the new phase. The theoretical understanding of the physical mechanism for the dynamics of the aggregation phenomenon has important implications for the more general task to realize relaxation phenomena in equilibrium and nonequilibrium complex systems [28].

The modelling of such supersaturated (metastable or unstable) systems is dominated by a very complicated energy landscape with many local minima and can be done by different means. Having in mind the homogeneous or heterogeneous (e. g. by electrical charges) nucleation and condensation process we divide the following theoretical and computational methods [14]:

1. Molecular dynamics
   Based on the classical many–particle Hamiltonian with interaction, e. g. of well–known Lennard Jones type, the integration of the equations of motions shows the clustering process quite obvious.
2. Master equation approach
   In this Monte Carlo simulation technique the master equation for the chemical reactions of the attachment and the detachment of elementary particles (monomers) on or out of clusters is solved with phenomenologically motivated transition probabilities. This method will be used later on.
3. Monte Carlo simulation technique
   The formation process of clusters on a two–dimensional lattice can be modelled by a so–called nucleation automaton with a well–explained energy function. The relaxation to a one–big–cluster–state as the equilibrium situation is a searching process on the energy landscape to find a minimum in a very high dimensional situation.
4. Cellular automata models
   Chemically reacting systems with phase separations like the well–known Schlögl model can be simulated by probabilistic lattice–gas automaton models where space, time and particle velocities are discrete. Each molecule (occupied site) can undergo elastic and reactive collisions and depending on the updating rule interesting spatio–temporal configurations of particles emerge.
5. Deterministic nonlinear dynamical systems
   The Becker–Döring cluster equations and other kinetic equations of such type provide a very famous dynamical model for the growth, shrinkage, coagulation and fragmentation of clusters (droplets) in a condensing vapour. The trends to equilibrium can be studied in dependence on the kinetic coefficients. Especially for the late Ostwald ripening regime a number of dynamical systems as coupled nonlinear sets of equations exists.

It should be mentioned that nonphysical systems, like traffic flow or economics, can also be described by methods of statistical thermodynamics. Such an approach, based on the use of state function entropy and its maximization principle, has been presented by Montroll, Reiss and co–workers [20,25,26]. Although our present investigation of traffic flow does not involve the entropy maximization principle, it clearly supports the idea that nonphysical systems like traffic flow and physical systems like supersaturated vapour can be treated similarly.

The aim of the present paper is to give a comparison between aggregation in supersaturated vapour and in traffic flow by providing unique description of these two phenomena within the stochastic master equation approach. We emphasize that phase transitions which are well known and investigated by statistical physics and thermodynamics take place in nonphysical systems too, as already pointed out by Montroll [20]. We concentrate here on vehicular traffic, but other fields like cancer growth and stock market development behave similarly.

2 Stochastic Master Equation Approach

Stochastic processes enter into many physical descriptions of nature. Historically first the motion of a heavy particle in a fluid of light molecules has been observed. The path of such Brownian particle consists of stochastic displacements due to random collisions. Such motion was studied by the Scottish botanist Robert Brown (1773 – 1858). In 1828 he discovered that the microscopically small particles into which the pollen of plants decay in an aqueous solution are in permanent irregular motion. Such stochastic process is called Brownian motion and can be interpreted as discrete random walk or continuous diffusion movement. The intuitive background to describe the irregular motion completely as stochastic process is to measure values $x_1, x_2, \ldots, x_n, \ldots$ at time moments $t_1, t_2, \ldots, t_n, \ldots$ of a well–defined time dependent random variable $x(t)$ and assume that a set of joint probability densities or distributions

$$p_n(x_1, t_1; x_2, t_2; \ldots; x_n, t_n) \quad \text{for} \quad n = 1, 2, \ldots \tag{1}$$

exists.

This stochastic description in terms of macroscopic variables is called mesoscopic. Typical systems encountered in the everyday life like gases, liquids, solids, biological organisms, human or technical objects consist of about 10^{23} interacting atoms. The macroscopic properties of matter are usually the result of collective behaviour of large number of atoms and molecules acting under the laws of quantum mechanics. To understand and control these collective macroscopic phenomena the complete knowledge based upon the known fundamental laws of microscopic physics is useless because the problem of interacting particles is much beyond the capabilities of the largest recent and future computers. The understanding of complex macroscopic systems consisting of many basic particles (in the order of atomic sizes: 10^{-10} m) requires the formulation of new concepts. One of the methods is the stochastic description taking into account the statistical behaviour. Since the macroscopic features are averages over time of a large number of microscopic interactions, the stochastic description links both approaches, the microscopic and the macroscopic one, together to give probabilistic results [3,8,10].

Speaking about a stochastic process from the physical point of view we always refer to stochastic variables (random events) changing in time. A realization of a stochastic process is a trajectory $x(t)$ as function of time.

A stochastic process without any dynamics (like a coin throw or any hazard game) is called a temporally uncorrelated process and the normalized distribution $p_1(x_1, t_1)$ describes the process totally.

Introducing dynamics via correlations between two different time moments, this assumption enables us to define the Markov process by two quantities totally, namely the first–order and the second–order probability density $p_2(x_1, t_1; x_2, t_2)$, or equivalently by the joint probability $p_1(x_1, t_1)$ and the conditional probability $p_2(x_2, t_2 \mid x_1, t_1)$ to find the value x_2 at time t_2, given that its value at previous time t_1 $(t_1 < t_2)$ is x_1. In contradiction to uncorellated processes, Markov processes are characterized by the following temporal relationship

$$p_2(x_1, t_1; x_2, t_2) = p_2(x_2, t_2|x_1, t_1)\, p_1(x_1, t_1) \; . \tag{2}$$

This Markov property enables us to calculate all higher–order joint probabilities p_n for $n > 2$. To determine the fundamental equation of stochastic processes of Markov type we start with the third–order distribution $(t_1 < t_2 < t_3)$

$$p_3(x_1, t_1; x_2, t_2; x_3, t_3) = p_2(x_3, t_3 \mid x_2, t_2)\, p_2(x_2, t_2 \mid x_1, t_1)\, p_1(x_1, t_1) \tag{3}$$

and integrate this identity over x_2 and divide both sides by $p_1(x_1, t_1)$. We get the following result for the conditional probabilities defining a Markov process

$$p_2(x_3, t_3 \mid x_1, t_1) = \int p_2(x_3, t_3 \mid x_2, t_2)\, p_2(x_2, t_2 \mid x_1, t_1)\, dx_2 \; , \tag{4}$$

called Chapman–Kolmogorov equation. As already stated the Markov process is uniquely determined through the distribution $p_1(x, t)$ at time t and the conditional probability $p_2(x', t' \mid x, t)$, also called transition probability from x at t to x' at later t', to determine the whole hierarchy p_n $(n \geq 3)$ by the Markov property (2).

The memory in a Markov process is very short, only one time interval from t to t' plays any role. If the trajectory has reached x at time t, the past is forgotten, and it moves toward x' at t' with a probability depending on x, t and x', t' only. The entire information relevant for the future is thus contained in the present. This gives sense to transform the Chapman–Kolmogorov equation (4) in an equivalent differential equation in the short time limit $t' = t + \tau$ with small τ tending to zero. The short time behaviour of the transition probability $p_2(\cdot \mid \cdot)$ should be written as series expansion with respect to time interval τ in the form

$$p_2(x, t+\tau \mid x'', t) = [1 - \bar{w}(x,t)\tau]\,\delta(x - x'') + \tau w(x, x'', t) + \mathcal{O}\left(\tau^2\right). \quad (5)$$

The new quantity $w(x, x'', t) \geq 0$ is the transition rate, the probability per time unit, for a jump from x'' to $x \neq x''$ at time t. This transition probability multiplied by the time step τ gives the second term in the series expansion describing transitions from an other state x'' to x. The first term (with the delta function) is the probability that no transitions takes place during time interval τ.

Inserting (5) in the right hand side of the Chapman–Kolmogorov equation (4) and integrating over x' (taking into account the normalization conditions for probabilities) we obtain the differential formulation of the Chapman–Kolmogorov equation

$$\frac{\partial}{\partial t} p_1(x,t) = \int w(x, x', t) p_1(x', t)\, dx' - \int w(x', x, t) p_1(x, t)\, dx' \quad (6)$$

called master equation in the (physical) literature [3,8,10]. The name 'master equation' for the above probability balance equation is used in a sense that this differential expression is a general, fundamental or basic equation. For a homogeneous in time process the transition rates $w(x, x', t)$ are independent of time t. The short time transition rates w have to be known from the physical context, often like a clever ansatz, or have to be formulated based on a reasonable hypothesis or approximation. With known transition rates w the master equation (6) gives the resulting evolution of the probability p_1 over a long time period.

The master equation can be written in different ways. Since we are particularly interested in the case where the stochastic variable has discrete values (natural numbers) within a finite range $0 \leq n \leq N$, the master equation for the time evolution of the probabilities $p_1(x,t)$, now $P(n,t)$, then is written as

$$\frac{dP(n,t)}{dt} = \sum_{n' \neq n} \left\{ w(n, n') P(n', t) - w(n', n) P(n, t) \right\}, \quad (7)$$

where $w(n', n) \geq 0$ are rate constants for transitions from n to other $n' \neq n$. Together with the initial probabilities $P(n, 0)$ ($n = 0, 1, 2, \ldots, N$) and the boundary conditions at $n = 0$ and $n = N$ this set of equations governing the time evolution of $P(n,t)$ from the beginning at $t = 0$ to the long–time limit $t \to \infty$ has to be solved. The meaning of both terms is clear. The first (positive) term is the inflow probability current to state n due to transitions from other states n', and the second (negative) term is the outflow current due to opposite transitions from n to n'.

Defining stationarity, sometimes called steady state, as a time independent distribution $P^{st}(n)$ by the condition $dP(n,t)/dt|_{P=P^{st}} = 0$, the stationary master equation is given by

$$0 = \sum_{n' \neq n} \{w(n,n')P^{st}(n') - w(n',n)P^{st}(n)\} \ . \tag{8}$$

This equation states the obvious fact, that in the stationary or steady state regime the sum of all transitions into any state n must be balanced by the sum of all transitions from n into other states n'. Based on the properties of the transition rates per unit time the probabilities $P(n,t)$ tend in the long–time limit to the uniquely defined stationary distribution $P^{st}(n)$. In open systems a constant probability flow is possible. This fundamental property of the master equation may be stated as

$$\lim_{t \to \infty} P(n,t) = P^{st}(n) \ . \tag{9}$$

Let us shortly discuss the question of equilibrium in a system without external exchange. The condition of equilibrium in closed isolated systems is much more stronger than the former condition of stationarity (8). Here we demand a balance between each pair of states n and n' separately. This so–called detailed balance relation is written for the equilibrium distribution $P^{eq}(n)$ as

$$0 = w(n,n')P^{eq}(n') - w(n',n)P^{eq}(n) \ . \tag{10}$$

Of course, each equilibrium state is by definition also stationary. If the initial probability vector $P(n,0)$ is strongly nonequilibrium, many probabilities $P(n,t)$ change rapidly as soon as the evolution starts (short–time regime), and then relax more slowly towards equilibrium (long–time behaviour). The final (thermodynamic) equilibrium is reached in the limit $t \to \infty$. As a particular example of such a behaviour, the relaxation process in atomic clusters has been studied in [19].

We are speaking about a one–dimensional stochastic process if the state space is characterized by one variable only. As in our investigations, this discrete variable is a particle number $n \geq 0$ describing the amount of molecules or the size of an aggregate. In chemical physics such aggregation phenomena like formation and/or decay of clusters are of great interest. Examples are the formation of a crystal or glass upon cooling a liquid or the condensation of a droplet out of a supersaturated vapour. To determine the relaxation dynamics of model clusters we take a particularly simple Markov process with transitions between neighbouring states n and $n' = n \pm 1$. This situation is called a one–step process. In biophysics, if the variable n represents the number of living individuals of a particular species, the one–step process is often called birth–and–death process to investigate problems in population dynamics. The random walk with displacements to left and right by one step is well known in physics and often plays a role as an introductionary example and has been recently revisited and applied to new fields like econophysics [23].

Setting the transition rates $w(n, n-1) = w_+(n-1)$, $w(n, n+1) = w_-(n+1)$, and therefore also $w(n+1, n) = w_+(n)$, $w(n-1, n) = w_-(n)$, the master equation (7) reads

$$\frac{dP(n,t)}{dt} = w_+(n-1)\, P(n-1,t) + w_-(n+1)\, P(n+1,t) \\ - \left[w_+(n) + w_-(n)\right]\, P(n,t) \; . \tag{11}$$

In general the forward and backward transition rates $w_+(n), w_-(n)$ are nonlinear functions of the random variable n; the physical dimension is one over time (s^{-1}). We pay attention to particles as aggregates in a closed box or vehicular jams on a circular road. Therefore in finite systems the range of the discrete variable n is bounded between 0 and N $(n = 0, 1, 2, \ldots, N)$.

The general one–step master equation (11) is valid for $n = 1, 2, \ldots, N-1$, but meaningless at the boundaries $n = 0$ and $n = N$. Therefore we have to add two boundary equations as closure conditions

$$\frac{dP(0,t)}{dt} = w_-(1)\, P(1,t) - w_+(0)\, P(0,t) \; , \tag{12}$$

$$\frac{dP(N,t)}{dt} = w_+(N-1)\, P(N-1,t) - w_-(N)\, P(N,t) \; . \tag{13}$$

In finite systems with $n = 0, 1, 2, \ldots, N$ one finds a situation which corresponds to steady state with a detailed balance relationship similar to (10). Therefore the stationary distribution $P^{st}(n)$ fulfils the recurrence relation

$$P^{st}(n) = \frac{w_+(n-1)}{w_-(n)} P^{st}(n-1) \; . \tag{14}$$

By applying the iteration successively and taking into account the normalization condition (the sum of all the probabilities must be equal to 1), the stationary probability distribution $P^{st}(n)$ in finite systems is finally written as

$$P^{st}(n) = \begin{cases} \dfrac{\displaystyle\prod_{m=1}^{n} \frac{w_+(m-1)}{w_-(m)}}{1 + \displaystyle\sum_{k=1}^{N} \prod_{m=1}^{k} \frac{w_+(m-1)}{w_-(m)}} & n = 1, 2, \ldots, N \\[2em] \dfrac{1}{1 + \displaystyle\sum_{k=1}^{N} \prod_{m=1}^{k} \frac{w_+(m-1)}{w_-(m)}} & n = 0 \; . \end{cases} \tag{15}$$

The obtained result is a unique solution for the stationary probability distribution in finite systems with closed boundaries. For an isolated system the stationary solution of the master equation P^{st} is identical with the thermodynamic equilibrium P^{eq}, where the detailed balance holds, which for one–step processes reads

$$w_-(n)\, P^{eq}(n) = w_+(n-1)\, P^{eq}(n-1)\ . \tag{16}$$

The condition of detailed balance states a physical principle. If the distribution P^{eq} is known from equilibrium statistical mechanics and one of the transition rates is also known (e. g. by a reasonable ansatz w_+), the equation (16) provides the opportunity to formulate the opposite transition rate w_- in a consistent way. By this procedure the nonequilibrium behaviour is adequately described by a sequence of (quasi–)equilibrium states. The relaxation from any initial nonequilibrium distribution tends always to the known final equilibrium. In physical systems the equilibrium distribution usually is represented in an exponential form

$$P^{eq}(n) \propto \exp\left[-\Omega(n)/(k_B T)\right] \tag{17}$$

where $\Omega(n)$ is the thermodynamic potential depending on the stochastic variable n, k_B is the Boltzmann constant, and T is the temperature [21].

3 Nucleation in Supersaturated Vapours

If we consider a vapour at equilibrium then a certain change of the thermodynamic parameters enables us to remove the system into a nonequilibrium state. The vapour becomes supersaturated. The basic quantity describing the situation is the cluster distribution **N** at time t

$$\mathbf{N}(t) = (N_0, N_1, N_2, \ldots, N_n, \ldots, N_N) \tag{18}$$

which gives the number of clusters N_n of size n. The free particles (molecules) are called monomers of size $n = 0$. It is supposed that N_1 molecules are excited. These molecules may be named as precluster of size $n = 1$. The bound states are clusters of size $n \geq 2$. Investigating a finite system the overall number of particles N_{total} as well as the volume V and the temperature T are fixed. The particle conservation law

$$N_{total} = N_0 + N_1 + \sum_{n=2}^{N} n N_n = \text{const} \tag{19}$$

takes into account that particles are either free, excited, or bounded in clusters. There is always some difficulty to describe the initial stage of formation of a cluster. We have introduced the precluster as an intermediate state between free and bounded states to provide an easy and unified description of the aggregation process which is valid both in supersaturated vapour and in traffic flow discussed later on.

A simple example of a system of many particles is a gas consisting of identical molecules called monomers. If the gas is dilute (the density as number of molecules per unit volume is small), the average separation length between the monomers is large and correspondingly their interaction is negligible. The gas is said to be ideal, if the average separation length is much larger than the de Broglie wave length. We treat the monomers as undistinguishable particles moving in a closed volume and making reactive collisions to form aggregates called molecular clusters. Furtheron, we consider a simplified case where only one single cluster of size n (i. e., $N_n = 1$) coexists with $N_0 = N_{total} - n$ unbounded (free) particles, which means that the cluster distribution (18) reduces to

$$\mathbf{N}(t) = (N_0, 0, \ldots, 0, N_n = 1, 0, \ldots, 0) \tag{20}$$

and the overall particle conservation (19) to

$$N_{total} = N_0 + n \cdot 1 = \text{const} , \tag{21}$$

where the stochastic variable $n = n(t)$ is the number of particles bounded in the cluster at time t. The nucleation box (volume V) embedded in a heat bath (temperature T) displaying the situation schematically is shown in Fig. 1.

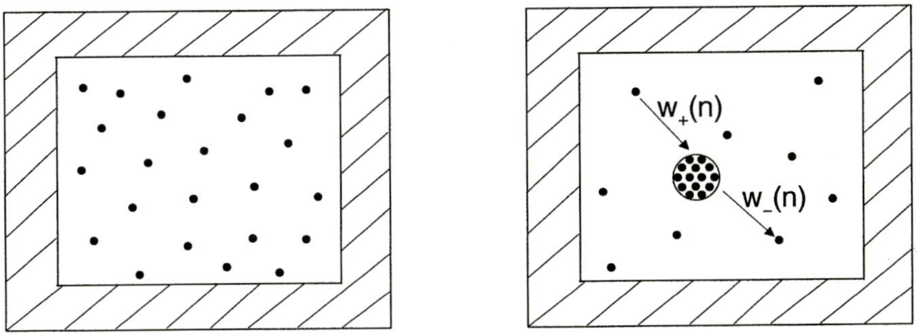

Fig. 1. Isothermal–isochoric nucleation in supersaturated vapour. Free molecules (black dots) called monomers in the initial stage (left) of the aggregation process form a cluster (spherical droplet) of certain size n, coexisting with gas of free molecules, afterwards (right).

The starting point is the one–dimensional one–step master equation (11) describing condensation and evaporation of a single particle on or from a molecular cluster, where $w_+(n)$ and $w_-(n)$ are transition rates of condensation and evaporation, respectively, which have to be formulated now.

The attachment probability per time of a monomer to a (spherical) cluster of size n is proportional to cluster surface $A(n)$ and to density of free monomers N_0/V, i. e.

$$w_+(n) = \alpha A(n) \left(N_{total} - n\right)/V \quad , \qquad 1 \leq n \leq N_{total} \, . \tag{22}$$

The coefficient α is up to now not specified (interpretation: velocity of sticking). The surface of spherical droplet is given by

$$A(n) = 4\pi r^2 = 4\pi \left(c_{clust} 4\pi/3\right)^{-2/3} n^{2/3} \sim n^{2/3} \tag{23}$$

with known incompressible particle density inside the cluster c_{clust} = const (liquid density as given experimental value). A special case is the formation of a precluster $n = 1$ out of an elementary particle ($n = 0$). The precluster can be understood as an excited monomer which is able to react with some other monomer to form a dimer ($n = 2$). In the free particle state $n = 0$ any of the N_{total} monomers can become excited, so that we can write

$$w_+(0) = \frac{p}{\tau} N_{total} \tag{24}$$

where the dimensionless parameter p in this case means the excitation probability per time multiplied by the time constant τ.

By using the detailed balance relation (16), the evaporation rate $w_-(n)$ of a monomer from a cluster of size n is calculated from the known attachment rate (22), as shown in Appendix, i. e.,

$$w_-(n) = \alpha A(n) \frac{1}{\lambda_0^3} \exp\left(\frac{f_n - f_{n-1}}{k_B T}\right) . \tag{25}$$

Here $f_n - f_{n-1}$ is the difference of binding energies between clusters of size n and $n-1$ and $\lambda_0(T)$ is de Broglie wave length of a monomer (particle of mass m) given by

$$\lambda_0 = h/(2\pi m k_B T)^{1/2} \approx 10^{-10} \text{ m} . \tag{26}$$

This is the wavelength of a quantum–mechanical free particle with energy $E = p^2/2m = \hbar^2 k^2/2m$, where k is the wave number related to the wavelength λ_0.

Clusters as bound states of elementary particles (monomers) have negative potential energy, which is the so–called binding energy. The potential function $f_n(T)$ is well known from atomic and nuclear theory and also the Bethe–Weizsäcker–formula [27–29] which in a simple nonlinear approximation reads

$$f_n(T) = \mu_\infty(T)\, n + \sigma A(n) . \tag{27}$$

The binding energy consists of a negative volume term ($\mu_\infty < 0$) and a positive surface contribution. The quantity $\mu_\infty(T)$ is the chemical potential of one monomer or, in other words, the energy necessary for taking away one elementary particle (monomer) from a cluster with a flat surface. The parameter σ can be understood as the surface tension of a flat surface. Ansatz (27) is a good approximation for large enough sizes n and provides also correct normalization condition $f_0 = 0$ for a free particle ($n = 0$). Substituting (27) into the detachment rate (25) we obtain the approximation

$$w_-(n) = \alpha A(n) \frac{1}{\lambda_0^3} \exp\left\{ \frac{\mu_\infty(T) + \sigma\left[A(n) - A(n-1)\right]}{k_B T} \right\}$$

$$\approx \alpha A(n) \frac{1}{\lambda_0^3} \exp\left(\frac{\mu_\infty(T)}{k_B T} \right) \exp\left(\frac{2\sigma k(n)}{c_{clust} k_B T} \right) . \tag{28}$$

This result is valid for large enough clusters (starting with $n \approx 10$) and contains the curvature $k(n)$ of a size–n–droplet

$$k(n) = 1/r = \left(c_{clust} 4\pi/3\right)^{1/3} n^{-1/3} . \tag{29}$$

The difference of surface areas $A(n) - A(n-1)$ in (28) can be evaluated by using a series expansion of $A(n-1)$ around n and retaining the leading term only, i. e.

$$n^{2/3} - (n-1)^{2/3} \approx n^{2/3} - n^{2/3}\left(1 - \frac{2}{3n}\right) \sim n^{-1/3} . \tag{30}$$

Taking into account the ideal gas model, the chemical potential μ_∞ is related in a simple way to the equilibrium density (concentration) $c_{eq}(\infty)$ of monomers for flat interface ($r \to \infty$) between liquid phase (droplet) and gaseous phase (free monomers). Thus, we have

$$\mu_\infty(T) = k_B T \ln\left[\lambda_0^3 c_{eq}(\infty)\right] . \tag{31}$$

However, in reality due to spherical droplets the interface is curved, therefore the concentration of free monomers in equilibrium is larger than $c_{eq}(\infty)$.

For large enough clusters, the detachment probability

$$w_-(n) = \alpha A(n) c_{eq}(\infty) \exp(\ell k(n)) \tag{32}$$

is obtained by inserting (31) into (28), where the length $\ell = \ell(T)$ defined as

$$\ell(T) = \frac{2\sigma}{c_{clust} k_B T} , \tag{33}$$

is explained and visualised in Fig. 2.

Considering the cluster sizes $n(t)$ as continuous variable which can be measured experimentally, the equation of motion dn/dt with a given velocity function $v(n)$, showing the time evolution of the cluster size, is well known. Putting forward phenomenological arguments like Fick's law the dynamical equation of reaction limited aggregation reads

$$\frac{dn}{dt} = v(n) \quad \text{with} \quad v(n) = \frac{D}{\ell} A(n) \left(c_{free} - c_{eq}(n)\right) , \tag{34}$$

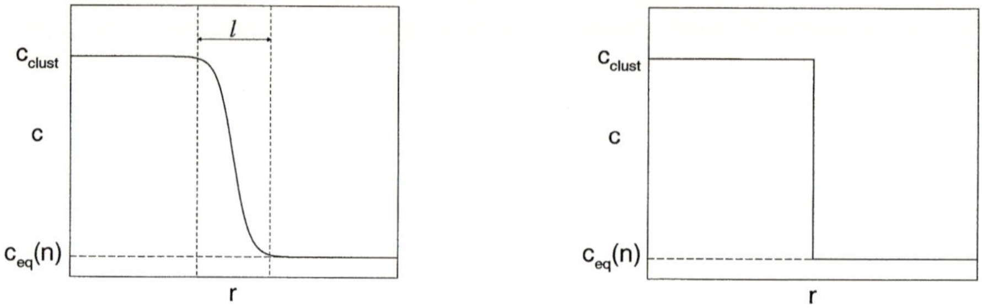

Fig. 2. Particle concentration c depending on the distance r from the center of cluster. The left picture shows a real density profile, whereas the right one approximates the profile corresponding to a model of cluster with sharp border. The length ℓ shows the width of the interface between the dense phase (c_{clust}) and the dilute surrounding ($c_{eq}(n)$) over a size–n–droplet.

where $c_{free} = N_0/V = (N_{total} - n)/V = c - n/V$ is the density of free particles, and $c = N_{total}/V$ is the total density of particles. The constant D is called diffusion coefficient; the coefficient ℓ called capillary length is a small interface thickness defined by (33).

In the stochastic approach an equation of the same type as (34) can be obtained for the average cluster size $\langle n \rangle$. Based on the master equation (11), we get a deterministic equation for the mean value

$$\frac{d}{dt}\langle n \rangle = \frac{d}{dt}\sum_n nP(n,t) = \langle w^+(n) \rangle - \langle w^-(n) \rangle \;, \tag{35}$$

which can be written in a certain approximation as follows

$$\frac{d}{dt}\langle n \rangle \approx w^+(\langle n \rangle) - w^-(\langle n \rangle) \;, \tag{36}$$

describing the time evolution of the average cluster size $\langle n \rangle$. According to the definitions of the transition frequencies (22) and (32), the time evolution of the mean cluster size (36) can be written in the same form as (34),

$$\frac{d}{dt}\langle n \rangle = \alpha A(\langle n \rangle) \left[\frac{N_{total} - \langle n \rangle}{V} - c_{eq}(\infty) e^{\ell k(\langle n \rangle)} \right] . \tag{37}$$

By comparing these two equations we find the up to now unknown coefficient α and the equilibrium concentration $c_{eq}(n)$, i. e.,

$$\alpha = D/\ell \tag{38}$$

and

$$c_{eq}(n) = c_{eq}(\infty)\, e^{\ell k(n)} \;. \tag{39}$$

The only difference between (34) and (37) is that in the latter case we always have the average value of the cluster size $\langle n \rangle$ instead of n. Rewritting (37) we get finally

$$\frac{d\langle n \rangle}{dt} = \frac{D}{\ell} A(\langle n \rangle) \left(c - \frac{\langle n \rangle}{V} - c_{eq}(\infty) e^{\ell k(\langle n \rangle)} \right) . \tag{40}$$

In the stationary state $d\langle n \rangle / dt = 0$ holds. Equation (40) is valid at large enough n only, whereas at $\langle n \rangle \to 0$ it should be modified to ensure that the transition rates and, therefore, the expression in the brackets does not diverge. Then we obtain three stationary solutions. The homogeneous situation without any cluster corresponds to $A(\langle n \rangle_{st}) = 0$, or to zero value of the stationary cluster size $\langle n \rangle_{st} = 0$. The other two solutions, describing the heterogeneous situation, originate from the identity $c_{free}(\langle n \rangle_{st}) = c_{eq}(\langle n \rangle_{st})$ or in extended version

$$c - \frac{\langle n \rangle_{st}}{V} = c_{eq}(\infty) \exp \left[\ell \left(c_{clust} 4\pi/3 \right)^{1/3} \langle n \rangle_{st}^{-1/3} \right] . \tag{41}$$

From this equation we can find the stationary cluster size $\langle n \rangle_{st}$ as function of the total density c. This is a nonlinear equation which cannot be solved analytically. However, it can be easily analysed (solved) graphically. In Fig. 3 terms on the l. h. s. (thick straight line) and on the r. h. s. (curved line)

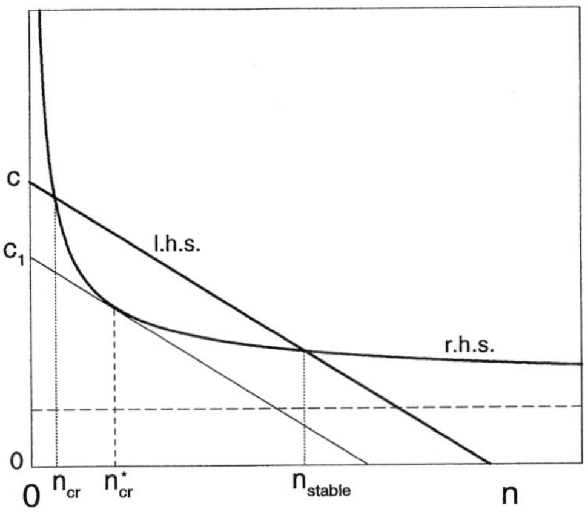

Fig. 3. Terms on the l. h. s. (straight lines) and on the r. h. s. (curved line) of (41) depending on the cluster size n. The two crossing points at $n = n_{stable}$ (stable cluster size) and $n = n_{cr}$ (unstable or critical cluster size) correspond to two different solutions of (41). The thin straight line related to the critical density c_1 has a single common point with the curved line at $n = n^*_{cr}$. The horizontal dashed line shows the value of equilibrium concentration $c_{eq}(\infty)$.

vs $\langle n \rangle$ are shown. The two crossing points $\langle n \rangle = n_{cr}$ and $\langle n \rangle = n_{stable}$ correspond to two different solutions of (41). The quantity n_{cr} is known as the critical cluster size in nucleation theory, whereas n_{stable} represents the stable stationary cluster size. Their meaning will be clarified in further discussion. The crossing points exist only if the total concentration c exceeds some critical value $c > c_1$ which corresponds to a bifurcation point where both solutions merge into one, as it is shown by thin straight line which has only one common point with the curved line at the marginal (largest possible) value of the critical cluster size $\langle n \rangle_{st} = n_{cr}^*$. At $c > c_1$ three different regions can be distinguished for the cluster size $\langle n \rangle_{st}$:

I. At $\langle n \rangle < n_{cr}$ we have $d\langle n \rangle/dt < 0$ which means that the cluster dissolves.
II. At $n_{cr} < \langle n \rangle < n_{stable}$ we have $d\langle n \rangle/dt > 0$ which means that the cluster growth until reaches the stable stationary size n_{stable}.
III. At $\langle n \rangle > n_{stable}$ we have $d\langle n \rangle/dt < 0$ which means that the cluster reduces its size (dissolves) to the stationary value n_{stable}.

According to this, the solution $\langle n \rangle = n_{stable}(c)$ corresponds to a stable cluster size, whereas the solution $\langle n \rangle = n_{cr}(c)$ to an unstable stationary cluster size. In Fig. 4 we have shown both solutions of (41) (branches $n_1(c)$ and $n_2(c)$),

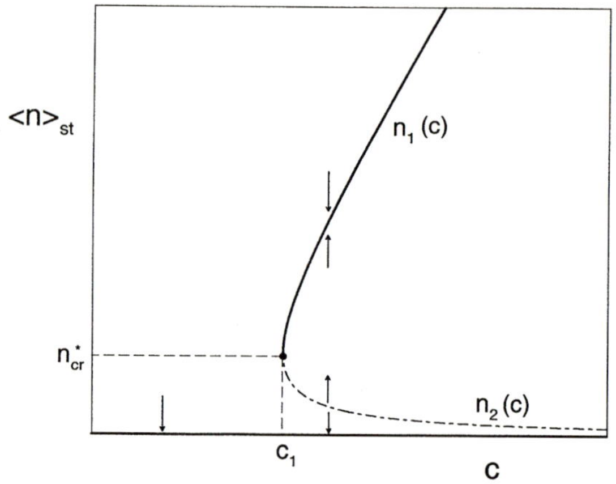

Fig. 4. The stationary cluster size $\langle n \rangle_{st}$ depending on the total density c of particles in a supersaturated vapour. The stable cluster size (the horizontal line and the branch $n_1(c)$) is shown by thick solid lines, whereas the unstable cluster size (the branch $n_2(c)$) is shown by dot–dashed line. Arrows indicate the time evolution of $\langle n \rangle$.

providing the stationary cluster size in a supersaturated vapour as a function of total density c. Branch $n_1(c)$ depicted by thick solid line corresponds to the stable cluster size, whereas branch $n_2(c)$ indicated by dot–dashed line correspond to the unstable (critical) cluster size. Several trajectories showing the time evolution of $\langle n \rangle$ to one of the stable stationary values (thick lines) are indicated by arrows.

Expansion of the exponent in (39), retaining the linear term only, yields

$$c_{eq}(\langle n \rangle) = c_{eq}(\infty)(1 + \ell k(\langle n \rangle)) . \tag{42}$$

By using this linearization around the critical cluster size n_{cr}, (40) can be written in the well–known form [28]

$$\frac{d\langle n \rangle}{dt} = Dc_{eq}(\infty)A(\langle n \rangle)\,[k(n_{cr}) - k(\langle n \rangle)] . \tag{43}$$

From this equation the above discussed property that clusters with an overcritical size ($n > n_{cr}$, i. e. $k(n) < k(n_{cr})$) grow, whereas those with an undercritical size ($n < n_{cr}$, i. e. $k(n) > k(n_{cr})$) dissolve, is obvious. In the actual bistable situation the growth from an undercritical to an overcritical cluster size cannot be described by deterministic equations of motion like (43). This phenomenon of noice–induced transitions over the critical value of cluster size can be treated in the stochastic approach only.

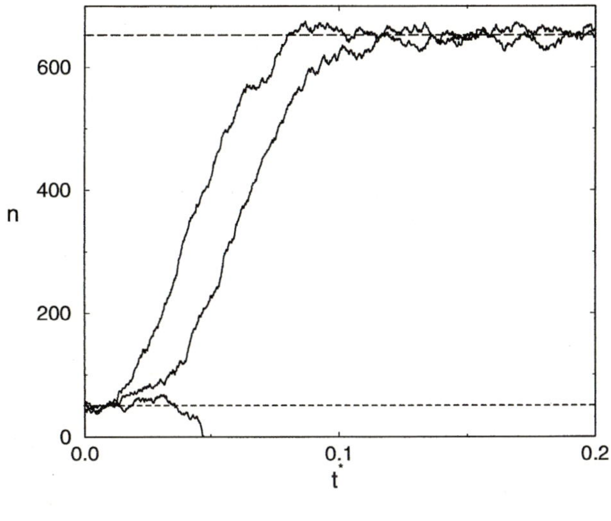

Fig. 5. Three different stochastic trajectories showing the time evolution of the cluster size n vs the dimensionless time t^*. The lower dashed line indicates the critical cluster size $n_{cr} = 54$; the upper dashed line represents the stable cluster size $n_{stable} = 650$.

We have simulated by Monte Carlo method and have presented in Fig. 5 three different stochastic trajectories showing the time evolution, i. e. the

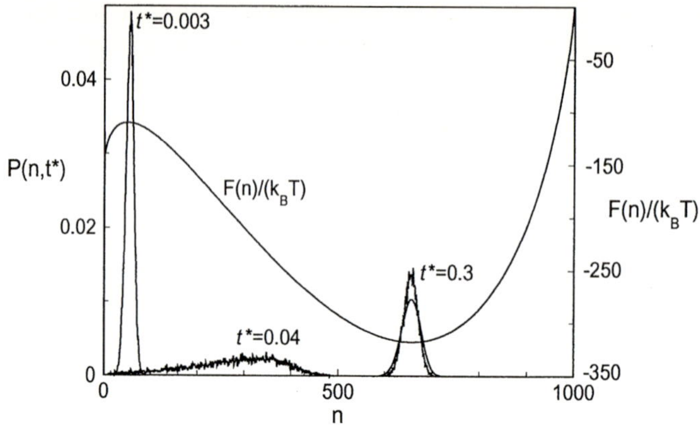

Fig. 6. The probability distribution $P(n, t^*)$ at three different dimensionless time moments $t^* = 0.003$ (left), $t^* = 0.04$ (middle), and $t^* = 0.3$ (right) calculated by averaging over 20 000 stochastic trajectories simulated by Monte Carlo method. The equilibrium distribution is shown by smooth solid line. The maximum of the equilibrium distribution corresponds to the minimum of free energy, i. e. $F(n)/(k_B T)$ (smooth curve).

cluster size n vs the dimensionless time $t^* = (\alpha A(1)/V) \cdot t$, of the system with $N_{total} = 1000$ particles starting with n values around the critical cluster size $n_{cr} \approx 54$ (the lower dashed line). The parameters of the system are choosen such that $\sigma A(1)/(k_B T) = 10$ and $V c_{eq}(\infty) = 160$. In one of the cases the cluster dissolves, whereas in other two cases it grows over the critical size. Besides, in distinction to the prediction of the deterministic equation (40), in one of the cases the growth up to the stable cluster size $n \approx 650$ (the upper dashed line) occurs starting with $n = 45 < n_{cr}$. The probability distribution at three different time moments, i. e. $t^* = 0.003$, 0.04, and 0.3, have been calculated by averaging over a large number of stochastic trajectories starting with the $n = n_{cr} = 54$. The results are shown in Fig. 6. The probability maximum moves towards larger values of the cluster size n with increasing time, and at $t^* = 0.3$ the probability distribution agrees approximately with the equilibrium distribution $P^{eq}(n) \propto \exp(-F(n)/(k_B T))$ (smooth solid line) given by (17). The maximum of the equilibrium distribution corresponds to the minimum of free energy, whereas the critical cluster size to the local maximum of free energy, as shown in Fig. 6.

4 Car Cluster Formation on Roads

The aggregation of particles out of an initially homogeneous situation is well known and it takes place in physical and in nonphysical systems, like traffic flow, and can be described similarly [12,15]. Traffic flow theory has a long history [24] and is still alive [7]. Theoretical approaches based on deterministic car following models [1] and particle hopping models [2] are well known. An experimental study of phase transitions in traffic flow has been presented by Kerner [13], the stochastic description is provided in [11,16–18].

In analogy to the droplet formation in supersaturated vapour, we consider a model of traffic flow on a one–lane road according to which N cars are moving along a circle of length L, as illustrated in Fig. 7 where two different regimes of traffic flow are shown, i. e., free flow (left) and congested traffic flow with two car clusters (right). For simplicity, we have restricted our further

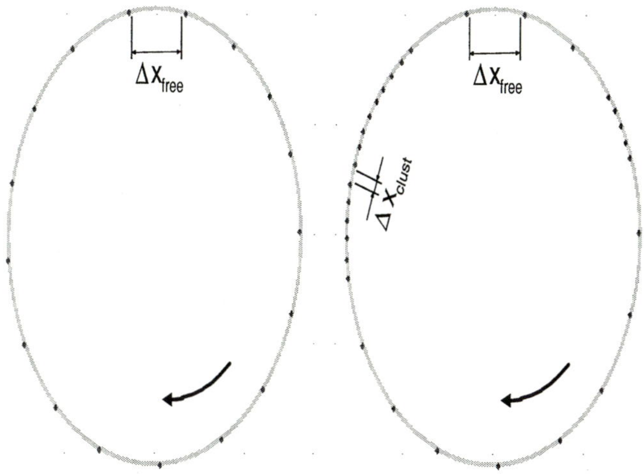

Fig. 7. Free traffic flow (left) and congested traffic flow (right) on the one–lane circular road. Each black dot represents a car and its direction of motion is indicated by arrow.

analysis to the simplest model where only one car cluster (jam) can exist. If a road is crowded by cars, each car requires some minimal space or length which, obviously, is larger than the real length of a car. We call this the effective length ℓ of a car. The maximal velocity of each car is v_{max}. Following the idea of Bando et al. [1], the behaviour of individual drivers is described by the optimal velocity model. The desired (optimal) velocity v_{opt}, depending on the distance between two cars Δx, is given in dimensionless variables $w_{opt} = v_{opt}/v_{max}$ and $\Delta y = \Delta x/\ell$ by the formula [16–18]

$$w_{opt}(\Delta y) = \frac{(\Delta y)^2}{d^2 + (\Delta y)^2} , \tag{44}$$

where the parameter $d = D/\ell$ is the interaction distance. D is the distance between two cars corresponding to the velocity value $v_{max}/2$.

Measurements on highways have been shown that the density c_{clust} of cars in congested traffic is independent of the size of the dense phase (jam). As a consequence the distance between jammed cars, the spacing $\Delta x_{clust} = \ell \Delta y_{clust} =$ const ≥ 0, is well known and has to be treated as a given measured quantity.

The length of the cluster (jam) depending on the number of congested cars n is defined by $L_{clust} = \ell n + (n-1)\Delta x_{clust}$. According to this, the average distance $\Delta x_{free} = \ell \Delta y_{free}$ between two cars outside the jam (or free cars) distributed over the free part of the road with length $L_{free} = L - L_{clust}$ is given by

$$\Delta y_{free}(n) = \frac{L/\ell - N - S(n)\Delta y_{clust}}{N - S(n)} , \tag{45}$$

where N is the total number of cars on the road (circle of length L) and $S(n) = n - 1 + \delta_{n,0}$ is the number of fixed spacings Δx_{clust} between cars in jam.

The traffic flow is described as a stochastic process where adding a vehicle to a car cluster of size n is characterized by a transition frequency (attachment probability per time unit) $w_+(n)$ and the opposite process by a frequency $w_-(n)$. The number n of cars in the cluster is the stochastic variable which may have values from 0 to N. The basic equation for the evolution of the probability distribution $P(n,t)$ to find a cluster of size n at time t with probability P is known as master equation (11). The one–dimensional stochastic equation reads

$$\begin{aligned} \frac{1}{\tau}\frac{dP(n,T)}{dT} &= w_+(n-1)P(n-1,T) - [w_+(n) \\ &\quad + w_-(n)]\, P(n,T) + w_-(n+1)P(n+1,T) \, . \end{aligned} \tag{46}$$

where $T = t/\tau$ is the dimensionless time. The time constant τ will be specified below.

The main task is to formulate expressions for both transition probabilities w_+ and w_-. We have assumed that the detachment frequency $w_-(n)$ or the average number of cars leaving the cluster per time unit is a constant independent of cluster size n. The ansatz for $w_+(n)$ is now formulated allowing for Δx_{clust} to be nonzero. Our general assumption is that a vehicle changes the velocity from $v_{opt}(\Delta x_{free})$ in free flow to $v_{opt}(\Delta x_{clust})$ in jam and approaches the cluster as soon as the distance to the next car (the last car in the cluster) reduces from Δx_{free} to Δx_{clust}. This assumption allows one to calculate the average number of cars joining the cluster per time unit or the attachment frequency to an existing cluster $w_+(n)$. The case $w_+(0)$ have to be considered separately. This is a probability per time of a stochastic event that a driver randomly reduces the velocity of his car to $v_{opt}(\Delta x_{clust})$ without obvious reason, thus forming a precluster or cluster of size $n = 1$. An ansatz for $w_+(0)$ similar to that we have used in the case of supersaturated vapour is appropriate. Thus, we have the ansatz which in dimensionless quantities reads

$$w_+(0) = \frac{p}{\tau} N \tag{47}$$

$$w_+(n) = \frac{b}{\tau} \frac{w_{opt}(\Delta y_{free}(n)) - w_{opt}(\Delta y_{clust})}{\Delta y_{free}(n) - \Delta y_{clust}} \tag{48}$$

$$w_-(n) = 1/\tau = \text{const} , \tag{49}$$

where $b = v_{max}\tau/\ell$ denotes a dimensionsless parameter and p is the stochasticity. The parameter τ is a time constant, which can be understood as the waiting time for the escape (detachment) of first car out of the jam into free flow [16].

The master equation (46) has been solved in Refs. [16–18] with an aim to extract from this solution information about the formation of traffic jams and about the various possible regimes of traffic flow depending on the parameters of the system, as well as to calculate the flux–density relationship, called fundamental diagramm of traffic flow. Our purpose is to study the dynamics of the cluster (jam) growth and dissolution by finding stable and unstable stationary states, and compare the results with aggregation in supersaturated vapour.

Like in the case of the cluster growth in supersaturated vapour, the deterministic equation (35) for growth of a single car cluster on the road is written as (36). The stationary cluster size $\langle n \rangle_{st}$ can be calculated from the condition $d\langle n \rangle/dt = 0$ or

$$\frac{w_+(\langle n \rangle)}{w_-(\langle n \rangle)} = b\frac{w_{opt}(\Delta y_{free}(\langle n \rangle)) - w_{opt}(\Delta y_{clust})}{\Delta y_{free}(\langle n \rangle) - \Delta y_{clust}} = 1 \tag{50}$$

consistent with the ansatz for transition probabilities (48) and (49). By using the definition (44) of the optimum velocity function $w_{opt}(\Delta y)$ we obtain the equation

$$b\left[\frac{(\Delta y_{free}(\langle n \rangle))^2}{d^2 + (\Delta y_{free}(\langle n \rangle))^2} - \frac{(\Delta y_{clust})^2}{d^2 + (\Delta y_{clust})^2}\right] = \Delta y_{free}(\langle n \rangle) - \Delta y_{clust} \tag{51}$$

which can be solved with respect to Δy_{free}. One solution of the third order equation (51) is trivial $\Delta y_{free} = \Delta y_{clust}$. The other two solutions, which have certain physical meaning, read

$$\Delta y_{free}^{(1,2)} = \frac{d}{2[d^2 + (\Delta y_{clust})^2]} \times \left\{bd \pm \sqrt{b^2 d^2 + 4b\Delta y_{clust}[d^2 + (\Delta y_{clust})^2] - 4[d^2 + (\Delta y_{clust})^2]^2}\right\} . \tag{52}$$

According to (52), the headway Δy_{free} between cars in a free flow coexisting with a single cluster has constant value depending merely on the control parameters of the model. Now, by means of (45), which states the relation between Δy_{free} and n, we are able to calculate the stationary cluster size $\langle n \rangle_{st}$. As already pointed out we set $S(n) = n - 1 + \delta_{n,0} \approx n$, assuming that the cluster (if it exists) contains a large number of cars. This leads to the equation

$$\frac{\langle n \rangle_{st}}{L/\ell} = \frac{c\left(1 + \Delta y_{free}^{(1,2)}\right) - 1}{\Delta y_{free}^{(1,2)} - \Delta y_{clust}} , \tag{53}$$

where $c = \ell N/L$ is the total density of cars and the term on the left hand side of the equation, in fact, is the relative part of the road crowded by cars. In this case $\Delta y_{free}^{(1,2)}$ has a constant value given by (52). Result (53) makes sense at large enough densities where it provides a positive value of $\langle n \rangle_{st}$. The solution with the largest value $\Delta y_{free}^{(1)}$ (positive sign in (52)) gives the average stationary size of a stable cluster depending on the total density c within the region $c > c_1$, where

$$c_1 = \frac{1}{1 + \Delta y_{free}^{(1)}} \tag{54}$$

is the critical density having the same meaning as in the case of cluster growth in supersaturated vapour. There is another critical density c_2, given by

$$c_2 = \frac{1}{1 + \Delta y_{free}^{(2)}} , \tag{55}$$

which defines the region $c > c_2$ where an unstable car cluster corresponding to the solution (53) with the smallest value $\Delta y_{free}^{(2)}$ of the headway can exist. In a special case of vanishing bumper–to–bumper distance in jam $\Delta y_{clust} = 0$ our result (52) reduces to

$$\Delta y_{free}^{(1,2)} = \frac{b}{2} \pm \sqrt{\frac{b^2}{4} - d^2} , \tag{56}$$

and (53) to

$$\frac{\langle n \rangle_{st}}{L/\ell} = c + \frac{c - 1}{\frac{b}{2} \pm \sqrt{\frac{b^2}{4} - d^2}} . \tag{57}$$

The value of $\langle n \rangle_{st}\, \ell/L$ with sign '+' in (57) corresponds to the stable, whereas that with sign '−' – to the unstable stationary cluster size. In Fig. 8 we have shown both branches $n_1(c)$ (stable) and $n_2(c)$ (unstable) as a function of total density c.

This result, obtained in the simple one–cluster model discussed here, remains true in the multi–cluster model, outlined in [12] and discussed in detail

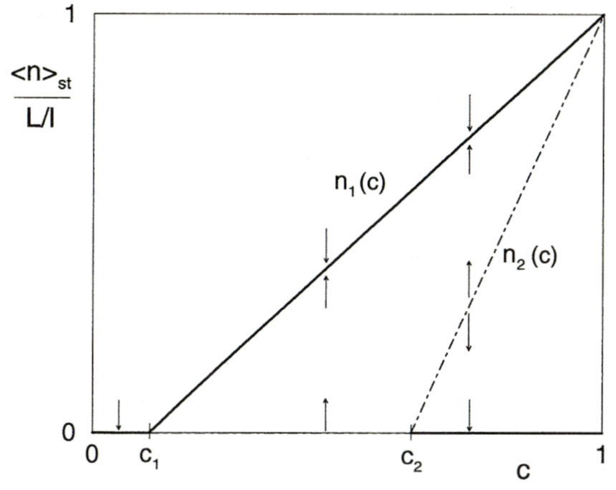

Fig. 8. The stationary cluster size $\langle n \rangle_{st}$ normalized to L/ℓ (i. e., the relative part of the road crowded by cars) depending on the total density of cars c. The stable cluster size (branch $n_1(c)$ and horizonta lines) is shown by thick solid lines, whereas the unstable cluster size (branche $n_2(c)$) and a horizontal line) is shown by dot–dashed lines. Arrows indicate the time evolution of $\langle n \rangle$. Parameters: $b = 8.5$, $d = 13/6$, and $\Delta y_{clust} = 0$.

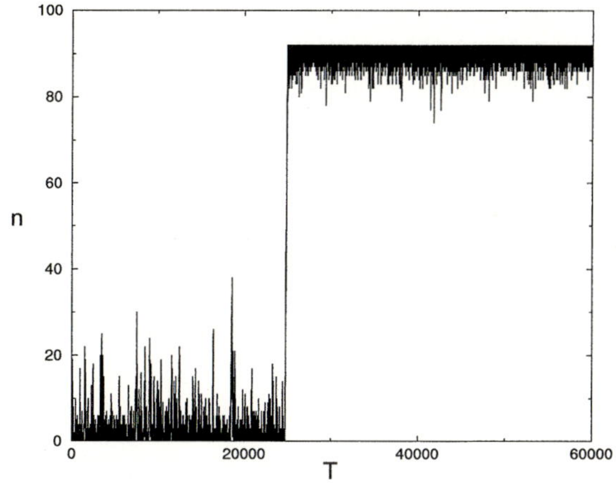

Fig. 9. A stochastic trajectory showing the total number of congested cars n vs dimensionless time T.

in [11], if the size of the system is large ($N \to \infty$) and stochasticity p is small. Note only that n, in general, is defined as the total number of congested cars.

Like in the case of supersaturated vapour, the growth of the car cluster starting with the undercritical (i. e., smaller than the unstable) cluster size can be described within a stochastic approach only. Fig. 9 presents a typical stochastic trajectory showing a noice–induced switching from free flow with n values fluctuating below the unstable cluster size (in this example $N = 92$, $c = 0.7886$, $b = 8.5$, $d = 13/6$, $\Delta y_{clust} = 1/6$, which corresponds to the critical cluster size about 53 and the stable cluster size about 91) to the state whith large stable cluster including almost all cars. Note, however, that we have two different kind of bifurcations at $c = c_1$ in Figs. 4 and 8. In distinction to the actual model of traffic flow (Fig. 8), in the case of the supersaturated vapour (Fig. 4) there is a jump in the cluster size. Besides, in the traffic flow we do not have the critical cluster size at intermediate densities $c_1 < c < c_2$, which means that even a very small car cluster at the initial time moment has tendency to grow up to the stationary size. At large densities ($c > c_2$) the critical cluster size $n_2(c)$ appears as well in our traffic flow model. The above discussed distinguishing feature disapears in an advanced traffic flow model discussed further on.

5 An Advanced Model of Car Cluster Formation

Here we propose an essential innovation in our traffic flow model: now the detachment frequency depends on the cluster size. In analogy to physical systems like droplets in supersaturated vapour, one can expect that smaller clusters dissolve easier, i. e., $w_-(n)$ increases remarkably when the cluster size becomes smaller than some characteristic value n_0. In traffic flow, this idea is based on a concept of higher irregularity of the cluster structure when the size is small. The increase in $w_-(n)$ might be related also to some multilane effect, i. e., due to several possible overtaking maneuvers, clusters consisting of few cars are particularly unstable. In some approximation our one–lane model effectively describes a multilane freeway. In this case n and N are the number of congested cars and the total number of cars per one lane. The above mentioned multi–lane effect is simulated by choosing an appropriate ansatz for $w_-(n)$. Our specific choice is to replace (49) with

$$w_-(n) = \frac{1}{\tau}\left[1 + \beta\left(\frac{n_0}{n+n_0}\right)^s\right] \qquad : \quad n \geq 1\,, \tag{58}$$

where β and s are positive constants. Assuming $n_0 \gg 1$, the parameter $\beta \approx (w_-(1) - w_-(\infty))/w_-(\infty)$ shows the relative increase in $w_-(n)$ when the cluster size n is reduced from large values to 1. The parameter s is responsible for the fastness of $w_-(n)$ convergence to its asymptotic value $w_-(\infty) = 1/\tau$ at $n \to \infty$. The specific form of $w_-(n)$ is choosen somewhat arbitrarily to get a simple but still realistic description.

The deterministic equation (51) for the the stationary mean cluster size in our advanced model becomes

$$b\frac{w_{opt}(\Delta y_{free}(\langle n\rangle)) - w_{opt}(\Delta y_{clust})}{\Delta y_{free}(\langle n\rangle) - \Delta y_{clust}} = 1 + \beta\left(\frac{n_0}{n+n_0}\right)^s \tag{59}$$

consistent with the ansatz for transition probabilities (48) and (58). Equation (59) has been solved together with (44) and (45) in the approximation $S(n) \approx n$,

$$\Delta y_{free}(n) = \frac{(L/\ell)\,(1-c) - n\,\Delta y_{clust}}{c\,(L/\ell) - n}\,, \tag{60}$$

at realistic values of the dimensionless control parameters $b = 8.5$, $d = 13/6$, $\Delta y_{clust} = 1/6$ estimated in [16]. Three different values of the parameter β, $\beta = 0$, 0.8, and 1.6, have been considered. Other parameters have been choosen $L/\ell = 500$, $s = 1$, and $n_0 = 10$. We have calculated the stationary mean cluster sizes $\langle n \rangle_{st}$ corresponding, in general, to different stationary solutions of these equations. Besides, the stable free flow state with $\langle n \rangle_{st} = 0$ has been identified as a state with $d\langle n \rangle / dT < 0$ at $\langle n \rangle \to 0$. The corresponding dimensionless fluxes

$$j = J\tau = b\left[w_{opt}(\Delta y_{clust})\frac{\ell\langle n \rangle_{st}}{L} + w_{opt}(\Delta y_{free}(\langle n \rangle))\left(c - \frac{\ell\langle n \rangle_{st}}{L}\right)\right] \tag{61}$$

have been calculated as well, where J is the flux averaged over the whole road or, which is the same, over an infinitely long time interval at a fixed coordinate. In this case we have neglected the stochastic fluctuations taking $n \equiv \langle n \rangle_{st}$. The results are shown in Figs. 10 and 11.

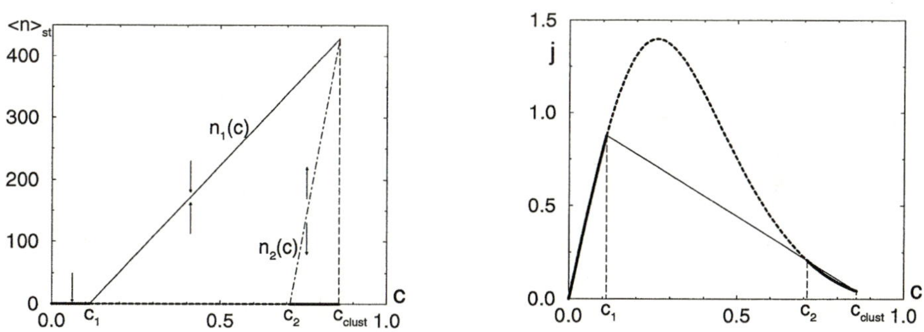

Fig. 10. Stationary cluster size (left) and flux (right) vs concentration calculated for finite road at dimensionless control parameters $\beta = 0, b = 8.5, d = 13/6, \Delta y_{clust} = 1/6, L/\ell = 500$. The thick and the thin solid lines correspond to stable free and congested states, respectively. Dashed lines represent an unstable free flow, dot–dashed line shows the unstable (critical) cluster size.

In Fig. 10 we have illustrated the behavior of the stationary cluster size $\langle n \rangle_{st}$ (left) and the corresponding flux (right) in a special case of parameter $\beta = 0$ where our advanced model reduces to the previous one discussed in Sec. 4. As a result, the stationary cluster size behaves in a similar way as

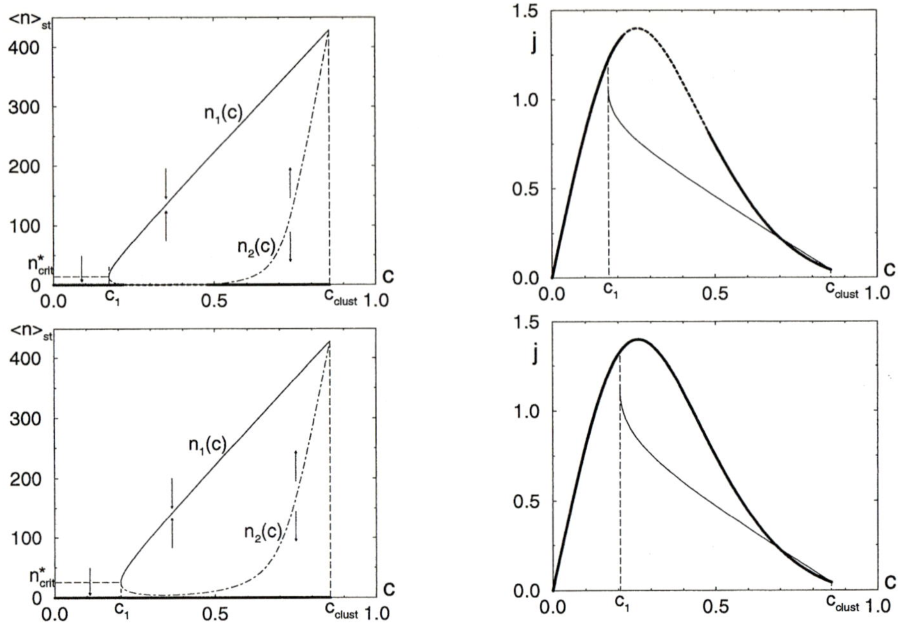

Fig. 11. Stationary cluster size (left) and flux (right) vs concentration calculated for finite road at $\beta = 0.8$ (top) and $\beta = 1.6$ (bottom). Other dimensionless control parameters are $n_0 = 10, s = 1, b = 8.5, d = 13/6, \Delta y_{clust} = 1/6, L/\ell = 500$. The meaning of solid, dashed, and dot–dashed lines is the same as in Fig. 10.

in Fig. 8, with the only essential difference that $\Delta y_{clust} > 0$. In this case the free flow with $\langle n \rangle_{st} = 0$ is stable up to some critical density $c = c_1$. Then a car cluster appears which growth linearly in size with increasing of c from c_1 to the maximum value $c_{clust} = 1/(1 + \Delta y_{clust})$ (the density in cluster), as shown in Fig. 10 by thin solid line. This corresponds to one of the stationary solutions $\langle n \rangle = n_1(c)$ of (59). At densities $c_2 < c < c_{clust}$ another positive solution $\langle n \rangle = n_2(c)$ exists (dot–dashed line) which corresponds to an unstable or critical cluster size. $(dn/dt > 0)$ if $n < n_1(c)$. The most probable time evolution of n, given by the deterministic equation (36), depending on the density c and initial conditions is indicated in Fig. 10 by arrows. Like in the supersaturated vapour, a growth of cluster starting from undercritical size $n < n_2(c)$ and finishing with the stable cluster size $n = n_1(c)$ is possible, as well, but only due to stochastic fluctuations.

In Fig. 11 we have shown the same pictures at $\beta > 0$. A distinguishing feature is that a nonzero critical cluster size appears already at $c = c_1$ where a phase transition takes place with a jump–like increase in the stable cluster size $\langle n \rangle_{st}$ from 0 to n^*_{crit}. This behavior is the same as in supersaturated vapour. At small values of β, however, a density region with vanishing critical cluster size still exists ($\beta = 0.8$, top in Fig. 11) which disapers at larger β values ($\beta = 1.6$, bottom in Fig. 11). Note also that at $\beta = 0.8$ the free flow becomes unstable somewhat above c_1 (see $j(c)$ plots in Fig. 11), while both free and congested states of the system are stable at $\beta = 1.6$. At positive β, a jump–like decrease of the flux j is observed at the critical point $c = c_1$ when switching from the free flow (thick solid line) to the congested flow (thin solid line). This is a finite–size effect which disapears in the limit $N \to \infty$, since $n^*_{crit}/N \to 0$.

6 Conclusion

To conclude this paper we recall the key points of the developed approach. It has been shown that both supersaturated vapour and traffic flow can be described by stochastic master equations. Solutions of the master equation in each of the cases have shown the existence of phase transition from initially homogeneous state to a heterogeneous state where some of the particles (molecules or cars) are bounded in a cluster. The cluster growth dynamics in supersaturated vapour and in traffic flow has been studied in view of both deterministic and stochastic theory. Finally, the comparison of results has been made showing that at certain conditions the nucleation in traffic flow is similar to that in supersaturated vapour. In particular, in the model where the detachment frequency increases remarkably at small cluster sizes $n \lesssim n_0$ (cf. Sec. 5) an undercritical bifurcation exists at some car density c_1. It means that, like in the supersaturated vapour, small, i. e. undercritical, clusters tend to dissolve, whereas those of overcritical size tend to grow if the density exceeds c_1.

7 Appendix: Derivation of Thermodynamic Potential and Corresponding Transition Rate

Our system is described by the cluster distribution $\mathbf{N}$ defined by (18). The total Hamiltonian $H(\mathbf{N})$ reads

$$H = \sum_{n=0}^{N} H_n \tag{62}$$

with the contribution H_n for the N_n clusters of size n at coordinates $\mathbf{r}_i^{(n)}$ and momentum $\mathbf{p}_i^{(n)}$ written as kinetic energy and interaction potential

$$H_n\left(\mathbf{p}^{(n)}, \mathbf{r}^{(n)}\right) = \sum_i \frac{\left(p_i^{(n)}\right)^2}{2m_i^{(n)}} + \sum_{i<j} U_{ij}^{(n,n)}\left(\mid \mathbf{r}_i - \mathbf{r}_j \mid\right) . \tag{63}$$

The mass $m_i^{(n)}$ of a cluster containing n monomers ($n \geq 1$) is given by

$$m_i^{(n)} \equiv m_n = n\, m \tag{64}$$

where $m \equiv m_0$ is the mass of one monomer. The canonical partition function, i. e. the statistical integral, is an integral over all space and momentum coordinates. In the semiclassical approximation it reads [21]

$$Z(T,V,\mathbf{N}) = \prod_{n=0}^{N} \frac{1}{N_n!\, h^{3N_n}} \int d^{3N_n}p\, d^{3N_n}q\, \exp\left(-\beta H_n\right) \,, \tag{65}$$

where $\beta = 1/(k_B T)$. This partition function can be divided in two factors, one of which represents an ideal part due to the kinetic energy

$$Z_{ideal}(T,V,\mathbf{N}) = \prod_{n=0}^{N} \frac{V^{N_n}}{N_n!\, h^{3N_n}} \left(\sqrt{2\pi\, m_n k_B T}\right)^{3N_n} \tag{66}$$

and the second part $Z_{binding}(T,V,\mathbf{N})$ is responsible for the energy stored in clusters. In a certain approximation, both terms together read

$$Z(T,V,\mathbf{N}) = \prod_{n=0}^{N} \frac{V^{N_n}}{N_n!} \left[\left(\frac{\sqrt{2\pi\, m_n k_B T}}{h}\right)^3 \exp\left(-\frac{f_n}{k_B T}\right)\right]^{N_n} \tag{67}$$

where the binding energy $f_n(T)$ is the minimum value of the potential energy searched over all spatial arrangements of the n bounded monomers

$$f_n(T) = \min_{\mathbf{r}} \sum_{i<j} U_{ij}^{(n,n)}\left(\mid \mathbf{r}_i - \mathbf{r}_j \mid\right) \,. \tag{68}$$

From the canonical partition function we can calculate the thermodynamic quantities using the relation to state function free energy

$$F(T,V,\mathbf{N}) = -k_B T \,\ln Z(T,V,\mathbf{N}) \,. \tag{69}$$

According to (67) and (69), in the isothermal–isochoric situation we obtain

$$F(T,V,\mathbf{N}) = k_B T \sum_{n=0}^{N} \left[N_n \ln\left(\lambda_n(T)^3/V\right) + \ln N_n!\right] + \sum_{n=0}^{N} N_n f_n(T) \tag{70}$$

with de Broglie wave length $\lambda_n(T) = n^{-1/2}\lambda_0(T) = n^{-1/2}\left(h^2/(2\pi m k_B T)\right)^{1/2}$ (at $n \geq 1$). Using Stirling formula $\ln N_n! \simeq N_n \ln N_n - N_n$ in (70) we obtain an approximation for large cluster numbers N_n [15]

$$F(T,V,\mathbf{N}) = k_B T \sum_{n=0}^{N} N_n \left[\ln\left(\lambda_n(T)^3 N_n/V\right) - 1\right] + \sum_{n=0}^{N} N_n f_n(T) \,. \tag{71}$$

In the case of one cluster of size n only (see (20)) and at $N_0 = N_{total} - n \to \infty$ (i. e., in the thermodynamic limit by expansion of $\ln N_0!$) the free energy (70) reads

$$F(T,V,n) = k_B T \left\{ (N_{total} - n) \left[\ln \left(\lambda_0(T)^3 (N_{total} - n)/V \right) - 1 \right] + (1 - \delta_{n,0}) \ln \left(\lambda_n(T)^3 / V \right) \right\} + f_n(T) \,. \quad (72)$$

For the equilibrium cluster distribution, the most probable value of the cluster size n corresponds to the minimum of the thermodynamic potential, in this case free energy F, as it is evident from (17). The latter equation (with $\Omega \equiv F$) combined with the detailed balance condition (16) allows to find the relation between transition rates $w(\mathbf{N}' \mid \mathbf{N})$ and $w(\mathbf{N} \mid \mathbf{N}')$ of opposite stochastic events (transition from state $\mathbf{N}$ to state $\mathbf{N}'$ and vice versa), i. e.

$$\frac{w(\mathbf{N}' \mid \mathbf{N})}{w(\mathbf{N} \mid \mathbf{N}')} = \exp \left(\frac{F(T,V,\mathbf{N}) - F(T,V,\mathbf{N}')}{k_B T} \right) \,. \quad (73)$$

In our special case, where only one cluster of size n is possible, (73) in the thermodynamic limit reduces to

$$\frac{w_-(n)}{w_+(n-1)} = \frac{V\,(1 - 1/n)^{3/2}}{\lambda_0^3(T)\,(N_{total} - n)} \exp \left(\frac{f_n(T) - f_{n-1}(T)}{k_B T} \right) : n \geq 2 \quad (74)$$

$$\frac{w_-(1)}{w_+(0)} = \frac{1}{N_{total}} \exp \left(\frac{f_1(T)}{k_B T} \right) \,. \quad (75)$$

From (74) we obtain an approximation

$$\frac{w_-(n)}{w_+(n)} = \frac{V}{\lambda_0^3(T)\,(N_{total} - n)} \exp \left(\frac{f_n(T) - f_{n-1}(T)}{k_B T} \right) , \quad (76)$$

consistent with Eqs. (22) and (25), which is true for large enough n. In a rough approximation, we have assumed that (76) can be extrapolated up to $n = 1$. Based on (75) the stochasticity parameter p in (24) can be written as

$$\frac{p}{\tau} = w_-(1) \exp \left(-\frac{f_1(T)}{k_B T} \right) \,. \quad (77)$$

By (22), (25), and (77) all the transition rates are well defined in a way consistent with the basic principles of statistical mechanics.

Acknowledgement: The author would like to thank Jevgenijs Kaupužs (Riga), Ihor Lubashevsky (Moscow) and Peter Wagner (Berlin) for useful comments, criticism and help.

References

1. Bando, M., Hasebe, K., Nakayama, A., Shibata, A., Sugiyama, Y.: Structure stability of congestion in traffic dynamics, Japan J. Indust. and Appl. Math. **11** (1994) 203

2. Chowdhury, D., Santen, L., Schadschneider, A.: Statistical physics of vehicular traffic and some related systems, Phys. Rep. **329** (200) 199
3. Gardiner, C. W.: *Handbook of stochastic methods* (Springer, Berlin, 1983). (1st ed.), 1985, 1990
4. Haberland H. (Ed.): *Clusters of atoms and molecules* (Springer, Berlin, 1994).
5. Haken, H.: *Synergetics. An introduction* (Springer, Berlin, 1978) (1st ed.), 1983
6. Heermann, D. W.: *Computersimulation methods in theoretical physics* (Springer, Berlin, 1990).
7. Helbing, D.: *Verkehrsdynamik. Neue physikalische Modellierungskonzepte* (Springer, Berlin, 1997).
8. Honerkamp, J.: Stochastische Dynamische Systeme, VCH Verlagsgesellschaft, Weinheim, 1990; Stochastic Dynamical Systems, VCH, New York, 1994
9. Jellinek, J. (Ed.): *Theory of atomic and molecular clusters* (Springer, Berlin, 1999).
10. van Kampen, N. G.: *Stochastic processes in physics and chemistry* (North–Holland, Amsterdam, 1981, 1992).
11. Kaupužs, J., Mahnke, R.: A stochastic multi–cluster model of freeway traffic, Europ. Phys. J. B **14** (2000) 793
12. Kaupužs, J., Mahnke, R.: Nucleation on roads, In: *Nucleation and Atmospheric Aerosols 2000* (Eds.: B. Hale and M. Kulmala), AIP Conf. Proceed., Vol. 535, p. 221
13. Kerner, B. S.: Phase transitions in traffic flow, In: *Traffic and Granular Flow '99* (Eds.: Helbing, D., Hermann, H. J., Schreckenberg, M., Wolf, D. E.), (Springer, Berlin, 2000) p. 253
14. Mahnke, R.: Aggregation phenomena to a single cluster regime under different boundary conditions, Z. Phys. Chem. **204** (1998) 85
15. Mahnke, R.: Nucleation in physical and non–physical systems, In: *Nucleation and atmospheric Aerosols 2000* (Eds.: B. Hale and M. Kulmala), AIP Conf. Proceed., Vol. 535, p. 229
16. Mahnke, R., Kaupužs, J.: Stochastic theory of freeway traffic, Phys. Rev. E **59** (1999) 117
17. Mahnke, R., Kaupužs, J.: Probabilistic description of traffic flow, In: *Special issue on traffic flow theory* (Ed.: Zhang, H. M.), Networks and Spatial Economics, Vol. 1, No. 1/2, March 2001, p. 103
18. Mahnke, R., Pieret, N.: Stochastic master–equation approach to aggregation in freeway traffic, Phys. Rev. E **56** (1997) 2666
19. Miller, M. A., Doye, J. P. K., Wales, D. J.: Structural relaxation in atomic clusters: Master equation dynamics, Phys. Rev. E **60** (1999) 3701
20. Montroll, E. W., Badger, W. W.: Introduction to quantitative aspects of social phenomena, Gordon and Breach Sci. Publ., New York, 1974
21. Münster, A.: *Statistical thermodynamics* Vol. I, (Springer, Berlin, 1969).
22. Newman, M. E. J., Barkema, G. T.: *Monte Carlo methods in statistical physics* (Clarendon Press, Oxford, 1999).
23. Paul, W., Baschnagel, J.: *Stochastic processes* From Physics to Finance, (Springer, Berlin, 1999).
24. Prigogine, I., Herman, R.: *Kinematic theory of vehicular traffic* (Elsevier, New York, 1971).
25. Reiss, H., Huang, C.: Statistical thermodynamic formalism in the solution of information theory problems, J. Stat. Phys. **3** (1971) 191

26. Reiss, H., Hammerich, A. D., Montroll, E. W.: Thermodynamic treatment of nonphysical systems: Formalism and an example (single–lane traffic), J. Stat. Phys. **42** (1986) 647
27. Schmelzer, J., Jr., Lembke, U., Kranold, R.: Nucleation and growth of AgCl clusters in a sodium borate glass: Numerical analysis and SAXS results, J. Chem. Phys. **113** (2000) 1268
28. Schmelzer, J., Röpke, G., Mahnke, R.: *Aggregation phenomena in complex systems* (Wiley–VCH, Weinheim, 1999).
29. Wu, D. T.: Nucleation theory, In: *Solid State Physics*, Vol. 50, (Eds.: Ehrenreich, H., Spaepen, F.), (Academic Press, San Diego, 1997) p. 37

Cellular Automata Simulation of Collective Phenomena in Pedestrian Dynamics

Andreas Schadschneider[1], Ansgar Kirchner[1,2], and Katsuhiro Nishinari[3,1]

[1] Institut für Theoretische Physik, Universität zu Köln, D-50937 Köln, Germany
[2] Theoretische Physik, Fakultät 4, Gerhard-Mercator-Universität Duisburg, D-47048 Duisburg, Germany
[3] Department of Applied Mathematics and Informatics, Ryukoku University, Shiga, Japan

Abstract. Pedestrian dynamics exhibits a variety of fascinating and surprising collective phenomena. A 2-dimensional cellular automaton model is presented which is able to reproduce these effects in a simple way. Inspired by the principles of chemotaxis the interactions between the pedestrians are mediated by a so-called *floor field*. This field has a similar effect as the chemical trace created e.g. by ants to guide other individuals to food places. In our case the floor field modifies the transition rates to neighbouring cells such that a motion in the direction of higher fields is preferred. Already the inclusion of only nearest-neighbour interactions allows to reproduce many of the collective effects and self-organization phenomena (lane formation, flow oscillations at doors etc.) encountered in pedestrian dynamics.

1 Introduction

Considerable research has been done on the topic of traffic flow using methods from physics during the last decade [1–4,6,5] where the focus has been especially on highway traffic. In contrast to the latter, pedestrian flow [7] is truely 2-dimensional and effects due to counterflow become important. In vehiclar traffic one usually deals with a well-defined ordering of vehicles. Interactions with nearest neighbours (i.e. the next car ahead) are most important. In pedestrian dynamics, however, the situation is more complex and interactions with other individuals (which might cross the path of walking) in a certain range have to be taken into account. This gives rise to several self-organization phenomena not observed in vehicular traffic.

1.1 Collective Effects

Many interesting collective effects and self-organization phenomena have been observed in pedestrian dynamics. Here we only give a brief overview and refer to [6–8] for a more comprehensive discussion and list of references.

Jamming: At large densities various kinds of jamming phenomena occur, e.g. when many people try to leave a large room at the same time where the flow is limited by e.g. a door (see Fig. 1) or narrowing. Therefore this kind of jamming phenomenon does not depend strongly on the microscopic dynamics of the particles. This clogging effect is typical for a bottleneck situation. It is important for practical applications, especially evacuation simulations. Other types of jamming occur in the case of counterflow where two groups

Fig. 1. Clogging near a bottleneck (door). Shown are typical stages of an evacuation simulation: **(a)** initial state ($t = 0$); **(b)** middle stages; **(c)** end stage with only a few particles left.

of pedestrians mutually block each other. This happens typically at high densities and when it is not possible to turn around and move back, e.g. when the flow of people is large.

Lane formation: In counterflow, i.e. two groups of people moving in opposite directions, a kind of spontaneous symmetry breaking occurs (see Fig. 2 and Sec. 4). The motion of the pedestrians can self-organize in such a way that (dynamically varying) lanes are formed where people move in just one direction [9]. In this way, strong interactions with oncoming pedestrians are reduced and a higher walking speed is possible.

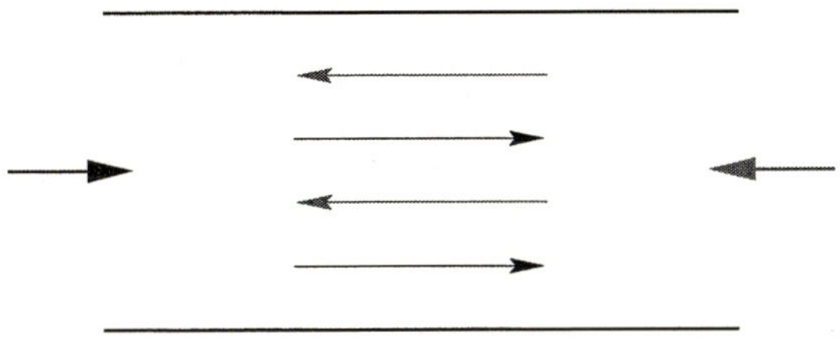

Fig. 2. Lane formation in counterflow in a narrow corridor.

Oscillations: In counterflow at bottlenecks, e.g. doors, one can observe oscillatory changes of the direction of motion. Once a pedestrian is able to pass the bottleneck it becomes easier for others to follow him in the same direction until somebody is able to pass (e.g. through a fluctuation) the bottleneck in the opposite direction (see Fig. 3).

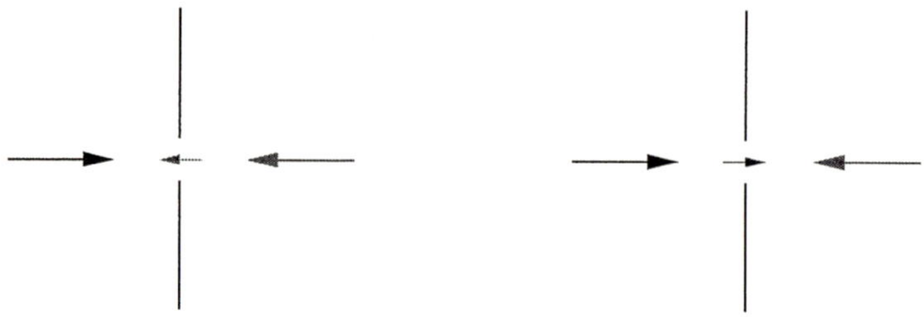

Fig. 3. Flow oscillations at a door with counterflow.

Patterns at intersections: At intersections various collective patterns of motion can be formed. A typical example are short-lived roundabouts which make the motion more efficient. Even if these are connected with small detours the formation of these patterns can be favourable since they allow for a "smoother" motion.

Panics: In panic situations, many counter-intuitive phenomena can occur. In the faster-is-slower effect [10] a higher desired velocity leads to a slower movement of a large crowd. In the freezing-by-heating effect [11] increasing the fluctuations can lead to a more ordered state. For a thorough discussion we refer to [8,10] and references therein.

1.2 Modelling Approaches

The number of models for pedestrian dynamics is much smaller than for vehicular traffic [4,6]. Nevertheless, several approaches [7] have been suggested:

Social force models: So far continuum models have been most successful in modelling pedestrian dynamics. An important example are the social force models (see e.g. [6,9,13] and references therein). Here pedestrians are treated as particles subject to long-ranged forces induced by the social behaviour of the individuals. The typical structure of the force between the pedestrian is described by [10]

$$\mathbf{f}_{ij} = \mathbf{f}_{ij}^{(soc)} + \mathbf{f}_{ij}^{(phys)}. \tag{1}$$

Here $\mathbf{f}_{ij}^{(soc)}$ is a repulsive social force modelling the tendency to keep a certain distance to other individuals. Typically it is long-ranged: $f_{ij}^{(soc)} \propto \exp(-r_{ij}/\xi)$. $\mathbf{f}_{ij}^{(phys)}$ is a physical force and describes friction and compression when pedestrians make contact. This idea leads then to equations of motion similar to Newtonian mechanics. There are, however, important differences since, e.g., in general the third law ("actio = reactio") is not fulfilled by social forces. It should be mentioned that the social force model is very successful in modeling the collective effects listed in Sec. 1.1.

Active-walker models: So-called active-walker models [12,14,15] have been used to describe the formation of human or animal trails etc. Here the walker leaves a trace by modifying the underground on his path. This modification is real in the sense that it could be measured in principle. For trail formation, vegetation is destroyed by the walker and some animals (like ants) leave a chemical trace which can be detected by other animals.

Cellular automata: Most cellular automata (CA) models for pedestrian dynamics proposed so far are rather simple [16–19] and can be considered as generalizations of the Biham-Middleton-Levine model for city traffic [20]. However, these models are not able to reproduce all the collective effects observed empirically. The same is true for more sophisticated discrete models, e.g. the one suggested by Gipps and Marksjös [21] or a disretized version of the social force model [22].

In [23–26] a new kind of CA model has been introduced which – despite its simplicity – is able to reproduce the observed collective effects. It takes its inspiration[1] from the process of chemotaxis (see [27] for a review) as used by some insects. They create a chemical trace to guide other individuals to food places. This is also the central idea of the active-walker models. In the approach of [23] the pedestrians also create a trace. In contrast to trail formation and chemotaxis, however, this trace is only virtual although one could assume that it corresponds to some abstract representation of the path in the mind of the pedestrians. Its main purpose is to transform effects of long-ranged interactions (e.g. following people walking some distance ahead) into a local interaction (with the "trace"). This allows for a much more efficient simulation on a computer.

2 Basic Principles of the Model

First we discuss some general principles applied in the development of the model [23,24]. To allow an efficient implementation for large-scale computer simulations a discrete model is preferable. Therefore a two-dimensional CA is used with a stochastic dynamics taking into account the interactions between the pedestrians. As mentioned above, similar to chemotaxis, we want to transform long-ranged interactions into local ones. This is achieved by the introduction of so-called *floor fields*. The transition probabilities for all pedestrians depend on the strength of the floor fields in their neighbourhood in such a way that transitions in the direction of larger fields are preferred.

Interactions between pedestrians are repulsive for short distances. One likes to keep a minimal distance to others in order to avoid bumping into

[1] Such "learning from nature" is the central idea of a field called *Bionics*.

them. This is taken into account through hard-core repulsion which prevents multiple occupation of the cells. For longer distances the interaction is often attractive. E.g. when walking in a crowded area it is usually advantageous to follow directly behind the predecessor. Large crowds may also be attractive due to curiosity and in panic situation often herding behaviour can be observed [10].

The long-ranged part of the interaction is implemented through the floor fields. We distinguish two kinds, a *static floor field* and a *dynamic floor field.* The latter models the dynamic interactions between the pedestrians, whereas the static field represents the constant properties of the surroundings.

The dynamic floor field corresponds to a virtual trace which is created by the motion of the pedestrians and in turn influences the motion of other individuals. Furthermore it has its own dynamics, namely through diffusion and decay, which leads to a dilution and finally the vanishing of the trace after some time.

The static floor field does not change with time since it only takes into account the effects of the surroundings. Therefore it exists even without any pedestrians present. It allows to model e.g. preferred areas, walls and other obstacles. A typical example can be found in Sec. 5 where the evacuation from a room with a single door is examined. Here the strength of the static field decreases with increasing distance from the door.

The introduction of the floor fields allows for a very efficient implementation on a computer since now all interactions are local. We have translated the *long-ranged spatial interaction* into a *local interaction with "memory"*. Therefore the number of interaction terms grows only linearly with the number of particles. Another advantage of local interactions can be seen in the case of complex geometries. Due to the presence of walls not all particles within the interaction range interact with each other. Therefore one needs an algorithm to check whether two particles "see" each other or whether the interaction is blocked by some obstacle. All this is not necessary here.

Furthermore we do not need to provide the pedestrians with some sort of "intelligence". The floor fields are sufficient to achieve the formation of complex structures and collective effects by means of self-organization. Even for an evacuation scenario [26] (see Sec. 5) the knowledge about the floor fields is sufficient to find the exit. Therefore the pedestrians behave like simple 'particles' in a field without explicit intelligence. Since we also not make detailed assumptions about the human behaviour it allows us to keep the model simple. Nevertheless it is able to reproduce many of the basic phenomena.

In contrast to vehicular traffic the time needed for acceleration and braking is negligible in pedestrian motion. The velocity distribution of pedestrians is sharply peaked [28]. These facts naturally lead to a model where the pedestrians have a maximal velocity $v_{\max} = 1$, i.e. only transitions to neighbour cells are allowed. Furthermore, a larger $v_{\max}$, i.e. pedestrians can move more than just one cell per timestep, would be harder to implement in two dimensions and reduce the computational efficiency.

For some applications it is useful to introduce another field called *matrix of preference* which encodes the preferred walking direction and speed of each pedestrian. From this direction, a 3×3 matrix is constructed which contains the probabilities for a move of the particle. The central element describes the probability for the particle not to move at all, the remaining eight correspond to a move to one of the neighbouring cells[2] (see Fig. 4). The probabilities can directly be *related to observable quantities*, namely the velocity and the longitudinal and transversal standard deviations (see [23,29] for details). In the simplest case the pedestrian is allowed to move in one direction only without fluctuations. In the corresponding matrix of preference only one element is one and all others are zero. In the following it is assumed that a matrix of preference is given at every timestep for each pedestrian. In principle, it can differ from cell to cell depending on the geometry and aim of the pedestrians. Therefore it can be combined with some model for route selection which assigns certain routes to each pedestrian.

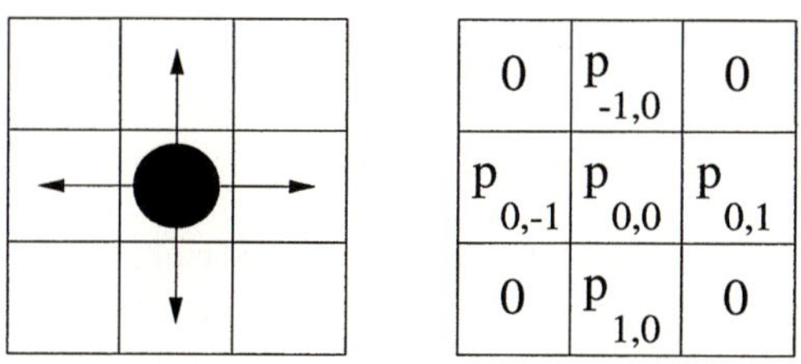

Fig. 4. A particle, its possible directions of motion and the corresponding transition probabilities p_{ij}.

3 Definition of the Model

CA are discrete in space, time and state variables. The area available for pedestrians is divided into cells of approximately $40 \times 40\ cm^2$ which is the typical space occupied by a pedestrian in a dense crowd [30]. Each cell can either be empty or occupied by exactly one particle (pedestrian). For special situations it might be desirable to use a finer discretization, e.g. such that each pedestrian occupies four cells instead of one, but we restrict to the simplest case which is sufficient for the applications studied here.

We now define the stochastic dynamics of the model by specifying the transition probabilities p_{ij} for a motion to a neighbouring cell in direction[3]

[2] In most cases motion in diagonal directions is not considered so that the matrix of preference has at most five nonzero elements.

[3] $i, j \in \{-1, 0, 1\}$, see Fig. 4.

(i, j). As explained in Sec. 2 the long-ranged interactions with other pedestrians and the surrounding is encoded in two fields, the dynamic floor field D and the static floor field S, respectively. Furthermore a matrix of preference M can be used to specify the walking direction (see Sec. 2), speed and fluctuations for each individual. The *transition probability* p_{ij} in direction (i, j) is then determined by all three contributions. Explicitly it is given by

$$p_{ij} = N e^{k_D D_{ij}} e^{k_S S_{ij}} M_{ij} (1 - n_{ij}) \xi_{ij}. \tag{2}$$

D_{ij} and S_{ij} are the strengths of the dynamic and static floor field at the target cell and M_{ij} is the matrix element of the matrix of preference for a motion in the direction (i, j). N is a normalization factor to ensure $\sum_{(i,j)} p_{ij} = 1$ where the sum is over the possible target cells. The factor $1 - n_{ij}$, where n_{ij} is the occupation number of the neighbour cell in direction (i, j), takes into account that transitions to occupied cells are forbidden. ξ_{ij} is a geometry factor (obstacle number) which is 0 for forbidden cells (e.g. walls) and 1 else. Finally, we have introduced two coupling constants k_D and k_S so that we can vary the coupling strengths to each field individually.

The actual values of the parameters k_D and k_S depend on the situation (see Sec. 5). A large k_S implies that the pedestrians choose their path mainly due to the surrounding without being distracted too much by other people. A large coupling k_D to the dynamic field, on the other hand, corresponds to a strong herding behaviour. Here the pedestrian tries to follow the lead of others, e.g. in the case of panics or insufficient knowledge about the surroundings. The influence of these couplings will be discussed in more detail in Sec. 5.

As mentioned before, the dynamic floor field is created by the motion of the pedestrians and corresponds to a virtual trace. At $t = 0$ it is zero everywhere. Whenever a particle moves from site (x, y) to one of its neighbours $(x + i, y + j)$, the field D_{xy} at the origin cell is increased by one ($D_{xy} \to D_{xy} + 1$). Thus D_{xy} has only non-negative integer values which can be interpreted as the number of 'bosons' located at site (x, y).

The dynamic floor field is not only changed by the motion of the pedestrians, but it is also subject to diffusion and decay which first leads a spreading and dilution of the trace and finally to its vanishing after some time. Diffusion and decay are controlled by two parameters $\alpha \in [0, 1]$ and $\delta \in [0, 1]$. In each time step of the simulation each boson of the dynamic field D decays with the probability δ and diffuses with the probability α to one of the neighbouring cells.

The update rules of the full model including the interaction with the floor fields then have the following structure:

1. The dynamic floor field D is modified according to its diffusion and decay rules.
2. Using equation (2), for each pedestrian the transition probabilities p_{ij} for a move to an unoccupied neighbour cell (i, j) are determined by the matrix of preference and the local dynamic and static floor fields.

3. Each pedestrian chooses a target cell based on the probabilities of the transition matrix $P = (p_{ij})$.
4. The conflicts arising by any two or more pedestrians attempting to move to the same target cell are resolved (see below).
5. The pedestrians which are allowed to move execute their step.
6. The pedestrians alter the dynamic floor field D_{xy} of the cell (x, y) they occupied before the move.

These rules have to be applied to all pedestrians at the same time (parallel dynamics). This introduces a timescale into the dynamics which can roughly be identified with the reaction time t_{reac} [23]. The existence of a timescale in the dynamics of the model is essential if one wants to make *quantitative predictions* for real processes. In the deterministic limit, corresponding to the maximal possible walking velocity in our model, a single pedestrian (not interacting with others) moves with a velocity of one cell per timestep, i.e. 40 *cm* per timestep. Empirically the average velocity of a pedestrian is about 1.3 m/s [30]. This gives an estimate for the real time corresponding to one timestep in our model of approximately 0.3 *sec* which is of the order of the reaction time t_{reac} and thus consistent with our microscopic rules. It also agrees nicely with the time needed to reach the normal walking speed which is about 0.5 *sec*.

One detail is worth mentioning. If a particle has moved in the previous timestep the boson created then is not taken into account in the determination of the transition probability. This prevents that pedestrians get confused by their own trace. One can even go a step further and introduce 'inertia' [23] which enhances the transition probability in the previous direction of motion.

Finally we like to comment on step 4, the resolution of conflicts which occur if $m \geq 2$ particles choose the same destination cell in step 3. This is relevant for high density situations. In order to avoid multiple occupancies of cells only one particle is allowed to move while the others keep their position. There are different ways to choose this particle. E.g. the winning particle can be picked at random with probability $1/m$ or according to the relative probabilities p_{ij} with which each particle chose their target [23]. For the problems studied here the details of the conflict resolution turned out to play no important role. The influence of conflicts will be discussed in more detail in Sec. 6.

4 Collective Phenomena

In this section we show that the model is indeed able to reproduce the collective effects mentioned in Sec. 1.1. As most prominent example we want to discuss lane formation out of a randomly distributed group of pedestrians. This corresponds to a spontaneous breaking of the symmetry of the particle number distribution in space. Simulations show that an even as well as an odd number of lanes may be formed. The latter corresponds to a spontaneous breaking of the left-right symmetry of the system.

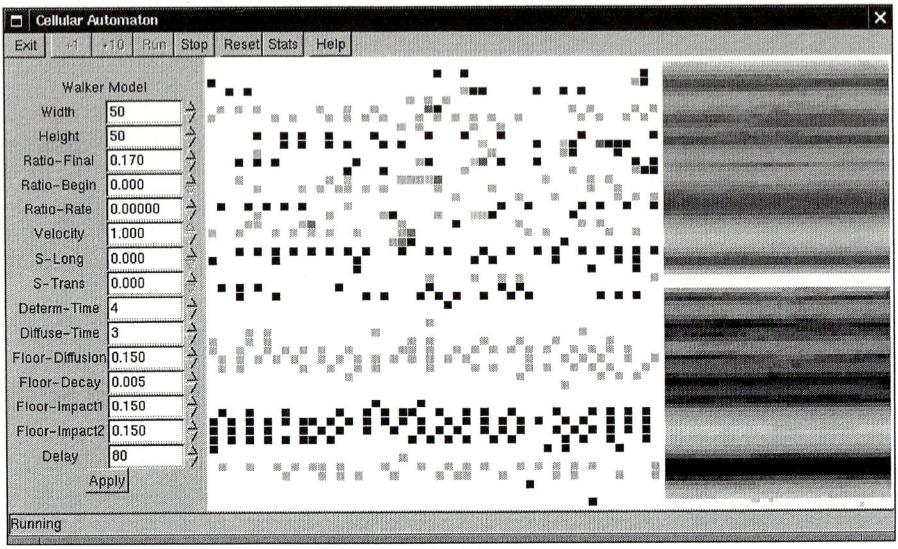

Fig. 5. Snapshot of a simulation of counterflow along a corridor. The left part shows the parameter control. The central window is the corridor and the light and dark squares are right- and left-moving pedestrians, respectively. The right part shows the floor fields for the two species.

Fig. 5 shows simulations of a rectangular corridor which is populated by two species of pedestrians moving in opposite directions. Parallel to the direction of motion the existence of walls is assumed. Orthogonal to the direction of motion both periodic and open boundary conditions have been investigated. With periodic boundary conditions, the density of pedestrians is fixed for each run. The numbers of left- and right-movers are equal and each species interacts with its own dynamic floor field. For open boundaries, we fix the rate at which pedestrians enter the system at the boundaries. The pedestrians leave the system as soon as they reach the opposite end of the corridor. Fig. 5 shows the graphical frontend running a simulation of a small periodic system. Lanes have already formed in the lower part of the corridor and can be spotted easily, both in the main window showing the cell contents and the small windows on the right showing the floor field intensity for the two species.

The formation of the lanes can also clearly be seen in the velocity profile (Fig. 6) which has been measured at a cross section perpendicular to the direction of flow. In a certain density regime, the lanes are metastable. Spontaneous fluctuations can disrupt the flow in one lane causing the pedestrians

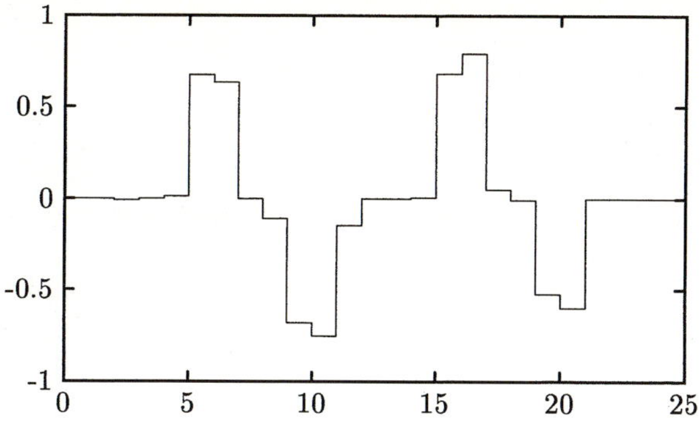

Fig. 6. Velocity profile of a periodic system with $\rho = 0.12$ exhibiting lane formation.

to spread and interfere with other lanes. Eventually the system can run into a jam by this mechanism.

Apart from lane formation we have also observed oscillations of the direction of flow at doors and the formation of roundabout-like flow patterns at intersections [23,25,29]. Therefore the model captures the main phenomena correctly which is important for practical applications, e.g. evacuation simulations or the optimization of escape routes.

5 Influence of the Floor Fields

In order to elucidate the influence of the coupling parameters k_S and k_D we investigated an evacuation process in a simple geometry, namely a large room with one door (see Fig. 1). In the following we do not use the matrix of preference. All information about the location of the exits is obtained from the floor fields. Fig. 7(a) shows the form of the static floor field S for such a situation. For each lattice site S is calculated using some distance metric [26] so that the field values increase in the direction of the door. Fig. 7(b) shows a 3-d plot of the dynamic field D in the final stages of an evacuation process.

The value of k_S, the coupling to the static field, can be viewed as a measure of the knowledge of the pedestrians about the location of the exit. A large k_S implies a motion to the exit on the shortest possible path. For vanishing k_S, on the other hand, the people will perform a random walk and just find the exit by chance. So the case $k_S \ll 1$ is relevant for processes in dark or smoke-filled rooms where people do not have full knowledge about the location of the exit.

The parameter k_D for the coupling to the dynamic field controls the tendency to follow the lead of others. A large value of k_D implies a strong herding behaviour which has been observed in the case of panics [10].

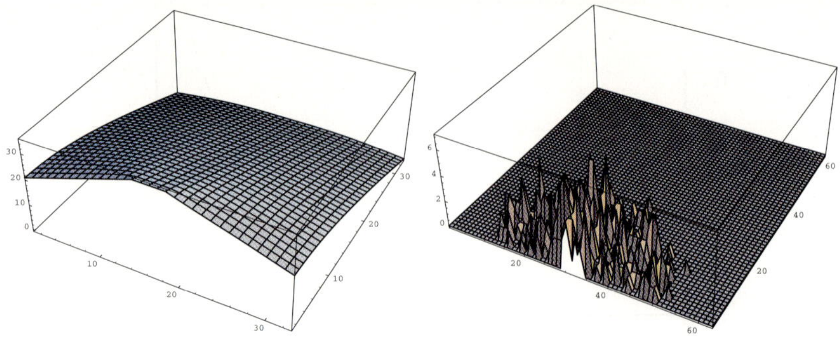

Fig. 7. Evacuation from a large room with a single door: **(a)** static floor field S; **(b)** dynamic floor field D in the final stages of the evacuation.

In the following we consider a grid of size 63×63 sites with an exit of one cell in the middle of one wall. The particles are initially distributed randomly and try to leave the room. The only information they get is through the floor fields. Fig. 1 shows typical stages of the dynamics for an initial particle density of $\rho = 0.3$, which means 1116 particles. In the middle picture of Fig. 1 a half-circle jamming configuration in front of the door is easy to spot. A typical feature of the dynamics is a radial motion of 'holes' created by particles escaping through the door.

Fig. 8(a) shows the evacuation times for fixed sensitivity parameter k_D of the dynamic field and variable sensitivity parameter k_S of the static field. For $k_S \to 0$ the pedestrians do not sense the strength of the field. Therefore they do not have any guidance through the surroundings and perform a pure random walk which leads to a maximal value of the evacuation time for $k_S = 0$. For $k_S \to \infty$ they have full information about the shortest distance to the door and the evacuation time converges towards a minimal value. The movement of the particles becomes almost deterministic. Therefore k_S can be interpreted as some kind of inverse temperature for the degree of information about the inanimate surrounding of the pedestrians.

In the same way the sensitivity parameter k_D of the dynamic field works as an inverse temperature for the information about the virtual trace. If k_S is turned on from zero to infinity, a non-zero value of k_D only means additional noise to the pedestrians, the evacuation times are increased for higher coupling strength to k_D (see Fig. 8(a)).

Much more interesting is the behavior for fixed k_S. In this case the evacuation times for various sensitivity parameters k_D are shown in Fig. 8(b). They converge to maximal values for growing sensitivity parameter k_D of the dynamic field. The most interesting point is the occurence of minimal evacuation times for non-vanishing small values of the sensitivity parameter k_D of the dynamic field. This effect also depends on the value of the diffusion

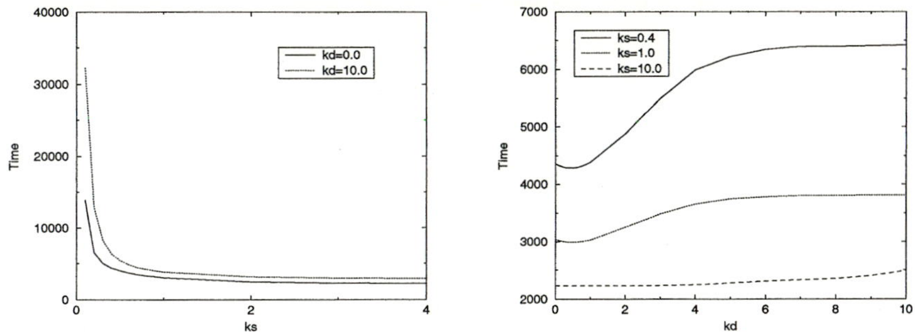

Fig. 8. Averaged evacuation times for a large room with an initial particle density of $\rho = 0.3$ and $\delta = 0.3$, $\alpha = 0.3$ for **(a)** fixed k_D, and **(b)** fixed k_S.

parameter α and becomes most pronounced in the limit $\alpha \to 0$. Therefore a small interaction with the dynamic field, which is proportional to the velocity-density of the particles, is of advantage. It represents some sort of minimal intelligence of the pedestrians. They are able to detect regions of higher local flow and minimize their waiting times.

If the coupling to the dynamic field is further increased, the evacuation times increase again from the small regime of the minimal times to maximal values. The interaction with other pedestrians becomes more and more unfavourable, one can compare this to the arising of a panic situation. The weaker the coupling to the static field S is, the higher are the evacuation times: the particles than have less information of the inanimate surrounding (for example they cannot find the way to the door because of smoke in a fire situation).

Three main regimes for the behaviour of the particles can be distinguished [26]. For strong coupling to k_S and very small coupling to k_D we find an *ordered regime* where particles only react to the static floor field and the behaviour than is in some sense deterministic. The *disordered regime* characterized by strong coupling to k_D and weak coupling to k_S leads to a maximal value of the evacuation time. The behaviour here is typical for panic situations. Between these two regimes an *optimal regime* exists where the combination of interaction with the static and the dynamic floor fields minimizes the evacuation time.

6 Friction Effects

In [23,26] the conflicts between pedestrians were solved in the following way: when $m > 1$ particles share the same target cell, one ($l \in \{1, \dots, m\}$) is chosen to move while its rivals for the same target keep their position. There are two main ways to determine the moving particle l:

1. According to the relative probabilities with which each particle chooses its target cell, i.e. the probability for the l-th particle to move is $\frac{p_{ij}^{(l)}}{\sum_{s=1}^{m} p_{ij}^{(s)}}$.
2. With equal probability, i.e. each particle moves with probability $\frac{1}{m}$.

Both methods yield very similar results in the simulations.

We now extend the basic model by a new friction parameter $\mu \in [0,1]$ (see [31] for details), in order to describe clogging effects between the pedestrians. Whenever two or more pedestrians try to attempt to move to the same target cell, the movement of all involved particles is denied with the probability μ, i.e. all pedestrians remain at their site (see Fig. 9). This means that with probability $1-\mu$ one of the individuals moves to the desired cell. Which particle actually moves is then determined by one of the rules for the resolution of conflicts described above. With this definition of μ and the extended up-

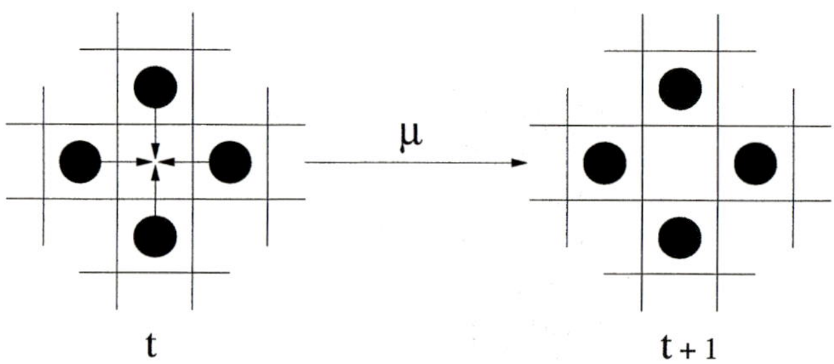

Fig. 9. Refused movement due to the friction parameter μ (for $m = 4$).

date rule it is easy to see that μ works as some kind of local pressure between the pedestrians. If μ is high, the pedestrians handicap each other trying to reach their desired target sites. This local effect can have enormous influence on macroscopic quantities like flow and evacuation time [31]. Note that the kind of friction introduced here only influences interacting particles, not the average velocity of a freely moving pedestrian.

As one can see the use of a parallel update is essential. Any other form of random- or ordered-sequential update will disguise the real number of arising conflicts between the pedestrians in the system.

Fig. 10(a) shows the influence of the friction parameter on the evacuation time T for the scenario described in Sec. 5. As expected, T is monotonically increasing with μ. The strongest effect can be observed in the ordered regime. Here the evacuation time T is mainly determined by the clogging at the door. For large values of μ, T increases strongly due to arching effects similar to those in granular materials.

For fixed μ and varying coupling strength k_S a surprising result can be observed (Fig. 10(b)). For $\mu = 0$ the evacuation time is monotonically decreasing with increasing k_S since for large coupling to the static field the

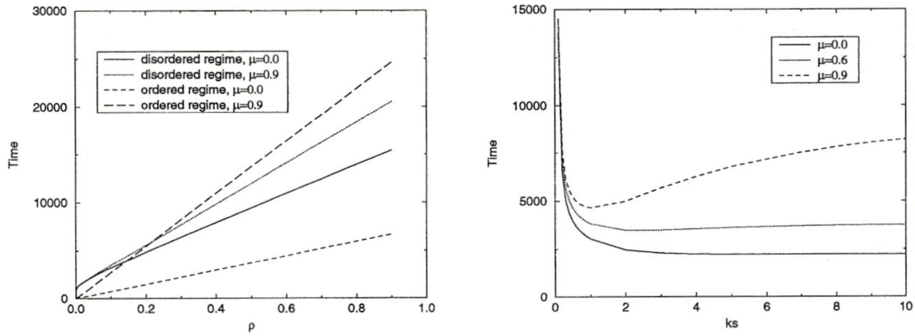

Fig. 10. Dependence of evacuation times on the friction parameter μ (a) in the ordered and disordered regimes and (b) as a function of k_S for $\rho = 0.3$.

pedestrians will use the shortest way to the exit. For large μ, however, $T(k_S)$ shows a minimum at an intermediate coupling strength $k_S \approx 1$. This is similar to the faster-is-slower effect described in Sec. 1.1: Although a larger k_S leads to a larger effective velocity in the direction of the exit, it does not necessarily imply smaller evacuation times if it increases the number of conflicts close to the door.

7 Conclusions

We have introduced a stochastic cellular automaton to simulate pedestrian behaviour[4]. The general idea in our model is similar to chemotaxis. However, the pedestrians leave a virtual trace rather than a chemical one. This virtual trace has its own dynamics (diffusion and decay) which e.g. restricts the interaction range (in time). It is realized through a dynamic floor field which allows to give the pedestrians only minimal intelligence and to use local interactions. Together with the static floor field it offers the possibility to take different effects into account in a unified way, e.g. the social forces between the pedestrians or the geometry of the building.

The floor fields translate spatial long-ranged interactions into non-local interactions in time. The latter can be implemented much more efficiently on a computer. Another advantage is an easier treatment of complex geometries. We have shown that the approach is able to reproduce the fascinating collective phenomena observed in pedestrian dynamics. As an example the formation of lanes in counterflow has been discussed in Sec. 4.

In Sec. 5 we have investigated the influence of the coupling strengths k_S and k_D to the static and dynamic floor fields on the dynamics, especially on the evacuation time in a simple scenario. Interesting nontrivial phenomena

[4] Further information and Java applets for the scenarios studied here can be found on the webpage `http://www.thp.uni-koeln.de/~as/as.html`.

could be observed, e.g. the nonmonotonic dependence of the evacuation time on the coupling to the dynamic floor field. Further surprising effects are caused by friction (Sec. 6), related to the resolution of conflict situations where several individuals want to occupy the same space.

It is surprising to note that the properties of the model are in many respects very similar to the social force model [9] although the interactions are very different. In our approach pedestrians interact with the velocity-density through an attractive coupling whereas the social-force model uses a repulsive density-density interaction. It would be interesting to get a deeper understanding of these similarities.

References

1. Wolf, D.E., Schreckenberg, M., Bachem, A. (Eds.): *Traffic and Granular Flow* (World Scientific, Singapore 1996)
2. Schreckenberg, M., Wolf, D.E. (Eds.): *Traffic and Granular Flow '97* (Springer 1998)
3. Helbing, D., Herrmann, H. J., Schreckenberg, Wolf, D. E. (Eds.): *Traffic and Granular Flow '99: Social, Traffic, and Granular Dynamics* (Springer 2000)
4. Chowdhury, D., Santen, L., Schadschneider, A.: Statistical physics of vehicular traffic and some related systems. Phys. Rep. **329**, 199 (2000)
5. Nagel, K., Esser, J., Rickert, M.: Large-scale traffic simulation for transportation planning. Annu. Rev. Comp. Phys. **7**, p. 151, ed. D. Stauffer (World Scientific, 2000)
6. Helbing, D.: Traffic and related self-driven many-particle systems. Rev. Mod. Phys. **73**, 1067 (2001)
7. Schreckenberg, M., Sharma, S.D. (Ed.): *Pedestrian and Evacuation Dynamics*, Springer 2001
8. Helbing, D., Farkas, I., Molnar, P., Vicsek, T.: Simulation of pedestrian crowds in normal and evacuation situations. in [7], p. 21
9. Helbing, D., Molnar, P.: Social force model for pedestrian dynamics. Phys. Rev. **E51**, 4282 (1995)
10. Helbing, D., Farkas, I., Vicsek, T.: Simulating dynamical features of escape panic. Nature **407**, 487 (2000)
11. Helbing, D., Farkas, I., Vicsek, T.: Freezing by heating in a driven mesoscopic system. Phys. Rev. Lett. **84**, 1240 (2000)
12. Helbing, D., Schweitzer, F., Keltsch, J., Molnar, P.: Active walker model for the formation of human and animal trail systems. Phys. Rev. **E56**, 2527 (1997)
13. Hoogendoorn, S.P.: Walker behaviour modelling by differential games. These proceedings.
14. Helbing, D., Keltsch, J., Molnar, P.: Modelling the evolution of human trail systems. Nature **388**, 47 (1997)
15. Chowdhury, D., Guttal, V., Schadschneider, A.: Cellular-automata model of ant-trail and vehicular traffic: similarities and differences. cond-mat/0201207
16. Fukui, M., Ishibashi, Y.: Self-organized phase transitions in cellular automaton models for pedestrians. J. Phys. Soc. Jpn. **68**, 2861 (1999)
17. Muramatsu, M., Irie, T., Nagatani, T.: Jamming transition in pedestrian counter flow. Physica **A267**, 487 (1999)

18. Muramatsu, M., Nagatani, T.: Jamming transition in two-dimensional pedestrian traffic. Physica **A275**, 281 (2000)
19. Klüpfel, H., Meyer-König, T., Wahle, J., Schreckenberg, M.: Microscopic simulation of evacuation processes on passenger ships. in *Theory and Practical Issues on Cellular Automata*, S. Bandini, T. Worsch (Eds.), Springer (2000)
20. Biham, O., Middleton, A.A., Levine, D.: Self-organization and a dynamical transition in traffic-flow models. Phys. Rev. A **46**, R6124 (1992)
21. Gipps, P.G., Marksjös, B.: A micro-simulation model for pedestrian flows. Math. and Comp. in Simulation **27**, 95 (1985)
22. Bolay, K.: Nichtlineare Phänomene in einem fluid-dynamischen Verkehrsmodell. Diploma Thesis, Stuttgart University (1998)
23. Burstedde, C., Klauck, K., Schadschneider, A., Zittartz, J.: Simulation of pedestrian dynamics using a 2-dimensional cellular automaton. Physica **A295**, 507 (2001)
24. Schadschneider, A.: Cellular automaton approach to pedestrian dynamics - Theory. in [7], p. 75
25. Burstedde, C., Kirchner, A., Klauck, K., Schadschneider, A., Zittartz, J.: Cellular automaton approach to pedestrian dynamics - Applications. in [7], p. 87
26. Kirchner, A., Schadschneider, A.: Simulation of evacuation processes using a bionics-inspired cellular automaton model for pedestrian dynamics. Physica **A** (in press)
27. Ben-Jacob, E.: From snowflake formation to growth of bacterial colonies. Part II. Cooperative formation of complex colonial patterns. Contemp. Phys. **38**, 205 (1997)
28. Henderson, L.F.: The statistics of crowd fluids. Nature **229**, 381 (1971)
29. Burstedde, C.: Simulation von Fußgängerverhalten mittels zweidimensionaler zellulärer Automaten. Diploma Thesis, Universität zu Köln (2001); available for download at `http://www.burstedde.de/carsten/diplom.html`
30. Weidmann, U.: Transporttechnik der Fussgänger. Schriftenreihe des IVT **80**, ETH Zürich (1992)
31. Kirchner, A., Nishinari, K., Schadschneider, A.: Friction effects and clogging in a cellular automaton model for pedestrian dynamics. in preparation.

Modeling, Simulation and Observations for Freeway Traffic and Pedestrian

Yuki Sugiyama[1] ⋆ and Akihiro Nakayama[2]

[1] City College of Mie, Tsu, Mie 514-0112, Japan
[2] Gifu Keizai University, Ohgaki, Gifu 503-8550, Japan

Abstract. We perform the simulations of Optimal Velocity Model with a bottleneck and investigate the new aspects of this model under such boundary conditions for the purpose of providing the interesting possibility of mathematical understanding of "synchronized flow", and compare our results with the observation. We also give the short review of extending OV model in 2-dimensional space for applying the several kinds of phenomena of collective behaviors, such as granular flow and pedestrian dynamics.

1 Introduction

The traffic flow provides many interesting aspects of physics for collective phenomena of non-equilibrium dynamics [1] [2]. Recently, the phenomenon of traffic flow in the bottleneck, so called "synchronized flow" affects many researchers [3]. It is challenging problem to provide the mathematical understanding on the view point of non-equilibrium physics. In this paper, we perform the simulations of Optimal Velocity Model with a bottleneck and investigate the new aspects of this model under such boundary conditions for the purpose of providing the interesting possibility of mathematical understanding of "synchronized flow".

2 Brief Review of OV Model (Traffic in Circuit)

Optimal Velocity Model is one of the successful models for describing the dynamics of traffic flow [4][5]. The model is formulated as

$$\frac{d^2}{dt^2}x_n(t) = a\left\{V(\Delta x_n) - \frac{d}{dt}x_n(t)\right\}. \tag{1}$$

x_n is the position of the nth car, and $\Delta x_n = x_{n+1} - x_n$ is the headway. a is a sensitivity constant. $V(\Delta x_n)$, so called OV-function, determines the optimal velocity according to the headway, which should have the form such as

$$V(\Delta x) = V_0\{\tanh(\Delta x - c) + const.\}. \tag{2}$$

⋆ The present address is School of Informatics and Sciences, Nagoya University, Nagoya 464-8601, Japan.

The model has the homogeneous flow solution; cars are moving uniformly distributed with the same velocity $V(1/k)$, where $k = N/L$ is the average vehicle density in the case of circuit (N is the number of cars and L is the length of the circuit). The homogeneous flow is linearly unstable under the condition $\frac{d}{dx}V(x)|_{x=1/k} > a/2$ In this case, the jam flow is formed as shown in Fig. 1. After relaxation all jam clusters are moving with the same velocity opposed to the direction of the vehicle. When we take a different

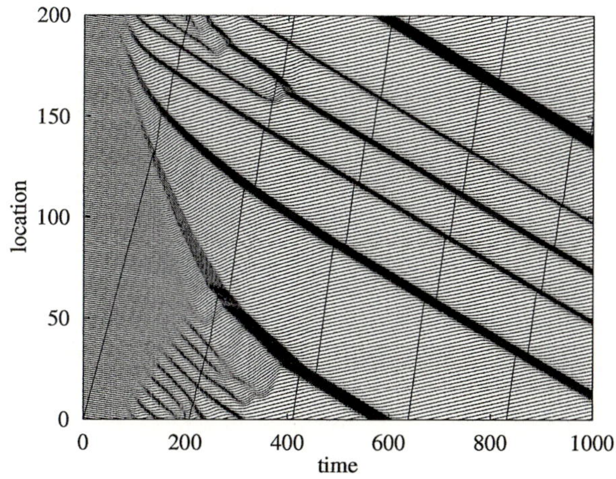

Fig. 1. The formation of traffic jam. The bold line is the trajectory of a sample car.

initial condition, we obtain a different figure of jam formation. But we always recognize the universal profile of jam flow solution by the trajectory of vehicles in the phase space of headway and velocity $(\Delta x_n, \dot{x}_n)$ in Fig. 2. In the jam flow solution all cars are moving along the specific closed curve, so called "hysteresis loop" [4], which is a kind of limit cycle. The size of the loop depends on a, not on k. As a becomes larger and exceeds the super-critical value a_c, the loop shrinks and disappears.

We show the fundamental diagram of OV model in Fig. 3, We have two Q-k (the flow rate - the car density) relations corresponding to the homogeneous flow and the jam flow solutions. We can observe the transition from the homogeneous flow state to the jam flow state as in Fig. 3. In the region-I the homogeneous flow is absolutely stable. The jam flow is absolutely stable in the region-III defined by $\frac{d}{dx}V(x)|_{x=1/k} > a/2$. In the region-II (the cross-over state), both flow-solutions are stable. The region-II is defined by the size of the hysteresis loop $\Delta x_F(a) - \Delta x_J(a)$ in Fig. 2.

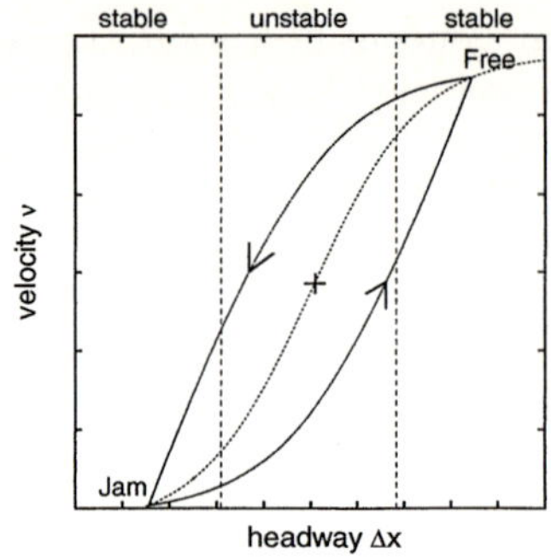

Fig. 2. The profile of jam flow solution. All vehicles move along the hysteresis loop in the direction of arrow.

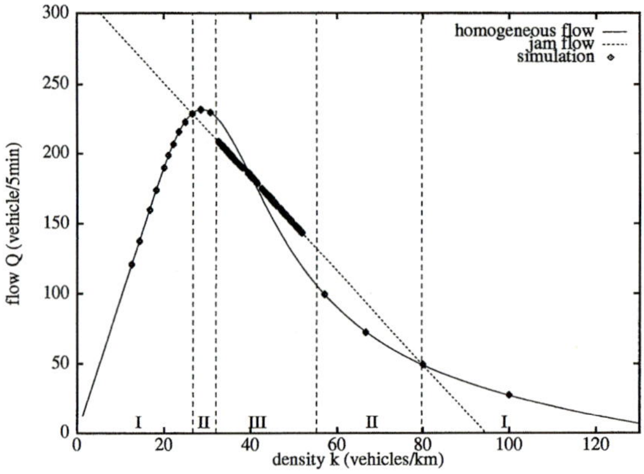

Fig. 3. *q*-*k* diagram of OV model is drawn by the homogeneous flow solution and limit cycle (jam flow) solution. The results of simulations are plotted together.

3 Observations in the Bottleneck

We first summarize some aspects of "synchronized flow". In the original meaning, the "synchronized flow" is the phenomena that the oscillation of velocities in traffic flow are synchronized between two parallel lanes in the existence of bottleneck [6]. The data observed by Koshi indicates such phenomena as a typical example [6][7]. Figure 4 shows the temporal sequential data for the oscillation of average velocities in two parallel lanes at the three different points (0.56km, 2.3km and 4.7km) in congested flow on the upstream of the bottleneck. It seems that the amplitude of oscillation is quite small at the immediate upstream of the bottleneck. From Fig. 5, the correlation of velocities between two lanes become clear as the upstream distance larger (than 1.2km).

This data indicates the synchronization appears in the larger distance from the bottleneck. As the correlation between two lanes become clear, the amplitude of velocities become larger. Actually, the synchronized flow is followed by jam flow in much larger distance. The synchronized flow occurs in some intermediate distance from the bottleneck, between the small-amplitude flow and the jam flow.

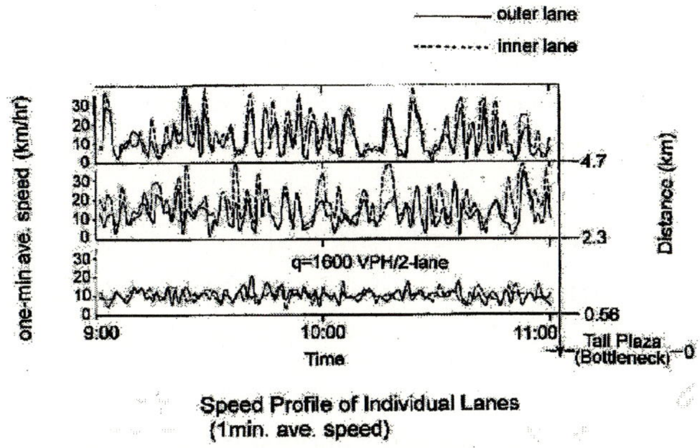

Fig. 4. The temporal sequence of the velocity-changing of two lanes at three points upstream from the bottleneck.

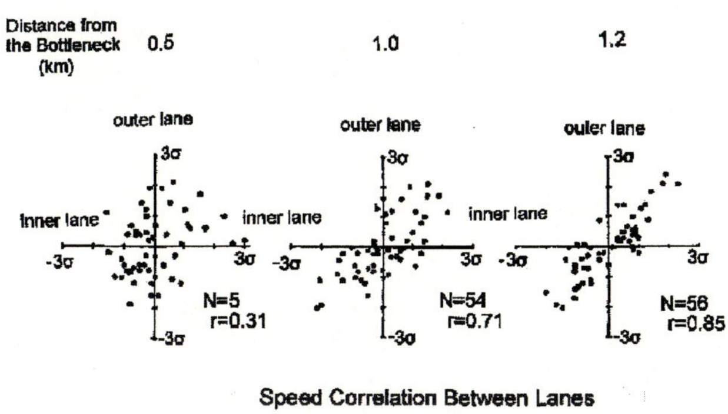

Fig. 5. The correlation of two lanes at three points upstream from the bottleneck.

4 OV Model with a Bottleneck

4.1 Modelling and Simulation

In this section, we perform the simulations of OV model with a bottleneck in one-lane traffic. In the original meaning, the simulation of two-lane traffic is needed for observation of synchronization. But, we can suppose that one of the most important aspects of "synchronized flow" is the structure of flow on the upstream of the bottleneck, which consists of three distinct regions of flow mentioned in the previous section. Such structure can be observed even in one-lane simulation with bottleneck. The characteristic behaviors of q-k diagram for "synchronized flow" can be also investigated.

In our simulation of OV model, we define the bottleneck using OV function as

$$V_{bottle}(\Delta x) = V_b\{\tanh(\Delta x - c) + const.\} \tag{3}$$

in the region of bottleneck, where the values of V_b and V_0 in (2) are chosen such that

$$V_{bottle}(\Delta x) < V(\Delta x), \tag{4}$$

as in Fig. 6. This definition means the optimal velocity in a bottleneck (Here, we suppose a tunnel as a bottleneck.) is reduced comparing with that outside a bottleneck. This condition is reasonable from the view point of psycholog-

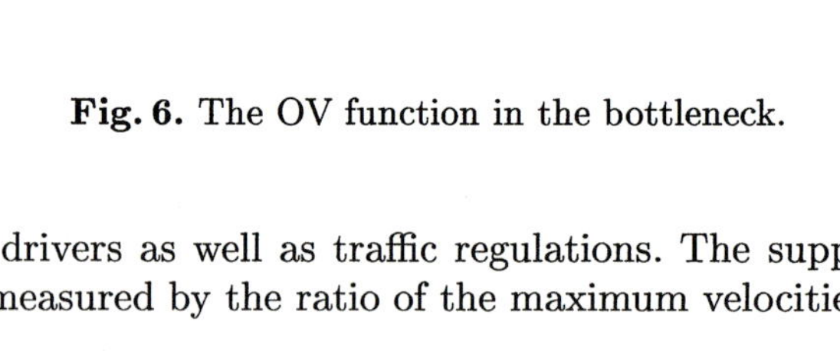

Fig. 6. The OV function in the bottleneck.

ical effect for drivers as well as traffic regulations. The suppression rate of bottleneck is measured by the ratio of the maximum velocities as

$$S_{bottle} \equiv \frac{V_{bottle}(\Delta x \to \infty)}{V(\Delta x \to \infty)} \equiv \frac{V^{max}_{bottle}}{V^{max}} < 1. \tag{5}$$

Here, we simply define $V_b = S_{bottle} V_0$ in the bottleneck.

The simulation starts in the condition that vehicles are homogeneously moving on the upstream of the bottleneck and no vehicles are on the downstream. The length of the bottleneck is 100m and the length of lane is enough long and the number of vehicles are enough not to suffer an artifact to the behavior of traffic flow around the bottleneck.

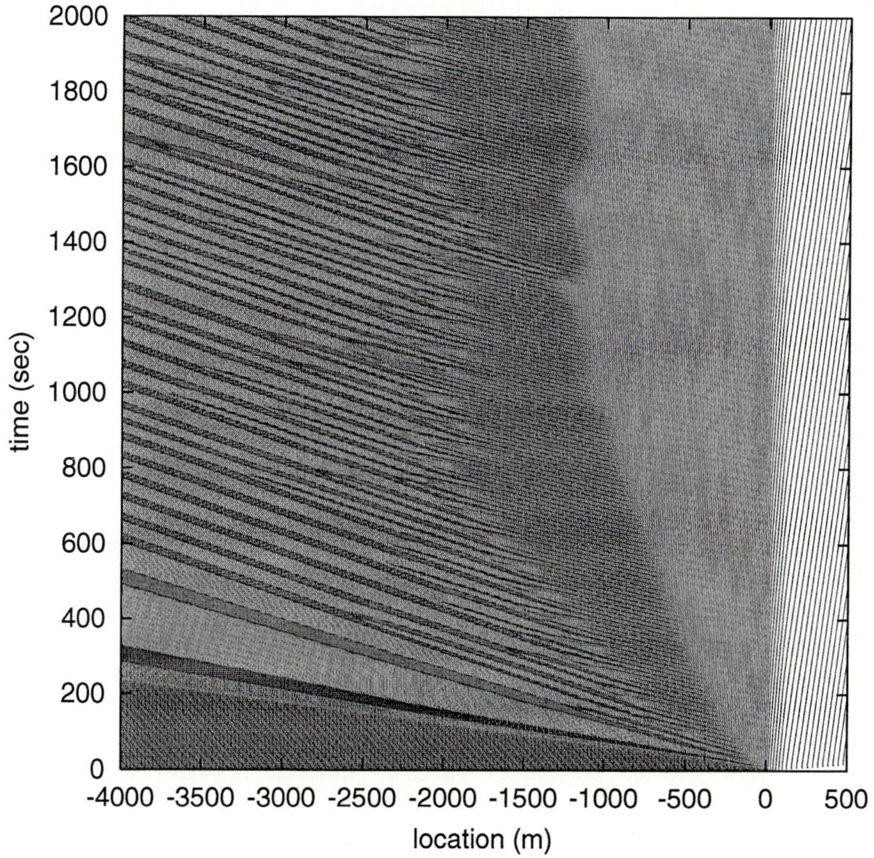

Fig. 7. The spatio-temporal plot on the upstream of the bottleneck for $S_{bottle} = 0.65$, $a = 1.6$. The bottleneck is located from $0m$ to $100m$. Vehicles move in the positive direction of location.

The figure 7 shows the formation of stable structure of traffic flow around the bottleneck. After relaxation three distinct spatial temporal patterns of flow are formed on the upstream of the bottleneck. In the immediate upstream the vehicles are distributed almost uniformly. The fine stripes are observed in the further upstream region, which means the vehicle density is oscillatory changing with small period in this region. The "oscillatory wave flow" is followed by the jam flow, where jam clusters appear clearly. The difference among these three patterns can be recognized more clearly in Fig. 8.

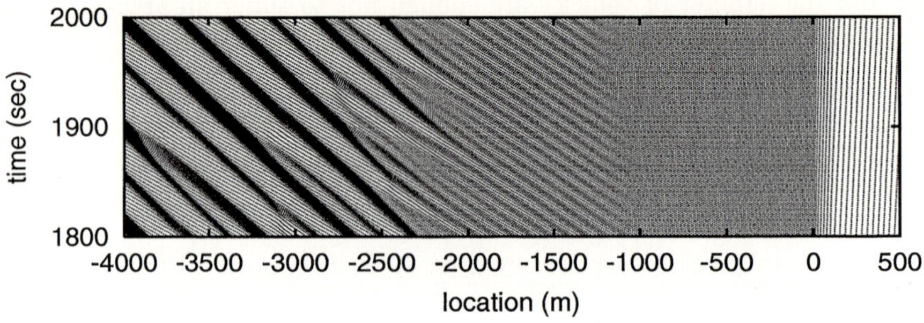

Fig. 8. The three stable patterns on the upstream of the bottleneck for $S_{bottle} = 0.65$, $a = 1.6$ after optimization.

4.2 The Structure of Flow Upstream of Bottleneck

Figure 9 shows the snapshots of velocity/headway distributions corresponding to Fig. 8. We can see the homogeneous flow with the identical velocity and headway in the region immediate upstream of the bottleneck. In the intermediate region 'the oscillatory wave' is observed for both velocity and headway. In the final region the kink-like cluster appears, which implies stop-and-go-wave as jam flow. Jam clusters propagates upstream, but the region of the oscillatory wave is located near but in some distance of the bottleneck, as you can see in Fig. 7.

Figure 10 is also the snapshots for the bottleneck of $S_{bottle} = 0.7$. In this case the size of each region of three patterns changes. The region of homogeneous flow shrinks and the region for oscillatory wave is spread. These sizes are determined by the strength of bottleneck. Moreover, the amplitudes of velocity/headway for the oscillatory waves are different between the two bottlenecks. While, the amplitudes for the homogeneous flow and the jam flow are originally characterized in the basic properties of OV model in circuit. Actually, the amplitude of jam flow is the same as that on the circuit with no bottleneck. In the contrast, the amplitude of the oscillatory wave is determined by the strength of the bottleneck, which is an additional element in the system.

The sequence of these three patterns on the upstream of the bottleneck can be qualitatively identified as the aspects of the flow accompanied with synchronized flow as mentioned in the previous section. The observed "synchronized flow" occurs in some intermediate distance from the bottleneck, between the regions of the small-amplitude flow and of the jam flow. In our simulation, the homogeneous flow is identified as the small-amplitude flow in the observation. And, the oscillatory wave appears in some intermediate distance from the bottleneck, which is followed by the jam flow. The wave has the characteristic amplitude, which is smaller than that of the jam flow. It can be understood the flow of oscillatory wave as the "synchronized flow".

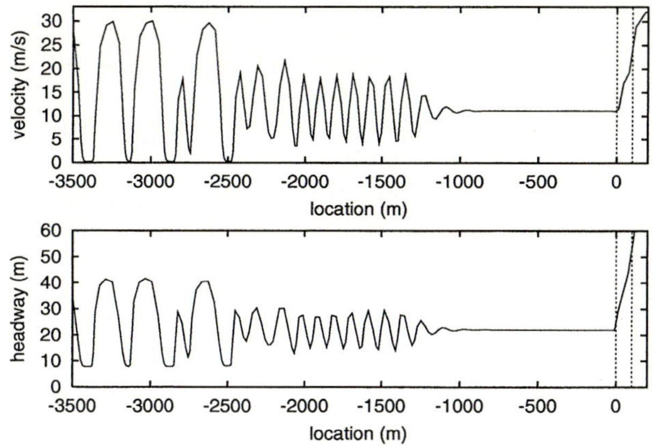

Fig. 9. The snapshots of velocity/headway distribution on the upstream of the bottleneck (the region between the dashed lines) of $S_{bottle} = 0.65$.

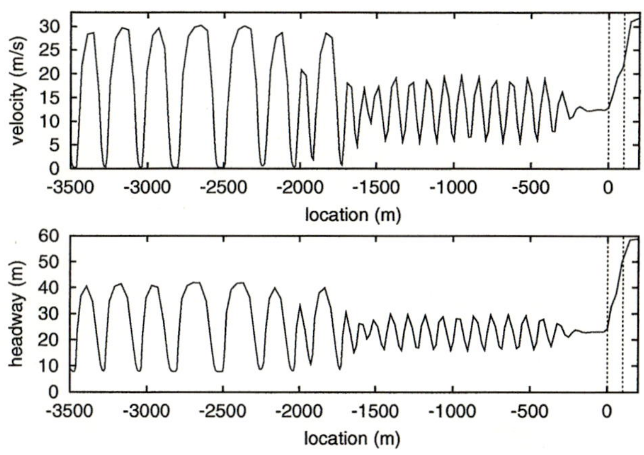

Fig. 10. The snapshots of velocity/headway distribution on the upstream of the bottleneck of $S_{bottle} = 0.7$.

Figure 11 is the headway-velocity plots observed at four different distances from the bottleneck in Fig. 10. Each figure has the vehicle-points within $100m$ of the observation point in 5 minutes. Figure 10(a) shows the transit-state from the homogeneous flow to the oscillatory wave.

Figure 10(b) shows the small closed loop as a clear trajectory in comparison with Fig. 10(a). The small closed loop can be understood as the profile of the oscillatory wave. Within the interval from $-500m$ to $-1400m$ the profile preserve its shape, that indicates the oscillatory wave is stable in this situation. This state is followed by the jam flow solution through the transit-state shown in Fig. 10(c). The profile of jam flow in Fig. 10(d) is just the same as that on the circuit with no bottleneck in Fig. 2. The small closed loop in Fig. 10(b) seems to be a kind of limit cycle solution as the jam flow solution is. This small limit cycle have not observed in the simulation on the circuit, that implies this solution is originally unstable. But this unstable solution is convectively stabilized by the bottleneck. In this sense the small loop solution can be called 'quasi-stable' solution which can be stabilized in some boundary conditions such as a bottleneck.

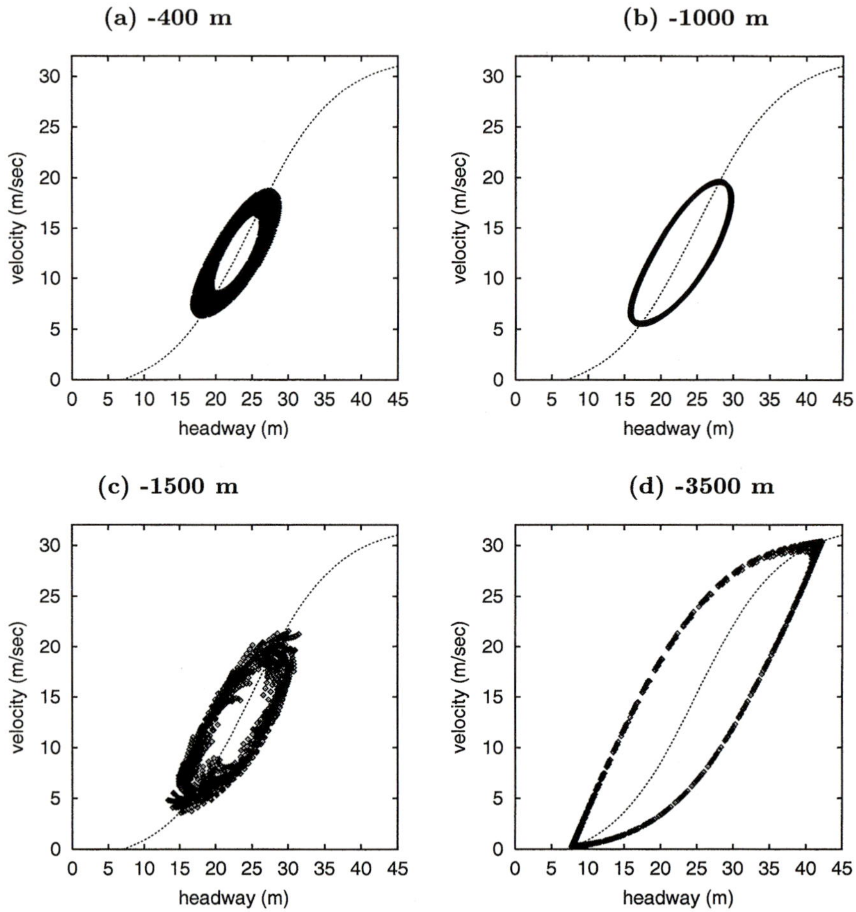

Fig. 11. The plots of vehicles-points in the phase space (headway, velocity) at four points on the upstream from the bottleneck of $S_{bottle} = 0.7$.

Figure. 12 is the plot of vehicle-points in q-k space corresponding to Fig. 11. The plots of 1 min. average data corresponding to Figs. 10 (b) and (d) in Fig. 12 are shown in Fig. 13. We take 1 min. average data every 10 m point from $-1000m$ to $-1200m$, and from $-3500m$ to $-37000m$. We compare the data with q-k relations for the homogeneous and jam flow solutions in the system on the circuit in Fig. 3, where the relations are universal for OV model. The data for $-1000m$-point are concentrate on the point between the curve of the homogeneous flow solution and the line of jam flow solution. The data for $-3500m$ are distributed along the line of jam flow solution.

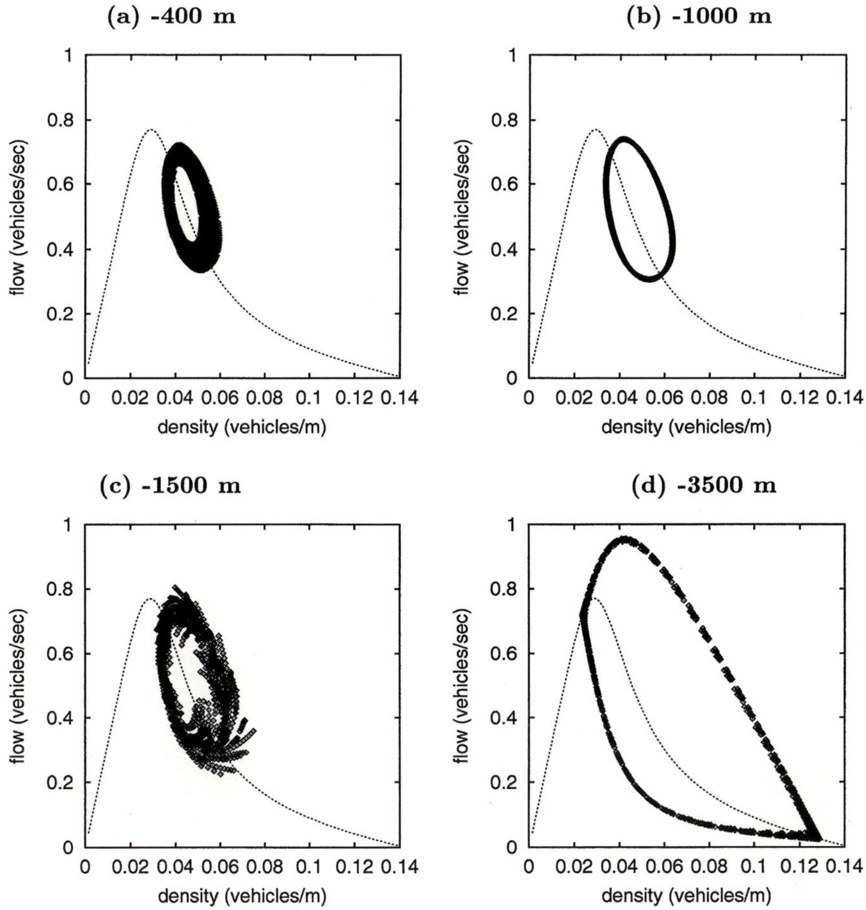

Fig. 12. The plots of vehicles-points in $q-k$ diagram (flow, density) at four points on the upstream from the bottleneck of $S_{bottle} = 0.7$. The dotted curve represents the $q-k$ relation of the homogeneous flow solution.

More clearly presenting the dynamical structure in the sequence of flow-patterns, we take 5 min. average data observed at several successive points upstream of the bottleneck, and plot the data on the q-k space together with the q-k diagram of homogeneous flow and jam flow solutions.

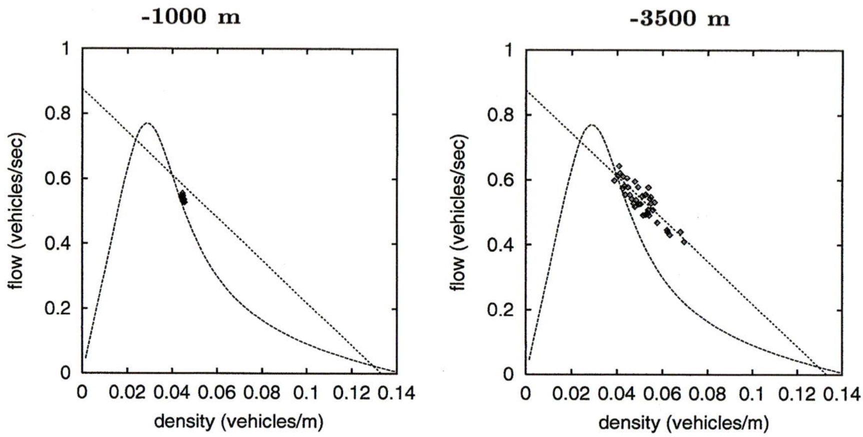

Fig. 13. The plots of 1 min. averaged data in $q-k$ diagram (flow, density) at two points on the upstream from the bottleneck of $S_{bottle} = 0.7$. The dashed curve represents the $q-k$ relation of homogeneous flow solution, and the dotted line represents that of jam flow solution.

Figure 14 shows the data from $0m$ to $-3500m$ for $S_{bottle} = 0.7$ and from $0m$ to $-5000m$ for $S_{bottle} = 0.65$. The data are taken along the sequence of three stable patterns of flow in Fig. 8. On the immediate upstream of the bottleneck there exits the homogeneous flow , as the data points are on the curve for the homogeneous flow solution. Originally, it is unstable in the region for the density (the region III in Fig. 3). The homogeneous flow is stabilized by the existence of the bottleneck. As the data is taken in upstream direction through the oscillatory wave, the data points transit from the homogeneous flow solution to the jam flow solution in q-k space. The jam flow is stable in this region. In contrast, the homogeneous flow and the oscillatory wave flow are originally unstable, but they are stabilized by the existence of the bottleneck. The data points of the upstream are aligned at the same flow rate with different densities in q-k space. The average density is increasing as the distance is larger through the three patterns upstream of the bottleneck. And this structure is convectively stabilized by the bottleneck.

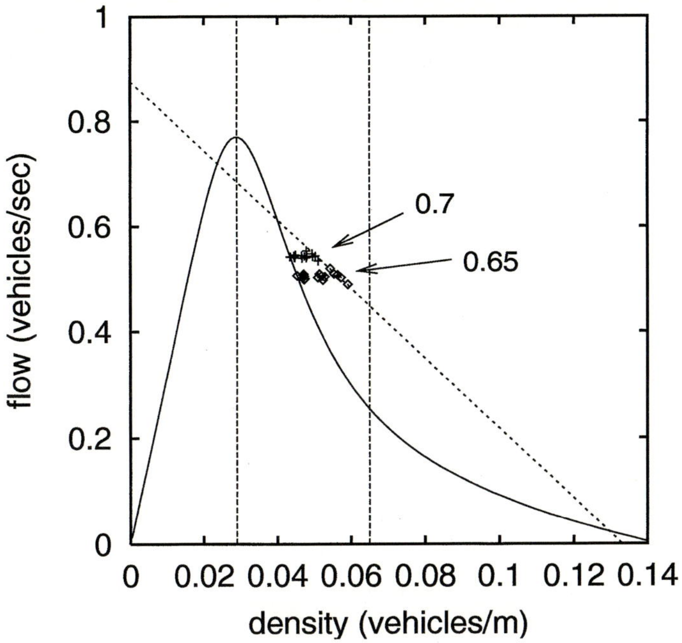

Fig. 14. The plots of 5 min.- averaged data for successive observed points on the upstream of the bottleneck. The solid curve and dotted line are the q-k relations for the homogeneous flow and jam flow solutions, respectively. In the region between dashed lines (the region III in Fig. 3), the homogeneous flow is unstable, and the jam flow is stable.

4.3 Summary and Conjecture

We have found the new solution of OV model, which appears as the flow of oscillatory wave on the upstream of the bottleneck. The amplitude of the wave is rather smaller than that of jam flow solution, and the velocity of the small cluster is different from the jam cluster. The properties of oscillatory wave are manifestly different form the jam flow and the homogeneous flow solutions. The flow of oscillatory wave is 'quasi-stable' solution, which is originally unstable but convectively stabilized by bottleneck. The similar oscillatory wave solution as ours, have been found by Mitarai and Nakanishi on the analysis of OV model in open boundary with the existence of localized perturbation [8]. They suggest the similar sequence of patterns on the flow upstream of the localized noise, which sustains the structure of their flow.

The flow of oscillatory wave has small amplitude comparing with jam flow and it is localized at some distance beyond the homogeneous flow upstream of the bottleneck. In flow-density plot, the average density varies with the same flow-rate. These properties indicate the possibility to understand the "synchronized flow" as the "oscillatory wave" solution of OV model with bottleneck. Our investigation in this paper is for one-lane traffic, but the above properties can be preserved for two-lane traffic. We hope the synchronization phenomena such as velocity changing for parallel two lanes, may be reproduced by two-lane simulation of OV model with the same situation studied in this paper.

5 Modeling Pedestrians in 2-dimensional OV Models

OV model can be extended to 2-dimensional version for applying the phenomena of several kinds of collective motions of interacting particles, such as granular flow in liquid, pedestrians, evacuation dynamics and collective bio-motions, etc. In principle, the two characteristics in OV model are independently appear in 2-dimension; the following behavior by the attractive force, and the exclusion behavior by the repulsive force.

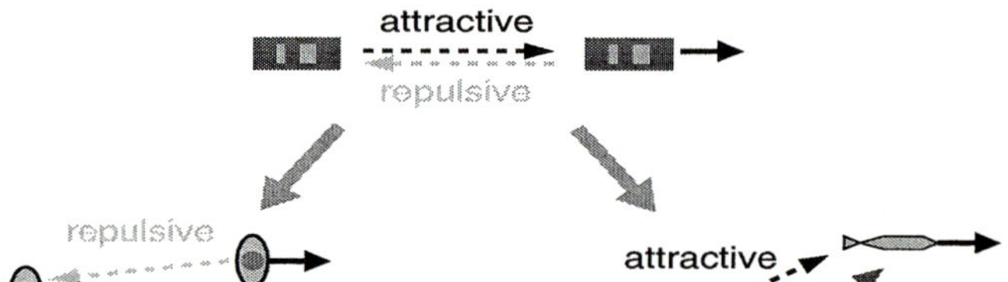

Fig. 15. The illustration of two behaviors in 2-d OV model.

In this paper we formulate the model expressed with the exclusive effect by the repulsive force and apply the pedestrian flow.

$$\frac{d^2}{dt^2}\mathbf{x}_n(t) = a\left\{V_c\mathbf{e}_x - \sum_{m}^{N(n)}[v_{max} - V(r_{mn}(t))]P(\theta_{mn}(t))\mathbf{e}_{mn}(t) - \frac{d}{dt}\mathbf{x}_n(t)\right\}. \quad (6)$$

$V_c\mathbf{e}_x$ presents the free-moving in x- direction of the desired velocity for each particle, which we set the same constant value for simplicity. The interaction between two particles (m, n) are dependent on the distance and direction as $V(r_{mn})$ and $P(\theta_{mn})$. For OV-function $V(r)$ we use the usual hyperbolic-tangent type function. The anisotropy of interaction is introduced by choosing $P(\theta_{mn}(t)) = \cos\theta_{mn}(t)$, because a pedestrian usually look forward in the direction of movement. The effect of each particle (mth) surrounding nth particle are collected.

For the simulation of pedestrian flow in passage, we introduce two kinds of particle, which have the different OV-function. In this case we can observe the fast particles overtake the slow particles in uni-direction flow. Fig. 16 shows the typical snapshots for different densities of pedestrians. In the low density

$N = 80$

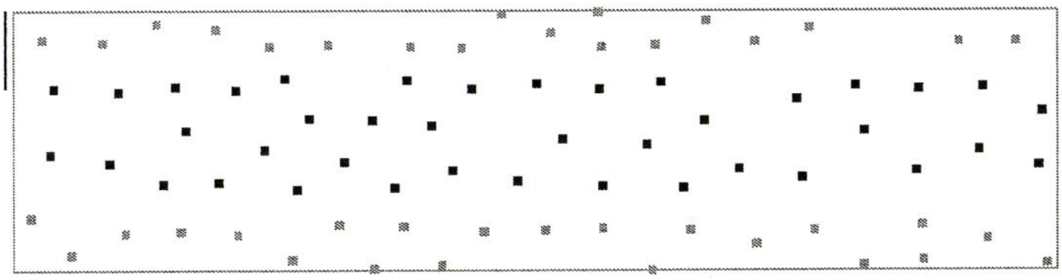

$N = 350$

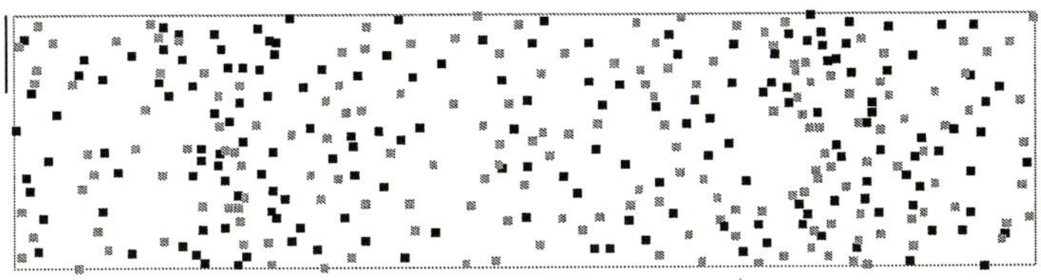

$N = 600$

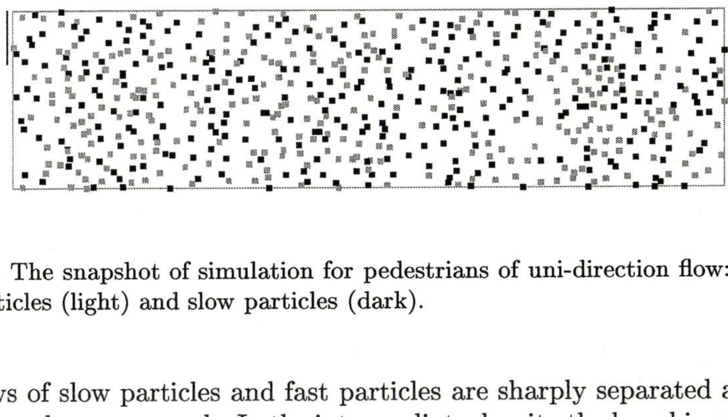

Fig. 16. The snapshot of simulation for pedestrians of uni-direction flow: $a = 0.5$, fast particles (light) and slow particles (dark).

the flows of slow particles and fast particles are sharply separated and each flow moves homogeneously. In the intermediate density the bunching appears, which is similar to the phenomena in traffic flow. This results in the formation of the 'island' of jam cluster, in which shape is qualitatively different from the case of granular. The jam of granular forms the 'band' perpendicular in the direction of the flow, which is caused by the difference of angular-dependence of sight, $P(\theta_{mn}) = 1 + \cos\theta_{mn}(t)$. In the high density no structure appears for too much particles.

Figure 17 shows the results of simulation for a counter flow of pedestrians. In the low density several thin streams in the opposite directions are formed. This is the specific properties for a counter flow of pedestrians. In the intermediate density the streams become unstable and grow wide and move winding. But their stability are very sensitively dependent to initial conditions. In the high density the jam islands are formed clearly, but several particles find their way to escape from the jam island to free-moving region.

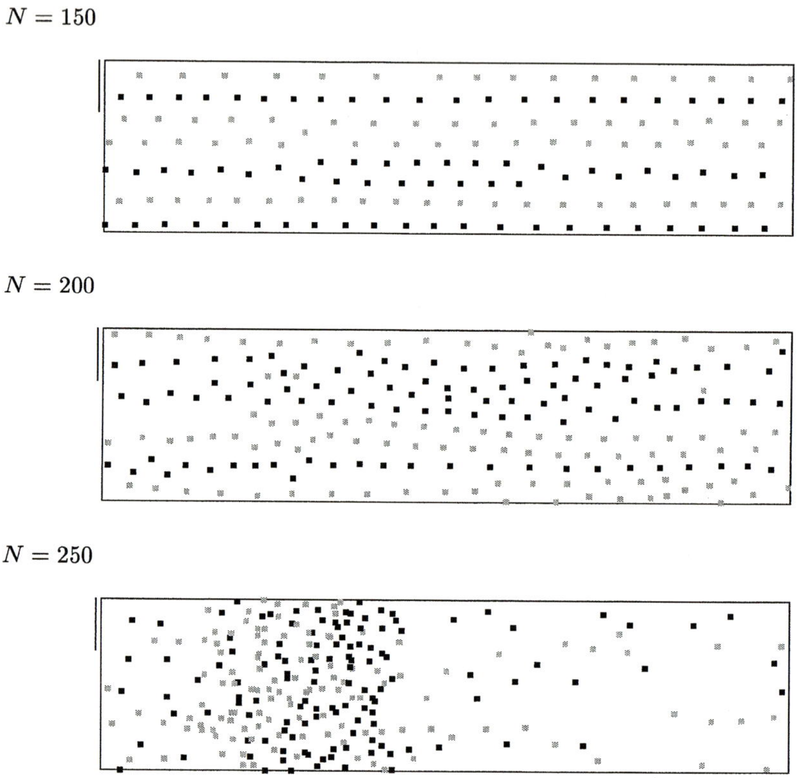

Fig. 17. The snapshot of simulation for pedestrians of counter flow: $a = 0.5$, fast particles (light) and slow particles (dark).

OV model is one of the successful model for describing the dynamics of traffic flow. OV model can be extended to several directions. The 2-dimensional OV models can develop to applications for the phenomena of several kinds of cooperative transportation system. We hope the models can provide the unified view of large class of such systems.

Acknowledgments

This work is partly supported by a Grant-in-Aid for Scientific Research (C) (No.12650065) and (C) (No.13640409) of the Japanese Ministry of Education, Science, Sports and Culture, and by Center for Integrated Research in Science and Engineering in Nagoya University.

References

1. Chowdhury, D., Santen, L., Schadschneider, A.: Phys. Rep. **329** 4-6, 199, Statistical Physics of Vehicular Traffic and Some Related Systems (2000)
2. Helbing, D.: Rev. Mod. Phys.**73**, 1067, Traffic and related self-driven many particle systems (2001)
3. Kerner, B. S., Rehborn, H.: Phys. Rev. E**53**, R4275; Kerner, B. S., Rehborn, H., Aleksic, A.: in Traffic and Granular Flow '99, edited by Helbing, D., Herrmann, H. J., Schreckenberg, M. and Wolf, D. E. (Springer, Berlin 2000)
4. Bando, M., Hasebe, K., Nakayama, A., Shibata, A., Sugiyama, Y.: Japan J. of Ind. and Appl. Math, **11**, 203 (1994); Phys. Rev. E**51**, 1035 (1995); Bando, M., Hasebe, K., Nakanishi, K., Nakayama, A., Shibata, A., Sugiyama, Y.: J. Phys. I France **5**, 1389 (1995).
5. Sugiyama, Y.: Wolf, D. E., Schreckenberg, M., Bachem, A.(Eds.), Traffic and Granular Flow'(World Scientific) 137 (1996); Schreckenberg, M., Wolf, D. E. (Eds.), Traffic and Granular Flow '97 (Springer) 301 (1998); Computer Physics Communications (in CCP99 conference) **121-122**,399 (1999).
6. Koshi, M., Iwasaki, M., Ohkura, I.: In Proceedings of the 8th International Symposium on Transportation and Traffic Flow Theory, ed. by Hurdle, V. F., Hauser, E., Stewart, G. N. (University of Toronto, Tronro, Ontario 1983) pp.403
7. Koshi, M.: In Traffic and Granular Flow '01 , Fukui, M., Sugiyama, Y., Schreckenberg, M., Wolf, D. E.: (Eds.)(Springer) to be published.
8. Mitarai, N., Nakanishi, H.: Phys. Rev. Lett. **85**, 1766-1769 (2000)
9. D. Helbing: Traffic and Granular Flow '97, M. Schreckenberg, D. E. Wolf (Eds.) (Springer) 21 (1998); D. Helbing, I.J. Frakas, T. Vicsek: Traffic and Granular Flow '99, Helbing, D., Herrmann, H. J., Schreckenberg, M., Wolf, D. E.:(Eds.) (Springer)(1998) 245; Vicsek, T., Czirók, A., Helbing, D.: 147 and references therein.

Testing Traffic Flow Models

Elmar Brockfeld and Peter Wagner

German Aerospace Centre, Institute for Transportation
Research,Rutherfordstrasse 2, 12489 Berlin, Germany, URL: http://ivf.dlr.de

Abstract. The huge number of different traffic flow models available calls for an attempt to classify, simplify and assess those models. In principle, this is simple to accomplish, since the principal algorithms for assigning quality measures such as χ^2 are well known. However, due to a lack of publicly available data, a commonly agreed upon benchmark could not be established up to now. This contribution shows (i) what can be done with such a data-sets, once they are available and (ii) tell about efforts to change the far-from-optimal situation related to the data.

1 So Much Models, so Little Time

The current state of the art of modelling traffic flow is far from optimal. There is a big market of different models of traffic flow that are thought to describe traffic flow into more or less detail. They provide the playing ground for theoreticians, each of which would like to convice the others, that (s)he has gained the biggest understanding of them all, and that her/his model describes the zoo of phenomena like jams, headway distributions, synchronized flow, and even the stock-market [2].

It is true, that most of the models can be taken as a certain approximation to the features observed in real life, and that the models are in a certain kind qualitatively correct. However, when it comes to more detailed questions it starts to become difficult.

Not surprisingly, the result of the usual reviews of those models states something like "for this and that application, this model may be used", but that's most of the time an excuse for being not exact. This small contribution to computational physics tries to go a very small step ahead, by assigning numbers for one given data-set and a number of different models.

2 Theory

Models may be classified into microscopic, mesoscopic and macroscopic, with several subclasses (micro can be subdivided further into ODE, map, and cellular automaton models, respectively). When it comes to comparison with reality, most models are tested at most on the level of the so called fundamental diagram. With fundamental diagram the plot of the flow versus density is meant, where both variables flow q_i and density k_i can be assigned in principle to an individual car. In this case, they are defined as:

$$q_i = \frac{1}{\tau_i} = \frac{\Delta x_i}{v_i} \tag{1}$$

$$k_i = \frac{1}{\Delta x_i} \tag{2}$$

i.e. the flow is defined either as the inverse of the time headway τ_i, or, alternatively, as the quotient between front-bumper-to-front-bumper-distance Δx_i and the velocity v_i of a car. Unfortunately, most counting devices record a temporal average of the velocity $\langle v\rangle_T$ and the flow $\langle q\rangle_T = \Delta N/T$ only, where T is the averaging interval. In most cases the density can be inferred only by using the questionable relationship $\langle k\rangle_T = \langle q\rangle_T/\langle v\rangle_T$. Any textbook comes up with the much better formula $\langle k\rangle_T = \langle q\rangle_T\langle \frac{1}{v}\rangle_T$.

This is discussed at some length in order to make clear, that this did not contain any idea about a specific model that may be hidden behind such a plot.

Sometimes, the fundamental diagram is refered to as a model $q = q(k)$ that may serve as a fixed point model for an underlying microscopic car following model. With fixed point it is meant, that if a car is following another car that drives with constant speed $v = v_{\text{lead}}$, this following car will approach the speed v_{lead} and keep an optimal distance $g = g(v_{\text{lead}})$. This correspondents to a homogeneous situation on the road, where all cars drive with the same speed and the same headway.

One of the interpretations of the fundamental diagram is shown in Fig. 1, in this case there is an additional regime called synchronized flow, where cars do not pick one particular fixed point. Instead, there are infinitely many marginally stable states, for more details read the presentation of Boris Kerner in this volume.

But compare this more or less theoretical fundamental diagram with an empirical one. The scatter in the microscopic data is so big, that there didn't seem any particular structure at all. Big question: can models applied successfully to those data?

3 Testing Dynamical Models

Quite in general, any dynamical system can be written as:

$$\dot{q}(t) = f(q(t), t; p) + \sum_j g_{ij}(q,t)\Gamma_j(t) \quad i = 1, \ldots, n \tag{3}$$

where the structure of the noise is given by

$$\langle \Gamma_j(t)\rangle = 0\,, \quad \langle \Gamma_j(t)\cdot \Gamma_k(t+\tau)\rangle = 2\delta_{jk} f(\tau)\ .$$

with $f(\tau)$ the noise correlation function, if $f(\tau) = \delta(\tau)$ it is a white noise process, and a coloured noise process otherwise. Here, the phase-space variable q is given by $q = \{(x_i, v_i, \ldots)\}_{i=1,\ldots,N}$, and the variable p denotes a set

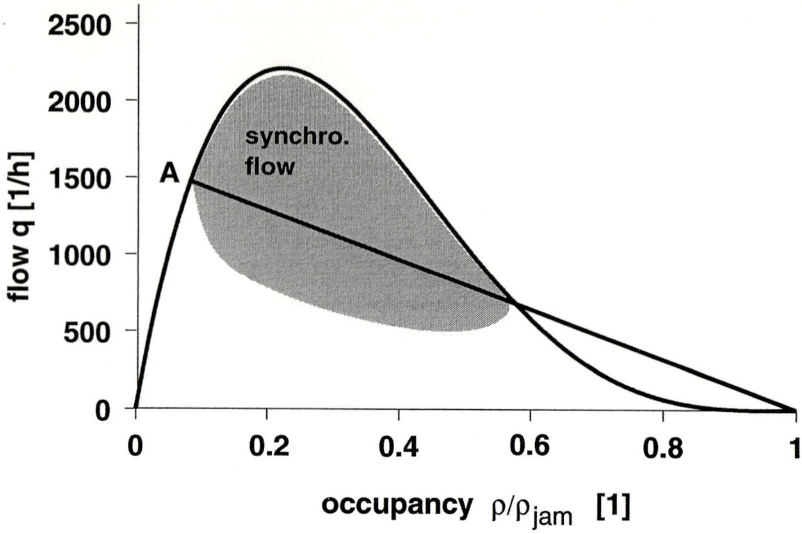

Fig. 1. Theoretical fundamental diagram. Shown is the line of virtual homogeneous states, and the illusive state of synchronized flow.

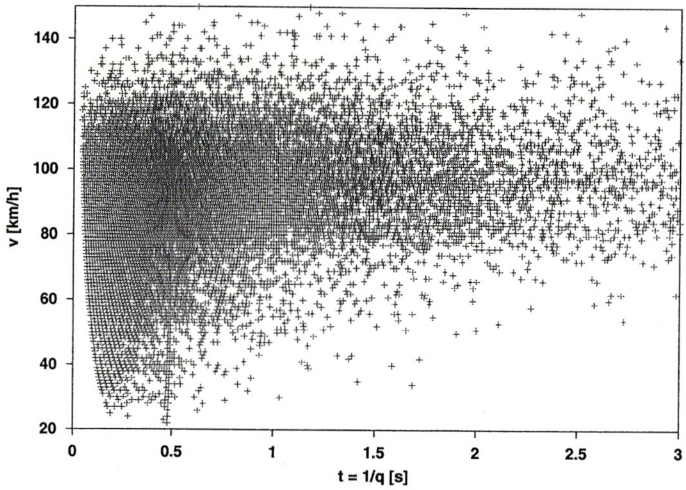

Fig. 2. Empirical fundamental diagram. It looks unfamiliar, since in this case the speed is plotted versus the time-headway, which is the inverse of the flow and can be understood as a scaled distance (distance divided by speed, see 1). The advantage: plotted here are only measured values, not computed ones like the density.

of parameters. (Of course, in case of a time-discrete dynamics anything can be carried over by just replacing $\dot{q}$ by the finite difference $(q(t+h)-q(t)/h)$.)

In principle, any car i may have its own set of parameters, which is true but would lead to a nightmare of parameters to determine. But note, that this is not just an excuse: certain parameters like the desired acceleration or the maximum speed are definitely different from driver to driver, and even for the same driver from time to time. Nevertheless, having a description such as those above at hand is physicist's dream, however, in most instances only a small part of a system can actually be observed:

$$Q(t) = G(q(t)) + \Gamma_o(t).$$

(Here, $G(q(t))$ is some map from the high-dimensional state space of the system into the low dimensional of the measurement, and $\Gamma_o(t)$ is the measurement noise.) Not surprisingly, in many cases, knowing Q is not enough to infer either the system's state q or it's dynamics f, g. This is especially true for traffic, and obtaining really good ideas about f and g for traffic flow constitutes an interesting research program in its own right.

Nevertheless, some work can still be done if a comparatively bad model is at hand, or even none at all, and this is what will be reported next. If there is a model at hand, two things can be done:

- use a Kalman filter to obtain q and the set of free parameters p, or, alternatively,
- fit the parameters p directly to the data, by some more or less sophisticated method.

In principle, it can be tried to infer the model directly from the data, an approach whose first results are reported in [1]. This is a different and very interesting avenue of research, however this contribution will stay with the simplest thing to be done, fitting the parameters directly. (The Kalman approach is a fairly complex one if it comes to testing microscopic models, which are the main focus of this contribution.)

Suppose that a certain data-set should be reproduced by a given model. That means that an appropriate error measure, for instance:

$$e = \frac{\langle |q_{\text{emp}} - q_{\text{sim}}| \rangle}{\langle q_{\text{emp}} \rangle} \tag{4}$$

should be minimized by searching for a set of parameters that minimizes e. (In this equation, q_{emp} are the observed values and q_{sim} are the simulated values. This measure gives an easy to understand error measure, the relative absolute error.) The fitting itself can be done by starting a simulation with a given set of parameters, obtaining e and looking then for a new set of parameters p' by means of one of the following algorithms:

- guessing (that is really done, and it is sometimes not as bad as it sounds),
- Nelder-Mead type algorithm (amoeba in Numerical Recipes),
- simulated annealing techniques,

– MJD Powell's COBYLA (implemented in IBM's NAO),

or any other means of doing a non-linear optimization when derivatives are not available. Of course, the business of doing such an optimization is an interesting and still fertile field of research in its own right, which would deserve another presentation as well. For an quick idea, visit [13].

4 The Trouble with the Data

Let's go! (i) Getting some data, using one of the algorithms above, this contribution will stick to Nelder Mead, and report the results. However, it turns out that it is not that simple to obtain meaningful data-sets, and even if one such data-set is at hand, sharing the data with others is not possible or not allowed. In order to remedy this situation, we are going to record them by our own devices and publish them on our homepage ivf.dlr.de/clearing. Successively we will add (stay tuned!):

– single car data from various locations,
– data from car-following experiments,
– video-based data, and their extracted traffic data,
– probably some aerial data,
– FCD-Data, recorded in Berlin (NDA required!).

together with some tests we are going to perform with those data and a number of models whose description is publicly available. Do you have a good idea what to measure? Please tell! And: please tell what you have done with the data. Watch out there for other goodies as well, e.g. the open source traffic simulation project SUMO (www.sourceforge.org/sumo).

5 Daganzo's Data

Here is an example, and what to do with it. Some years ago, Carlos Daganzo recorded a nice data-set [4] which will be used in the following to perform model testing. Along a one-lane road where overtaking was not possible, eight observers were positioned, each of them equipped with a lap-top to record the times a certain car passes him/her. One lead car that drove several times along this road provided the initial car number, therefore traffic flow is recorded by so called $N(t)$–curves where N is the number of cars that passed a certain position up to a certain time t. Typical recording errors are of the order of 5 cars out of 600 by which the counts of the different observers differed. In the following a data-set has been used that has been deliberately corrected for those small mistakes, we will make it available on our homepage as well. The time accuracy can be assumed as of the order of the one second resolution that is reported in Daganzo's data. The Fig. 3 shows a sketch of the recoding site, in the following some additional informations are given:

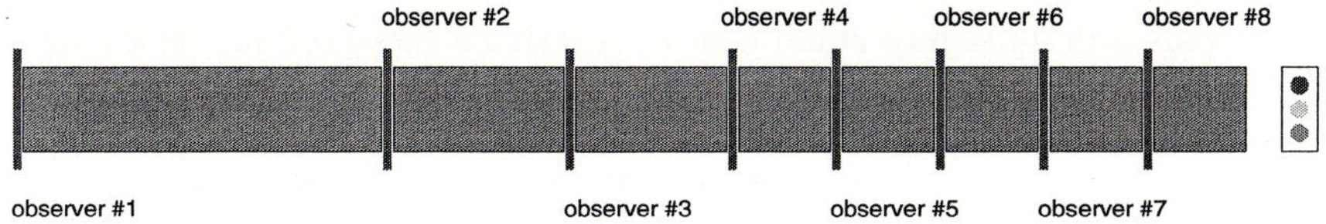

Fig. 3. Sketch of the Pablo Dam Road where the data used in this article have been recorded.

- The total length of the one-lane road is about 6.4 km,
- there is a traffic light at the end, controlling the flow into another highway,
- a speed-limit of about 25 m/s is imposed on this road,
- two days of data have been recorded, each of them from 7 to 9 am.

The traffic light at the end causes a congestion, when demand rose during the morning rush-hour. This can be seen most easily in the Fig. 4 where the travel times on segment one and the last segment six are plotted. Only six segments have been used, since the data from observer # 5 are partly missing, so they were discarded alltogether in order to simplify the analysis.

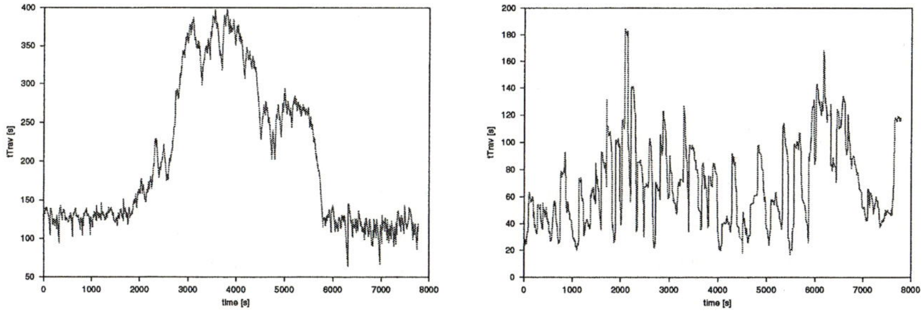

Fig. 4. Travel time as function of the number of seconds that have passed since starting of the record. It can be seen very clearly how the small jams created by the traffic light travel backwards. Note however, that they merge more or less into one big jam wave that starts to flatten out at the first observer.

6 The Contest

6.1 Building a Reference Model

How well can models describe this situation? In order to have some model for comparison, a very simple reference model will be constructed. Since the data are from Carlos Daganzo, a variant of his cell transmission model will be used [5]. This is basically a very simple queueing model, which takes care of continuity & capacity constraints. The model is specified further by:

- The road is divided into small cells of length 13 cars (100 m), where any cell has a maximum occupancy (N_i), an actually occupancy n_i and a maximum flow out $c_{\max}$.
- Cars can pass to the next cell only if there is enough room, and the number of cars that pass from one cell i to the next one $i+1$ during one time-step are computed as $\Delta N(i \to i+1) = \min\{vn, c_{\max}, w(N_{i+1} - n_{i+1}\}$.
- The size of a time-step h is just the minimum travel time, i.e. $h = L_i/v_{\max}$.

In order to compare this model directly to the data, some more microscopic details have been added. After some try and error, it converged to the following set of rules:

- Upon entering the cell, any car i gets assigned an assumed leaving time $t_i^{\text{exit}} = t_i^{in} + t_{\min}$, where $t_{\min}$ is the minimum travel time.
- Each cell is modelled as a short first-in-first-out queue, where the capacity is implemented by noting that a flow is just an inverse time headway. Therefore, cars can leave only if the car that has left the cell before is at minimum $\tau \approx 1/c_{\max}$ away, where $c_{\max}$ is the maximum possible flow.
- The entry into the next cell can be delayed if space is scarce there. This is being done by adding a penalty time before allowing to leave cell i that is proportional to n_{i+1}.

This cell transmission and its microscopic counterpart are well-known for the fact that jams are not stable, fortunately they decay fairly slowly. In order to render the jam-waves stable, the entry into the next cell has to be delayed a little if $n_{>}n_{i+1}$, compared to the approach above [15].

In the presentation given at the comphys02-conference, a travel time function has been used that was a function of the number of cars in the cell. Surprisingly, this basically has no effect on the outcome of a simulation of the queue-model, therefore it is not used anymore. It should be admitted, that there still a better way to translate cell transmission into a microscopic queueing model can be found, see for instance [14].

This simple model has four parameters $\tau_{\min}$, $c_{\max}$, α and β, where α and β are just the proportionality constants in the computation of the penalty term, $T_{\text{pen}} = \alpha n_{i+1}$, the same for β instead of α.

6.2 Technicalities

In order to actually run a simulation, some technical problems have to be addressed. These are basically the boundary conditions, which are complicated by the fact that the observers have not recorded the speeds. So, for the inflow, velocities were set to the so called safe velocity of the corresponding microscopic model, which may cause in principle travel times that are too short. The outflow condition was modelled at the position of observer # 8 by a fictitious traffic light that is switched by the number of cars that have left the road. The light is on red if $n_{out}^{(\text{sim})} \geq n_{out}^{(\text{emp})}$, and it switches to green once this inequality is false. This forces the system to the correct boundary condition based upon the number of cars, obviously, this is not the best possible solution, since the models have to deal with the discontinuity of switching the traffic light.

The reference model, it turns out, gives reasonable results. Best values obtained by the Nelder Mead algorithm was about 18 % error for the travel times between the observers, some more details are discussed in the figure caption of Fig. 5.

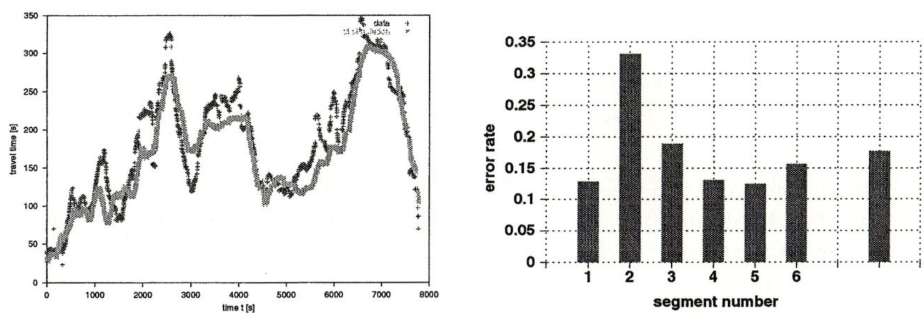

Fig. 5. Left figure: travel time as function of time for the simulation, compared to the data. It can be seen, that even the simple cell transmission model fits the data quite well. Right figure: The distribution of the errors for the different segments. It can be seen, that the second segment has the largest error, a pattern that is robust for all of the other models.

6.3 The Rest of the Pack

Interestingly, no model could do much better. Even the much more sophisticated models did not yield an error rate substantially better than the 17 % of the cell transmission model. Since a winner must be named, we do that in the following, note however, that the differences to the second best model is very small and may subject to change: optimization is sometimes more an art than a science, and 1 % decrease in the error rate may be achieved e.g. by making minor changes to the simulation set-up. The model that performed best is, we are really sorry about that, our own. (Remember what we told in the introduction!) It is a variant of the Gipps-model [12], invented some years ago by Stefan Krauß [9]. The basic idea of the model is still fairly simple: in order to avoid crashes, brake realistically, and try to keep the speed always smaller than a certain maximum safe velocity v_{safe}. If braking happens to occur with constant deceleration, then v_{safe} can be computed exactly:

$$v_{\text{safe}} = -b\tau + \sqrt{b^2\tau^2 + \tilde{v}^2 + 2bg}. \tag{5}$$

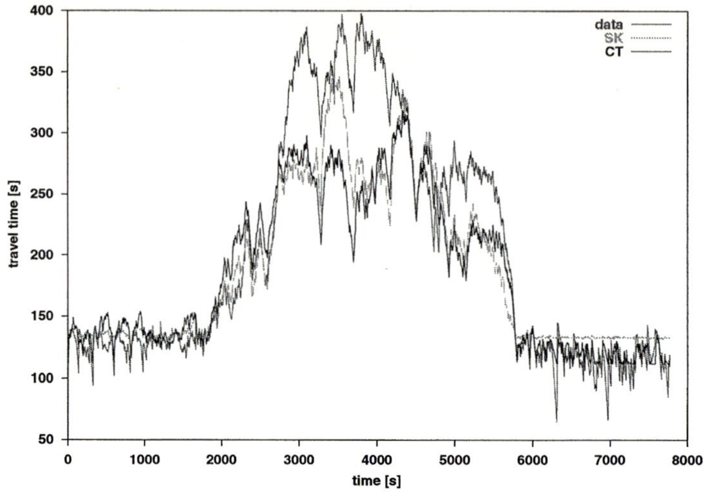

Fig. 6. Travel times for the SK-model below on the first segment, in comparison with the data and the best values obtained by the cell transmission model. Note, that there are very few differences in the performance of the two models.

Note, that this model is, like the CA–model, a kind of time-discrete optimum velocity model (OVM), this time however with an optimal velocity that additionally depends on the speed of the leading vehicle. It can be understood as the special case of an optimal velocity model differential equation which has been time-discretized with the (too big) time step $h = \tau$:

$$\tau \dot{v} = v_{\text{safe}} - v \tag{6}$$

Unfortunately, even this result came with a drawback: the optimal parameters of the model found need a very large acceleration of around $a \approx 1.8 m/s^2$ which are not very realistic, and which are different from the parameters Krauß has found [8].

6.4 Finally, the Rest (Preliminary Results!!!)

Below, the results of all the models tested so far are shown. We re-iterate, that these results are preliminary, for the reasons stated above. The abbreviations used in the plots are translated as follows: IDM = intelligent driver model [11], CA = cellular automaton model [7], SK is the Gipps-like model mentioned above, CT is the cell transmission model above, VDR = velocity dependent randomization, a variant of the CA-model [16] with so called slow–to–start rule, OVM = optimal velocity model [10] (see also [15] for the reference to a similar model). We failed up to now to implement the recent model of

Kerner [3] without any bugs (which is by no means surprisingly, given its enormous complexity), the same is true for the new CA–model with braking lights [6]. The results will be reported elsewhere. Note, that our best value obtained so far is 16.3 %, but note also that only one of us (EB) obtained this result. By using a different program, the other author (PW) gets 16.9 % error only for this model.

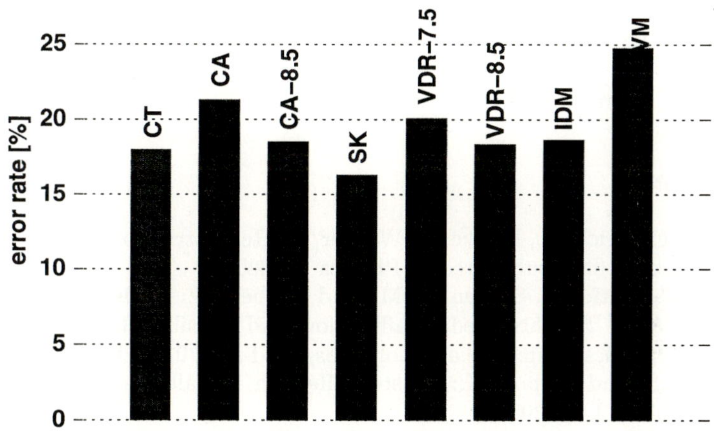

Fig. 7. Comparison of the different models.

7 Conclusions & Further Plans

Physicists can measure certain numbers with an accuracy better than 10^{-12}, but in this case the models just display, that either there are no good models actually available, or that there is an unexplainable rest of about 20 % in the behaviour of the drivers that is completely random, or far beyond our current description. It remains to be seen whether this is true for other data-sets and other models as well. Note, that the situation provided here was a very simple one, that favoured the cell transmission model. In other set-ups, it probably isn't as good as in this application.

Another point is that we have to utilize the second day of data as well just to discuss how well a given model is capable of generalizing. In order to improve the models, a number of suggestions will close this contribution:

- add some psycho facts, e.g. about recognition, learning, comfort, planning,
- people are sometimes planning their behaviour,

- acceleration usually adopts on time-scales of 1...2 second ⇒ use $\dot{a}$–models?,
- people drive sloppy, a phenomena that is hard to model.

Acknowledgements

We would like to thank Carlos Daganzo for putting this beautiful data-set on his web-site. Additionally, we would like to thank Chris Cassir, Georg Hertkorn, Boris Kerner, Kai Nagel, Michael Schreckenberg and Richard Woesler for discussions, and the Federal German Ministry of Education and Research for financial support.

References

1. Kriso, S., Friedrich, R., Peinke, J., Wagner, P.: Reconstruction of dynamic equations for traffic flow, submitted to Physica A (2002).
2. Helbing, D., Batic, D., Schoenhof, M., and Treiber, M.: Modelling Widely Scattered States in 'Synchronized' Traffic Flow and Possible Relevance for Stock Market Dynamics, http://de.arxiv.org/abs/cond-mat/0108548 (2001).
3. Kerner, B.S. and Osipov, X.: Cluster Effect in Initially Homogeneous Traffic Flow, **35** L31 – L43 (2002)
4. The data of Carlos Daganzo are publicly available on his web-site www.ce.berkeley.edu/~daganzo.html
5. Daganzo, C. F.: The cell transmission model: a simple dynamical representation of highway traffic, Transp. Res. B **28** 269 (1994).
6. Knospe, W., Santen, L., Schadschneider, A., and Schreckenberg, M.: Towards a realistic microscopic description of highway traffic, J. Phys. A33, L477 (2000).
7. Nagel, K., and Schreckenberg, M.: J. Physique I **2** 2221 (1992).
8. Krauß, S.: Microscopic modelling of traffic flow: Investigation of Collision Free Vehicle Dynamics, University of Cologne, 1998.
9. Krauß, S., Wagner, P., and Gawron, C.: Metastable states in a microscopic model of traffic flow, Phys. Rev. E, **55** 5597 – 5605 (1997).
10. Bando, M, Hasebe, K., Nakayama, A., Shibata, A., and Sugiyama, Y.: Dynamical model of traffic congestion and numerical simulation, Phys. Rev. E, **51** 1035 – 1042 (1995).
11. Treiber, M., Hennecke, A., Helbing D.: Derivation, Properties, and Simulation of a Gas-Kinetic-Based, Non-Local Traffic Model, Physical Review E, **59** 239 – 253 (1999).
12. Gipps, P. G.: A behavioural car following model for computer simulation, Transp. Res. B **15** 105 – 111 (1981).
13. Surf to http://plato.la.asu.edu/guide.html.
14. Heidemann, D.: A Queueing theory model of nonstationary traffic flow, Transportation Science **35**, 405 – 412 (2001).
15. Newell, G. F.: Theories of instability in dense highway traffic, J. Oper. Res. Soc. Japan, **5**, 9 – 54 (1962).
16. Barlovic, R., Santen, L., Schadschneider, A., and Schreckenberg, M.: Metastable States in Cellular Automata for Traffic Flow, Europ. Journ. Phys. B **5**, 793 (1998)

Editorial Policy

§1. Volumes in the following three categories will be published in LNCSE:

i) Research monographs
ii) Lecture and seminar notes
iii) Conference proceedings

Those considering a book which might be suitable for the series are strongly advised to contact the publisher or the series editors at an early stage.

§2. Categories i) and ii). These categories will be emphasized by Lecture Notes in Computational Science and Engineering. **Submissions by interdisciplinary teams of authors are encouraged.** The goal is to report new developments – quickly, informally, and in a way that will make them accessible to non-specialists. In the evaluation of submissions timeliness of the work is an important criterion. Texts should be well-rounded, well-written and reasonably self-contained. In most cases the work will contain results of others as well as those of the author(s). In each case the author(s) should provide sufficient motivation, examples, and applications. In this respect, Ph.D. theses will usually be deemed unsuitable for the Lecture Notes series. Proposals for volumes in these categories should be submitted either to one of the series editors or to Springer-Verlag, Heidelberg, and will be refereed. A provisional judgment on the acceptability of a project can be based on partial information about the work: a detailed outline describing the contents of each chapter, the estimated length, a bibliography, and one or two sample chapters – or a first draft. A final decision whether to accept will rest on an evaluation of the completed work which should include

- at least 100 pages of text;
- a table of contents;
- an informative introduction perhaps with some historical remarks which should be accessible to readers unfamiliar with the topic treated;
- a subject index.

§3. Category iii). Conference proceedings will be considered for publication provided that they are both of exceptional interest and devoted to a single topic. One (or more) expert participants will act as the scientific editor(s) of the volume. They select the papers which are suitable for inclusion and have them individually refereed as for a journal. Papers not closely related to the central topic are to be excluded. Organizers should contact Lecture Notes in Computational Science and Engineering at the planning stage.

In exceptional cases some other multi-author-volumes may be considered in this category.

§4. Format. Only works in English are considered. They should be submitted in camera-ready form according to Springer-Verlag's specifications. Electronic material can be included if appropriate. Please contact the publisher. Technical instructions and/or TEX macros are available via http://www.springer.de/math/authors/help-momu.html. The macros can also be sent on request.

General Remarks

Lecture Notes are printed by photo-offset from the master-copy delivered in camera-ready form by the authors. For this purpose Springer-Verlag provides technical instructions for the preparation of manuscripts. See also *Editorial Policy*.

Careful preparation of manuscripts will help keep production time short and ensure a satisfactory appearance of the finished book.

The following terms and conditions hold:

Categories i), ii), and iii):
Authors receive 50 free copies of their book. No royalty is paid. Commitment to publish is made by letter of intent rather than by signing a formal contract. Springer-Verlag secures the copyright for each volume.

For conference proceedings, editors receive a total of 50 free copies of their volume for distribution to the contributing authors.

All categories:
Authors are entitled to purchase further copies of their book and other Springer mathematics books for their personal use, at a discount of 33,3 % directly from Springer-Verlag.

Addresses:

Timothy J. Barth
NASA Ames Research Center
NAS Division
Moffett Field, CA 94035, USA
e-mail: barth@nas.nasa.gov

Michael Griebel
Institut für Angewandte Mathematik
der Universität Bonn
Wegelerstr. 6
D-53115 Bonn, Germany
e-mail: griebel@iam.uni-bonn.de

David E. Keyes
Computer Science Department
Old Dominion University
Norfolk, VA 23529–0162, USA
e-mail: keyes@cs.odu.edu

Risto M. Nieminen
Laboratory of Physics
Helsinki University of Technology
02150 Espoo, Finland
e-mail: rni@fyslab.hut.fi

Dirk Roose
Department of Computer Science
Katholieke Universiteit Leuven
Celestijnenlaan 200A
3001 Leuven-Heverlee, Belgium
e-mail: dirk.roose@cs.kuleuven.ac.be

Tamar Schlick
Department of Chemistry
Courant Institute of Mathematical Sciences
New York University
and Howard Hughes Medical Institute
251 Mercer Street
New York, NY 10012, USA
e-mail: schlick@nyu.edu

Springer-Verlag, Mathematics Editorial IV
Tiergartenstrasse 17
D-69121 Heidelberg, Germany
Tel.: *49 (6221) 487-8185
e-mail: peters@springer.de
http://www.springer.de/math/peters.html

Lecture Notes in Computational Science and Engineering

Vol. 1 D. Funaro, *Spectral Elements for Transport-Dominated Equations.* 1997. X, 211 pp. Softcover. ISBN 3-540-62649-2

Vol. 2 H. P. Langtangen, *Computational Partial Differential Equations.* Numerical Methods and Diffpack Programming. 1999. XXIII, 682 pp. Hardcover. ISBN 3-540-65274-4

Vol. 3 W. Hackbusch, G. Wittum (eds.), *Multigrid Methods V.* Proceedings of the Fifth European Multigrid Conference held in Stuttgart, Germany, October 1-4, 1996. 1998. VIII, 334 pp. Softcover. ISBN 3-540-63133-X

Vol. 4 P. Deuflhard, J. Hermans, B. Leimkuhler, A. E. Mark, S. Reich, R. D. Skeel (eds.), *Computational Molecular Dynamics: Challenges, Methods, Ideas.* Proceedings of the 2nd International Symposium on Algorithms for Macromolecular Modelling, Berlin, May 21-24, 1997. 1998. XI, 489 pp. Softcover. ISBN 3-540-63242-5

Vol. 5 D. Kröner, M. Ohlberger, C. Rohde (eds.), *An Introduction to Recent Developments in Theory and Numerics for Conservation Laws.* Proceedings of the International School on Theory and Numerics for Conservation Laws, Freiburg / Littenweiler, October 20-24, 1997. 1998. VII, 285 pp. Softcover. ISBN 3-540-65081-4

Vol. 6 S. Turek, *Efficient Solvers for Incompressible Flow Problems.* An Algorithmic and Computational Approach. 1999. XVII, 352 pp, with CD-ROM. Hardcover. ISBN 3-540-65433-X

Vol. 7 R. von Schwerin, *Multi Body System SIMulation.* Numerical Methods, Algorithms, and Software. 1999. XX, 338 pp. Softcover. ISBN 3-540-65662-6

Vol. 8 H.-J. Bungartz, F. Durst, C. Zenger (eds.), *High Performance Scientific and Engineering Computing.* Proceedings of the International FORTWIHR Conference on HPSEC, Munich, March 16-18, 1998. 1999. X, 471 pp. Softcover. 3-540-65730-4

Vol. 9 T. J. Barth, H. Deconinck (eds.), *High-Order Methods for Computational Physics.* 1999. VII, 582 pp. Hardcover. 3-540-65893-9

Vol. 10 H. P. Langtangen, A. M. Bruaset, E. Quak (eds.), *Advances in Software Tools for Scientific Computing.* 2000. X, 357 pp. Softcover. 3-540-66557-9

Vol. 11 B. Cockburn, G. E. Karniadakis, C.-W. Shu (eds.), *Discontinuous Galerkin Methods.* Theory, Computation and Applications. 2000. XI, 470 pp. Hardcover. 3-540-66787-3

Vol. 12 U. van Rienen, *Numerical Methods in Computational Electrodynamics.* Linear Systems in Practical Applications. 2000. XIII, 375 pp. Softcover. 3-540-67629-5

Vol. 13 B. Engquist, L. Johnsson, M. Hammill, F. Short (eds.), *Simulation and Visualization on the Grid.* Parallelldatorcentrum Seventh Annual Conference, Stockholm, December 1999, Proceedings. 2000. XIII, 301 pp. Softcover. 3-540-67264-8

Vol. 14 E. Dick, K. Riemslagh, J. Vierendeels (eds.), *Multigrid Methods VI.* Proceedings of the Sixth European Multigrid Conference Held in Gent, Belgium, September 27-30, 1999. 2000. IX, 293 pp. Softcover. 3-540-67157-9

Vol. 15 A. Frommer, T. Lippert, B. Medeke, K. Schilling (eds.), *Numerical Challenges in Lattice Quantum Chromodynamics.* Joint Interdisciplinary Workshop of John von Neumann Institute for Computing, Jülich and Institute of Applied Computer Science, Wuppertal University, August 1999. 2000. VIII, 184 pp. Softcover. 3-540-67732-1

Vol. 16 J. Lang, *Adaptive Multilevel Solution of Nonlinear Parabolic PDE Systems.* Theory, Algorithm, and Applications. 2001. XII, 157 pp. Softcover. 3-540-67900-6

Vol. 17 B. I. Wohlmuth, *Discretization Methods and Iterative Solvers Based on Domain Decomposition.* 2001. X, 197 pp. Softcover. 3-540-41083-X

Vol. 18 U. van Rienen, M. Günther, D. Hecht (eds.), *Scientific Computing in Electrical Engineering.* Proceedings of the 3rd International Workshop, August 20-23, 2000, Warnemünde, Germany. 2001. XII, 428 pp. Softcover. 3-540-42173-4

Vol. 19 I. Babuška, P. G. Ciarlet, T. Miyoshi (eds.), *Mathematical Modeling and Numerical Simulation in Continuum Mechanics.* Proceedings of the International Symposium on Mathematical Modeling and Numerical Simulation in Continuum Mechanics, September 29 - October 3, 2000, Yamaguchi, Japan. 2002. VIII, 301 pp. Softcover. 3-540-42399-0

Vol. 20 T. J. Barth, T. Chan, R. Haimes (eds.), *Multiscale and Multiresolution Methods.* Theory and Applications. 2002. X, 389 pp. Softcover. 3-540-42420-2

Vol. 21 M. Breuer, F. Durst, C. Zenger (eds.), *High Performance Scientific and Engineering Computing.* Proceedings of the 3rd International FORTWIHR Conference on HPSEC, Erlangen, March 12-14, 2001. 2002. XIII, 408 pp. Softcover. 3-540-42946-8

Vol. 22 K. Urban, *Wavelets in Numerical Simulation.* Problem Adapted Construction and Applications. 2002. XV, 181 pp. Softcover. 3-540-43055-5

Vol. 23 L. F. Pavarino, A. Toselli (eds.), *Recent Developments in Domain Decomposition Methods.* 2002. XII, 243 pp. Softcover. 3-540-43413-5

Vol. 24 T. Schlick, H. H. Gan (eds.), *Computational Methods for Macromolecules: Challenges and Applications.* Proceedings of the 3rd International Workshop on Algorithms for Macromolecular Modeling, New York, October 12-14, 2000. 2002. IX, 504 pp. Softcover. 3-540-43756-8

Vol. 25 T. J. Barth, H. Deconinck (eds.), *Error Estimation and Adaptive Discretization Methods in Computational Fluid Dynamics.* 2003. VII, 344 pp. Hardcover. 3-540-43758-4

Vol. 26 M. Griebel, M. A. Schweitzer (eds.), *Meshfree Methods for Partial Differential Equations.* 2003. IX, 466 pp. Softcover. 3-540-43891-2

Vol. 27 S. Müller, *Adaptive Multiscale Schemes for Conservation Laws.* 2003. XIV, 181 pp. Softcover. 3-540-44325-8

Vol. 28 C. Carstensen, S. Funken, W. Hackbusch, R. H. W. Hoppe, P. Monk (eds.), *Computational Electromagnetics.* Proceedings of the GAMM Workshop on "Computational Electromagnetics", Kiel, Germany, January 26-28, 2001. 2003. X, 209 pp. Softcover. 3-540-44392-4

Vol. 29 M. A. Schweitzer, *A Parallel Multilevel Partition of Unity Method for Elliptic Partial Differential Equations.* 2003. V, 194 pp. Softcover. 3-540-00351-7

Vol. 30 T. Biegler, O. Ghattas, M. Heinkenschloss, B. van Bloemen Waanders (eds.), *Large-Scale PDE-Constrained Optimization.* 2003. approx. 344 pp. Softcover. 3-540-05045-0

Vol. 31 M. Ainsworth, P. Davies, D. Duncan, P. Martin, B. Rynne (eds.) *Topics in Computational Wave Propagation.* Direct and Inverse Problems. 2003. VIII, 399 pp. Softcover. 3-540-00744-X

Vol. 32 H. Emmerich, B. Nestler, M. Schreckenberg (eds.) *Interface and Transport Dynamics.* Computational Modelling. 2003. XV, 432 pp. Hardcover. 3-540-40367-1

Vol. 33 H. P. Langtangen, A. Tveito (eds.) *Advanced Topics in Computational Partial Differential Equations.* Numerical Methods and Diffpack Programming. 2003. XIX, 659 pp. Softcover. 3-540-01438-1

Texts in Computational Science and Engineering

Vol. 1 H. P. Langtangen, *Computational Partial Differential Equations.* Numerical Methods and Diffpack Programming. 2nd Edition 2003. XXVI, 855 pp. Hardcover. ISBN 3-540-43416-X

Vol. 2 A. Quarteroni, F. Saleri, *Scientific Computing with MATLAB.* 2003. IX, 257 pp. Hardcover. ISBN 3-540-44363-0

For further information on these books please have a look at our mathematics catalogue at the following URL: `http://www.springer.de/math/index.html`

Printing: Mercedes-Druck, Berlin
Binding: Stein+Lehmann, Berlin